STORMWATER MANAGEMENT

STORMWATER MANAGEMENT

Martin P. Wanielista, P.E., Ph.D.
Yousef A. Yousef, P.E., Ph.D.

JOHN WILEY & SONS, INC.
New York · Chichester · Brisbane · Toronto · Singapore

In recognition of the importance of preserving what has been written, it is a policy of John Wiley & Sons, Inc., to have books of enduring value published in the United States printed on acid-free paper, and we exert our best efforts to that end.

Copyright © 1993 by John Wiley & Sons, Inc.

All rights reserved. Published simultaneously in Canada.

Reproduction or translation of any part of this work beyond that permitted by Section 107 or 108 of the 1976 United States Copyright Act without the permission of the copyright owner is unlawful. Requests for permission or further information should be addressed to the Permissions Department, John Wiley & Sons, Inc.

This publication is designed to provide accurate and authoritative information in regard to the subject matter covered. It is sold with the understanding that the publisher is not engaged in rendering legal, accounting, or other professional services. If legal advice or other expert assistance is required, the services of a competent professional person should be sought. *From a Declaration of Principles jointly adopted by a Committee of the American Bar Association and a Committee of Publishers.*

Library of Congress Cataloging in Publication Data:

Wanielista, Martin P.
    Stormwater management / by Martin P. Wanielista, Yousef A. Yousef.
      p. cm.
    Includes index.
    ISBN 0-471-57135-0
    1. Storm sewers. 2. Storm sewers—Mathematical models. 3. Water quality management. 4. Water quality management—Mathematical models. I. Yousef, Yousef A. II. Title.
TD665.W36 1992
628′.21—dc20   92-6914

Printed in the United States of America

10 9 8 7 6 5 4 3 2 1

PREFACE

The study of stormwater management principles, concepts, and ideas has progressed from one of an individual understanding of traditional disciplines (i.e., hydrology and hydraulics) to an interdisciplinary study of subject areas commonly found in civil engineering, environmental engineering, economics, and planning. Stormwater management is part of many disciplines and as such, in academic curricula, it may be found in water resources studies. Specifically, it can be classified as a water resources engineering study area.

Recently, federal, state, and local regulations have been announced for stormwater management. The regulations are causing an explosion of and the potential for additional work in the public and private sectors. Regulators are actively developing laws and rules to govern stormwater work. Consultants are engaged in contracts for feasibility studies, structural designs, operation and maintenance procedures, and setting up stormwater utilities. This book with its many practical problems will help regulators and consultants understand and solve some of the many stormwater related problems.

Many stormwater management efforts follow a multidisciplinary approach and also require an understanding of the physical features of stormwater management systems. In addition to the traditional goals of peak discharge management and flood control, the contemporary goals of water quality control and reuse are presented. We present a multidisciplinary, many-objective approach for stormwater management. To accomplish this, we begin with philosophies and reasons for stormwater management and progress to individual subject areas for a complete multidisciplinary approach. Examples of conventional hydrologic and hydraulic techniques are enhanced with optimization and fiscal procedures to manage water quality and quantity.

ORGANIZATION

All eleven chapters are usually taught to fourth-year undergraduate and first-year graduate engineering and science students or advanced-level graduate students of different academic backgrounds. The first chapter is introductory material that is important to establish reasons for stormwater management. Usually, proportionally less time is spent on it relative to the remaining chapters.

Chapters 2 and 3 may be a review for those with in-depth knowledge of hydrology and especially, hydrograph generation procedures. However, the probability distributions and hydrograph procedures are fundamental and should be understood before progressing through the remaining chapters. Both students of the subject and those who practice on a daily basis will find the material of Chapters 2 and 3 very beneficial from a technical viewpoint. In Chapter 4 popular computer-based models for the control of peak discharge and volume are presented.

The book is decidedly different from others in the sense that it includes extensive coverage of water quality issues. The measures of water quality are presented in Chapter 5 along with event mean concentrations and mass loadings. This is followed by water quality impacts and assessment procedures in Chapter 6. Next, management alternatives for water quality improvement are presented in Chapters 7 and 8. Popular conventional methods, such as sedimentation, off-line infiltration systems, on-line wet-detention ponds, and swales, are discussed with regard to design and operation.

When reading the subject matter of Chapter 9, the applications are important. Both economic and fiscal feasibility measures are as important as technical feasibility. Once technical and economic feasibility are determined, the best alternative is one that is fiscally responsible. Therefore, minimum-cost criteria must be examined along with who pays what and when they pay.

The optimization procedures of Chapter 10 are presented to aid in the selection of the best stormwater management system as judged by some political and social criteria that are translated into mathematical models. Minimum-cost alternatives are evaluated knowing that technical feasibility can be achieved. Thus optimization, water quantity, water quality, and economic feasibility are used together. In Chapter 11, erosion and sediment control for rural areas are examined.

There are problems and computer-assisted problems at the end of each chapter that can be assigned to enhance knowledge. It is helpful to solve the example problems before attempting the end-of-chapter problems. The computer diskette provided with the book can be used to solve the computer-assisted problems and allow solutions to otherwise time-consuming problems. A special feature of the book is the inclusion of practical problems and case studies. Appendix H includes solution to some of the case studies and practical problems* with the purpose of illustrating solutions to otherwise more complex problems. A solution manual is available for all the problems to those who adopt the book or otherwise purchase in bulk quantities.

---

*Boxed problem numbers in the end-of-chapter problems indicate solutions available in Appendix H.

ACKNOWLEDGMENTS

The authors appreciate the manuscript reviews provided by the publisher. Also, we thank Dr. Robert D. Kersten for his valuable review and suggestions. Special thanks are extended to Greg Harper, Marc Gauthier, and Ron Eaglin, who helped develop the computer programs. The programs have been tested by both students and engineers practicing stormwater management.

The text materials were reviewed by students in a water resources course at the University of Central Florida. Special recognition of Sara Mayo, Betty Wanielista, Colleen Pedersen, Lisa Guthrie, Carol Ann Pohl, Deb Keely, and Cheryl Brooks is extended for their work on manuscript preparation.

MARTIN P. WANIELISTA
YOUSEF A. YOUSEF

Maitland, Florida
Winter Park, Florida
July, 1992

CONTENTS

Preface		v
1. Introduction		**1**
1.1	Organization of Book and Related Disciplines	1
1.2	Magnitude of Stormwater Problems	2
1.3	Hydrologic Cycle	4
1.4	Stormwater Management Objectives and Limitations	8
1.5	Stormwater Management Policies	10
1.6	Feasibility Considerations	11
1.7	Stormwater Mathematical Models	13
	1.7.1 Types of Models	13
	1.7.2 Stick Diagram	14
1.8	Computation Aids and Computer Programs	19
1.9	Summary	20
1.10	Problems	20
1.11	Computer-Assisted Problems	22
1.12	References	22
2. Probability and Statistical Methods for Hydrologic Events		**24**
2.1	Introduction	24
	2.1.1 Frequency–Intensity–Duration Curves	25
	2.1.2 Methods for Estimating Extreme Events	26
2.2	Assumptions for Statistical Analysis	27
	2.2.1 Statistical Sampling	27
	2.2.2 Independent Events	28
	2.2.3 Rainfall Volume	28
2.3	Regression Formulas	30
2.4	Theoretical Frequency Distributions	34
	2.4.1 Parameter Estimation	35
	2.4.2 Confidence Intervals	39
	2.4.3 Incomplete Records and Extreme Events	39
	2.4.4 Example Graphical Applications	42
	2.4.5 Example Computer Applications	44

		2.4.6	Example Calculations for the Log-Pearson Type III Probability Distribution	48
		2.4.7	Rainfall-Runoff Events and Probability Models	51
	2.5	Summary		59
	2.6	Problems		60
	2.7	Computer-Assisted Problems		62
	2.8	References		63

3. Hydrographs 65

	3.1	Synthetic Hydrographs		65
	3.2	Rainfall Excess		66
		3.2.1	Site-Specific Measure for Soil Infiltration Rates	66
		3.2.2	A Constant Rate	68
		3.2.3	Mass Balance Methods	68
		3.2.4	SCS Soil Complex Curves Method	70
	3.3	Hydrograph Procedures		72
		3.3.1	Rational Method	72
		3.3.2	SCS Procedure	73
		3.3.3	Santa Barbara Urban Hydrograph	76
		3.3.4	Continuous Convolution	77
		3.3.5	Computer Programs	77
	3.4	Hydrograph Attenuation		78
		3.4.1	Inventory Equation	79
		3.4.2	Muskingum Method	82
	3.5	Summary		83
	3.6	Problems		84
	3.7	Computer-Assisted Problems		87
	3.8	References		88

4. Management Models for Flow Rate and Volume Control 90

	4.1	Reasons for Models		90
	4.2	Some Existing Models		91
	4.3	Basic Concepts		92
		4.3.1	Runoff Quantity	92
		4.3.2	Runoff Quality	94
		4.3.3	Storage, Treatment, and Overflow	95
		4.3.4	Soil Erosion	95
	4.4	Application of HEC-1		96
	4.5	Application of SMADA		98
		4.5.1	Rainfall Data	98
		4.5.2	Watershed Data	98
		4.5.3	Nodal Network Data	99
		4.5.4	Peak Attenuation and Water Quality Control	103

4.6	Summary	104
4.7	Problems	105
4.8	Computer-Assisted Problems	105
4.9	References	106

5. Stormwater Quality — 108

5.1	Water Quality Parameters	110
5.2	Dustfall and Quality of Precipitation	112
5.3	Stormwater Sampling, Event Mean Concentration, and Loading	116
	5.3.1 Stormwater Sampling	116
	5.3.2 Event Mean Concentration	117
	5.3.3 Event Loading	120
5.4	Rural Nonpoint Sources	120
5.5	Urban Nonpoint Sources	123
5.6	Runoff Water Quality	127
5.7	Mass Loadings	135
	5.7.1 Mass Loadings and Flow Rates	137
	5.7.2 Other Mass Versus Flow Relationships	139
	5.7.3 Calculation Steps for Loading Rates from Flow Rates	139
5.8	Mathematical Modeling	140
	5.8.1 Transported Rates Related to Rainfall Excess	146
5.9	Summary	150
5.10	Problems	151
5.11	Computer-Assisted Problems	155
5.12	References	155

6. Receiving Water Quality — 159

6.1	Water Quality Assessment	159
6.2	Separating Point from Nonpoint Sources	163
6.3	Combined Sewers and Comparison to Separate Sewers	166
6.4	Toxicity	169
	6.4.1 Bioassay	170
	6.4.2 Algal Assays	173
	6.4.3 Toxicity of Heavy Metals in Stormwater	174
	6.4.4 Deicing	176
6.5	Dissolved Oxygen Impacts	176
	6.5.1 DO Mass Balance	177
	6.5.2 Critical DO Conditions	184
	6.5.3 Photosynthesis, Respiration, and Benthic Demand	186
	6.5.4 Impoundment Dissolved Oxygen	186

	6.6	Sediment Accumulation	188
	6.7	Eutrophication and Trophic Analysis	190
	6.7.1	Shannon–Brezonik Trophic State Index	191
	6.7.2	Vollenweider Model	192
	6.7.3	Dillon Model	193
	6.7.4	Larsen–Mercier Model	194
	6.8	Mathematical Modeling for Receiving Waters	195
	6.8.1	Reaction Kinetics	195
	6.8.2	Types of Models	199
	6.8.3	Dead Volume	203
	6.9	Summary	207
	6.10	Problems	208
	6.11	References	211

7. Stormwater Management Alternatives for Water Quality Improvement — 216

	7.1	Philosophies for Stormwater Quality Management	216
	7.2	Alternative Stormwater Quality Management Practices	218
	7.3	Fundamental Efficiency Considerations	220
	7.3.1	Mass Balances	220
	7.3.2	Rainfall Processes and Runoff Diversion	220
	7.3.3	Diversion Based on the Number of Rainfall Events per Year	221
	7.3.4	Diversion Based on the Volume of Runoff per Year	222
	7.4	Off-Line Retention (Diversion) Systems	225
	7.5	Sedimentation	231
	7.5.1	Sedimentation Ponds: Physical Configuration	232
	7.5.2	Settling Velocities	232
	7.5.3	Design Storm Sizing	235
	7.6	Removal of Dissolved Contaminants in Stormwater	237
	7.6.1	Activated Carbon	237
	7.6.2	Coagulation	239
	7.7	Swales	240
	7.7.1	Swale Hydraulics and Hydrology	240
	7.7.2	Swale Design	244
	7.7.3	Berms (Swale Blocks)	248
	7.7.4	Water Quality Considerations	249
	7.8	Summary	254
	7.9	Problems	254
	7.10	Computer-Assisted Problems	256
	7.11	References	258

CONTENTS   xiii

8. Wet-Retention / Detention Ponds   262

   8.1 Design Considerations   264
       8.1.1 Temporary Storage Removal by Infiltration and Irrigation   264
       8.1.2 Temporary Storage Removal by Direct Surface Discharge   265
   8.2 Modeling Concepts for Wet-Detention Ponds   266
       8.2.1 Mass Balance for Concentration Changes   267
       8.2.2 Temporary Pond Volume in a Wet-Detention Pond   268
       8.2.3 Permanent Pool Volume in a Wet-Detention Pond   270
       8.2.4 Wet-Pond Construction Details   273
       8.2.5 On-Line Infiltration Pond Volume   274
       8.2.6 Probability Models   278
       8.2.7 Fate of Pollutants   283
   8.3 Wet-Detention Pond as a Holding Tank and a Reuse Pond   292
       8.3.1 Holding Pond with No Discharge for a Minimum Interevent Dry Period   294
       8.3.2 Reuse Ponds Operating on a Schedule   295
   8.4 Summary   306
   8.5 Problems   306
   8.6 Computer-Assisted Problems   309
   8.7 References   310

9. Economic and Fiscal Feasibility   313

   9.1 Comprehensive Stormwater Management   313
   9.2 Economic Measures   315
       9.2.1 Costs   315
       9.2.2 Benefits   317
       9.2.3 Benefit–Cost Comparisons   319
   9.3 Fiscal Responsibility and Financial Planning   321
       9.3.1 Interest Rate and Planning Horizon   322
       9.3.2 Types of Financial Plans   322
       9.3.3 Plan Equivalency   322
       9.3.4 Stormwater Financing Alternatives   327
   9.4 Uncertainty and Risk   328
       9.4.1 Flood Damage Analysis   329
       9.4.2 Expected Values with Benefits and Cost   330
       9.4.3 Expected Values with Total Cost   332
   9.5 Summary   337
   9.6 Problems   337
   9.7 References   342

xiv CONTENTS

10. Optimization — 343

- 10.1 Introduction — 343
  - 10.1.1 Formulating the Problem — 344
  - 10.1.2 Constructing a Model — 344
  - 10.1.3 Deriving a Solution — 345
  - 10.1.4 Testing the Model and Solutions — 345
  - 10.1.5 Implementing the Solution — 346
- 10.2 Linear Programming — 346
  - 10.2.1 Objective Function and Constraints — 347
  - 10.2.2 Problem Formulation/Graphical Solution — 348
  - 10.2.3 Regional Stormwater Facilities Operation and Design — 351
  - 10.2.4 Unit Hydrograph Estimation — 352
  - 10.2.5 Optimal Load Reduction Model — 354
- 10.3 Multiobjective Programming — 357
- 10.4 Other Optimization Techniques — 359
  - 10.4.1 Classical Optimization — 360
  - 10.4.2 Nonlinear Optimization — 361
  - 10.4.3 Piecewise Linear Approximations — 362
  - 10.4.4 Dynamic Programming — 364
- 10.5 Cost-Effectiveness — 371
  - 10.5.1 Cost-Effectiveness Defined — 371
  - 10.5.2 Broom Sweeping of Mall Areas — 372
  - 10.5.3 Combined Sewer Flushing — 373
  - 10.5.4 Cost-Effectiveness of Diversion Systems — 373
  - 10.5.5 Alternative Comparisons — 378
- 10.6 Optimization of Stormwater Management Practices — 381
  - 10.6.1 Linear Programming Formulation — 381
  - 10.6.2 Two Subwatershed Linear Programming Form — 382
- 10.7 Summary — 385
- 10.8 Problems — 386
- 10.9 Computer-Assisted Problems — 391
- 10.10 References — 395

11. Rural Area Stormwater Management — 397

- 11.1 Erosion and Sedimentation — 397
- 11.2 Universal Soil Loss Equation — 399
  - 11.2.1 Rainfall Factor (R) — 399
  - 11.2.2 Soil Erodibility Factor (K) — 400
  - 11.2.3 Slope Length (L) and Slope Gradient (S) Factor — 400
  - 11.2.4 Cropping Management Factor (C) — 404

		11.2.5 Erosion-Control Practice Factor (P)	407
		11.2.6 Conclusions on USLE	410
	11.3	Water Yield Model	410
	11.4	Solids Disposal	414
		11.4.1 Land Sprinkling	416
		11.4.2 Storage Reservoirs	417
	11.5	Summary	419
	11.6	Problems	420
	11.7	References	420
Appendix A	**Notation**		423
Appendix B	**Metric Units With English Equivalents**		430
Appendix C	**Nondimensional Rainfall and Frequency–Intensity–Duration Curves**		435
Appendix D	**Statistical Tables**		443
Appendix E	**Computer Programs, Descriptions, and Example Outputs**		464
Appendix F	**Selected Hydrologic Soil Classifications and Curve Numbers**		504
Appendix G	**Description of Common Stormwater Management Practices**		528
Appendix H	**Solutions to Selected Case Studies**		536
Index			573

STORMWATER MANAGEMENT

CHAPTER 1

Introduction

Stormwater management is knowledge used to understand, control, and utilize waters in their different forms within the hydrologic cycle. A majority of work is done on surface and groundwaters resulting from precipitation events. Streamflow appearing during and immediately after a precipitation event is from groundwater and surface runoff; however, groundwater flow and volume storage over a longer period are related to precipitation events. The applications of the concepts, design methods, and maintenance principles of stormwater management are evident in many diverse areas, such as agricultural drainage, water supply, flood control, lake management, urban runoff, forest management, and ecological impact studies. Indeed, the applications of stormwater management are vast with as many definitions of this subject as its areas of applications. The subject content can be found in the educational disciplines of hydrology, hydraulics, environmental engineering, and water resources engineering. In fact, stormwater management is both an expansion of each of these four disciplines and an integration of the discipline concepts.

1.1 ORGANIZATION OF BOOK AND RELATED DISCIPLINES

Each chapter starts with opening remarks and introduces the purpose of the material in that chapter. The purpose of the material in this chapter is to provide an overview of stormwater management and to introduce some of the common goals and objectives for stormwater management. The limitations on achieving these objectives along with policies for obtaining a feasible stormwater management program are discussed. Five tests for project feasibility are presented: (1) engineering, (2) economic, (3) financial, (4) political, and (5) social. Engineering feasibility depends to a great extent on a knowledge of subject matter related to hydrology, hydraulics, and water quality. The remaining chapters introduce hydrology, water quality, and engineering feasibility measures with economic and fiscal considerations as related to stormwater management.

Example problems are solved in each chapter, and other computation problems are stated at the end of each chapter. The solved example problems should assist in the application of concepts as applied to many of the problems at the end of each chapter. Summary statements are listed before the problem sections of each chapter to highlight important ideas, issues, and concepts.

We have attempted to keep a consistent set of notation and have listed all in Appendix A. Also conversions between the English and System International (SI) systems are necessary. Thus Appendix B was constructed to facilitate these conversions.

An understanding of the principles and concepts from hydrology and hydraulics are very important for the design and operation of stormwater management systems. Hydrology is the study of waters and their occurrence on, above, and below the earth's surface. Methods to quantify the hydrologic processes of precipitation, evaporation, transpiration, rainfall excess, runoff, and infiltration should be known if we are to control flooding and pollution problems in a technically efficient way.

The study of hydraulics aids in explaining and quantifying the movement of waters on or below the surface. This requires basic scientific knowledge similar to that required for hydrology. Thus there is some overlap of areas of study. We cannot conduct stormwater management work without an integrated use of both hydrologic and hydraulic concepts.

We must also consider the quality of waters. This is important from a public health viewpoint, and thus environmental engineering chemistry and biology concepts and principles are required. Environmental engineering applies basic biological, hydrologic, hydraulic, and chemical principles for water pollution control as well as other human–environment interactions. Certainly, the designer and operator of stormwater management systems should be concerned with the interaction these systems have with the environment.

Finally, the engineering aspects of stormwater management requires an examination of the economic, social, and political impacts of all projects. Thus effective stormwater management requires an interdisciplinary body of knowledge for the planning, design, and operation of waters.

1.2 MAGNITUDE OF STORMWATER PROBLEMS

If a water-related problem exists in the minds of many people, the technical community develops specific measures for the problem and searches for the reasons the problem exists. The cause-and-effect relationship should be identified before a reasonable low-cost and technically effective stormwater management program is begun. The traditional problem of stormwater management is flooding. If flooding of an area exists, the high-water "mark," rates of flow, and storage volumes can be measured as the definition of the

1.2 MAGNITUDE OF STORMWATER PROBLEMS 3

problem. Why the flood occurred and its frequency of occurrence are common questions of the stormwater management area.

The two most common causes of flooding are heavy-volume rainfall and rapidly melting snow that can also be mixed with rainfall. Since rainfall and snowmelt cannot be controlled, designs, and operation of transport and storage system must be done to minimize the flooding. About 7% of the land in the United States (almost the size of Texas) is in the floodplain. Floodplains are low areas adjacent to streams, lakes, and oceans that are subject to flooding once every 100 years. A 100-year flood is one likely to be equaled or exceeded on the average only once every 100 years. This statement is meaningful only over long periods of time (centuries). It is possible that the "big" floods will occur at shorter durations. Failures of upstream controls (dams, ponds, etc.) also cause flooding. Floods in the United States cause damages totaling over $2.2 billion per year, and in the 1970s, flood-related deaths were 200 per year, with another 80,000 people being forced from their homes per year (U.S. Water Resources Council, 1981).

Since the 1960s the problems of water quality in lakes, estuaries, and rivers that receive only stormwater have presented many opportunities (challenges) to those interested in solving the problem of a degrading environment. In the past 50 to 100 years, water quality enhancement has been achieved by managing the readily identified sewage and industrial treatment systems. In the United States alone, the cost to manage these sewage and industrial waste systems between 1974 and 1988 was about $200 billion (Novotny and Bendoricchio, 1989). The money was primarily from federal sources. The other source of pollution (stormwater) is being managed primarily by local programs (state, regional, city, county levels) at a much lower funding level. The discharges from sewage and industrial systems are relatively easy to identify on a map because they appear as a "point" and thus are called *point sources*, whereas stormwater sources are spread out on a map and are called *nonpoint sources*.

Nonpoint sources are atmospheric fallout, surface runoff that immediately follows rainfall, low-flow longer-duration groundwater infiltration, and residual chemicals and sediment that release pollutants to the water column over longer periods of time. The damaging effects of these long-term and time-variable sources may be of greater magnitude than the short-term surface runoff events. However, the surface runoff may carry with it the chemicals that cause long-term effects. The International Joint Commission on the Great Lakes North America reported that nonpoint sources, including atmospheric fallout, account for over 50% of the measured organic chemicals and about 98% of the suspended solids in the Great Lakes (International Joint Commission on the Great Lakes, 1978; Sonzogni et al., 1978). In combined sewered areas with secondary treatment, storm-generated runoff accounted for 40 to 80% of the annual oxygen-consuming load (Pisano, 1976). For separate systems in Florida, Livingston and Cox (1989) reported that 85 to

90% of the metals, 90% of the oxygen-consuming material, over 50% of the nutrients, and 99% of the suspended solids are from nonpoint sources. Suspended solids concentration from nonpoint sources far exceed those from point sources (Soderlund and Lektinen, 1972).

Point versus nonpoint pollution problems were assessed by the Association of State and Interstate Water Pollution Control Administrators (Savage, 1985). In the United States, 43% of the states indicated that their current nonpoint-source problems are greater in magnitude than their point source problems. All states have some water quality problems and economic impacts due to nonpoint sources. Also, there was indicated a continuing need for hydrologic and hydraulic data to define more clearly the magnitude of surface and groundwater problems. It is hoped that a greater understanding of stormwater management will reduce the financial burden.

This fiscal hardship is not confined to the United States; other countries have similar problems. In the United Kingdom, the Water Resources Act of 1963 established new stream gaging stations to provide additional data for stormwater management. International organizations have focused on the solution of water-related problems through stormwater management. This is evident by creation of the International Drinking Water Supply and Sanitation Decade (1980s) by the United Nations. International bodies are financing hydrologic and hydraulic studies to provide data useful for stormwater management.

1.3 HYDROLOGIC CYCLE

The *hydrologic cycle* is a water balance that accounts for the endless circulation of the earth's water. Energy to operate the cycle in its present form is derived from the sun and gravity. The sun's heat vaporizes water from water surfaces (i.e., lakes, oceans, ponds, rivers, and soil pores). Water also is lost to the atmosphere from plants that use water available through the plant's root zone. This is called transpiration and requires the energy of the sun. Gravity causes movement from higher to lower elevations on the earth's surface. Also, water is lost from the upper surficial aquifer to deeper aquifers because of gravity and pressure differentials. However, geologic formations and pressure differentials may provide opportunities for groundwater to move from lower to higher aquifers (Hammer and MacKickan, 1981).

The hydrologic cycle is vital for sustaining life as we currently understand it. Human beings require directly a definite quantity and quality of water for consumption, and indirectly, the hydrologic cycle is necessary to maintain human food sources and social values. Once rainfall on a region is lost by surface discharge to other regions, it may not be economically recovered, political boundaries may make it lost forever, or salt water and other forms of contaminants may make it unfit for human uses. Then it should be clear that those who are interested in the hydrologic cycle must focus attention on

1.3 HYDROLOGIC CYCLE 5

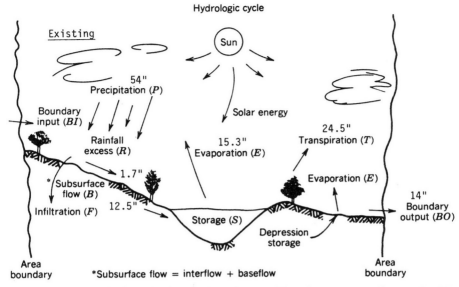

Figure 1.1 Hydrologic cycle for rural sandy area (data from an actual watershed in southwest Florida).

methods to measure and calculate runoff rates and rainfall excess (runoff volume).

A *water budget* is an accounting in mathematical terms (a quantification) for waters of the hydrologic cycle within a fixed boundary. A water budget requires measurements or estimates of the hydrologic variables, some of which are precipitation, evaporation, transpiration, and surface flows. Also, for some regions, water flow from the surficial to deep aquifers or flows in the reverse direction may be measured or estimated from other data.

A common situation that uses stormwater management concepts and principles is the duplication of an existing hydrologic cycle after a land-use change. A water budget for an existing rural area and then another budget for a proposed land development change will be used (Figures 1.1 and 1.2) to illustrate that water can be lost from a region and water tables are lowered after a watershed is paved. The data are representative of an area of southwest Florida. Boundary output is measured as streamflow, and streamflow is the sum of runoff (overland flow) plus groundwater that has infiltrated into the stream. Stormwater management that conserves water within the watershed may be necessary.

The water budget for Figure 1.1, assuming no change in groundwater and surface water storage and no deep aquifer exchange during the time period (1 year), is:

$$\text{inputs} - \text{outputs} = \text{storage change} = 0 \qquad (1.1)$$

6 INTRODUCTION

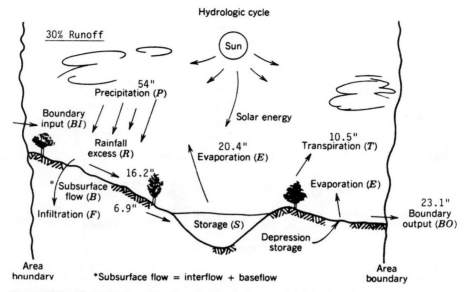

Figure 1.2 Hydrologic cycle after developing a rural sandy area (data from an actual watershed in southwest Florida).

Inputs:

$$\begin{aligned}\text{Precipitation} &= 54 \text{ in.}\\ \text{Boundary inputs} &= \underline{0 \text{ in.}}\\ \text{Total inputs} & 54 \text{ in.}\end{aligned}$$

Outputs:

$$\begin{aligned}\text{Evaporation} &= 15.3 \text{ in.}\\ \text{Transpiration} &= 24.5 \text{ in.}\\ \text{Boundary outputs} &= \underline{14.2 \text{ in.}}\\ \text{Total outputs} & 54.0 \text{ in.}\end{aligned}$$

The boundary output is from a streamflow gage for a 60.2-mi^2 rural watershed in southwest Florida and thus measures overland flow (runoff) and groundwater. For a rural area with soils that have rapid infiltration rates and high storage capacity, very low surface runoff can be expected. If the soils have slow infiltration rates and low potential storage capacity, more rainfall excess and higher runoff rates can be expected.

Next, this rural area land use is changed and is partially paved over, with 30% of the impervious area being directly connected impervious areas going directly to surface-water bodies. Since some vegetation is removed to provide building space for the buildings, driveways, streets, and the like, the volume of water from transpiration is reduced.

However, stormwater management requires additional ponding areas for flood and water quality control; thus evaporation increases. The water budget for 30% directly connected impervious areas (30% runoff) is shown in Figure 1.2. The boundary output consists of runoff and groundwater infiltrating the surface waters. The rainfall excess in volume terms (also called runoff volume and defined as the volume of overland flow available for discharge from the property) is

$$R = CP \tag{1.2}$$
$$= 0.3(54)$$
$$= 16.2 \text{ in.}$$

Based on remaining water and land areas with the same evaporation and transpiration rates used for the water budget of Figure 1.1, the water budget for the developed case of Figure 1.2 is completed.

A comparison of Figures 1.1 and 1.2 reveals that rainfall excess and streamflow have increased while groundwater flow (water table) has decreased. The rate at which rainfall excess occurs (runoff) also increased (see Figure 1.3). Once the impervious area of a region increases, more water is

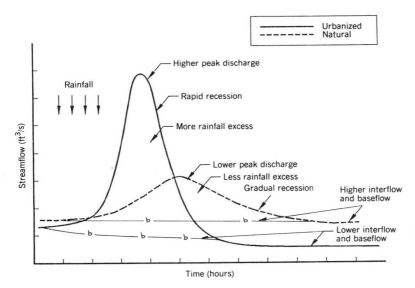

Figure 1.3 Comparison of hydrographs from natural and urbanized areas with high groundwater storage.

8 INTRODUCTION

lost to the surface water bodies at a faster rate than before the land-use change. The construction of holding ponds can return rainfall excess to the ground by infiltration and irrigation. This is an approximate duplication of the predevelopment hydrologic cycle that could benefit not only the existing plant and animal life but also prevent pollution and flooding of downstream regions. Thus it is important to understand how to calculate rainfall excess and runoff rates from a watershed for all different watershed characteristics.

1.4 STORMWATER MANAGEMENT OBJECTIVES AND LIMITATIONS

Objectives are specific end results that are formulated from general goals. The goals frequently are statements of a particular problem coming from the general public. Also, special-interest groups present problem areas that become part of the general goals. Stormwater management goals are usually related to economic development, historical preservation, and environmental protection. However, there generally exists a conflict between economic and other goals. If one statement is a goal, the others may become a limitation or a minimum level of achievement. An example would be the storage of streamflow waters to generate hydropower, but the storage cannot exceed a fixed elevation because of a need to protect wildlife and save historical areas. The needs of special-interest groups and the general public welfare must be examined to set goals and objectives. A listing of some of the more common objectives of stormwater management is introductory to new students of the subject and reinforcing to practitioners. Each of these objectives can be stated separately but more frequently are combined, especially in the management of large watersheds.

1. Prevent the flooding of an area from a specific frequency event, such as the one-in-50-year runoff event. Perhaps this is the historical goal of stormwater management. There is a probable economic impact expressed as loss of property and a cost associated with the structures that prevent the flooding.

2. Reduce land loss by a fixed percentage due to erosion and sedimentation resulting from rainfall. If the land is in agricultural use, the goal is to help improve the economic base of the community.

3. Provide a reservoir of specific size for irrigation and municipal supply. Streamflow and rainfall records with possible losses due to evapotranspiration and infiltration must be examined for each site. The goal is primarily one of economic value to a community.

4. Implement a combination of programs to reduce by 80% the mass loading of chemicals and suspended solids to a receiving water. The goal is to help preserve the environmental quality of the receiving water.

1.4 STORMWATER MANAGEMENT OBJECTIVES AND LIMITATIONS

5. Reduce postdevelopment peak discharge to predevelopment conditions. The goals are to protect downstream users from economic harm and to maintain environmental values.

6. Maintain on-site a fixed percentage of rainfall excess. The goals are to recharge groundwater and provide for passive recreation opportunities. Also, in an estuary environment, a goal may be to reduce freshwater inputs.

A limitation is a restriction or a narrowing of options available for the achievement of an objective. One of the objectives listed in the preceding section may also be a limitation when another objective is used. Examples of some of the more commonly used limitations are:

1. *Cost.* An organization responsible for stormwater management has a monetary budget that limits both capital (construction) and operating expenses. Furthermore, emphasis is placed on cost-effective systems, which are those that accomplish the stated objectives, such as peak flow reduction or pollution removal, at a minimum cost. Any one of the objectives listed in the preceding section can also have the mention of cost in the objective statement.

2. *Site Feasibility.* The impact on land amenities may be substantial. Minimum acceptable changes in water area and landscaping should be specified. Some controls may appear unsightly, provide no other direct benefits, and are plagued by nuisance complaints. Some stormwater ponds may be called attractive hazards to children because of their desire for water activities. Thus some ponds must be fenced. In addition, tests to verify hydrologic variables at the site should be done. As an example, consider infiltration rates as a hydrologic variable affecting the size of a pond. Sometimes sites are picked for infiltration ponds based on published general soil data without a site investigation to estimate the rate of infiltration. This practice of using generalized data can lead to false results.

3. *Environmental Impact.* The removal of sediment, chemicals, and biological organisms that cause a detrimental effect on water quality may be specified. In some situations, pollutant removal to enhance water quality is essential to achieve desired benefits from a receiving water body and its adjacent lands. Stormwater management methods that reduce quantities of water and pollutants should be used along with the traditional controls of sewage and industrial discharges.

4. *Potential Reuse.* Stormwaters released to the oceans have lost their reuse potential. Reservoirs and ponds can be constructed to store fresh water for future uses. A certain percentage or quantity can be specified for a beneficial use. An example would be a stormwater pond that must provide a minimum quantity of water for irrigating a golf course.

5. *Floodplains and Wetlands.* The area adjacent to a stream or lake may need to be preserved for storage and transport of stormwater and preserva-

tion of wildlife. In addition, the floodplain may provide space for recreation, agricultural development, space for important and diverse plants and animals, and open space for aesthetic purpose.

6. *Labor and Maintenance.* Some control methods are more labor intensive and present and future laborers are not available. Besides the number and quality of maintenance people, the cost should be included in any analysis.

7. *Institutional Preferences.* The legal and political entities may dictate the bounds on any stormwater management system. This is frequently translated into land, personnel, and budget terms and may be expressed in unclear or somewhat cryptic statements. A written statement of facts and preferences should be encouraged.

1.5 STORMWATER MANAGEMENT POLICIES

A *policy* is a definite course or statement of action selected from many alternatives that is used to guide and determine present and future decisions. Each policy is reviewed periodically for relevance to the community. All existing and proposed policies should be readily available to the public and the professional community. A number of policies are necessary for effective stormwater management. Some of these are:

1. *Criteria.* These are statements relative to specific details used for design, operation, and maintenance that are written as regulations or rules. These details are necessary to ensure proper compliance with objectives and limitations. Some examples related to hydrologic variables are:

a. Rainfall volume used for sizing flood control structures is based on the one-in-100-year, 24-h-duration storm event. The frequency–intensity–duration curves for an area can be used to determine this volume.

b. Peak flow rate discharged to adjacent water bodies after development cannot exceed predevelopment peak rates.

c. Plant life in a lake must not produce an evapotranspiration from the lake area which exceeds evaporation by more than 10%.

d. Sediment and oxygen-demanding materials in stormwaters must be reduced by an amount equivalent to that from point sources (e.g., 80%).

e. On-line wet detention ponds must be provided to store the runoff waters from 3 in. of rainfall, and the release must occur over a 5-day period.

f. The rainfall excess from the first inch of rainfall must be diverted to an off-line pond and prevented from direct discharge to a receiving water body.

Development of the foregoing policies is complex and time consuming, but similar criteria have been determined and published for regional and statewide use (State of Maryland, 1987; State of Florida Department of Environmental Regulation, 1985; Austin, Texas, 1986; South Florida Water Management District, 1987; Washington Metropolitan Water Resources Planning Board, 1987). It should be noted that criteria may vary from one region to another. Criterion (a) above varies with the type of system and the risk involved in the failure of the system. Some roadway drainage systems on low-traffic roads specify as low as a one-in-5-year rainfall volume and intensity associated with watershed time of concentration. On the other extreme, the floor (pad) elevation for most residential structures must be set at the elevation resulting from floodwaters from the one-in-100-year storm event. This is a policy of the National Flood Insurance Program. The floodwater areas are delineated using synthetic hydrograph and flood routing procedures with statistical analyses of streamflow records.

2. *Drainage Plans.* For each watershed, the boundaries should be determined and existing conduits and storage areas delineated. A policy on interbasin transfers must be determined. The greater the details on land use, pipe sizes, invert elevations, storage volumes, slopes, and so on, the better the analysis for hydrologic parameters, environmental impacts, and future designs.

3. *Financial Plan.* This establishes how much money is available, who pays, and when to spend it on a stormwater system. Methods to assess the users of the system must be determined, such as taxing districts, or the income must come from the general revenue base. The basis for establishing a monetary rate for services must be determined. One equitable approach may be to establish a rate based on a directly connected impervious area. Another is based on equivalent residential units. Equivalency is in terms of runoff volume or peak discharge. The means to obtain money for capital improvement must be identified and operating budgets specified. The financial plan is a major policy issue that must be determined if the stormwater system is to be built and operated in a cost-effective manner.

1.6 FEASIBILITY CONSIDERATIONS

There are at least five feasibility tests that should be used before a stormwater management plan is found acceptable. The first is engineering feasibility. The remaining four feasibility tests are: economic, financial, political, and social. The engineering tests must prove that the project is physically, biologically, and chemically capable of performing its intended function. For this, a basic fundamental understanding of hydrology, hydraulics, and water quality is necessary. Most fundamental principles of hydrology and hydraulics help establish the physical feasibility of stormwater systems and have been presented in other publications (Wanielista, 1990; Ackers et al., 1978). Applications of these principles to the design of stormwater management systems and an understanding of the remaining feasibility tests are needed for an effective stormwater system. The system simply must perform as designed and operated.

Economic feasibility is determined by comparing benefits of a project (in addition to those without the project) to the cost of the project. If benefits exceed costs, the project is judged economically feasible. Traditionally, economic decisions consider the production, distribution, and consumption of a commodity. Stormwater, if managed, is a commodity that is stored, distributed, and used. Decisions on the size of a stormwater-related project must involve income, expenditures, and other economic impacts of the project.

The overall cost of a water-related project is frequently not considered or is not obvious. A project may cause an effect outside its boundaries. The cost associated with "off-site" effects are given the term "indirect costs." The indirect costs can be substantial and affect many people. An example is a set of water controls within one county that cause floods in another county. Other examples are related to pollution and are not always well defined or obvious. It is important to include the proper identification of benefits and cost when completing a stormwater plan.

Economic analysis of a project provides a thought process that helps identify direct and indirect costs. As an example, surface and groundwater quality can become altered by stormwater chemicals and sediment. Sediment resulting from soil erosion can be a carrier for chemicals. Instream and adjacent water damages due to soil erosion were calculated to be approximately $3.2 billion (1980) per year (Clark, 1985). The instream damages were those caused by sediments and nutrients and included damages to water-based recreation, navigation, property, reservoirs, and commercial fishing. Adjacent stream damages were estimated from sediment-related damages, drainage-related repairs, and electric power maintenance activities. In 1979, one of every 15 jobs in the United States was related to leisure activities (Clark, 1985). Loss of water-based recreation can cause significant loss of employment.

Financial participation in large stormwater projects requires an identification of both on-site and off-site costs. The closer one can make an economic study to the real cost, the more likely it is that a financial plan can be developed and supported by all affected parties. The economic, hydrologic, and environmental plans should be used to specify a financial plan. The financial plan is one that specifies sources of money for the cost of construction and operation. The financial plan may be the final measure of what gets done. A plan may be economically feasible but financially infeasible because the benefits may be intangible or distributed among so many people for payment to be impractical. Thus governments may wish to participate in a financial way in many regional stormwater management projects.

Political feasibility is determined by legal and public support of a project. Politicians measure the needs and wishes of the people and translate feelings into actions. Social acceptance is related to cultural habits and change. How will a stormwater management project change the lives of people? The social acceptance is measured by how well people accept the change.

1.7 STORMWATER MATHEMATICAL MODELS

Stormwater mathematical models are mathematical and logical equations that relate stormwater objectives with hydraulic, hydrologic, engineering, and economic data. Data include measures of precipitation, evaporation, infiltration, transpiration, runoff, storage, wind velocities, solar radiation, quality changes, and other parameters too numerous to provide an extended list. There is a vast quantity of data from both hydrology and hydraulic disciplines and other technical and nontechnical areas required to describe stormwater systems. The mass of data forms important inputs and outputs for equations in mathematical models. Mathematical models reproduce system performance that otherwise would be impossible to reproduce in a relatively short time period. As an example, consider estimating runoff for a 100-year-return-period rainfall. If we had to wait for this event, it may not occur in our lifetime.

1.7.1 Types of Models

Mathematical models for surface flows are abstractions of the surface-water systems in equation form. A mass balance relating runoff to precipitation and storage is a simple example. Mathematical models for surface waters can be divided into at least three basic types: (1) single event, (2) multiple or continuous events, and (3) economic.

Single-event models usually are based on design storms and many are derived from empirical approaches using uncomplicated relationships among watershed parameters, experience factors, and judgment. One example of a single-event mathematical model is the rational formula for estimating a single event (peak discharge) as a function of rainfall intensity, watershed area, and a coefficient related to land use and other ground-cover conditions. The rational formula will continue to be very popular among designers because of its ease of application and the credible results obtained on "small" watersheds (those areas with time of concentration short enough so that rainfall intensity can be considered to be constant during that time period). Rainfall intensity and the runoff coefficient are input parameters for which the designer can exercise judgment; thus a "safe" design can be obtained for "small" watersheds.

A single-event model usually has parameters that are fixed for a given period. That time period can be as short as a few hours or as long as one year. An example of a single-event model capable of using different time scales is an empirical model for sediment studies called the universal soil loss equation (USLE). It is the product of six factors, one of which is the distribution of rainfall with time. Soil loss to surface waters in terms of mass and concentration of sediment is an important quality consideration. This model is discussed further in Chapter 11.

Examples of multiple-event models are some of the hydrograph models, such as unit hydrograph approach, the Environmental Protection Agency Stormwater Management Model (SWMM), the U.S. Hydrologic Engineering Center hydrograph model (HEC-1), and the Stormwater Management Design Aid (SMADA) model of the University of Central Florida. The models can be used for both single events and multiple events. When used for multiple events, flow rates and storage volumes can be calculated for longer periods of time or larger watersheds can be analyzed for nonindependent conditions. As an example, synthesized rainfall and flow records can be generated based on probability distributions for long time periods, producing various combinations of rare hydrologic events. These models are generally computer executed and useful in design and operation of large multiuse water resources systems. Other statistical methods are used to estimate extreme conditions of the hydrograph (floods and droughts). Multiple-event models vary in details and, as such, the input data can range from minimum to extensive.

The last model types are classified as economic because they incorporate capital and/or operating costs. Essentially, economic models are useful for allocation of economic-related resources subject to technical and institutional constraints. On large multiuse reservoir storage and downstream use models, economic benefits and costs are assigned to each water user. Then, based on present-value calculations or other suitable economic models, water is allocated to each user. A community may also be interested in the least-cost location of wastewater treatment facilities and the least cost of flood control. Again, the hydrologic or flow models are integrated with economic criteria for the "best" or least-cost combination of treatment consistent with hydrologic principles.

1.7.2 Stick Diagram

Complex water resources systems are simplified for mathematical modeling using a schematic of the system known as a *stick diagram* (Figure 1.4). Consider the allocation of surface waters among many users. Example complex systems for the transport of waters are found in northern California to southern California, the allocation of waters from the Rio-Colorado, Colorado River, Nile River, Lake Okeechobee, and many others. The issue

1.7 STORMWATER MATHEMATICAL MODELS

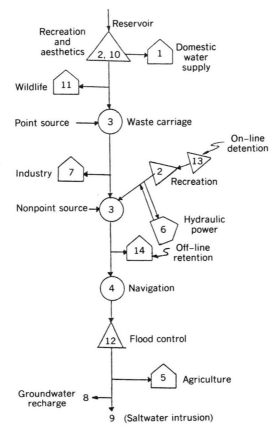

Figure 1.4 Schematic of a water allocation system (otherwise known as a stick diagram). Numbers refer to beneficial water uses; arrows for flow direction.

can simply be stated as what levels of water quality and quantity can be provided for each user within ecological, institutional, and technical constraints. Many users of water may be in competition. This could force reduced consumption, additional storage capacity, or basin boundary transfers in order to satisfy yearly needs. These uses are illustrated in Figure 1.4 and may be categorized as (1) domestic water supply, (2) recreation, (3) waste carriage, (4) navigation, (5) irrigation (agricultural), (6) power generation, (7) industry, (8) groundwater recharge, (9) prevention of saltwater intrusion, (10) aesthetics, (11) wildlife, (12) flood control, (13) on-line detention, and (14) off-line retention. (These last three refer primarily to stormwater practices or methods to manage the quantity and quality of stormwaters.)

The stick diagram is a simplification of the "real-world" condition. To design the transport systems that move water from storage to a user requires

16 INTRODUCTION

the use of a single-event or hydrograph-type model. Maximum or peak flows determine the size of conduits for small watersheds, while the addition of hydrographs would be required for determining peak flows when multiple transport systems blend into one transport system at different geographic locations (Figure 1.4). Water quality may also be considered. Dissolved oxygen quality predictions are possible using existing models for various flow and wastewater discharge conditions. The design of the reservoir storage of Figure 1.4 may be done with the aid of stochastic models to predict flood events. Similarly, the stochastic models are useful for predicting drought conditions.

Some stormwater management practices control both water quantity and quality. Here we will use three of these to demonstrate the construction of a stick diagram. Nonpoint sources can be controlled at three locations, as shown on the nonpoint-source diagram of Figure 1.5. The question of how much water should be controlled, at what cost, and of what quality relate to stormwater management. To answer these questions, on-line detention, off-line retention, or on-line retention is available and the source diagram of Figure 1.5 is translated to a stick diagram as shown in Figure 1.6.

Note that the stick diagram is a simplification of the real situation. *Retention* refers to holding stormwater for long periods of time or preventing it from direct discharge to a receiving water. *Detention* refers to on-line stormwater ponds that detain water for short periods of time (hours to days). They are usually designed to reduce peak flow rates but can also be used for limited water quality improvement. *On-line retention* refers to ponds that retain water for long periods of time (days to months). On-line detention has both inlet and outlet structures on the same drainage system. Off-line structures have only an inlet structure for diverted waters. The diverted water evaporates, infiltrates, or goes to other watersheds, but does not return, except possibly by groundwater flow, to the surface waters of the originating drainage system.

The foregoing discussion of surface-water models illustrates the use of mathematical models in hydrologic studies. Single-event models are usually used to estimate one event (peak flow or sediment load), multiple-event models are used for a continuous record (quantity or quality with time), and economic models are used for the allocation of resources associated with hydrologic events. These classes of mathematical models provide the majority of information necessary to plan, design, and manage water systems.

Example Problem 1.1 Historical records on pond levels are available for a rectangular holding pond (10 m by 20 m) for the storage and infiltration of stormwater. The hourly pond levels for one storm are shown in Table 1.1. Precipitation volume equals the evapotranspiration volume during the period of measurement. Calculate the rainfall excess (input volume) to the pond if the infiltration rate for the pond is constant at 4 cm/h and the pond storage

1.7 STORMWATER MATHEMATICAL MODELS 17

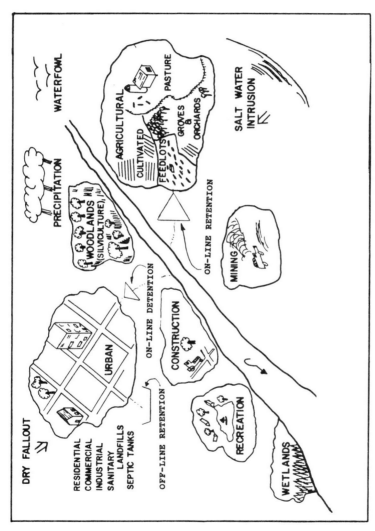

Figure 1.5 Nonpoint sources with three possible stormwater management location sites.

18 INTRODUCTION

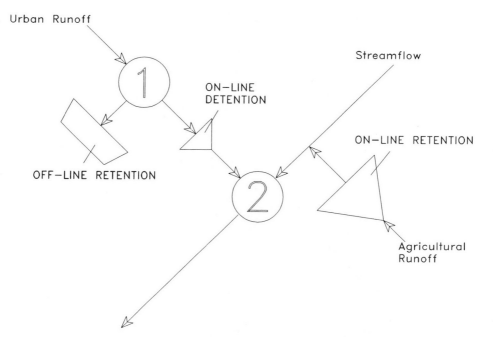

Figure 1.6 Stick diagram of a stormwater management system.

TABLE 1.1 Pond Levels for Example Problem 1.1

Pond Level[a] (cm)	Time (h)
0	0
10	1
16	2
13	3
10	4
6	5
2	6
0	7

[a]Measurements at the end of the hour.

volumes per time period are shown in Table 1.2. Assume that the rate of infiltration is through the total pond bottom.

SOLUTION: Use a mass balance equation with units of depth (volume) = cm:

$$\text{ending storage} = \text{beginning storage} - \text{infiltration} + \text{excess}$$
$$S(t + 1) = S(t) - F(\Delta t) + R(\Delta t)$$

TABLE 1.2 Solution Table for Example Problem 1.1

Given:					Find:	
Time	$S(t)$	$S(t+1)$	$F(\Delta t)$	$R(\Delta t)$	\multicolumn{2}{c}{ΣR}
(h)	(cm)	(cm)	(cm)	(cm)	(cm)	(m³)[a]
0	0	10	4	14	14	28
1	10	16	4	10	24	48
2	16	13	4	1	25	50
3	13	10	4	1	26	52
4	10	6	4	0		
5	6	2	4	0		
6	2	0	2	0		

[a]$R(m^3) = R$ cm × 10 m × 20 m × 1 m/100 cm.

Solve for $R(\Delta t)$:

$$R(\Delta t) = S(t+1) - S(t) + F(\Delta t)$$

It is assumed that infiltration rates are constant.

1.8 COMPUTATION AIDS AND COMPUTER PROGRAMS

Problems in stormwater management frequently require repetitive calculations. Some are solved by trial-and-error methods. Others require repeating the basic solution algorithm many times. Many aids for solving these problems have been developed over the years. Nomographs and coaxial graphs were popular about 25 years ago (1950–1965) and are still used today. They make the computation procedure less complex and save time. Starting in the 1960s, calculators of various types further streamlined the repetitive and routine calculations. Also, around the same time, computers were gaining popularity. In the early 1980s the personal computer (PC) gained greater recognition as an aid to solving complex repetitive problems in hydrology. As more software developed, the availability of PCs increased and cost decreased. In the late 1980s, many researchers, planners, engineers, and hydrologists increased their dependency on computers.

To understand some of the complex problems that the principles of stormwater management can aid in solving, it is frequently necessary to use computer programs. Many computer programs are available. Computer programs are provided with this book and can be used for educational or professional purposes. The programs are resident on a computer diskette. These programs should be installed on your computer and the menus reviewed. These programs are similar to the programs used by professionals with regard to input data and computation procedures. As with any computational aid, all programs are specific to certain types of problems and input data identification is very important.

1.9 SUMMARY

Stormwater management is a body of knowledge used to understand, control, and utilize waters in different forms within the hydrologic cycle. The fundamental knowledge for stormwater management is found in the subject material of hydrology, hydraulics, economics, and disciplines related to water quality. Before methods are developed in the remaining chapters to understand, design, and operate stormwater systems, additional fundamental knowledge of objectives, limitations, and policies for stormwater management should be understood. Additional fundamental concepts from hydrology with example problems are presented in the next chapters. Highlights of some of the ideas and concepts from this chapter are:

- Stormwater appearing during and immediately after a precipitation event are primarily surface runoff waters. However, groundwater infiltration to surface waters over a longer period can also be a significant amount of the total streamflow.
- Frequently specified objectives for stormwater management relate to (1) flood control, (2) erosion control, (3) reuse storage, (4) pollution reduction, (5) peak discharge reduction, and (6) groundwater recharge.
- Many limitations to the attainment of objectives do exist. A partial listing would include the objectives plus (1) cost, (2) site feasibility, (3) environmental impact, (4) potential reuse, (5) floodplain preservation, (6) labor, and (7) institutional preferences.
- A policy should be developed that specifies criteria for design and operation, existing drainage details, and financial plans. Economic plans help determine a stormwater management plan, while a financial plan aids in determining funding sources for the construction and operation.
- Nonpoint sources of pollution are in general dispersed over an area and typically not permitted by governmental bodies to discharge pollution. As the economic and environmental significance of stormwater discharges are defined, additional governmental controls should be expected. Rare hydrologic events and the hydrologic outcomes of projected land-use changes require the use of predictive techniques for which mathematical models are primarily used.

1.10 PROBLEMS

1. Formulate a goal with objective(s) and policies for a stormwater management program.
2. For the objective(s) of Problem 1, list some constraints (written and mathematical) to obtaining these objectives.

3. Which of the following may be considered as nonpoint sources of pollution?
   a. Gasoline leaks from underground tanks at a gas station
   b. A permitted solid waste landfill
   c. Runoff from fertilized agricultural land
   d. Treated wastewater from an industrial plant

   Name two more possible nonpoint sources of pollution.

4. For the following hydrologic processes, discuss a control measure and what you hope to achieve.
   a. Precipitation
   b. Evapotranspiration
   c. Rainfall excess
   d. Runoff
   e. Infiltration

5. State engineering quantities for your region and an objective for each for the following criteria.
   a. One-in-25-year rainfall in 24 h
   b. One-in-100-year rainfall in 24 h
   c. Flow rate control
   d. Yearly evapotranspiration
   e. Pollution control volume

6. A retention pond is designed to infiltrate stormwater runoff at a rate of 2.0 in./h. The area of the pond is 1 acre and consider the area to be constant with depth. The watershed area is 50 acres. If the average rainfall intensity of a one-in-100-year storm is 1.9 in./h and the storm lasts 4 h, what is the depth of water in the pond after 2 days? Assume that 30% of the rainfall is available for runoff into the pond and that the pond is empty before the storm. Evapotranspiration from the pond is 3 mm/day.

7. What area must the retention pond in Problem 6 have for it to empty in 1 day? Neglect losses from evapotranspiration. Assume that all runoff has reached the pond.

8. Explain in your own words two problems commonly used to illustrate the magnitude of the stormwater problem.

9. Construct a hydrologic balance on a yearly basis for a 40% directly connected impervious area in Florida and determine the expected streamflow in a year that can result from a 50-in./yr rainfall. First, list all your assumptions on the percentage of land area in water and in vegetation. Next, list your assumptions for evaporation and transpiration rates. Assume that there is no surface storage within the watershed, but the surficial aquifer contributes to streamflow and no exchange of waters with the deeper aquifer.

1.11 COMPUTER-ASSISTED PROBLEMS

1. Install the computer program diskette provided with this book. Review the menus of programs and locate at least one chapter of this book where the programs can be used. This exercise will begin to familiarize one with the computer programs and the subject material.

2. Using your own disk, develop a computer program to convert any input data from pounds/acre-day to kilograms/hectare-year; from acre-feet to million gallons; from cubic feet/second to cubic meters/second; from inches/hour to millimeters/hour; and from liter/second to gallons/minute. Document the program and make it user friendly. Print out the input data as well as the converted data along with the appropriate units.

1.12 REFERENCES

Ackers, P., White, W. R., Perkins, J. A., and Harrison, A. J. M. 1978. *Weirs and Flumes for Flow Measurement*, Wiley, New York.

Austin, Texas. 1986. "The Evolution of Erosion and Sedimentation Control Regulations and Enforcement," presented by G. E. Oswald at the *State of Maryland Sediment and Stormwater Management Conference*, Salisbury, Md., July.

Clark, Edwin H. 1985. "The Off-Site Costs of Soil Erosion," *Journal of Soil and Water Conservation*, Vol. 40, No. 1, pp. 19–22.

Hammer, M. J., and MacKickan, K. A. 1981. *Hydrology and Quality of Water Resources*, Wiley, New York.

International Joint Commission on the Great Lakes. 1978. *Environmental Management Strategies for the Great Lakes*, International Joint Commission, Windsor, Ontario.

Livingston, Eric, and Cox, John. 1989. *Florida Development Manual*, Vol. I, Florida Department of Environmental Regulation, Tallahassee, Fla.

Novotny, Vladimir, and Bendoricchio, Guiseppe. 1989. "Linking Nonpoint Pollution and Deterioration," *Water Environment and Technology*, Nov., pp. 400–407.

Pisano, M. 1976. "Nonpoint Sources of Pollution: A Federal Perspective," *Journal of Environmental Engineering Division*, ASCE, Vol. 102, pp. 555–565.

Savage, Roberta. 1985. "State Initiatives in Nonpoint Source Control," *Journal of Soil and Water Conservation*, Vol. 1, No. 1, pp. 1 and 17.

Soderlund, G., and Lehtinen, H. 1972. "Comparison of Discharges from Urban Stormwater Runoff, Mixed Storm Overflow and Treated Sewage," *IAWPR International Conference*, Jerusalem, Israel, Pergamon Press, Oxford, England.

Sonzogni, W. C., Monterth, T. M., Bach, W. N., and Hughes, V. G. 1978. *United States Great Lakes Tributary Loadings*, International Joint Commission, Windsor, Ontario.

South Florida Water Management District. 1987. *Permit Information Manual*, Vol. IV, West Palm Beach, Fla.

State of Florida Department of Environmental Regulation. 1985. *Stormwater Management*, Chapter 17.25 F.A.C., Tallahassee, Fla., Apr.

State of Maryland. 1987. *Guidelines for Constructing Wetland Stormwater Basins*, Maryland Department of National Resources, Baltimore, Mar.

U.S. Water Resources Council. 1981. *Floodplain Management Handbook*, U.S. Government Printing Office, Washington, D.C.

Wanielista, M. P. 1990. *Hydrology and Water Quantity Control*, Wiley, New York.

Washington Metropolitan Water Resources Planning Board. 1987. *Controlling Urban Runoff: A Practical Manual for Planning and Designing Urban BMP's*, T. R. Schueler, Washington, D.C.

CHAPTER 2
Probability and Statistical Methods for Hydrologic Events

Hydrology is the study of the occurrence and distribution of water on, above, and beneath the surface of the earth. Any study of stormwater management must include an understanding of the methods of probability and statistics to define hydrologic events. *Extreme events* such as floods and droughts are those that occur infrequently. Examples of extreme events are the peak streamflow value in a year or the maximum rainfall over 24 h given a specific interevent dry period. Commonly used methods to estimate extreme events are statistical or probabilistic. These methods use data for an event (rainfall volume, runoff rate) to estimate some extreme value for that event. The major questions addressed in this chapter are (1) how to quantify rainfall and runoff and (2) how to use existing data with the methods of probability and statistics to estimate hydrologic events.

2.1 INTRODUCTION

Flooding and water quality deterioration are related to the precipitation process of the hydrologic cycle. Thus it is convenient to initiate hydrologic analyses with a discussion of the precipitation process. Precipitation is a random event over time and can be defined by volume, intensity, and interevent dry periods. A graph of the process is shown in Figure 2.1. Intensity (rate of precipitation) is plotted on the vertical axis and is displayed below the horizontal axis to indicate a "falling" condition. The plot of intensity versus time indicates a changing intensity with time and is called a *hyetograph*. The typical units of intensity are inches per hour and millimeters per hour. On a continuous time scale, the intensity is variable but frequently is expressed as a constant value per time period because of some reporting times for precipitation measurements. Also, some gages record continuously but interpretation of the data is more convenient and accurate if a time-interval accumulation is used. During the duration of a rainfall event, there may be time periods during which there is little or no rainfall.

The definition of a rainfall event has to be consistent with the use of the data. When rainfall is recorded on a daily basis, it is convenient to report a

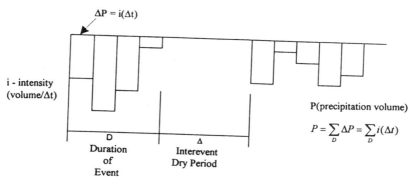

Figure 2.1 Precipitation process.

rainfall event with a duration of rainfall in days. Then the minimum *interevent dry period* (period of no rainfall) is 1 day. However, the choice of the minimum interevent dry period should be made such that the precipitation events are independent of each other. Typical interevent dry periods for rainfall independence range from 4 to 6 h (Hvitved-Jacobsen and Yousef, 1987; Arnell, 1982; Wenzel and Voorhees, 1981; Driscoll et al., 1989).

There exist other situations for the choice of other interevent dry periods. First, for an urban watershed, the minor conveyance system (underground pipes) can be designed to transport the runoff from a 6-h storm event, and for this situation the runoff hydrograph takes 12 h to pass through the pipes. Thus, to maintain independence between runoff events, at least a 12-h interevent dry period is required for analysis. For the minimum 12-h interevent dry period, the volume for each rainfall event is tabulated and an empirical frequency distribution developed. For this empirical distribution, the extreme rainfall event is determined. Another example is the choice of a rainfall event volume to design a stormwater detention pond. If the pond drains in 72 h after the storm event, the minimum interevent dry period to tabulate rainfall volume is 72 h. A conditional empirical probability distribution results. From this distribution, an extreme event volume for design is determined.

2.1.1 Frequency – Intensity – Duration Curves

Design of structures to transport runoff is a common task for the control of floods. The rate of transport in a gutter, ditch, or pipe is related to the intensity of precipitation. Once the total watershed area drains, the rate of precipitation (input) is equal to maximum runoff (output). Thus the average rate of precipitation for specified frequencies (return periods, such as once every 5 years, once every 10 years, etc.) and storm duration are developed from precipitation event data. The storm duration must be long enough for the total watershed area to drain in order to achieve maximum runoff for a stated return period. The *return period* is the average time period for an extreme event to occur. It is calculated from the frequency distribution for the event. By knowing the exceedence probability and the number of events per time period, the return period is calculated using

$$T_r = \frac{M/N}{1 - \Pr(X \leq x)} \qquad (2.1)$$

where T_r = return period, months or years
$\Pr(X \leq x)$ = probability of occurrences
M = number of months or years in the record
N = number of rainfall events in the record

Using maximum yearly data (i.e., one event per year), a return period of 100 years means an exceedence probability of 0.01. The return periods for floods are given as yearly values, and the return periods used for water quality violations have monthly or less frequent values.

A conditional probability distribution for average rainfall intensity given a geographical area, storm duration, and interevent dry period is developed from rainfall event data. A distribution is developed for many durations, such that a smooth curve relating average intensity of rainfall to averaging time (duration) for a specific frequency is developed. Typical curves are shown in Appendix C and are called FID or IDF curves.

The use of the FID curve requires information on the design frequency, maximum watershed drainage time, and the drainage area. The frequency of overflow (occurrence) is specified based on economic considerations. The duration is a time over which rainfall is averaged, and volume is calculated by multiplying the event duration by average rainfall intensity. For short durations (maximum drainage time of 20 min) conventional design procedures generally use the average intensity to calculate the runoff rate. For a larger watershed, a time distribution for rainfall intensity is used to better represent the changes in rainfall intensity over time. "Time of concentration" is the term used when estimating the maximum watershed drainage time.

2.1.2 Methods for Estimating Extreme Events

If it is desired to predict an extreme event for a location, generally one of the following two categories is used:

1. *Data Sites:* those at or near a gaging station (i.e., streamflow or precipitation) where the record is fairly complete, accurate, and of sufficient length to provide precise estimates of an event.
2. *Nondata Sites:* no data are available; however, synthetic predictive methods can be used, such as synthetic hydrograph generation procedures.

In general, the more data from a gage station, the higher the reliability and confidence in the estimate of return frequency and associated magnitude. Thus the confidence in estimation is high, with verified, relatively complete data.

For estimating extreme events, the following methods are available:

1. Probable maximum floods (PMF)
2. Regression analysis
3. Frequency distribution analysis
4. Synthetic hydrographs

The PMF method relies primarily on the use of meteorological data, indirect measures, and fluid flow equations, but will not be discussed in this book. Regression analysis and frequency distributions are estimation methods where probabilities can be assigned. These methods require a greater understanding of statistical analysis. When data are available, these are the preferred methods. Synthetic hydrograph generation is preferred when few or no hydrologic data are available for a site. This is common for smaller watersheds and for sites where land-use changes are anticipated.

2.2 ASSUMPTIONS FOR STATISTICAL ANALYSIS

Statistical analyses of hydrologic data start with concern about reliability and representative data. Were the data collected and reported with an acceptable degree of error (or is it reliable)? The measuring instrumentation should be calibrated and checked frequently to ensure reliability. A person cannot simply mandate reliability. Measurement errors are usually the greatest during maximum events, such as floods and hurricanes. Calibration of data and honesty in reporting must be checked. At one time period in our history it was difficult to find water quality data in rivers during runoff conditions. Afterall, who wants to sample stream flow during rainy conditions? The data must be representative of a particular situation and include those watershed and meteorological conditions of interest.

2.2.1 Statistical Sampling

It is important to review the concepts and terminology used in forecasting before studying extreme event models. Investigators collect some data from an area, then attempt to estimate probability distributions and numerical values associated with the probabilities. Therefore, care must be exercised to sample only those data from the identified event, thus minimizing the possibility of mixed meteorological and watershed conditions. As an example, over a few years, watershed land use may change and thus the runoff rates

28 PROBABILITY AND STATISTICAL METHODS FOR HYDROLOGIC EVENTS

and volumes may change. The population from which the sample is taken controls the characteristics of the sample. A completely impervious watershed produces one shape of a frequency distribution, while a sandy soil with a low water table produces another shape. Mixing land uses may produce a false estimate for a distribution. Thus the investigator must ensure that the sample is representative of the same land use, producing many events (population). For the analysis of hydrographs and extreme events for mixed land uses, it may be desirable to monitor runoff from each land use separately.

Over time, land-use activities on watersheds may change. This is relatively to document given the area information from local residents. Examples include cover alterations such as urbanization, agricultural crop rotation, and drainage. Changing watershed conditions may change runoff flow rates and volume. This should be documented during the period of measurements. Mixed populations caused by different covers on the watershed may be present in the sample. However, an investigator may wish to determine changes in hydrologic data caused by watershed changes.

At some locations, more than one meteorological force may be responsible for the variability in an event. Again, mixed populations may be present. An example is streamflow caused by snowmelt, rainfall, or a combination of both. Another example is rainfall produced by frontal storms and rainfall produced by convective storms. Records may not always be homogeneous and may have to be separated based on time of year or other factors to increase the confidence of an estimate.

2.2.2 Independent Events

An event is said to be *independent* of another event when the occurrence of one does not affect the occurrence of another. Streamflow peak rates separated by a long period of time may be independent, but two peaks close to one another may not be independent. This is true when the recession limb of the first hydrograph becomes part of the rising limb of the next hydrograph.

Most meteorological characteristics are considered to be independent unless one is calculated from the other, such as intensity of rainfall from rainfall volume. Duration, volume, and interevent time are generally assumed to be independent of one another. Independence means that joint probabilities can be calculated. The joint probability of these characteristics is useful in practical design problems. For example, intensity and duration are important for flood determination and precipitation and interevent times are important for storage pond sizing and water quality performance.

2.2.3 Rainfall Volume

Volume of rain has been described by a histogram and a frequency distribution as shown in Figure 2.2 (Wanielista, 1990). Thus if we were interested in

2.2 ASSUMPTIONS FOR STATISTICAL ANALYSIS

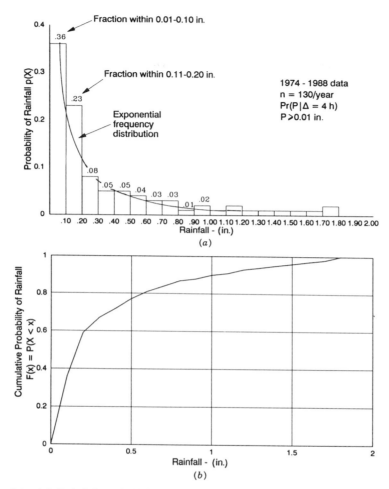

Figure 2.2 (*a*) Rainfall probability histogram for Orlando Jetport for interevent dry period of 4 h; (*b*) rainfall cumulative probability distribution for Orlando Jetport.

the percentage of storms associated with a rainfall volume, a frequency distribution of rainfall volumes can be used. An example problem will help explain use of the data.

Example Problem 2.1 (a) Obtain a frequency distribution on rainfall volume for your area or use Figure 2.2*a* or *b*. If your budget limits rainfall event sampling to 20 times but you wish to get a reasonable representation of the mass of nitrogen in all rainfalls, how many rainfall events greater than or equal to 25.4 mm (1 in.) should be measured to ensure proportionality with respect to the distribution?

(b) Also, assume that an empirical frequency distribution for interevent dry periods is available. From this distribution, the probability of a dry period following a storm event can be estimated (e.g., the probability of an interevent time less than 72 h is 0.70). How many rainfall events of 1 in. or more are followed by an interevent dry period of less than 72 h? What is the return period?

SOLUTION: (a) From Figure 2.2a, only 10% of the rainfalls are greater than or equal to 1 in.; thus about 2 out of 20 events should be sampled with a volume of 1 in. or greater. Almost every rainfall event is sampled, but some of the samples will not be analyzed because the volume of rainfall associated with a sample has already been analyzed. This problem assumes that the sampling must be consistent with the volume distribution.

(b) If the two events are independent, the joint probability is

$$\Pr(P \geq 1 \text{ in.} \cap \Delta \leq 72 \text{ h}) = 0.10(0.70) = 0.07$$

and the number per year is $(0.07)(130) = 9.1$ (call this 9 per year). The return period is

$$T_r = \frac{1}{\Pr(x \geq x_0)N} \quad \text{or} \quad \frac{1}{0.07(130)} = \tfrac{1}{9} \text{ year}$$

where M is the number of years in the record, $M = 1$, and N is the number of rainfall events in the record, $N = 130$.

2.3 REGRESSION FORMULAS

Regression analysis is a statistical procedure that minimizes the variability in estimating a variable which is dependent on other independent variables. It may be the choice for extreme event estimation when a general equation is required for an area. Various watershed characteristics and meteorological conditions are measured and related to a hydrologic variable. Then estimates of the hydrologic variable can be made from the easily measured characteristics. A simple equation derived from statistical linear regression is (Strahler, 1964)

$$Q_p = cA^n \tag{2.2}$$

where Q_p = peak streamflow, m³/s (ft³/s)
A = watershed area, km² (mi²)
c, n = regression analysis constants

This type of equation is not applied to other watersheds because it does not take into account important watershed variables (most notably, channel slope and storage area) and is most likely for specific regional climate conditions. Thus like most statistical formulas, its use is limited. Coefficients for the formula have been developed and used for specific locations in India, Wales, England, and the United States.

When many data points are available, regression analysis with the aid of computer programs is helpful to reduce the time it takes to obtain an answer. Such computer programs are provided with this book. Other common equation forms are:

Linear:	$Y = a + bX$	(2.3)
Exponential:	$Y = ae^{bx}$	(2.4)
Log:	$Y = a + b \cdot \ln x$	(2.5)
Power:	$Y = aX^b$	(2.6)
Polynomial:	$Y = a_0 + a_1(X) + \cdots + a_n(X^n)$	(2.7)

All of the equations above can be made linear so that the parameters can be estimated using linear regression. Also available are nonlinear regression routines to estimate the parameters of equations 2.3 through 2.7.

As an example for making a nonlinear equation into a linear one, consider the power equation and take logarithms to the base e of both sides:

$$\ln Y = \ln a + b \ln X \qquad (2.8)$$

Thus by transforming Y's to $\ln Y$ and X's to $\ln X$, a linear equation is developed.

Equations 2.3 through 2.7 have been used to estimate peak flows and mass loadings. Other equation forms, including equations 2.3 through 2.7, have been used for rainfall analysis. Consider estimating the parameters of the typical rainfall intensity (i) and duration (D) equation:

$$i = \frac{1}{a + b(D)} \qquad (2.9)$$

This equation can be made linear simply by letting $Y = 1/i$; thus

$$Y = a + b(D) \qquad (2.10)$$

which as a linear form is the same as equation 2.3.

Almost all regression equations for predicting peak flows include the watershed area as an independent variable. For the multivariable equations, other watershed characteristics, such as channel slope, shape factors, land use, average land slope, detention storage, percent impervious, and many others, are included. Thus multiple regression procedures can be used to estimate the coefficients to obtain a best-fit equation. The usual criterion for *best fit* is to minimize the variability in the estimate of measured versus predicted peak flows (dependent variable).

The U.S. Geological Survey, the Federal Highway Administration, and state highway departments have developed equations for peak flow estimates valid throughout the United States (U.S. Department of Transportation, 1984). Some examples of these equations with their standard errors (S_T) are:

For region 5 in Texas (Schroeder and Massey, 1977):

$$Q_2 = 4.82 A^{0.779} S_0^{0.966} \quad S_T = 62.1\%$$
$$Q_{25} = 180 A^{0.776} S_0^{0.544} \quad S_T = 41.3\% \quad (2.11)$$
$$Q_{100} = 399 A^{0.782} S_0^{0.497} \quad S_T = 44.1\%$$

where Q_T = peak flow for specific return period, ft^3/2
A = watershed area, mi^2
S_0 = average streambed slope between points 10 and 85% along the length of the main stream, ft/mi
S_T = standard error for T return period

It is important to note the range of areas and slopes used for the equations. For equations 2.11, the areas were between 1.08 and 1947 mi^2, and the slope was between 9.15 and 76.8 ft/mi.

The standard error is a measure of statistical reliability:

$$S_T = \frac{\delta S}{(n)^{1/2}} \quad (2.12)$$

where S = standard deviation
n = sample size
δ = parameter dependent on type of distribution

Equation 2.12 expresses the error in the parameter choice due to the record length. Shorter records have a higher standard error than longer records for a given type of distribution.

Kite (1985) analyzed various estimation methods for extreme-value distributions. He has provided tables for the parameter δ. These values were tabulated and are shown as Tables D.4, D.5, D.6, and D.7 in Appendix D for the normal, log-normal, Gumbel, and log-Pearson type III distributions, respectively. For the normal distribution, δ is a function of the return period; for the log-normal distribution, δ is a function of the return period and the logarithm of the coefficient of variation ($s/\bar{X}$). Using the Gumbel distribution, δ is a function of the sample size and return period (skewness is fixed at 1.140). For the log-Pearson type III distribution, δ is a function of skewness and return period.

Example Problem 2.2 For the Texas watershed represented by the flow rate equations 2.11, estimate the 25-year flood for a 100-mi^2 watershed with a slope of 10 ft/mi.

SOLUTION: The 25-year flood estimate is

$$Q_{25} = 180(100)^{0.776}(10)^{0.544} = 22{,}453 \text{ ft}^3/\text{s } (636 \text{ m}^3/\text{s})$$

Example Problem 2.3 For the estimate of the data of Example Problem 2.2, what is the range of estimates for 68% of the values, assuming a normal distribution about the prediction (± 1 standard deviation)?

SOLUTION: The equation for a 25-year return period and its standard error from Equations 2.11 are used. The standard error is the square root of the variance estimator. It is a measure of accuracy. The range is 22,453 ft^3/s $\pm$ (41.3/100)(22,453) or 13,180 to 31,726 ft^3/s. Note that the standard error of 41.3% is from Equation 2.11 for Q_{25}.

Other equations for natural areas in Florida were developed by Bridges (1982) using 182 gaging stations. Three independent variables are specified. For two of three regions in Florida, the form of the equation is

$$Q_T = aA^{b_1}(S_0)^{b_2}(LK + 3.0)^{b_3} \tag{2.13}$$

where Q_T = peak runoff rate for return period T, ft^3/s
 a, b_1, b_2, b_3 = regression coefficients
 S_0 = channel slope (difference between elevation at the 10 and 85% points), ft/mi
 A = watershed area, mi^2
 LK = lake area, % of total watershed area

Table 2.1 lists the equation constants and standard errors for region A in Florida. Also shown are the correlation coefficients.

Sauer et al. (1983) developed a seven-parameter equation for urban conditions. It related the urban peak discharge to (1) the rural discharge for similar watershed area, (2) watershed area, (3) rainfall intensity, (4) channel slope, (5) watershed storage, (6) basin development factor, and (7) percent imperviousness.

TABLE 2.1 USGS Regression Equations for Natural Flow Conditions in Florida's Region A[a]

Peak Runoff Equation[b]	R^2	Standard Error (%)
$Q_2 = 93.4 DA^{0.756} S_0^{0.268} (LK + 3)^{-0.803}$	0.868	42.6
$Q_5 = 192 DA^{0.722} S_0^{0.255} (LK + 3)^{-0.759}$	0.858	42.4
$Q_{10} = 274 DA^{0.708} S_0^{0.248} (LK + 3)^{-0.738}$	0.843	44.2
$Q_{25} = 395 DA^{0.696} S_0^{0.240} (LK + 3)^{-0.717}$	0.821	47.3
$Q_{50} = 496 DA^{0.690} S_0^{0.234} (LK + 3)^{-0.705}$	0.803	50.0
$Q_{100} = 609 DA^{0.685} S_0^{0.227} (LK + 3)^{-0.695}$	0.784	52.9
$Q_{200} = 779 DA^{0.674} S_0^{0.205} (LK + 3)^{-0.694}$	0.763	55.8
$Q_{500} = 985 DA^{0.668} S_0^{0.196} (LK + 3)^{-0.687}$	0.738	59.7

Source: Bridges (1982).

Florida Region A

[a]Basin Characteristic	Range of Applicability
Drainage area	1,170 acres to 3,066 mi^2
Slope	0.15 to 24.2 ft/mi
Lake area	0 to 28.16%

[b]Q_T, peak runoff rate for return period of T years, ft^3/s; DA, drainage area, mi^2; S_0, channel slope between points at 10 and 85% of total channel length, ft/mi; LK, lake area, % of total watershed area.

Regression equations for other areas can be determined. There are many computer programs that can be used to aid in determining the parameters for the bivariate or multivariate cases.

2.4 THEORETICAL FREQUENCY DISTRIBUTIONS

The estimation of the probability of some future extreme event of specified magnitude and its recurrence frequency is often required when only limited data are available. To facilitate this analysis, it is necessary to understand the concepts underlying parameter estimation for known theoretical frequency distributions. The theoretical (true) frequency distribution chosen is based on an empirical frequency distribution.

2.4.1 Parameter Estimation

The problem of estimating extreme events is simplified if the period of record is at least as long as the return period of interest. If 10 years of data are available, the one-in-100-year estimate is somewhat simplified because the choice of the true distribution is more apparent. For shorter periods of record, the theoretical distribution is not always evident and the extreme events or "tail" of a distribution are not well defined and must be estimated. In fact, the empirical distribution from a short record may indicate one type of theoretical distribution when another is more accurate. Sometimes, several theoretical distributions appear to "fit" the empirical distribution by graphical methods or statistical tests, such as the chi-square test. However, the basic problem of which one of the distributions is the "true" one cannot, in general, be determined except with the use of longer record periods.

Another problem in determining the proper theoretical distribution is the sampling error. The sampling error can be both human and equipment related. These errors can be reduced by proper attention to such details as equipment maintenance and personnel matters. There are four commonly used methods for distribution parameter estimation:

1. Graphical
2. Regression (least squares)
3. Method of moments
4. Maximum likelihood method

An empirical distribution for the data is developed first. The data are arranged in order of magnitude. If a less than or equal to distribution is required, the data are arranged from smallest to largest values. Then a plot position that is an estimate of the "true" cumulative probability is determined. There are at least four widely used plot position formulas: Weibull, California, Foster, and exceedence. The following example problem presents the plot position formulas and calculations. The Weibull is used most frequently in stormwater management because of the need to estimate probabilities near 0 and 1. With a small number of data points an empirical distribution should not be expected to reflect the certainty of a zero or 1 probability; therefore, the Weibull or Foster are favored for a low number of data points.

Example Problem 2.4 Using four different plot position formulas, develop the empirical probability distribution for rainfall at Bushnell, Florida using the limited but ordered rainfall data in Table 2.2, which are the maximum values for a year.

TABLE 2.2 Rainfall Data for Example Problem 2.4

Daily Rainfall Volume (in.)	m Plot Position	Weibull $m/(n+1)$	California m/n	Foster $(2m-1)/2n$	Exceedence $(m-1)/n$
1.55	1	0.06	0.07	0.03	0.00
2.00	2	0.13	0.13	0.09	0.07
2.80	3	0.19	0.20	0.17	0.13
3.00	4	0.25	0.27	0.23	0.20
3.08	5	0.31	0.33	0.30	0.27
3.11	6	0.38	0.40	0.37	0.33
3.50	7	0.44	0.46	0.43	0.40
3.82	8	0.50	0.53	0.50	0.46
3.97	9	0.56	0.60	0.57	0.53
4.13	10	0.63	0.67	0.63	0.60
4.70	11	0.69	0.73	0.70	0.67
5.15	12	0.75	0.80	0.77	0.73
5.27	13	0.81	0.87	0.83	0.80
7.60	14	0.88	0.93	0.90	0.87
9.08	15	0.94	1.00	0.97	0.93

SOLUTION: The data points are listed in Table 2.2.

The *graphical method* has the advantage of simplicity with visual appeal. Once the data are plotted as an empirical distribution, a line is fit to the data points and the judgment of the user is beneficial. Unfortunately, many people will draw a different line, and thus different estimates of extreme events will result. The graphical method is highly subjective and is usually not reproducible.

As an example of a graphical presentation, consider the yearly tide data above or equal to 6.0 ft MSL (mean sea level) for a location on the Gulf coast of North America. The data are shown along with coincidental rainfall data for the same location in Table 2.3. The distribution is skewed to the left. A graphical plot would indicate the skewness. What is a best fit to these data? Is it a straight line, a curved line in the tail of the distribution, or some other line? A cumulative distribution graphical plot is shown later in the chapter as Figure 2.7. The "true" theoretical distribution can be estimated by graphical means.

The *least-squares method* (Yevjevich, 1972) uses mathematical formulas to determine the parameters of an empirical distribution, such as the slope and intercept of the distribution. The results are reproducible among users. A best fit is achieved when the sum of squares of all deviations between the observed point and some theoretical function is minimized. The function is calculated for each data point, and then the difference between the observed and calculated is squared such that the sum is minimized. This method has gained in popularity and is especially useful if the theoretical function can be made linear. Such is the case for the Weibull distribution.

2.4 THEORETICAL FREQUENCY DISTRIBUTIONS

TABLE 2.3 Maximum Tidal and Associated Rainfall Data

Year	Tide Date[a]	Tide (ft)	Rainfall[b] (in.)
1947	5/21	6.7	0.00
1948	12/30	6.6	0.73
1949	8/28	7.0	1.68
1950	9/5	8.4	6.13
1951	11/3	7.2	0.58
1952	12/3	6.5	0.00
1953	9/27	7.4	4.25
1954	10/30	6.3	2.50
1955	6/19	6.5	1.90
1956	9/25	6.9	1.70
1957	6/9	7.2	4.86
1958	1/7	7.0	1.02
1959	6/18	7.2	6.84
1960	7/29	7.4	7.95
1961	11/23	6.5	1.33
1962	12/6	6.6	0.09
1963	9/29	7.9	0.31
1964	9/12	6.5	1.30
1965	9/9	7.8	0.26
	7/29	7.0	3.47
1966	6/9	7.3	2.93
1967	5/23	6.7	0.51
1968	10/18	8.1	1.50
1969	7/28	6.6	0.20
1970	10/20	7.0	0.00
1971	11/7	6.5	0.05
1972	12/22	7.3	1.88
1973	4/4	7.25	2.42
1974	6/25	7.57	5.67
1975	9/23	6.93	0.64
1976	5/15	7.03	3.90
1977	6/2	6.5	0.40
1978	5/4	6.9	2.00
1979	9/12	7.3	0.38
1980	8/8	6.5	0.00
1981	11/17	6.92	0.00
1982	6/18	9.15	4.95
	12/16	6.9	0.00
1983	3/24	7.07	0.78
	11/8	7.02	0.00
1984	3/29	6.79	0.03
1985	9/1	10.5	3.80

[a] Date of highest tide.
[b] Cumulative for day of tide and preceding storm event.

The *method of moments* also is used for parameter estimation. It is similar in concept to moments as described and used in basic physics; that is, the kth moment about the origin is defined using a probability distribution as

$$u_k = x^k f(x) \tag{2.14}$$

and the kth moment around the mean is

$$u_k = (x - u)^k f(x) \tag{2.15}$$

Thus the mean value of a distribution is the first moment about the origin, or in terms of grouped discrete data

$$u = \sum_{\text{all } X} Xf(x) \tag{2.16}$$

and the variance is the second moment about the mean, or

$$\sigma^2 = \sum_{\text{all } X} (X - u)^2 f(x) \tag{2.17}$$

The third moment measures the skewness of a distribution and is very useful when examining the shape of the empirical data. The least calculated moment, the fourth, is called kurtosis of a distribution, which is a measure of the peakedness.

The *method of maximum likelihood*, which is preferred by most statisticians when a large number of data points are available, also is an explicit method. However, the number of computation steps is frequently large, and in some cases it is difficult to converge on an acceptable estimate for the distribution parameters. The objective is to maximize the likelihood of obtaining a given value of the variable (X), which is proportional to the probability given the distribution parameters. The procedure is done to calculate the parameters of the distribution that maximizes the probability of obtaining the proportional samples. Thus a product function is formed as an increasing value of X, so the base e logarithm (ln) function is used. The total probability is the product of all individual probabilities.

$$\ln L = \prod_{L=1}^{n} \ln \Pr(X_i; \mu; \sigma^2) \tag{2.18}$$

where $\ln L$ = likelihood function
   X = variable value
   μ and σ^2 = parameters of the distribution

By taking the partial derivative of equation 2.18 with respect to each of the parameters and setting it equal to zero, estimates of μ and σ^2 are obtained.

These estimates have minimum variation and are identified as statistically efficient solutions.

2.4.2 Confidence Intervals

A *confidence interval* establishes the upper and lower limits on an estimated value of a return period. Since the estimated value is based on a single sample from a population, it is highly probable that another sample of equal size from the same population would produce a different estimate. Thus the confidence interval defines the range within which these frequency estimates could be expected.

For given confidence levels (90%, 95%), confidence intervals are defined. Equations have been developed for confidence intervals for various distributions. For the Gumbel distribution, Kite (1985) presents an estimation equation for the upper and lower 95% confidence limits as

$$U_T = X_T + 1.96 S_T \tag{2.19}$$
$$L_T = X_T - 1.96 S_T \tag{2.20}$$

where U_T = upper bound estimate for return period T using Gumbel distribution
L_T = lower bound estimate using Gumbel distribution
X_T = point estimate using Gumbel distribution
S_T = standard error using Gumbel distribution

For other distributions, similar formulas can be developed and in general have the following formulas for limits:

$$U_T = \frac{K_{G,T} + K_{G,T}^2 - ab}{a} \tag{2.21}$$

$$L_T = \frac{K_{G,T} - K_{G,T}^2 - ab}{a} \tag{2.22}$$

where $K_{G,T}$ = confidence limit deviate values (see Table D.8 for normal and log normal)
$a = 1 - Z/2(n-1)$
$b = K_{G,T}^2 - Z^2/n$
Z = standard normal deviate at zero skewness (from normal distribution)

2.4.3 Incomplete Records and Extreme Events

Hydrologic data are frequently not complete, for reasons of human or equipment error. Gages may be removed for budgetary reasons or destroyed by weather. The rainfall data of Table 2.4 provide an example of missing data. The cumulative distribution for these data is shown in Table 2.5. If we

TABLE 2.4 Listing by Month, Day, and Year for Maximum Daily Storms for Each Year at Bushnell

Year	Date	Rainfall[a] (in.)
1918	9/27	2.00
1937	7/30	3.00
1938	6/22	3.86
1939	7/07	2.81
1940	7/04	2.90
1941	4/03	3.57
1942	2/24	1.55
1943	4/09	3.00
1944	10/19	7.60
1945	6/24	8.32
1946	6/27	2.53
1947	10/24	3.65
1948	7/02	3.50
1949	4/05	3.90
1950	9/06	9.08
1951	11/16	3.04
1952	5/20	3.09
1953	12/23	2.69
1954	7/26	3.94
1955	11/10	3.09
1956	10/16	4.13
1957	12/26	4.70
1958	3/02	3.08
1959	7/17	2.70
1960	7/29	5.27
1961	2/07	1.73
1962	5/28	2.80
1963	11/10	3.10
1964	9/11	3.82
1965	8/05	3.11
1966	5/08	3.02
1967	6/11	3.29
1968	10/19	3.79
1969	3/17	3.17
1970	2/03	3.97
1971	2/08	4.00
1972	3/31	6.65
1973	2/15	2.41
1974	6/25	6.90
1975	10/29	2.98
1976	6/05	2.98
1984	4/04	2.58
1985	8/18	5.15

[a] Average is 3.78 in.; standard deviation is 1.66 in.

2.4 THEORETICAL FREQUENCY DISTRIBUTIONS

TABLE 2.5 Empirical Cumulative Distribution Function

$X(I)$ (in.)	Plot Position m	$\Pr = m/(N+1)$ $\leq$ Probability	Exceedence Probability $Q = 1 - \Pr$	Return Period Tr (yrs)
1.55	1	0.0227	0.977	1.023
1.73	2	0.0455	0.955	1.048
2.00	3	0.0682	0.932	1.073
2.41	4	0.0909	0.909	1.100
2.53	5	0.1136	0.886	1.128
2.58	6	0.1364	0.864	1.158
2.69	7	0.1591	0.841	1.189
2.70	8	0.1818	0.818	1.222
2.80	9	0.2045	0.795	1.257
2.81	10	0.2273	0.773	1.294
2.90	11	0.2500	0.750	1.333
2.98	12	0.2727	0.727	1.375
2.98	12	0.2727	0.727	1.375
3.00	14	0.3182	0.682	1.467
3.00	14	0.3182	0.682	1.467
3.02	16	0.3636	0.636	1.571
3.04	17	0.3864	0.614	1.630
3.08	18	0.4091	0.591	1.692
3.09	19	0.4318	0.568	1.760
3.09	19	0.4318	0.568	1.760
3.10	21	0.4773	0.523	1.913
3.11	22	0.5000	0.500	2.000
3.17	23	0.5227	0.477	2.095
3.29	24	0.5455	0.455	2.200
3.50	25	0.5682	0.432	2.316
3.57	26	0.5909	0.409	2.444
3.65	27	0.6136	0.386	2.588
3.79	28	0.6364	0.364	2.750
3.80	29	0.6591	0.341	2.933
3.82	30	0.6818	0.318	3.143
3.90	31	0.7045	0.295	3.385
3.94	32	0.7273	0.273	3.667
3.97	33	0.7500	0.250	4.000
4.00	34	0.7727	0.227	4.400
4.13	35	0.7955	0.205	4.889
4.70	36	0.8182	0.182	5.500
5.15	37	0.8409	0.159	6.286
5.27	38	0.8636	0.136	7.333
6.65	39	0.8864	0.114	8.800
6.90	40	0.9091	0.091	11.000
7.60	41	0.9318	0.068	14.667
8.32	42	0.9545	0.045	22.000
9.08	43	0.9773	0.023	44.000

are interested in extreme events and a break in the record is not related to extreme events (destroyed by a flood), no augmentation is needed. When the record is related to extreme events, it may be possible to estimate values of the extreme events from nearby gages (rainfall) or highwater marks (streamflow). These estimates are made part of the record and the frequency analysis is performed normally.

Low or zero flows in a stream present another type of problem since the log transformation may be necessary for the log-Pearson and maximum likelihood estimators. The U.S. Water Resources Council (1981) in their Bulletin 17B recommends an augmentation based on conditional probability concepts. The augmentation is reasonable if not more than 25% of the sample is eliminated. The procedure is to calculate a relative frequency above which the annual flow will exceed a truncation level.

$$\text{Pr}_a = \frac{M_T}{N} \tag{2.23}$$

where M_T is the number of flows above a truncation level and N is the total record period. Next, the exceedence probability for the original distribution (nontruncated), Pr_n, is computed as a conditional probability using

$$\text{Pr}_n = \text{Pr}_a \text{Pr}_x \tag{2.24}$$

where Pr_x is the selected probability on truncated distribution.

Example Problem 2.5 The exceedence probability for a design is the one-in-50-year event and the truncated distribution is given as a relative frequency of one-fourth. What is the exceedence probability when the truncated distribution is used?

SOLUTION

$$\text{Pr}_n = \text{Pr}_a \text{Pr}_x$$

$$\text{Pr}_x = \frac{1/50}{1/4} = \frac{2}{25} \tag{2.25}$$

or the two-in-25-year exceedence probability value used on the truncated distribution to estimate the one-in-50-year design value from the nontruncated distribution.

2.4.4 Example Graphical Applications

The U.S. Department of Transportation (1984), using streamflow data from the Medina River near San Antonio, Texas, developed a graphical plot of the empirical data and then imposed theoretical distributions on each plot.

2.4 THEORETICAL FREQUENCY DISTRIBUTIONS

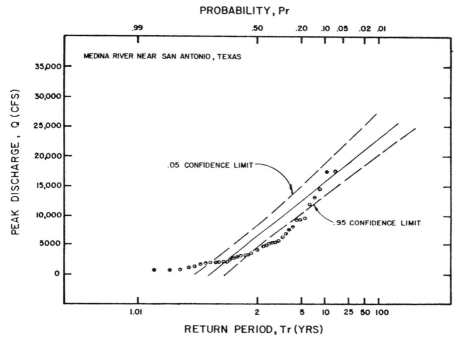

Figure 2.3 Normal distribution with confidence limits, Medina River, Texas. (From *Hydrology*, Circular 19, Federal Highway Administration, USDOT.)

Figures 2.3, 2.4, 2.5, and 2.6 are the plots on probability paper for the normal, Gumbel, log-normal, the log-Pearson type III distributions. First, note that the normal and Gumbel distributions do not appear to be a good fit to the data, while the log-normal and log-Pearson type III distributions appear to fit better. Note that the normal and log-normal distributions would plot as straight lines on probability and log probability paper, respectively.

Confidence limits are also shown on Figures 2.3 through 2.6. The confidence interval increases at the extremes. Based on the confidence limits, it appears that the log-Pearson type III and the log-normal are acceptable distributions. A high degree of correspondence exists between the empirical distribution and the theoretical distributions. The data are within the confidence limit over the entire range. The log-Pearson confidence limit is narrower or tighter than the log-normal, and the standard error of the estimate is less. Thus a log-Pearson type III distribution is more appropriate and preferred for this case.

Similar analyses can be performed for other data sets and using other distributions. As an example, the three-parameter log-normal, truncated normal, and Pearson distribution with standard error, residual, and parameter estimation can also be executed. (See the computer package available

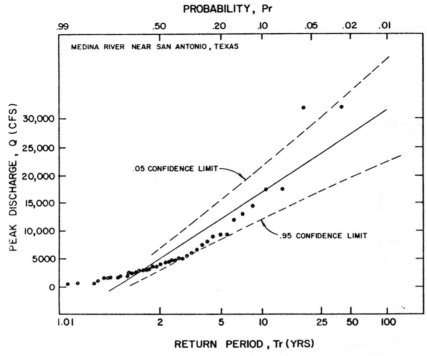

Figure 2.4 Gumbel extreme value distribution with confidence limits, Medina River, Texas. (From *Hydrology*, Circular 19, Federal Highway Administration, USDOT.)

with this book.) It uses the same basic data input formats as those of many other program packages.

2.4.5 Example Computer Applications

The purpose of this section is to illustrate the use of the computer programs provided with this book that aid in the selection of the theoretical or best-fit distribution. The computer programs are explained in Appendix E and are (1) Pearson, (2) two-log-normal and Gumbel, (3) error estimation, and (4) statistical menu on the stormwater management diskette. Selection of the theoretical distribution can be based on graphical comparison between the empirical and theoretical and by the use of distribution statistics. Some of these statistics that measure the precision of the fit (lower values indicate better fit) are:

1. Residuals or differences between empirical and theoretical values over the probability range of interest
2. Standard error of the estimate or a measure of the vertical deviation of the sample points from the best-fit line at a given probability and for a given distribution

2.4 THEORETICAL FREQUENCY DISTRIBUTIONS

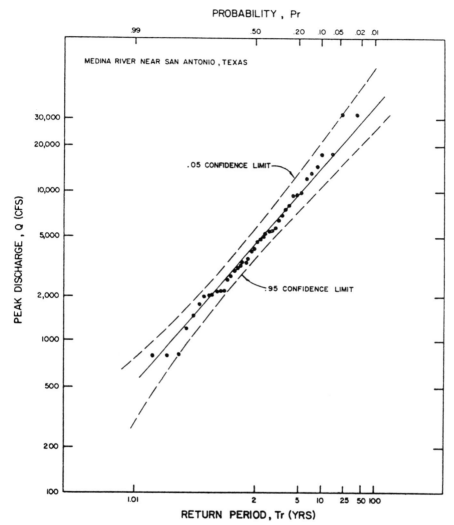

Figure 2.5 Log-normal distribution with confidence limits, Medina River, Texas. From *Hydrology*, Circular 19, Federal Highway Administration, USDOT.)

3. Random test as a measure of the randomness of the data (also called a runs test)
4. Chi-square as a measure of variance or deviation of the entire distribution
5. Kolmogorov–Smirnov as a nonparametric test for differences between two cumulative distributions

A data file on maximum yearly rainfall for Tampa, Florida, is on the diskette and is labeled as MYDTAM.DAT. Using this file, statistical data

46 PROBABILITY AND STATISTICAL METHODS FOR HYDROLOGIC EVENTS

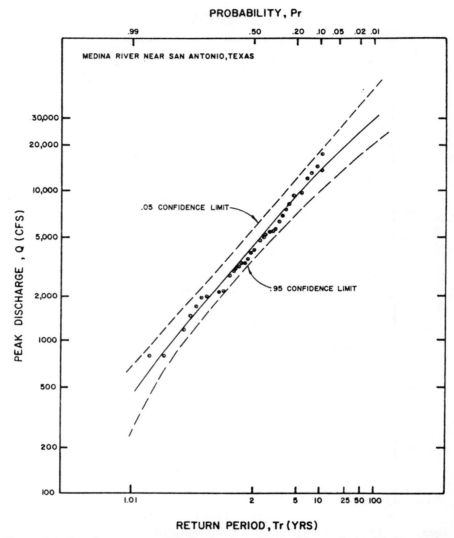

Figure 2.6 Log-Pearson type III distribution with confidence limits, Medina River, Texas. (From *Hydrology*, Circular 19, Federal Highway Administration, USDOT.)

and graphical analyses between the empirical and theoretical distributions can be developed. Plots of actual computer screen captures using the Leasts computer program show the empirical and theoretical values in Figures 2.7 and 2.8. Note that the theoretical distributions fit so well that the empirical data can only be seen at very high probabilities (high return periods). All other extreme-value distributions can be tested.

2.4 THEORETICAL FREQUENCY DISTRIBUTIONS

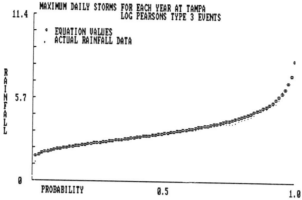

Figure 2.7 Empirical versus theoretical distribution for Tampa using log-Pearson type III distribution.

For the empirical frequency distribution of rainfall at Bushnell, Florida (Table 2.5), the frequency distribution that best fits the data can be determined. The three-parameter log-normal and the log-Pearson type III appear to have the best fits. Comparing the statistical parameters for Busnell, the following is noted:

The tidal data of Table 2.3 can be analyzed for statistical measures using computer programs. The log-Pearson type III statistics are:

	Mean tide	7.17 ft (MSL)
	Standard deviation	0.80 ft
	Skewness	2.23
100-year prediction		10.50 ft
Empirical data:	Mean	3.78 in.
	Standard deviation	1.66 in.
	Skewness	1.63
Theoretical distributions:		
Standard error for distribution	Two-log-normal	0.65 in.
	Gumbel	0.63 in.
	Log-Pearson	0.48 in.
	Three-log-normal	0.48 in.
Residual 100-year	Two-log-normal	1.64 in.
(Equation—empirical)	Gumbel	2.02 in.
	Log-Pearson	1.17 in.
	Three-log-normal	1.19 in.
Residual 25-year	Two-log-normal	1.11 in.
(Equation—empirical)	Gumbel	1.33 in.
	Log-Pearson	1.01 in.
	Three-log-normal	1.01 in.

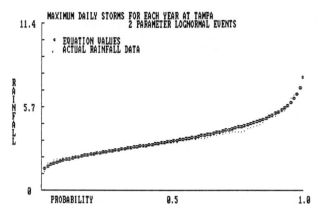

Figure 2.8 Empirical versus theoretical distribution for Tampa using log-normal distribution.

For estimating the 100- and 25-year rainfall volumes, either the log-Pearson or the three-log-normal distribution would provide the best estimate. All the distributions underestimate the empirical estimate. However, the log-Pearson may be "best" from among the four based on the graphical plots and the statistical analysis. Also, it may be possible to weight a distribution to place emphasis on extreme events and then solve for the distribution parameters.

2.4.6 Example Calculations for the Log-Pearson Type III Probability Distribution

An application of an extreme probability distribution will be presented to illustrate its use in water quality studies and is not intended to be an exhaustive exercise but primarily an example of what can be done. Nationwide studies have been initiated to determine water quality impacts, but extensive water quality data are not available. However, work has been completed on methods to estimate flow frequencies. The U.S. Water Resources Council (1981) extended and updated standard procedures for peak annual discharges with their exceedence probabilities.

Mass for a time period is defined as the product of flow and concentration. If concentration is random, it may be appropriate to use the same distribution for mass as that used for flow over a time period. Since there is no procedure that will always produce well-defined estimates, an element of risk and uncertainty is present. An investigator would be wise to consider a number of theoretical distributions and select the one that best fits the empirical distribution. In most applications of annual flows or mass discharge, the distribution of choice is usually the log-Pearson type III, gamma, or log-normal distribution.

2.4 THEORETICAL FREQUENCY DISTRIBUTIONS

Annual mass discharges are assumed to be a succession of random events that may be described by the log-Pearson type III distribution with the parameters mean, standard deviation, and skew coefficient.

Example Problem 2.6 The Little Econ River has recorded phosphorus discharge records for 35 years. The annual mass discharges are shown in Table 2.6. What are estimates for mass per year associated with seven probabilities of exceedence assuming a log-Pearson type III distribution and use of Hardison's (1974) generalized skew coefficients?

SOLUTION: Let $X = \log(\text{flow})$ with the transformed values shown in Table 2.6. Then using computation formulas (Wanielista, 1990),

$$\bar{X} = 3.596 \qquad S = 0.425 \qquad G = 0.20 \qquad n = 35$$

A generalized skew coefficient for flow rate data can be obtained from Hardison (1974). Using mass data, a general coefficient for the area is -0.5. The weighted skew coefficient (U.S. Water Resources Council, 1981) is

$$G = \frac{N-25}{75}G + \left(1 - \frac{N-25}{75}\right)G_g \qquad (2.26)$$

$$= \left(\frac{35-25}{75}\right)(0.2) + \left(1 - \frac{35-25}{75}\right)(-0.5) = -0.40$$

The computations for the frequency curve coordinates (for plotting) requires the determination of an appropriate standard deviate for the assumed distribution. An example computation for an exceedence of 0.01 is

$$X = \log Q = \bar{X} + KS \qquad (2.27)$$

and from the U.S. Water Resources Council (1981)

$$K = \frac{2}{G}\left[\left(K_n - \frac{G}{6}\right)\left(\frac{G}{6}\right) + 1\right]^3 - \frac{2.0}{G} \qquad (2.28)$$

when $K_n = 2.326$. Thus $K = 2.03$ and

$$\log Q = 3.596 + 2.03(0.425)$$
$$= 4.459$$
$$Q = 28{,}757 \text{ kg/year}$$

TABLE 2.6 Annual Mass Discharge of Phosphorus (kg/yr)

Year	Peak	X	X^2	X^3
1951	1,400	3.146	9.897	31.137
1952	7,445	3.872	14.992	58.049
1953	6,281	3.798	14.425	54.786
1954	1,394	3.144	9.886	31.082
1955	2,065	3.315	10.988	36.428
1956	4,048	3.607	13.012	46.935
1957	1,458	3.164	10.009	31.669
1958	3,568	3.552	12.620	44.825
1959	1,299	3.114	9.695	30.189
1960	1,717	3.235	10.464	33.851
1961	1,362	3.134	9.823	30.785
1962	5,324	3.726	13.885	51.735
1963	3,239	3.510	12.323	43.254
1964	4,519	3,655	13.359	48.828
1965	8,322	3.920	15.368	60.243
1966	18.580	4.269	18.224	77.798
1967	6,240	3.795	14.403	54.661
1968	5.655	3.752	14.081	52.831
1969	13,333	4.125	17.015	70.187
1970	2,950	3.470	12.039	41.778
1971	4,730	3.675	13.505	49.629
1972	17,250	4.237	17.950	76.056
1973	1,988	3.298	10.880	35.881
1974	830	2.919	8.521	24.873
1975	7,160	3.855	14.860	57.287
1976	7,520	3.876	15.025	58.237
1977	730	2.863	8.199	23.473
1978	3,810	3.581	12.823	45.919
1979	25,558	4.408	19.430	85.647
1980	2,900	3.463	11.988	41.515
1981	22,050	4.343	18.862	81.918
1982	1,915	3.282	10.773	35.356
1983	975	2.989	8.934	26.704
1984	14,330	4.156	17.272	71.782
1985	4,093	3.612	13.047	47.124
		125.86	458.983	1692.452

The results for seven exceedence probabilities are shown in Table 2.7. Procedures for adjusting the curves for very high or very low flow rates and the determination of these values are presented elsewhere (Gumbel, 1945). This example was presented for the purpose of illustrating the required basic calculations.

TABLE 2.7 Example Problem 2.5 Exceedence Calculations

Exceedance Probability	Appendix D (Table D.3) $-0.4P$	log Q	Q (kg/yr)
0.01	2.029	4.459	28,757
0.05	1.523	4.243	17,510
0.10	1.231	4.119	13,158
0.50	0.066	3.624	4,208
0.90	−1.317	3.036	1,087
0.95	−1.750	2.852	712
0.99	−2.615	2.485	305

2.4.7 Rainfall-Runoff Events and Probability Models

Rainfall-runoff and pollutant concentrations can be represented by random processes (Athayde, 1979). A schematic of the rainfall-runoff processes and detention pond operation is shown in Figure 2.9. Commonly used terms associated with the schematic are defined as follows.

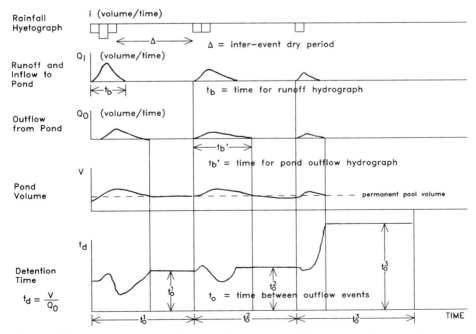

Figure 2.9 Schematic of rainfall runoff processes with wet-detention pond volume and detention time.

Rainfall Volumes per Storm Event. Rainfall volume is the sum of rainfall during a rainfall event, and the rainfall event is one separated by a period of no rainfall (interevent dry period). Generally, 4 hours is the minimum interevent time for storms to be considered independent (Hvitved-Jacobsen and Yousef, 1987; Wenzel and Vorhees, 1981) with an exponential frequency distribution used to describe rainfall volume per storm event (Wanielista, 1990). Knowing the mean rainfall volume ($\bar{P}$), the exponential frequency distribution is

$$f\left(x;\frac{1}{\bar{P}}\right) = \frac{1}{\bar{P}}e^{-x/\bar{P}} \quad \text{for } x \geq 0, \quad \bar{P} > 0 \quad (2.29)$$

$$F\left(x;\frac{1}{\bar{P}}\right) = 1 - e^{-x/\bar{P}} \quad \text{for } x \geq 0, \quad e^{-x/\bar{P}} > 0 \quad (2.30)$$

Example Problem 2.7 The average rainfall volume for storms producing runoff in a region of West Virginia is about 0.50 in. What percentage of storm events is less than or equal to 1 in., assuming an exponential distribution for rainfall volumes per storm event?

SOLUTION: Using the cumulative exponential distribution (equation 2.30) gives

$$F(1; 0.5) = 1 - e^{-1/0.5} = 0.865$$

In other words, 86.5% of the storm events have volume less than 1 in.

Number of Events. The number of storm events per time period has been shown (Athayde, 1979) to follow a Poisson distribution with x = number of events in that fixed time period. If t is the time period and $\bar{\Delta}$ is the average interevent dry period, the Poisson distribution can be used to represent the probability of a number of events in a time period:

$$f(x; \lambda) = \frac{\lambda^x e^{-\lambda}}{x!} \quad \text{for } x = 0, 1, 2, \ldots, \lambda > 0 \quad (2.31)$$

where x = number of events per time period
λ = average number in the time period = $t/\bar{\Delta}$
$\bar{\Delta}$ = average interevent dry period

Example Problem 2.8 For a summer month in Dallas, Texas, the average number of storms that exceed 0.04 in. of rainfall is 10. What is the probability of exactly 9 events and at most 20 events?

2.4 THEORETICAL FREQUENCY DISTRIBUTIONS

SOLUTION: For exactly 9 events, use the frequency estimate given by equation 2.31.

$$f(9; 10) = \frac{10^9 e^{-10}}{9!} = 0.125$$

For at most 20, use the cumulative distribution function or Table D.9 in Appendix D.

$$F(20; 10) = \sum_{x=0}^{20} f(x; 10) = 0.998$$

With very high probability, the number of rainfall events in the month will be 20 or less.

Interevent Times. The interevent dry period is a time with no rainfall or rainfall depth lower than specified minimum value and is considered to be a random variable. An interevent time probability distribution can be defined by an exponential as

$$F(t; \bar{\Delta}) = \Pr\{t \leq \bar{\Delta}\} = 1 - e^{-t/\bar{\Delta}} \qquad (2.32)$$

where $\bar{\Delta}$ is any average interevent dry period and

$$\Pr\{t > \bar{\Delta}\} = e^{-t/\bar{\Delta}} \qquad (2.33)$$

The mean of this exponential curve is $\bar{\Delta}$. Analysis of rainfall records usually indicate that the exponential distribution closely fits the empirical distribution (Athayde, 1979). Note that the coefficient of variation (standard deviation divided by the mean) for the exponential distribution is approximately equal to 1 if the exponential distribution fits the empirical data. If not, a gamma distribution may be used.

Rainfall Excess and Runoff Time. Runoff rates and runoff time have been shown by investigators (DiToro and Small, 1979; Athayde, 1979) to be independent random events. The runoff volume (rainfall excess) is the product of runoff rate and time of runoff. Thus rainfall excess may also be represented by an independent random variable. This assumption may not be entirely accurate, especially when summer and winter sotrm events are mixed together. However, a gamma distribution has been used for rainfall excess and duration before (DiToro and Small, 1979) and is expressed as

$$F(x) = \left(\frac{k}{\bar{X}}\right)^k \frac{x^{k-1}}{\Gamma(k)} \exp\left(\frac{-kx}{\bar{X}}\right) \quad \text{for } x > 0 \qquad (2.34)$$

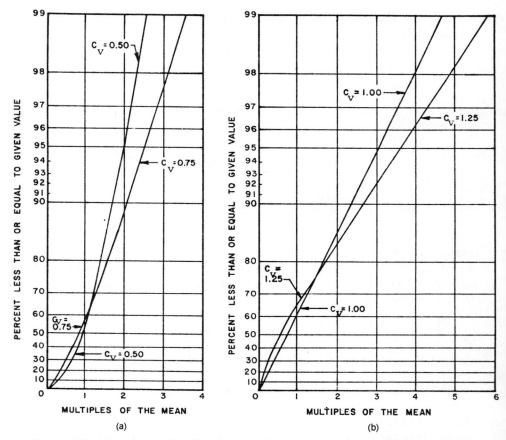

Figure 2.10 Cumulative distribution function for gamma distribution: (a) C_v = 0.50 and 0.75; (b) C_v = 1.00 and 1.25; (c) C_v = 1.50 and 2.0; (d) C_v = 2.50 and 3.0. (From Athayde, 1979.)

where $\quad k = 1/(C_{vx})^2$

C_{vx} = coefficient of variation for variable x, fraction and equal to standard deviation divided by the average
x = value for rainfall excess (R) or for runoff duration (t_b)
$\overline{X}$ = average rainfall excess $(\overline{R})$ or runoff duration (t_b)
$\Gamma(k)$ = gamma function defined by $(k-1)(k-1)!$ for $k > 0$

The cumulative gamma function (less than type) distributions are shown as a function of coefficient of variation in Figure 2.10 and can be used to determine the percent of excess volume and runoff time less than or equal to a given value for a given coefficient of variation. If a person were interested in the value of runoff volume for which 10% of all storm values would be

2.4 THEORETICAL FREQUENCY DISTRIBUTIONS

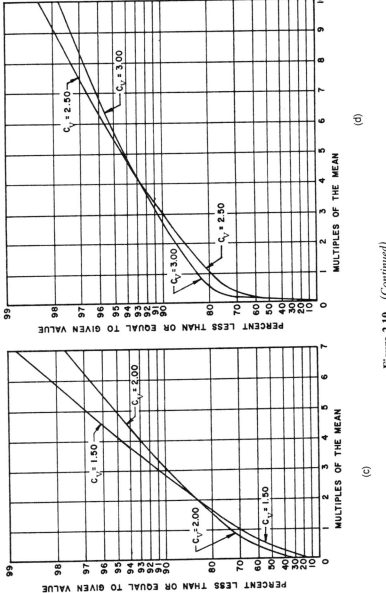

Figure 2.10 (*Continued*)

exceeded and the coefficient of variation were 2.00, Figure 2.10c can be used and the multiplier is 3.0. Thus the 90% value of runoff would be three times greater than the mean.

Mass of Pollutants. The mass of runoff pollutants for each storm event is the product of concentration and rainfall excess. If event mean concentration is used, a gamma function would probably result because rainfall excess per event is multiplied by a constant, or

$$M = \overline{C}R \qquad (2.35)$$

where M = mass per event, mg
$\overline{C}$ = event mean concentration, mg/L
R = rainfall excess per event, L

Regional Variability. Statistical analysis of hourly rainfall by the U.S. Environmental Protection Agency (1986) provided estimates of the average and coefficient of variation for volume of rainfall (P), duration of event (D), average storm intensity (i), and interval between event midpoints (δ). A minimum interevent dry period of 3 to 4 h was used. The minimum rainfall producing runoff was 0.10 in. The analysis of rainfall events indicated that these storm parameters can be approximated by a gamma distribution. Representative regional values are shown for the United States in Figure 2.11. These are regional values and should be further justified for site-specific areas. The coefficient of variation ranges from 0.84 to 1.77. For specific exceedence probabilities, the value of the statistic can be estimated using Figure 2.10 knowing the mean and coefficient of variation.

Runoff-Producing Rainfall Events. At the start of a rainfall event, depression storage, evaporation from heated surface, and vegetation surfaces account for an initial abstraction of rainfall. For urban residential areas, initial abstraction has been estimated at 1 to 3 mm (0.04 to 0.12 in.) of rainfall (Wanielista and Shannon, 1977; Overton and Meadows, 1976; Schuler, 1987). For parking lot areas with almost 100% directly connected surfaces, the initial abstraction is lower, about 1 mm (0.04 in.) (Wanielista and Shannon, 1977). Thus statistical analyses of rainfall data producing runoff can exclude events that do not produce runoff. Driscoll (1989) calculated these statistics using a minimum rainfall volume of 2.5 mm (0.10 in.), with a minimum interval between rainfall midpoints of 6 h (see Table 2.8). The statistics for average rainfall event volume and duration will increase as the minimum interevent dry period increases. The stormwater management system may require more than 4 to 6 h to return to its initial assumed design condition or to treat the pollutants in the stormwater; thus the minimum interevent time is longer and set equal to the treatment time or the time to return to design

2.4 THEORETICAL FREQUENCY DISTRIBUTIONS

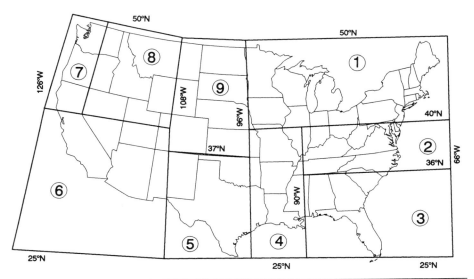

Zone	Period[a]	Volume P (in.) Mean	C_v	Intensity i (in./h) Mean	C_v	Duration D (h) Mean	C_v	Interval δ (h)[b] Mean	C_v
1	Annual	0.26	1.46	0.051	1.31	5.8	1.05	73	1.07
	Summer	0.32	1.38	0.082	1.29	4.4	1.14	76	1.07
2	Annual	0.36	1.45	0.066	1.32	5.9	1.05	77	1.05
	Summer	0.40	1.57	0.101	1.37	4.2	1.09	77	1.08
3	Annual	0.49	1.47	0.102	1.28	6.2	1.22	89	1.05
	Summer	0.48	1.52	0.133	1.34	4.9	1.33	68	1.01
4	Annual	0.58	1.46	0.097	1.35	7.3	1.17	99	1.00
	Summer	0.52	1.54	0.122	1.35	5.2	1.29	87	1.06
5	Annual	0.33	1.74	0.080	1.37	4.0	1.07	108	1.41
	Summer	0.38	1.71	0.110	1.39	3.2	1.08	112	1.49
6	Annual	0.17	1.51	0.045	1.04	3.6	1.02	277	1.48
	Summer	0.17	1.61	0.060	1.16	2.6	1.01	425	1.26
7	Annual	0.48	1.61	0.024	0.84	20.0	1.23	101	1.21
	Summer	0.26	1.35	0.027	1.11	11.4	1.20	188	1.15
8	Annual	0.14	1.42	0.031	0.91	4.5	0.92	94	1.39
	Summer	0.14	1.51	0.041	1.13	2.8	0.80	125	1.41
9	Annual	0.15	1.77	0.036	1.35	4.4	1.20	94	1.24
	Summer	0.18	1.74	0.059	1.44	3.1	1.14	78	1.13

[a]Summer includes June, July, August and September.
[b]Interval between rainfall midpoints.

Figure 2.11 Representative regional values for rainfall statistics. (From U.S. Environmental Protection Agency, 1986.)

TABLE 2.8 Typical Values of Storm Event Statistics with Minimum Between Rain Midpoints of 6 Hours and Minimum Precipitation Producing Runoff of 0.1 Inch

Rain Zone	Annual Statistics				Independent Storm Event Statistics							
	Number of Storms		Precipitation (in./yr)		Duration		Intensity		Volume		Mid point Time	
	Avg.	C_v	Avg.	C_v	Avg.	C_v	Avg.	C_v	Avg.	C_v	Avg.	C_v
Northeast	70	0.13	34.6	0.18	11.2	0.81	0.067	1.23	0.50	0.95	126	0.94
Northeast coastal	63	0.12	41.4	0.21	11.7	0.77	0.071	1.05	0.66	1.03	140	0.87
Mid-Atlantic	62	0.13	39.5	0.18	10.1	0.84	0.092	1.20	0.64	1.01	143	0.97
Central	68	0.14	41.9	0.19	9.2	0.85	0.097	1.09	0.62	1.00	133	0.99
North central	55	0.16	29.8	0.22	9.5	0.83	0.087	1.20	0.55	1.01	167	1.17
Southeast	65	0.15	49.0	0.20	8.7	0.92	0.122	1.09	0.75	1.10	136	1.03
East Gulf	68	0.17	53.7	0.23	6.4	1.05	0.178	1.03	0.80	1.19	130	1.25
East Texas	41	0.22	31.2	0.29	8.0	0.97	0.137	1.08	0.76	1.18	213	1.28
West Texas	30	0.27	17.3	0.33	7.4	0.98	0.121	1.13	0.57	1.07	302	1.53
Southwest	20	0.30	7.4	0.37	7.8	0.88	0.079	1.16	0.37	0.88	473	1.46
West inland	14	0.38	4.9	0.43	9.4	0.75	0.055	1.06	0.36	0.87	786	1.54
Pacific south	19	0.36	10.2	0.42	11.6	0.78	0.054	0.76	0.54	0.98	476	2.09
Northwest inland	31	0.23	11.5	0.29	10.4	0.82	0.057	1.20	0.37	0.93	304	1.43
Pacific central	32	0.25	18.4	0.33	13.7	0.80	0.048	0.85	0.58	1.05	265	2.00
Pacific northwest	71	0.15	35.7	0.19	15.9	0.80	0.035	0.73	0.50	1.09	123	1.50

Source: Driscoll (1989).

TABLE 2.9 Comparison of Rainfall Statistics for 4- and 72-Hour Minimum Interevent Dry Periods by Geographic Region with a Minimum Precipitation Producing Runoff of 0.04 in.

Region	4-h Minimum Interevent Dry Period		72-h Minimum Interevent Dry Period	
	Mean Volume (in.)	Mean Duration (h)	Mean Volume (in.)	Mean Duration (h)
Apalachicola	0.63	5.5	1.54	49.9
Daytona Beach	0.53	4.8	1.34	51.0
Fort Myers	0.54	2.6	1.82	70.3
Gainesville	0.50	2.9	1.34	51.8
Inglis	0.55	3.0	1.45	47.2
Jacksonville	0.52	5.3	1.34	56.7
Key West	0.45	3.8	1.13	45.8
Lakeland	0.50	3.1	1.52	61.3
Melbourne	0.46	2.7	1.25	46.5
Miami	0.46	3.9	1.71	83.1
Moore Haven	0.49	4.1	1.35	58.4
Niceville	0.62	3.8	1.70	51.7
Orlando	0.49	4.1	1.41	64.0
Parrish	0.52	2.6	1.49	55.6
Tallahassee	0.67	5.2	1.74	56.5
Tampa	0.54	4.4	1.32	54.7
West Palm Beach	0.53	4.3	1.71	68.9
Average	0.53	3.9	1.48	57.3

flood control levels. An example of the volume and duration changes for two different interevent dry periods for different areas of region 3 are given in Table 2.9 (Wanielista et al., 1991). When using runoff-producing rainfall, the average statistics increase. Thus empirical distributions based on rainfall statistics at specific exceedence probabilities and interevent dry periods must be calculated.

2.5 SUMMARY

Probability distributions are used to estimate the relative frequency of an event. Empirical data are ranked into a frequency distribution and various theoretical probability distributions are "fit" to the empirical one. The goodness of fit can be determined using a graphical examination or statistical tests. Also, regression analysis has been used to estimate the parameters for the equations used for extreme events. Additional specific information from this chapter is as follows:

- A rainfall event is defined by the volume and intensity of rainfall given a minimum interevent dry period. The minimum interevent dry period is picked so that resulting use of the precipitation data produces either independent rainfall or runoff events. An example is the calculation of a runoff hydrograph that requires the initial flow rate not to be affected by previous flow rates. Another example is the requirement to retain runoff for at least 48 h to achieve infiltration or otherwise to remove its pollutants.
- Frequency–intensity–duration curves are used to specify average intensity for a given averaging time and frequency of return.
- The frequency of events used for flood prediction has a return period of years, while the frequency used for pollution control is on the order of months.
- Empirical data collection must follow the concepts of statistical sampling. Thus the data must be independent and not biased.
- Some nonlinear equations (equations 2.4 through 2.7) can be transformed to a linear relationship, then linear regression procedures can be used to estimate the parameters of the equation. Nonlinear regression procedures use a gradient search routine on untransformed data to estimate a "best fit." The nonlinear procedures usually produce more accurate estimates.
- Regression equations for hydrologic data are specific to the area where the data were collected. Extrapolation to other areas without calibration is poor practice.
- The method of moments, maximum likelihood method, graphics, and regression analysis are used to estimate parameters of frequency distributions.
- Rainfall and runoff processes can be described by probability distributions. The gamma, exponential, and log-Pearson have been used.
- To improve an understanding of extreme events, the reader should develop, by hand calculations, and plotting on probability paper, an empirical probability distribution. Then computer programs can be used for parameter estimation and sensitivity analysis. Computer programs are available with this book to aid in the selection of theoretical probability distributions.

2.6 PROBLEMS

1. Explain in your own words how to sample streamflow with the purpose to estimate flows from snowmelt. When do you sample? What is the population?

2. Using Figure 2.2, how often would you sample rainfall intensities for a storm during 1 year if you wish to determine runoff volumes due to storms with volume greater than or equal to 1 in.? Assume that there are 130 storms per year and that you have a continuously measuring rainfall gage.

3. Show mathematically how the exponential curve of equation 2.4 is made linear.

4. For equation 2.11, estimate the value of the peak flow for return periods of 25 years and 100 years if the streambed slope is 2 m/km and the watershed area is 8 km². Be careful with units. Now, estimate the range of values for 95% of the estimates (2 standard deviations). Comment on the values for slope.

5. Using the equation from Table 2.1 for a 4000-acre watershed and the basin characteristics (slope and lake area) that will produce the greatest peak and be within the range of applicability, estimate the greatest peak for the 100-year event.

6. If you wish to attenuate in half (reduce by 50%) the peak estimate of Problem 5 using equations from Table 2.1 for the same slope and watershed area, how much lake area should you build into the system? Comment on the expected attenuation results from the viewpoint of statistical accuracy.

7. For the following flow rate data, develop an empirical distribution (use either Weibull, California, or Foster plot position). Comment on the result using probability paper.

Year	Peak Flow (m³/s)	Year	Peak Flow (m³/s)
1970	602	1979	178
1971	214	1980	249
1972	106	1981	365
1973	312	1982	250
1974	280	1983	912
1975	143	1984	404
1976	190	1985	136
1977	236	1986	101
1978	737		

8. Using the maximum daily rainfall data of Table 2.4, construct a histogram of rainfall. Comment on the shape of the histogram and compare to the cumulative plot of data from Table 2.5.

9.* A cumulative probability distribution for rainfall intensity (i), duration (D), and interevent times (Δ) are used for the following estimates.

$$\Pr\{i \leq 0.40 \text{ in.}/\text{h}\} = 0.98$$
$$\Pr\{D \geq 6 \text{ h}\} = 0.05$$
$$\Pr\{\Delta \geq 92 \text{ h}\} = 0.10$$

   a. If there were 100 events per year, calculate the return period and the number of events per year for each hydrologic measure.
   b. What is the return period and number of events per year for a storm of intensity greater than 0.40 in./h and a duration greater than 6 h?

10. Using 2 months of rainfall data measured and reported on an hourly basis, develop an empirical frequency distribution for two interevent dry periods. Use an interevent dry period commonly used for independent rainfall and the other for a runoff hydrograph that is estimated to be 24 h long.

11. A mass distribution for solids into a detention pond follows a gamma distribution with mean equal to 200 mg/L and a coefficient of variability of 1.25. What is the concentration for which 10% of the time, one can expect a higher value?

12. If the average rainfall volume producing runoff is 0.45 in. and the distribution of these rainfall events follows an exponential distribution, what percentage of storm events are less than or equal to 1 in.? If the distribution is a gamma with a coefficient of variation similar to summer storms in region 3 of Figure 2.11, what percentage of the storms are less than 1 in.?

2.7 COMPUTER-ASSISTED PROBLEMS

1. For the tidal data of Table 2.3, estimate the coefficients for the log-Pearson distribution using the method of moments and a computer program. Then calculate the 100-year return period tidal depth using the log-Pearson method.

2. Use the least-squares curve-fit program to determine a mathematical relationship between the tidal and rainfall data of Table 2.3. Comment on the best fit.

*Denotes a problem that is solved in Appendix H.

3. Using a computer program, determine the rainfall for the 2-, 25-, and 100-year return using a log-Pearson type III distribution for the data of Table 2.4.

4. Using a computer program and the data of Problem 7, Section 2.6, determine the parameters of a log-Pearson type III distribution and comment on the goodness of fit. Change the 1983 flow rate from 912 to 312, and comment on the change.

5. Using the data of Table 2.5 (maximum daily storms for each year at Bushnell), develop an empirical distribution graph and comment on the goodness of fit for a two-parameter log-normal versus a Gumbel theoretical distribution.

2.8 REFERENCES

Arnell, V. 1982. *Rainfall Data for the Design of Sewer Pipe Systems*, Report Series A: 8, Department of Hydraulics, Chalmers University of Technology, Gothenburg, Sweden.

Athayde, D. 1979. *A Statistical Method for the Assessment of Urban Stormwater*, EPA 440/3-79-023, U.S. EPA Nonpoint Source Branch.

Bridges, W. C. 1982. *Techniques for Estimating Magnitude and Frequency of Floods on Natural-Flow Streams in Florida*, Water Resources Report 84-4012, U.S. Geological Survey, Tallahassee, Fla.

DiToro, D. M., and Small, M. J. 1979. "Stormwater Interception and Storage," *Journal of the Environmental Engineering Division*, ASCE, Vol. 105, No. EE1.

Driscoll, E. D. 1989. *Analysis of Storm Event Characteristics for Selected Rainfall Gages Throughout the United States*, draft report prepared for U.S. Environmental Protection Agency, Woodward-Clyde Consultants, Oakland, Calif.

Driscoll, E. D., Palhegye, G. E., Strecher, E. W., and Shelley, P. E. 1989. *Analysis of Storm Event Characteristics for Selected Rainfall Gages Throughout the United States*, draft report to Environmental Protection Agency by Woodward-Clyde Consultants, Oakland, Calif.

Gumbel, E. J. 1945. "Floods Estimated by the Probability Method," *Engineering News Record*, Vol. 134; pp. 833–837.

Hardison, C. H. 1974. "Generalized Skew Coefficient of Annual Floods in the United States," *Water Resources Research*, Vol. 10, No. 5, pp. 745–752.

Hvitved-Jacobsen, T. H., and Yousef, Y. 1987. "Analysis of Rainfall Series in the Design of Urban Drainage Control Systems," *Water Research*, Vol. 22, No. 4, pp. 491–496.

Kite, G. W. 1985. *Frequency and Risk Analyses in Hydrology*, Water Resources Publications, Littleton, Colo.

Overton, D. E., and Meadows, M. E. 1976. *Stormwater Modeling*, Academic Press, New York.

Sauer, V. B., Thomas, W. O., Stricker, V. A., and Wilson, K. U. 1983. *Flood Characteristics of Urban Watersheds in the United States*, Water Supply Paper 2207, U.S. Geological Survey, Washington, D.C.

Schroeder, E. E., and Massey, B. C. 1977. *Technique for Estimating the Magnitude and Frequency of Floods in Texas*, Water Resources Investigations, U.S. Geological Survey, Washington, D.C., pp. 77–110.

Schuler, T. B. 1987. *Controlling Urban Runoff: A Practical Manual for Planning and Designing Urban BMPs*, Washington Metropolitan Water Resources Planning Board, Washington, D.C.

Strahler, A. N. 1964. "Geology, Part II", *Handbook of Applied Hydrology*, McGraw-Hill, New York.

U.S. Department of Transportation. 1984. *Hydrology Circular 19*, Federal Highway Administration, McLean, Va.

U.S. Environmental Protection Agency. 1986. *Methodology for Analysis of Detention Basins for Control of Urban Runoff Quality*, EPA 440/S-87-001, U.S. EPA, Washington, D.C.

U.S. Water Resources Council, Hydrology Committee. 1981. *Guidelines for Determining Flood Frequencies, Bulletin 17B*, U.S. Water Resources Council, Washington, D.C.

Wanielista, M. P. 1990. *Hydrology and Water Quantity Control*, Wiley, New York.

Wanielista, M. P., and Shannon, E. 1977. *An Evaluation of Best Management Practices for Stormwater*, East Central Florida Regional Planning Council. Winter Park, Fla.

Wanielista, M. P., Sommerville, P. N., Cooper, G. E., Dendy, J. S., and Thompson, E. 1986. *Rainfall Analyses of Southwest Florida*, Southwest Florida Water Management District, Brooksville, Fla.

Wanielista, M. P., Yousef, Y. A. and Lineback, T. 1991. *Precipitation and Inter-event Dry Periods for Selected Areas of Florida*, State Department of Environmental Regulation, Tallahassee, Fla.

Wenzel, H. G., and Voorhees, M. L. 1981. *An Evaluation of the Urban Design Storm Concept*, Research Report VILU-WRC-81-0164, Water Resource Center, University of Illinois at Urbana–Champaign, Ill.

Yevjevich, V. 1972. *Probability and Statistics in Hydrology*, Water Resources Publications, Fort Collins, Colo.

CHAPTER 3

Hydrographs

Stormwater management includes objectives and policies that require the calculation of rainfall excess, peak rate discharge, and runoff rate over time. The volume of rainfall available for runoff or intentional on-site storage is called *rainfall excess*. It has volume measurement units and is different from runoff rates, which have measurement units of volume per time period. An expression of runoff rate over time is called a *runoff hydrograph*, while a streamflow hydrograph is the sum of runoff and soil resident water that infiltrates from the ground into the stream. The highest rate of discharge is called the *peak discharge*.

Watershed changes (i.e., increase in directly connected impervious areas) can increase the rainfall excess and peak discharge relative to an existing land use. The major questions addressed in this chapter are related to methods for calculating rainfall excess, peak discharges, hydrograph shapes, and detention pond volume for hydrograph attenuation.

3.1 SYNTHETIC HYDROGRAPHS

A *hydrograph* is an expression (usually graphical) for flow rate at a point over time. Synthetic hydrographs are predictions of flow rate based on watershed characteristics and assumed rainfall intensities over a period of time. Few if any streamflow (discharge) data are available to aid in developing the synthetic hydrograph. Synthetic hydrograph procedures are used more often than procedures based on streamflow data because each time a land-use change is anticipated, or no stream gage is in place for the watershed, synthetic hydrograph procedures to "generate" the hydrograph must be used. The development of a synthetic hydrograph requires estimates for runoff volume (rainfall excess) and hydrograph shape. The shape of a runoff hydrograph can be specified by the time to peak flow and time for recession flow if rainfall excess is given.

3.2 RAINFALL EXCESS

Rainfall excess is the volume of water from rainfall that is available for planned constructed storage on the watershed, or if not stored, it becomes runoff from the site. When expressed as runoff it is a rate term (volume per unit time).

Four estimation procedures are in common use: namely, site-specific and variable infiltration rate measures, a constant rate, mass balances, and the soil-complex-cover method (SCS-CN). All four methods are used to estimate that portion of rainfall that infiltrates on site in addition to that available for runoff.

If infiltration is to occur, there must be precipitation and the soil must have the potential to infiltrate the rainfall. There are three conditions relating potential infiltration rate and volume to precipitation rate and volume as shown in Figure 3.1. Precipitation intensity and potential infiltration rate are denoted by $i(t)$ and $f(t)$, respectively, with P, F, and R used for precipitation volume, actual infiltration volume, and rainfall excess. This assumes there is no initial abstraction. During a rainfall event when the precipitation rate is less than the potential rate of infiltration and the total infiltration volume is less than the potential infiltration volume all the rainfall will infiltrate (case 1). When the rate of precipitation exceeds the potential infiltration rate and there is no on-site surface storage, rainfall excess will result (case 2), and maximum potential infiltration volume will not be obtained. For case 3, precipitation rate exceeds the potential infiltration rate and the maximum groundwater storage was exceeded before the end of the storm event.

3.2.1 A Site-Specific Measure for Soil Infiltration Rates

The rate of infiltration into the soil is generally greater at the start of infiltration and decreases with time until a relatively constant rate is obtained assuming that surface water is available. This is usually reflected by an exponential relationship (Horton, 1939) and valid when the potential infiltration rate is greater than or equal to rate of surface water supply, such as rainfall intensity.

$$f(t) = f_c + (f_0 - f_c)e^{-Kt} \qquad (3.1)$$

where $f(t)$ = infiltration rate, cm/h or in./h
f_c = final or constant rate, cm/h or in./h
f_0 = initial infiltrate rate, cm/h or in./h
K = recession constant, h^{-1}
t = time, h

3.2 RAINFALL EXCESS   67

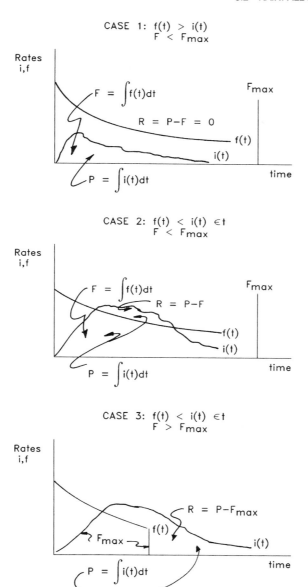

Figure 3.1 Rainfall excess related to potential infiltration rates and rainfall intensity.

Assuming surface water is available for infiltration, integration of equation 3.1 produces a total volume of infiltrate (F), or

$$F = f_c(t) + \frac{f_0 - f_c}{K}(1 - e^{-Kt}) \tag{3.2}$$

Thus by knowing the rate of infiltration and the volume of infiltrate at any time during precipitation, the rate of rainfall excess or rainfall excess volume can be calculated. If there were any depression storage or other abstraction by vegetation, the rate of discharge must be reduced by the rate at which depression storage and abstraction is occurring.

3.2.2 A Constant Rate

A constant rate of infiltration is assumed. It is sometimes called the Φ index method. It is calculated by assuming that the rate of infiltration is constant with time using the formula

$$\Phi = \frac{P - R}{D} \qquad (3.3)$$

where Φ = index, in./h
R = runoff volume or rainfall excess, in.
P = rainfall volume, in.
D = duration of the storm, h

Rainfall intensity minus the Φ index is the rate of rainfall excess.

3.2.3 Mass Balance Methods

Mass balance methods are used to equate rainfall intensity to runoff rate. For a completely impervious watershed, the volume of rainfall excess is equal to the volume of precipitation. Precipitation intensity (rainfall rate) onto a watershed of contributing area (CA in acres) must equal the instantaneous rate of discharge (Q_I in ft^3/s) from the watershed using appropriate conversion factors and sufficient travel time for the total area to contribute runoff.

Instantaneous runoff rate = precipitation intensity

or

$$Q_I = i(\text{CA})(1.008) \qquad (3.4)$$

where Q_I = instantaneous runoff rate, ft^3/s
i = precipitation rate, in./h
CA = contributing area, acres
1.008 = conversion factor, (ft^3/s)-h/acre-in.

The constant (1.008) frequently can be dropped from the equation. When the contributing area increases, runoff rate increases for constant rainfall rate.

The original work of Kuichling (1889) over 100 years ago in an urban watershed showed that the ratio of runoff rate to precipitation rate is equal to the contributing area at a time into the storm when approximately the total impervious area was drained. Multiplying the runoff rate by this time results in a runoff volume [i.e., flow rate $(L^3/t) \times$ time (t) = volume (L^3)] and multiplying rainfall intensity by time and by area yields a rainfall volume [i.e., rainfall intensity $(L/t) \times$ time $(t) \times$ area (L^2) = volume (L^3)], or from Kuichling (1889):

$$\frac{Q}{i} = CA \tag{3.5}$$

Integrating over time (t) gives

$$\frac{Qt}{iAt} = C = \frac{R}{P} \tag{3.6}$$

Kuichling (1889) concluded that the ratio of Q/i is the rational value that can be used for the design of urban sewer systems and the value of C is equal to the extent of the impervious surfaces divided by the total area assuming no drainage from the pervious surfaces. Also, C is identified as the runoff coefficient and can be calculated using two formulas: a ratio of rainfall excess to precipitation volume (equation 3.6) or a ratio of peak discharge to the product of rainfall intensity and area (equation 3.5). Since rainfall intensity is rarely constant over the time it takes to drain the watershed area, average intensity is used in equation 3.5, while maximum total precipitation for the time of drainage is related to the maximum volume discharge in equation 3.6.

Rainfall excess results from both the directly connected impervious areas and from the pervious areas when rainfall intensity exceeds infiltration rate or the soil becomes saturated. The volume of runoff that results from the directly connected impervious area is about equal to the volume of water from rainfall. However, there could exist depression storage within the impervious area, which reduces the volume of runoff. For watersheds divided into hydraulically (directly) connected impervious areas and pervious areas, the runoff coefficient is 1.0 for the directly connected areas. To aid designers of sewer systems and stormwater ponds, composite C factors have been published and are used in practice (see Table 3.1). Conservative (larger) values are generally used for design. However, Table 3.1 should be used with care because the runoff coefficient may change with the volume of rainfall and intensity of rainfall. Frequently, it is more accurate to identify the directly connected impervious area and the pervious area and sum the rainfall excess from both areas. Then the runoff coefficient is calculated by dividing the rainfall excess by the rainfall volume.

TABLE 3.1 Runoff Coefficients C Recurrence Interval $\leq$ 10 Years

Description of Area	Runoff Coefficient	Character of Surface	Runoff Coefficient[a]
Business		Pavement	
Downtown	0.70–0.95	Asphalt or concrete	0.70–0.95
Neighborhood	0.50–0.70	Brick	0.70–0.85
Residential		Roofs	
Single family	0.30–0.50	Lawns, sandy soil	
Multiunits, detached	0.40–0.60	Flat, 2%	0.05–0.10
Multiunits, attached	0.60–0.75	Average, 2 to 7%	0.10–0.15
Residential, suburban	0.25–0.40	Steep, 7% or more	0.15–0.20
Apartment	0.50–0.70	Lawns, heavy soil	
Industrial		Flat, 2%	0.13–0.17
Light	0.50–0.80	Average, 2 to 7%	0.18–0.22
Heavy	0.60–0.90	Steep, 7% or more	0.25–0.35
Parks and cemeteries	0.10–0.25		
Railroad yard	0.20–0.35		
Unimproved	0.10–0.30		

Source: Design and Construction of Sanitary and Storm Sewers, ASCE Manual of Practice No. 37, 1970. Revised by D. Earl Jones, Jr.

[a] For 25- to 100-year recurrence intervals, multiply coefficient by 1.1 and 1.25, respectively; the product cannot exceed 1.0.

3.2.4 SCS Soil Complex Curves Method

The SCS-CN method (U.S. Department of Agriculture, 1986) depends on knowledge of the hydrologic classification of soils and the vegetation cover. Through experimentation with over 3000 soil types and cover crops, an empirical relationship was derived relating maximum watershed storage to a curve number that reflects soil type and vegetative cover. The relationship is

$$\text{Metric} \qquad \text{English}$$
$$S' = \frac{25{,}400}{\text{CN}} - 254 \qquad S' = \frac{1000}{\text{CN}} - 10 \tag{3.7}$$

where S' is the maximum storage (mm, in.) and CN is the curve number $\leq$ 100. When the curve number is equal to 100, there is no storage. For impervious areas, the curve number is usually specified as 98 or 100. For pervious areas, curve numbers that are commonly used are shown in Table 3.2. Other data on soil types, hydrologic classification, and curve numbers are available in Appendix F (U.S. Department of Agriculture, 1986). Judgment on the type of land use has to be exercised and a conservative estimate (higher CN values) is prudent when doubt exists.

3.2 RAINFALL EXCESS

TABLE 3.2 Runoff Curve Numbers for Pervious Areas

Land Use	Hydrologic Soil Class			
	A	B	C	D
Bare ground	77	86	91	94
Natural desert landscaping	63	77	85	88
Gardens or row crop	72	81	88	91
Good grass (cover on > 75% of the pervious area)	39	61	74	80
Fair grass (cover on 50–75% of the pervious area)	68	79	86	89
Lightly wooded area	36	60	73	79
Good pasture and range land	39	61	74	80

Source: Adapted from U.S. Department of Agriculture (1986).

Using the SCS-CN procedure, rainfall excess calculations are a function of rainfall volume and curve number. Assuming that storage at any time is proportional to maximum storage and rainfall excess is proportional to precipitation volume, the following equations result (U.S. Department of Agriculture, 1986) for no initial abstraction ($I_A = 0$) and for an initial abstraction (I_A) expressed as a fraction of maximum storage:

$$R = \frac{P^2}{P + S'} \quad \text{for } I_A = 0 \tag{3.8}$$

$$R = \frac{(P - I_A S')^2}{P + (1 - I_A)S'} \quad \text{for } P > I_A S' \tag{3.9}$$

Otherwise, $R = 0$.

The typical SCS initial abstraction factor is 0.2, but other values can be used.

Example Problem 3.1 During a period of intense rainfall for a duration equal to the time it takes to drain the watershed completely (time of concentration), calculate the rainfall excess volume and the instantaneous runoff rate from an impervious area of 12 acres and a pervious area of 6 acres. The period of intense rainfall is 30 min at a rate of 5 in./h. The runoff coefficients for the pervious desert landscape and impervious areas are 0.30 and 1.00 respectively for the rainfall conditions and are assumed constant over time.

Next, using the SCS-CN method, calculate the rainfall excess from the pervious area if the area is in hydrologic soil classification A and desert landscaping with an initial abstraction equal to zero.

SOLUTION: The first part of the problem is solved using mass balances:

1. Volume of rainfall:

$$(5 \text{ in./h})(30/60) = 2.5 \text{ in.}$$

2. Rainfall excess volume:
   Impervious area: $R = CP = 1.0(2.5) = 2.5$ in.
   Pervious area: $R = CP = 0.3(2.5) = 0.75$ in.
   Total rainfall excess $= [2.5(12) + 0.75(6)]/18 = 1.92$ in.
3. The instantaneous runoff rate for:
   Impervious area: $Q_I = i(CA) = 5(12) = 60 \text{ ft}^3/\text{s}$
   Pervious area: The contributing area (CA) is equivalent to the runoff coefficient times the area; thus

$$Q_P = iCA = 5(0.3)(6) = 9 \text{ ft}^3/\text{s}$$

Next, the land use and hydrologic classification is given. Use the SCS-CN procedure and from Table 3.2, the CN is 63; thus $S' = (1000/63) - 10 = 5.87$ in. Using equation 3.8, the rainfall excess is $R = (2.5)^2/(2.5 + 5.87) = 0.75$ in. from the pervious area. Thus the rainfall excess is calculated as the same value (0.75 in.) using either the mass balance or SCS-CN method.

3.3 HYDROGRAPH PROCEDURES

Four commonly used hydrograph generation procedures are presented in this section: the rational method, the Soil Conservation Service procedure, the Santa Barbara urban hydrograph, and a functional form for continuous convolution.

3.3.1 Rational Method

The rational formula is traditionally used to calculate peak discharge and can be derived using a mass balance for precipitation rate and runoff rate (see equation 3.4). The contributing area is divided into two parameters: the runoff coefficient C and the watershed area A. As developed by Mulvaney (1851) and used by Kuichling (1889), equation 3.5 is restated as

$$Q_P = CiA \qquad (3.10)$$

where Q_p = peak discharge, ft³/s
  C = runoff coefficient
  i = precipitation rate, in./h
  A = watershed area, acres

The conversion factor (1.008) is dropped from the equation. Use of the equation must consider the following assumptions:

1. Rainfall intensity is constant over the time it takes to drain the watershed (time of concentration).
2. The runoff coefficient remains constant during the time of concentration.
3. The watershed area does not change.

These assumptions are reasonable for watersheds with short time of concentration (about 20 min). The intensity is relatively constant for travel time below 20 min. From frequency–intensity–duration curves, intensity of rainfall is selected. The intensity is associated with a rainfall of specific frequency and with a duration of storm equal to time of travel for the watershed time of concentration. Appendix C has example frequency–intensity curves.

As developed by Williams (1950), Mitchi (1974), Pagan (1972), Wanielista (1990), and others, a hydrograph shape can be assumed for the rational method. It is essentially an isosceles triangle with the hydrograph base equal to twice the time of concentration (travel time) when the storm duration equals time of concentration.

3.3.2 SCS Procedure

The Soil Conservation Service (SCS) hydrograph procedure (U.S. Department of Agriculture 1986) specifies either a triangular hydrograph shape or a curvilinear one. The triangular shape can be expressed as a function of the shape obtained using the rational formula (Wanielista, 1990). Equations 3.11 and 3.12 are used to calculate the base and peak discharge for the SCS hydrograph shape.

$$t_b = t_p + xt_p \tag{3.11}$$

$$Q_P = KCiA \tag{3.12}$$

where t_b = hydrograph base time, h
  t_p = time to peak, h
  $x = (1291/K) - 1$ (with A in square miles)
    $= (2/K) - 1$ (with A in acres)
  Q_P = peak discharge, ft³/s
  K = peak attenuation factor
  i = average intensity per time interval, in./h
  A = watershed area, acres or mi²

HYDROGRAPHS

TABLE 3.3 Triangular Hydrograph Attenuation Factors

General Description	Falling Limb Factor, x	Peak Attenuation Factor, K^a	
		A (mi^2)	A (acres)
Rational formula shape	1.00	645	1.00
Urban, steep slopes	1.25	575	0.89
Typical SCS	1.67	484	0.75
Mixed urban/rural	2.25	400	0.62
Rural, rolling hills	3.33	300	0.47
Rural, flat slopes	5.50	200	0.31
Rural, very flat	12.00	100	0.16

[a] Includes the 1.008 conversion factor.

Guidance for the selection of K is given in Table 3.3 and Figure 3.2. The calculation of peak discharge requires an estimate of time of concentration or a general description of the land use. When doubt exists, larger values of the attenuation factor should be used which will calculate larger peak discharges. However, every effort should be made to determine a correct attenuation factor. The effects of the attenuation factor on triangular hydrograph shapes are shown in Figure 3.3. The rainfall excess is the same for each triangular hydrograph.

Example Problem 3.2 Calculate the peak discharge from a 200-acre mixed urban/rural watershed with a time of concentration (t_c) equal to 50 min and a runoff coefficient of 0.30 for a 5-in. rainfall event over 1 h in SCS type II distribution area. Do the calculations using the SCS attenuation factor (K) from Table 3.3 and Figure 3.2. The initial abstraction is about 0.5 in. Also, use the rational formula and compare the three peak discharges. Assume that intensity is constant.

SOLUTION: Using Table 3.3:

$$K = 0.62 \text{ (mixed urban/rural)}$$
$$Q_P = KCiA = 0.62(0.30)(5)(200) = 186 \text{ ft}^3/\text{s}$$

Using Figure 3.2:

$$K = 400 \text{ at } t_c = \frac{50}{60} = 0.83 \text{ h} \quad \text{and} \quad \frac{I_A}{P} = 0.10$$

R (rainfall excess) $= 0.30(5) = 1.5$ in.

$$Q_P = KAR = 400\left(\frac{200}{640}\right)(1.5) = 187.5 \text{ ft}^3/\text{s}$$

3.3 HYDROGRAPH PROCEDURES   75

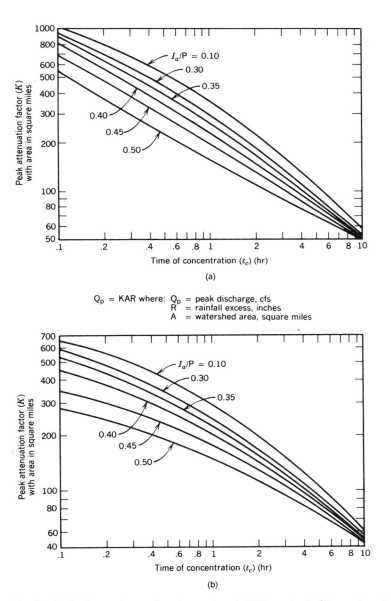

Figure 3.2 Peak discharge factor for the updated SCS method. (Reproduced from U.S. Department of Agriculture, 1986.)

76 HYDROGRAPHS

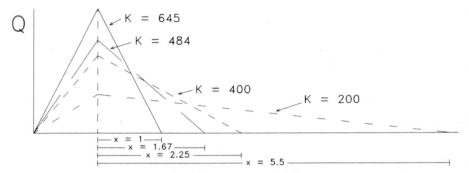

Figure 3.3 Triangular hydrograph related to hydrograph peak reduction factor.

Using the rational formula:

$$Q_P = CiA = 0.30(5)(200) = 300 \text{ ft}^3/\text{s}$$

Since the rational formula attenuation factor is equal to 1, the peak discharge calculated using the rational formula relative to the SCS should be higher.

3.3.3 Santa Barbara Urban Hydrograph

The Santa Barbara urban hydrograph (SBUH) procedure provides equations for calculating the instantaneous discharge from both the directly connected impervious area and the pervious area (Stubchaer, 1975). The sum of these two discharges are routed using

$$Q(t + 1) = Q(t) + K_r[I(t) + I(t + 1) - 2Q(t)] \qquad (3.13)$$

where $Q(t + 1)$ = routed flow in time $(t + 1)$, ft^3/s
$Q(t)$ = routed flow in time (t), ft^3/s
$I(t)$ = instantaneous flow rate in time (t), ft^3/s
   = $R(t)(A)(1.008)/\Delta t$
where $R(t)$ = sum of rainfall excess from both impervious and pervious areas, in.
A = watershed area, acres
Δt = time interval, h
1.008 = conversion factor
$I(t + 1)$ = instantaneous flow rate in time $(t + 1)$, ft^3/s
$K_r = \Delta t/(2t_c + \Delta t)$

The choice of K_r is critical because it establishes the peak discharge or attenuation. Since t_c is assumed as constant, the time increment for analysis (Δt) must be picked to represent a reasonable hydrograph shape, and Δt must be less than the time to peak (Wanielista, 1990).

3.3.4 Continuous Convolution

From unit hydrograph theory (Sherman, 1932) the product of hydrograph shape (unit hydrograph) and rainfall excess is the hydrograph for a specific watershed and rainfall hyetograph. Thus if rainfall excess is calculated and a hydrograph shape (routing function) is known, the resulting hydrograph can be calculated by simple multiplication. In its general form, the convolution integral is

$$Q_t = \int R(\tau) g(t - \tau) \, d\tau \tag{3.14}$$

where $R(t)$ = rainfall excess rate as a function of time, ft^3/s
$g(t - \tau)$ = routing function offset in real time by τ

Solving equation 3.14 assuming an exponential routing function and constant rainfall excess per unit time, Wanielista (1990) developed the following equation:

$$Q_t = R(t)(1 - e^{-kt}) \qquad 0 \leq t \leq D \tag{3.15}$$

$$Q_t = R(t) e^{-kt}(e^{kD} - 1) \qquad t > D \tag{3.16}$$

where $R(t)$ = rainfall excess rate, ft^3/s
 k = routing or storage coefficient, min^{-1}
 t = time, min
 D = time for each rainfall excess interval

Equations 3.15 and 3.16 were developed from basic mass balances with the routing coefficient shown (Wanielista, 1990) to be related to the inverse of time of concentration and watershed area (Figure 3.4). Because of the strong relationship to time in the watershed, the routing coefficient is also called the storage coefficient.

$$k = \frac{1}{t_t} \tag{3.17}$$

where k = routing or storage, min^{-1}
 t_t = travel time, min

3.3.5 Computer Programs

The extensive time-consuming calculations to solve a complete hydrograph requires the use of some type of computation aids. The computer programs included with this book are useful in reducing the time for obtaining

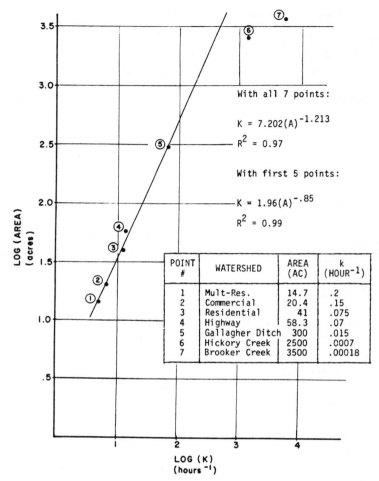

Figure 3.4 Relationship between storage coefficient k and drainage area for the seven watersheds.

solutions and aid in understanding hydrographs and in conducting sensitivity analyses (see Appendix E for more details).

3.4 HYDROGRAPH ATTENUATION

Hydrograph attenuation is the process of reducing peak discharges using storage ponds or a river segment. This can be accomplished by increasing the watershed time of concentration primarily by adding water storage in the transport system. As time of concentration increases, the hydrograph attenuation factor decreases; thus peak discharges decrease. For large storage areas, it is not always possible to estimate time of concentration; thus storage volume–outflow rate relationships are used.

There are in common use at least two hydrograph attenuation methods: the inventory equation and Muskingum formula. Both assume a relationship between inflow and outflow hydrographs, with the outflow being dependent on previous outflow and inflow.

3.4.1 Inventory Equation

The continuity equation (conservation of mass) can be written as

$$\frac{\Delta A}{\Delta t} + \frac{\Delta Q}{\Delta X} = q \tag{3.18}$$

where A = flow area, ft^2
t = time, s
Q = flow rate, ft^3/s
X = longitudinal distance, ft
q = discharge per unit length, ft^3/s-ft

The flow rate change with distance $(\Delta Q/\Delta X)$ is the change in rate per unit channel length or inflow minus outflow; thus

$$\frac{\Delta Q}{\Delta X} = \frac{I - O}{\Delta X} \tag{3.19}$$

where I = inflow, ft^3/s
O = outflow, ft^3/s
ΔX = distance, ft

The change in channel storage volume is the change in area multiplied by the channel length, or

$$\frac{\Delta A}{\Delta t} \frac{\Delta X}{\Delta X} = \frac{\Delta S}{\Delta t \Delta X} \tag{3.20}$$

where ΔS = storage volume, ft^3
Δt = time interval, s

Substituting equations 3.19 and 3.20 into equation 3.18 yields a form of the continuity equation after multiplying by ΔX and Δt recognizing that ΔS

80 HYDROGRAPHS

includes $q\Delta X\Delta t$:

$$\bar{I}(\Delta t) - \bar{O}(\Delta t) = \Delta S \qquad (3.21)$$

where ΔS = storage change, L^3
$\bar{I}$ = average inflow in period (Δt), $L^3 t^{-1}$
$\bar{O}$ = average outflow in period (Δt), $L^3 t^{-1}$
Δt = time period for analysis, t

Letting $\bar{O} = (O_1 + O_2)/2$ and arranging equation 3.21 with all known values on the right side of the equation, equation 3.21 is rewritten, with N_2 defined as the known quantity.

$$S_2 + \frac{O_2}{2}\Delta t = S_1 - \frac{O_1}{2}\Delta t + \bar{I}(\Delta t) = N_2 \qquad (3.22)$$

and in general,

$$N = S + \frac{O}{2}\Delta t \qquad (3.23)$$

Since outflow is a function of storage, substitute that function, $S = f(O)$, into equation 3.23:

$$N = f(O) + \frac{O}{2}\Delta t \qquad (3.24)$$

Thus outflow can be calculated if a storage–discharge relationship, previous outflow, and an influent hydrograph are available.

Example Problem 3.3 Calculate a discharge hydrograph using a rainfall intensity of 4 in./h for a 30-acre urban watershed with a time to peak equal to 20 min, the typical SCS hydrograph shape, and a runoff coefficient of 0.6. Use 10-min increments and a triangular shape for calculations. Next, use a pond volume discharge relationship of 1 ft^3/s per 5000 ft^3 of storage and calculate the pond dishcarge hydrograph.

SOLUTION: Use the SCS peak attenuation factor of 484, and calculate the triangular discharge hydrograph. The time from recession limb (peak to no flow) is $1.67(20 \text{ min}) \cong 33$ min. The peak flow is

$$Q_p = KCiA = 0.75(0.6)(4)(30) = 54 \text{ ft}^3/\text{s}$$

3.4 HYDROGRAPH ATTENUATION   81

TABLE 3.4   Calculations for Example Problem 3.3[a]

Time (min)	I (ft³/s)	I_{i-1} (ft³/s)	$\bar{I}$ (ft³/s)	$\bar{I}\Delta t$ (ft³)	S_{i-1} (ft³)	O_{i-1} (ft³/s)	$O_{i-1}\left(\dfrac{\Delta t}{2}\right)$ (ft³)	$N_i = \bar{I}\Delta t + S_{i-1} - O_{i-1}\left(\dfrac{\Delta t}{2}\right)$ (ft³)	O_i (ft³/s)	S_i (ft³)
0	—	0	0	0	0	0	0	0	0	0
10	27	0	13.5	8,100	0	0	0	8,100	1.528	7642
20	54	27	40.5	24,300	7,642	1.528	458	31,484	5.940	29,702
30	38	54	46.0	27,600	29,702	5.940	1,782	55,520	10.475	52,377
40	22	38	30.0	18,000	52,377	10.475	3,143	67,234	12.686	63,428
50	5	22	13.5	8,100	63,428	12.686	3,806	67,772	12.777	63,889
60	0	5	2.5	1,500	63,889	12.777	3,833	61,556	11.614	58,072
70	0	0	0	0	58,072	11.614	3,484	54,588	10.300	51,498
80	0	0	0	0	51,498	10.300	3,090	48,408	9.134	45,668
90	0	0	0	0	45,668	9.134	2,740	42,928	8.100	40,498
100	0	0	0	0	40,498	8.100	2,430	38,068	7.183	35,913
110	0	0	0	0	35,913	7.183	2,155	33,758	6.369	31,847
120	0	0	0	0	31,847	6.369	1,911	29,936	5.648	28,242
130	0	0	0	0	28,242	5.648	1,694	26,548	5.009	25,045

[a] Assume that at time zero, 0 ft³ of water is in storage. $\Delta t = 10$ min $= 600$ s. Calculations continue until the outflow rate or storage is below an acceptable small number or the volume of inflow equals the volume of outflow plus an acceptable error.

and

Time (min)	Q/Q_p	Flow Rate Q
0	0	0
10	0.5	27
20	1.0	54
30	0.7	38
40	0.4	22
50	0.1	5
53	0	0

The storage–outflow relationship is

$$S = 5000(O)$$

and the N–O relationship using equation 3.23 is

$$N = S + \frac{O}{2}\Delta t = 5000(O) + \frac{O}{2}\Delta t$$

and for $\Delta t = 10$ min or 600 s,

$$N = 5000(O) + 300(O) = 5300(O)$$

or

$$O = \frac{N}{5300}$$

The calculations are shown in Table 3.4 and the hydrograph presented in Figure 3.5.

3.4.2 Muskingum Method

The Muskingum method, developed by the Muskingum Conservancy Flood Control District in 1930 (McCarthy, 1938), is based on a power relationship between flow rates and depth. The storage within a stream reach is a weighting of both the input and output volumes over a period of time, or

$$S = K[cI + (1-c)O] \qquad (3.25)$$

where S = storage volume, L^3
K = storage time constant, t
c = weighting factor, $0 \le c \le 0.5$
I = inflow volume, L^3
O = outflow volume, L^3

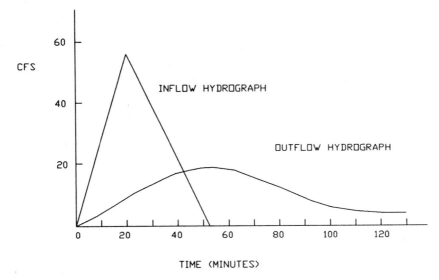

Figure 3.5 Detention pond inflow and outflow hydrographs for Example Problem 3.3.

An acceptable value of c is one that gives a linear relationship between discharge and storage, and K is thus the reciprocal of the slope, or

$$K = \frac{\Delta S}{\Delta O} \tag{3.26}$$

3.5 SUMMARY

Specific information from this chapter is:

- Rainfall excess is the volume of water from rainfall that is available for intentioned storage or surface discharge from the watershed.
- There are four estimation procedures for rainfall excess: site-specific infiltration measures, constant rate, mass balance, and the soil-complex-cover method.
- One method for rainfall excess and runoff rate based on a mass balance and presented in 1889 (Kuichling) has withstood the test of time and is still used on small watersheds defined by a maximum travel time of about 20 min or less. It is called the rational formula.
- Synthetic runoff hydrographs are used to estimate peak discharge and runoff volume for watersheds with little or no streamflow data.

84 HYDROGRAPHS

- The rational method is used for hydrograph generation when the rainfall intensity can be assumed constant for a time period equal to the time of concentration for the watershed. This usually limits the use of the rational formula to those watersheds having a time of concentration equal to or less than about 20 min.
- The SCS hydrograph generation procedure has a shape factor that is related to the watershed time of concentration.
- The Santa Barbara urban hydrograph procedure has a shape factor that is related to the time step for calculations and the watershed time of concentration.
- The continuous convolution method has a shape factor (storage coefficient) that has been related to watershed area and time of concentration.
- Hydrograph generation procedures are in general more accurate if directly connected impervious areas are separated from pervious areas.
- Detention volume calculations for "live" storage are dependent on a pond input hydrograph and a pond stage–storage discharge relationship.

3.6 PROBLEMS

1. What is rainfall excess? Discuss your answer using both volume and rate terms.

2. The total rainfall from a 2-h storm had an average intensity of 1.5 in./h. The initial abstraction on the watershed plus the infiltration volume was 1.0 in. What are the Φ index and runoff coefficient for the watershed? Also, the total watershed is 12 acres. What is the instantaneous rate of discharge?

3. An exponential infiltration rate results from a double-ring infiltrometer producing final rate of 1 in./h, an initial rate of 3 in./h with a recession constant of 2 per hour. What is the total infiltration volume (acre-feet) from a 24-acre pond after $\frac{1}{2}$ and 2 h?

4. Estimate rainfall excess (volume as inches) for a vegetated pasture land covered over with dense grass on about 98% of the area. Compare the estimates to areas using a 5-year 1-h design storm of 3 in. if the hydroologic soil classification for one area is C and the other is A. Assume that initial abstraction is 20% of maximum storage.

5. A land-use change is proposed for a vegetated area in hydrologic soil class A with 98% of the area covered with dense grass. The area is 48 acres in size. A developer plans to pave and build over 24 of the 48 acres. The 24 impervious acres will be directly connected. The water

management district requires that the runoff from 3 in. of rainfall should be stored in a flood plain area adjacent to the site. What-size (acre-feet) pond is needed for storage if the runoff volume is calculated using a composite curve number for the total area and by separating the directly connected impervious area from the pervious area? Assume that equation 3.9 with initial abstraction of 0.2 is appropriate for use. Comment on the difference.

6. For a 24-acre impervious area that is directly connected and has a runoff coefficient of 0.95, what is the peakk discharge for a rainfall intensity of 6 in./h? Also, size a pipe with a design velocity of 7 ft/s.

7. Using the typical SCS hydrograph shape, calculate the peak discharge from a 120-acre watershed with a runoff coefficient of 0.5 and an average intensity of 3 in./h for 1 h.

8. In a region for which the SCS type II rainfall distribution is used, the initial abstraction is about 30% of the total rainfall. The watershed is 240 acres with a 2-h time of concentration. What is the peak discharge for a 2-in. rainfall excess storm? Also, what is the base time for the hydrograph?

9. Compare the peak discharges using $K_r = 0.10$ and $K_r = 0.40$ in the SBUH procedure if the instantaneous discharges in the first two periods are 20 and 10 ft^3/s. Do the calculations for six time periods.

10. For a rainfall excess rate of $\frac{1}{2}$ in. in 30 min on a watershed of 50 acres with a time of concentration of 30 min, estimate the peak discharge using convolution.

11. An urban watershed has a 20-min time of concentration for a 50-acre area with a 60% directly connected impervious area. The pervious area is assumed not to contribute runoff for rainfalls less than 3 in. For a 2-in. rainfall in 20 min, what is the peak discharge using continuous convolution and the rational method? Assume a constant rainfall intensity.

12. A 100-acre completely impervious watershed in Orlando, Florida, produces a peak discharge after 30 min of rainfall (type II rain). For a 10-year return period, what are estimates of the peak using the rational formula, the SCS method with $I_a/P = 0.10$, and the Santa Barbara method with a routing parameter of 0.33 and $\Delta t = 30$ min? *Note:* It only rains for 30 min. Also, what volume of pond (cubic feet) do you recommend to store the rainfall excess?

13. a. Determine the Muskingum storage time constant for a river segment that has a relatively constant cross-sectional area and slope if

inflow and outflow hydrographs are available and the weighting factor (c) is equal to 0.3.

b. Governments wish to reduce the peak outflow and dredge the river, producing a $K = 1.5$. With $c = 0.2$, what is the new peak discharge given the same inflow? Were they successful? The available hydrographs are:

Day	I (ft^3/s)	O (ft^3/s)
0	0	0
1	18	5
2	47	25
3	30	30
4	16	25
5	10	15
6	5	12
7	2	7
8	0	5
9	0	2
10	0	0

14. Recalculate Example Problem 3.3 and comment on the results (peak discharge after detention and size of pond) if the input hydrograph flow rate values decrease by one-half.

15. What is the peak discharge from a 100-acre residential watershed ($\frac{1}{4}$-acre lot sizes) with soils exhibiting a curve number of 33 using the discrete unit convolution procedure? Use a constant rainfall intensity of 4 in./h for a duration of 1 and assume that the residential area is affected by storms similar to those in SCS region II. The time of concentration for the watershed is 60 min. Reference your numbers that you use and the equations.

16. Calculate the runoff hydrograph at 30 min after rainfall for a 50-acre residential area with a directly connected impervious area of 20 acres using the Santa Barbara procedure. The pervious area has a curve number of 40. The time of concentration is 20 min. Use a computation interval of 10 min. The rainfall for each 10-min interval is 0.2, 0.6, 1.0, 0.6, 0.4, and 0.2 in.

17. For the discharge hydrograph generated using the rational formula for a 4-acre parking lot in Orlando with a time of concentration of 10 min

and the one-in-50-year storm event, what size (maximum water storage volume) will you build if a permanent pool of 50,000 ft³ is used and the pond discharge relationship is 1 ft³/s per 10,000 ft³ of storage above the permanent pool? Use a 5-min computation interval.

3.7 COMPUTER-ASSISTED PROBLEMS

1. For a 3-h rainfall distribution recorded every 30 min with volumes of 0.4, 0.8, 1.1, 0.9, 0.6, and 0.2 in., calculate the runoff hydrograph from a 100-acre residential site with 40% directly connected impervious area, a pervious area curve number of 75, and time of concentration of 120 min. Use the SBUH procedure. Next, divert the first $\frac{1}{2}$ in. of runoff for pollution control and comment on the hydrograph shape.

2. a. Generate and compare, using the SBUH and SCS procedures, hydrograph shape and peak discharge from a 58.3-acre highway right-of-way that has the following characteristics.

Percent imperviousness	60
Impervious area	35 acres
Percent directly connected	100
Curve number pervious area	60

Use a 6-h/6-in.-volume storm event and, distribute the rainfall using SCS type II curve.

Time of concentration = 14.3 min (0.238 h)

For SCS method, assume that $I_a/P = 0.30$.

b. Compare these peak results with a peak estimate using the rational formula.

3. For the hydrograph of Problem 2 using the SCS procedure, what size detention pond would you recommend to attenuate the hydrograph given a storage–discharge relationship? Use only the stage–storage–discharge data given below. Storage at the 5-ft-deep control elevation is 1.10 acre-ft which is also storage at time zero. Also, what is the pond peak discharge?

The storage discharge data are:

Stage (ft)	Storage (acre-ft)	Discharge ft^3/s
50	1.10	0
51	2.50	2
52	5.00	6
53	8.00	14
54	12.00	22
55	16.00	32
56	20.00	50
57	30.00	8

Note: Create pond and hydrograph data, save data, exit and restart computer program before going to routing.

4. a. Compare hydrographs generated using the Santa Barbara versus the SCS-484 method using a rainfall interval of 5 min: namely, rainfall of 0.12, 0.13, 0.13, 0.38, 0.38, 0.38, 0.63, 0.63, 0.63, 0.13, 0.13, and 0.13 in. Use a 10-acre impervious watershed with a 30 min line of concentration.

b. Change the rainfall data to 15-min increments: namely, to change the Santa Barbara computation interval and routing factor to develop a hydrograph using the Santa Barbara methods for watershed with $K_r^B = 15/[2(30) + 15] = 0.20$. The 15-min rainfall volumes are 0.38, 1.14, 1.89, and 0.39 in. Compare these results to those of part (a) obtained using the Santa Barbara method. Comment on the differences.

c. Using a region with a type II rainfall distribution and an attenuation factor of 484, calculate the hydrograph for the area using $\Delta t = 5$ min and the SCS curvilinear hydrograph procedure.

5. There are two watersheds discharging to an open channel and then a pipe. The relevant hydrologic data are shown in the stick diagram of Figure 3.6. Using a rainfall distribution for a 24-h storm event with a 50-year return period and a 60-min computation interval, size the open channel and pipe and estimate the hydrograph shapes at nodes 1, 2, and 3. Use 8 inches of

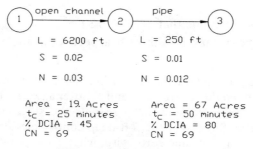

Figure 3.6 Stick diagram.

runoff and a type II rainfall distribution with the SCS hydrograph method ($K = 484$).

3.8 REFERENCES

Horton, R. E. 1939. "An Approach Toward a Physical Interpretation of Infiltration Capacity," *Transactions of the American Geophysical Union*, Vol. 20, pp. 693–711.

Kuichling, Emil. 1889. "The Relation Between the Rainfall and the Discharge of Sewers in Populous Areas," *Transactions of the American Society of Civil Engineers*, Vol. 20, pp. 1–56.

McCarthy, G. T., 1938. "The Unit Hydrograph and Flood Routing," unpublished paper at *U.S. Army Corps of Engineers North Atlantic Division Conference*.

Mitchi, C. 1974. "Determine Urban Runoff the Simple Way," *Water Wastes Eng.*, January.

Mulvaney, T. J. 1851. "On the Use of Self-Registering Rainfall and Flood Gauges," *Institute of Civil Engineers Transactions (Ireland)*, Vol. 4, No. 2, pp. 1–8.

Pagan, A. R. 1972. "Rational Formula Needs Change and Uniformity in Practical Applications," *Water Sew. Works*, October.

Sherman, L. K. 1932. "Stream-Flow from Rainfall by the Unit-Graph Method," *Engineering News Record*, Vol. 108, Apr., pp. 501–505.

Stubchaer, J. M. 1975. "The Santa Barbara Urban Hydrograph Method," *Proceedings of the National Symposium of Hydrology and Sediment Control*, Univ. of Kentucky, Lexington.

U.S. Department of Agriculture. 1986. *Urban Hydrology for Small Watersheds*, Technical Release 55, USDA Soil Conservation Service, Washington, D.C.

Wanielista, M. P. 1990. *Hydrology and Water Quantity Control*, Wiley, New York.

Williams, G. R. 1950. "Hydrology," in *Engineering Hydraulics*, H. Rouse, Ed., Wiley, New York.

CHAPTER 4

Management Models for Flow Rate and Volume Control

The evaluation of stormwater management practices can be performed with the aid of mathematical models. It is not practical or economical to build a physical stormwater system, evaluate it, and then change. Mathematical models are abstractions of real-world situations using mathematical equations. However, total reliance on the results of models to solve problems is neither realistic nor desirable. Many other technical, economic, social, and political factors may need to be considered, but sometimes they cannot be included. A review of some mathematical models as applied to the urban environment is presented in this chapter. The purpose of the review is to present information relative to the type of input and output data formats and the usefulness of such models. Some rural models are reviewed in Chapter 11.

4.1 REASONS FOR MODELS

Some goals of a stormwater management program are (1) estimates of peak flow rates, (2) generation of hydrograph shapes, (3) specification of management methods, (4) prediction of water quality, and (5) socioeconomic evaluation of management strategies to determine the best management practice. The acceptable levels of water volume and peak flows are some of the first data requirements needed for stormwater management. Basic mathematical relationships among the more important causative factors help identify the sensitivity and accuracy of related stormwater practices. Usually, mathematical models are used. As these models are continually updated with additional data, a greater certainty in predicting water quantity effects develops. These predictions are, where feasible, calibrated and verified using field-observed data. This calibration and verification will increase the credibility of any model.

The most common abstractions of stormwater problems into mathematical language are simple-to-use equations. These abstractions provide a basis for examining flow or quality. A good example is the rational formula (Wanielista, 1990). Other formulas for peak flows, based on statistical evaluation of data, also provide a basis for evaluating management methods. In this chapter we present example applications of three models: (1) U.S. Army Corps of

Engineers Stormwater Management Model (STORM) (U.S. Army Corps of Engineers, 1974), (2) Hydrologic Engineering Center Flood Hydrograph Program HEC-1 (U.S. Army Corps of Engineers, 1985), and (3) Stormwater Management and Design Aid (SMADA) (Wanielista and Eaglin, 1992). These models use many of the fundamental principles of hydrology and hydraulics presented in previous chapters.

The intended use of a model will determine the level of detail and required accuracy. All models have some usefulness. This utility is frequently based on the availability of data, facilities, personnel, and budget. Models used for water quantity may also include consideration of water quality. Some models for water quality and quantity estimation with abatement are examined for their utility. Pollution transport mechanisms are used in some models. Of interest in these models are (1) precipitation in all forms, (2) dry fallout, (3) surface water flows such as runoff and streamflow, (4) infiltration, (5) groundwater flow, (6) sediment transport, (7) quality changes, and (8) miscellaneous activities such as waterfowl, nitrogen fixation, and volatilization. It should be emphasized that the land or groundwater in one watershed may contribute to nonpoint-source pollution in another watershed. This may be significant in large urban areas where air pollution may drift, producing dry fallout and bulk precipitation in another watershed. Other more subtle watershed transfers will occur with groundwater flows. If the transport mechanisms were defined in mathematical terms, quantitative models for prediction would be valuable for illustrating relative effects and for predicting levels of water quality. If not, loading rate data can be used.

4.2 SOME EXISTING MODELS

A most complete comparative analysis of urban stormwater models was done by Nix (1990). Also, simplified models (Lager, 1975), water quality models (James and James, 1985), and an overview of several models have been published (Stahre and Urbonos, 1990). Here, the intent of the review is to report on some details related to input/output and model development. The use of any one model will depend on the desired end results, availability of input and calibration data, and the knowledge of the users. Following is a partial list of models with references noted at the end of this chapter: Corps of Engineers STORM (U.S. Army Corps of Engineers, 1974); U.S. Environmental Protection Agency Stormwater Management Model (SWMM) (Metcalf and Eddy, 1971; Huber et al., 1988); Illinois Urban Drainage Area Simulator ILLUDAS (Terstriep and Stall, 1974); British Road Research Laboratory Model (Terstriep and Stall, 1969); Corps of Engineers HEC-1 (U.S. Army Corps of Engineers, 1985); Environmental Protection Agency (EPA) Pesticide Transport and Runoff (PTR) Model (Crawford and Conigian, 1973); USDA Soil Conservation Service TR55 and TR20 (U.S. Department of Agriculture, 1986) models; Federal Highway Administration HYDRAIN, highway drainage software to include (HY8) Culvert Design Program (Federal

92 MANAGEMENT MODELS FOR FLOW RATE AND VOLUME CONTROL

Highway Administration, 1985); Stormwater Management and Design Aid (Wanielista, and Eaglin, 1992); Texas Water Board Water Yield Model (Williams and LaSeur, 1976); QUALHYMO, a continuous quality quantity simulation model (Wisner and Rowney, 1985); and other spreadsheet models for solving long-term time-related simulations.

4.3 BASIC CONCEPTS

Among the basic concepts of a stormwater management model, a rainfall-runoff relationship must exist. One of the earliest computer programs to use a basic rainfall-runoff mathematical relationship was called STORM (U.S. Army Corps of Engineers, 1974). The concept of the overall STORM model is shown in Figure 4.1 and includes the rainfall runoff, quality, and treatment concepts.

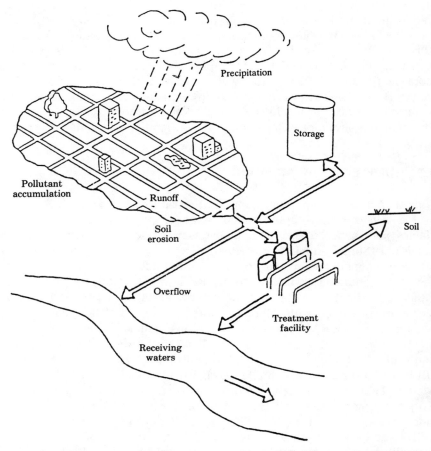

Figure 4.1 Concept of watershed as seen by STORM. (From U.S. Army Corps of Engineers, 1974.)

Stormwater management computer programs use mathematical relationships to calculate runoff rates, rainfall excess, storage volumes, and water quality. The program STORM was applied to two river basins: Palm Beach Gardens (Ingram, 1975) and Magnolia Ranch (Izzo, 1975) (Econlockhatchee River) to evaluate the mathematical relationships and then present some details in this chapter.

To understand any computer program, divide it into its major computation areas. The STORM program is used as an example with computation areas identified as (1) runoff quantity; (2) runoff quality; (3) treatment, storage, and overflow; and (4) soil erosion.

4.3.1 Runoff Quantity

The quantity of runoff is calculated from

$$R = C(P - I) \tag{4.1}$$

where R = runoff
P = precipitation
C = composite urban runoff coefficient
I = available depression storage, a function of the rainfall record and evapotranspiration rates

Urban and nonurban runoffs are combined by area weighting for the total watershed runoff.

The urban runoff coefficient in STORM is given as

$$C = C_P + (C_I - C_P) \sum_{i=1}^{L} X_i RA_i \tag{4.2}$$

where C = composite urban runoff coefficient
C_P = runoff coefficient for pervious surfaces
C_I = runoff coefficient for impervious surfaces
X_i = area$_i$/total area
RA_i = impervious area$_i$/area$_i$
L = number of urban land uses

4.3.2 Runoff Quality

Urban area pollutant washoff rates (M, lb/h of pollutant) are calculated from the following set of equations:

Suspended solids:

$$M_{SUS} = (0.057 + 1.4R^{1.1})(F_{SUS}DD_L N_D + P_{PO})(t)(\text{EXPT})$$

Settleable solids:

$$M_{SET} = (0.028 + R^{1.8})(F_{SET}DD_L N_D + P_{PO})(t)(\text{EXPT})$$

Biological oxygen demand:

$$M_{BOD} = (F_{BOD}DD_L N_D + P_{PO})(t)(\text{EXPT}) + 0.1 M_{SUS} + 0.02 M_{SET} \quad (4.3)$$

Nitrogen:

$$M_{NIT} = (F_{NIT}DD_L N_D + P_{PO})(t)(\text{EXPT}) + 0.045 M_{SUS} + 0.01 M_{SET}$$

Phosphorus:

$$M_P = (F_P DD_L N_D + P_{PO})(t)(\text{EXPT}) + 0.0045 M_{SUS} + 0.001 M_{SET}$$

where EXPT = $(1 - e^{-ER})$
  F = pounds of pollutant per pound of dust and dirt
  DD_L = dust and dirt accumulation, lb/day
  N_D = number of days without runoff
  P_{PO} = remaining pounds of pollutant on land at the end of previous storm
  t = time, 1-h fixed interval
  E = washoff decay coefficient
  R = runoff rate from impervious surfaces

The initial mass of pollutant P_p on the nonurban watershed at the beginning of a storm is computed as

$$P_P = PA_p A_n N_D + P_{po} \quad (4.4)$$

where P_P = total pounds of pollutant P on the nonurban area A_n at the beginning of the storm
  PA_p = accumulation rate for pollutant P, lb/day/acre
  A_n = nonurban area, acres
  N_D = number of days without runoff since the last storm
  P_{po} = total pounds of pollutant P remaining on the nonurban area at the end of the last storm

The washoff of nonurban pollutants is a function only of the nonurban runoff rate and the amount of pollutants on the watershed. The expression used to compute the rate at which pollutants are washed off the nonurban watershed is

$$MN_P = P_P(1 - e^{-E_n r_n \Delta t}) \quad (4.5)$$

where MN_p = pounds washoff of pollutant P after nonurban diversion
  E_n = nonurban washoff decay rate
  r_n = rate of nonurban runoff, in./h

No analysis of the availability of pollutants is made for the nonurban pollutants as done in the urban case. Nonurban runoff is also subject to diversion losses, and the pollutants are estimated using a fixed percent runoff diversion quantity, such as 1 in. of runoff.

Concentration is calculated using the pollutant washoff rates multiplied by the runoff duration for the total pounds of washoff and multiplied by the runoff volume. The following equation is used to calculate concentration:

$$C = 453{,}600 \sum_{i=1}^{n} M_i(\Delta t)_i / (R_i) \qquad (4.6)$$

where
C = concentration, mg/L
453,600 = conversion factor, mg/lb
M_i = pollutant mass in time period i, lb/h
Δt = time interval, h
n = total number of time periods
R = rainfall excess per time period, L

The total pounds of washoff and concentrations are calculated as before

4.3.3 Storage, Treatment, and Overflow

The program has the capability to analyze the computed runoff rate for overflow of a given storage capacity when being treated at an assigned treatment rate. The use of the storage facility is a function of the accumulation of runoff and the constant output of the treatment plant. The program can plot the storage utilization versus time curve.

4.3.4 Soil Erosion

The universal soil loss equation is used to calculate the loss of soil from a plot of ground. This equation is discussed in greater detail in Chapter 11, but the equation is

$$\text{soil erosion rate} = RK(L*S)CP \qquad (4.7)$$

where
R = rainfall factor
K = soil erodibility factor
$L*S$ = length–slope factor
C = cropping management factor
P = erosion-control practice factor

4.4 APPLICATION OF HEC-1

HEC-1 has many of the same input data requirements that the other models have and essentially many of the output specifications. However, there is one unique feature that merits some additional consideration, the parameter optimization routine. Whenever possible, any mathematical model results should be compared to observed hydrograph data. If gaged precipitation and runoff data are available, parameter optimization can be done. Optimization is achieved by minimizing an objective function that is the square root of the weighted squared differences between an observed and estimated hydrograph, or

$$\min \sum_{i=1}^{n} \left[\frac{(Q_i^\circ - Q_i)^2 WT_i}{n} \right]^{1/2} \quad (4.8)$$

where Q_i° = observed hydrograph flow rate at ordinate i, ft^3/s
Q_i = estimated hydrograph flow rate at ordinate i, ft^3/s
$WT_i = (Q_i^\circ - Q)/2Q$ or weighted flow
Q = average flow rate
n = total number of hydrograph ordinates

A search procedure is used to find the minimum value. Constraints consist of the physical and numerical limitations on parameter selection. The search procedure does not guarantee a global optimum. Thus the user can view graphical comparisons and make adjustments. An improvement in the objective function might be possible by specifying different parameter initial values.

An example of the optimization algorithm is presented to illustrate procedure and selected input and output format. This example was taken from the HEC-1 Users' Manual (U.S. Army Corps of Engineers, 1985). The procedure is:

1. Obtain observed flow rates, precipitation, and watershed data.
2. Specify the Clark unit hydrograph procedure.
3. Fix one or more parameters of the unit hydrograph and loss rate function (equation 4.9).

   Four parameters are optimized: unit hydrograph parameters t_c (time of concentration), RS (storage coefficient), and the exponential loss infiltration parameters STRKR and DLTKR in the following equations:

   $$\text{ALOSS} = (\text{AK} + \text{DLTKR})\text{ERAIN} \quad (4.9)$$

where ALOSS = potential rain loss rate, in./h
AK = STRKR/(RTIOL)0.1CUML
STRKR = starting value of loss coefficient
RTIOL = ratio of rain loss on exponential loss curve to 10 in. of accumulated loss
CUML = cumulative rain loss, in.
DLTKR = initial accumulated rain loss
i = precipitation intensity, in./h
ERAIN = watershed coefficient (0 < ERAIN < 1.0)

STRKR reflects watershed characteristics, such as soil type, land use, and vegetation cover, while DLTKR is a function of antecedent soil moisture deficiency and is storm dependent. For this example, the other parameters are known from other locations and are considered as constants. Some default values are shown in Table 4.1. The unit hydrograph parameters are displayed as sums and ratios

TABLE 4.1 HEC-1 Default Values for Unit Graph and Loss Rate Parameters for Optimization Using $t_c + R$ Method

Description	Parameter	Initial Value
Unit graph		
Time of concentration[a]	t_c	(area)$^{1/2}$
Storage coefficient (RS)[b]	$R/(t_c + R)$	0.5
Exponential loss		
Starting value	STRKR	0.2
Initial rain loss	DLTKR	0.5
Rain loss ratio	RTIOL	2.0
Watershed coefficient	ERAIN	0.5
SCS-CN		
Initial rain abstraction	$0.2S$	1.08
Curve number	CN	65

[a] Area of watershed in square miles.
[b] R is the routing parameter constant for each watershed (time units).

4. The hydrograph response is calculated. The volume of the simulated hydrograph is adjusted to within 1% of the observed value.
5. Each parameter of Table 4.1 is adjusted in the order shown in the table. Parameters that do not improve the objective function are held constant and the search technique proceeds.
6. The final results are shown in Table 4.2.

4.5 APPLICATION OF SMADA

Runoff hydrograph generation and the design of a detention pond will be illustrated. As with the other models, the input data requirements must be met to ensure proper output results.

TABLE 4.2 Optimization Results

Unit graph	Time of concentration	$t_c = 3.16$ h
	Storage coefficient	$RS = 3.88$
	Peak	$Q_p = 4332$ ft^3/s
	Time to peak	$t_p = 3.0$ h
Exponential loss	Starting value	STRKR = 0.49
	Initial rain loss	DLTKR = 0.50
	Rain loss ratio	RTIOL = 1.0
	Watershed coefficient	ERAIN = 0.5
Standard error	Root-mean-squared sum between observed and computer = 270	
Objective function	Equation 4.8 = 284	

SMADA can evaluate single or multiple land-use watersheds. For the same rainfall event, each specified subwatershed is evaluated for runoff effects and hydrograph prediction. The options of pre- versus post-, pollutant-retention, and/or peak-flow detention-type analyses are available for use on each subwatershed.

There are three main data groups: (1) rainfall, (2) watershed, and (3) nodal network.

4.5.1 Rainfall Data

Ranfall data requires the input of storm duration in hours and the calculation time increment in minutes. This time increment will naturally be also used to define the hydrograph(s). The program allows for up to 96 time increments for both rainfall entries and hydrograph generation. As prompted, one enters the rainfall volume in inches for each time increment.

For user convenience, the rain data inputted are stored automatically in a data file and can be reused in subsequent computer runs. If new rainfall data input are desired, the user simply answers the questions appropriately.

4.5.2 Watershed Data

Each land area to be evaluated requires the following data. The data are stored automatically in a file that can be recalled.

1. Drainage area (acres)
2. Time of concentration (minutes)
3. Impervious area (acres)
4. Percentage of impervious that is directly drained
5. Additional abstraction for impervious area (inches)
6. Additional abstraction for pervious area (inches)
7. Curve number for pervious area or Horton's limiting infiltration rate (in./h)
8. Horton's initial infiltration rate (in./h)
9. Horton's depletion coefficient (h^{-1})

4.5.3 Nodal Network Data

The nodal network is defined by entering the following data concerning the transmission system between the nodes:

1. Distance between nodes (feet)
2. Cross section (circle or trapezoid)
3. Slope of proposed drainage
4. Manning function coefficient

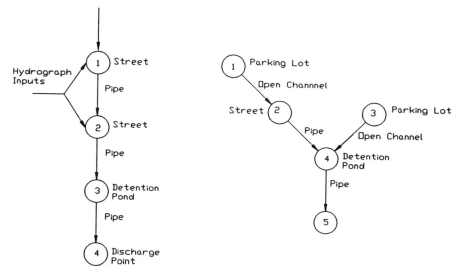

Figure 4.2 Number system schematic.

When considering a watershed with more than one land use, more than one entry point into the conveyance system exists. A network of nodes appropriately arranged allows for defining the system. Node 1 is placed at the farthest upstream point. The remaining nodes are numbered such that the numbers of upstream nodes are progressively less than the given node. See the example in Figure 4.2.

4.5.4 Peak Attenuation and Water Quality Control

Peak attenuation is defined as the reduction of peak flow rates caused by intentional (man-made) storage of water or the routing of a hydrograph through a natural body of water. SMADA is one of the computer programs that can be used to generate and route hydrographs through a storage basin or detention pond. The routing is done in the nodal network section. The network nodes (hydrograph inputs, tributary, and inlets areas) must be defined. Each node is numbered so that the most upstream node is numbered 1, and the most downstream node (discharge from a reservoir or detention pond or discharge to receiving waters) is the highest number. Always, a smaller number must flow into a larger number (see Figure 4.2).

Water quality control can be achieved using on-line (detention) or off-line (retention) ponds. The on-line ponds increase holding time for sedimentation and other transformations, while the off-line ponds promote the removal of pollutants, usually by infiltration.

A wet detention pond attenuates hydrograph peak discharges and has a permanent storage of water that exists during most, if not all, of the year.

TABLE 4.3 Example Stage–Storage–Discharge Relationship

Stage (ft)	Storage (acre-ft)	Discharge (ft^3/s)
50	1.15	0
51	2.43	7.0
52	5.42	19.0
53	6.80	25.0

The permanent storage is necessary to provide a large volume sufficient to increase detention time and thus improve the removal of sediment and chemical/biological activity. A way of approximating residence time is to approximate the holding period by a term called *detention time* (see Chapters 7 and 8).

For each storage area, a stage–storage–discharge relationship must be specified. If a permanent storage (volume below the pond control elevation) is available, it must be incorporated into the storage–discharge relationship. An example stage–storage–discharge relationships for a detention pond at the discharge of a 50-acre residential area is shown in Table 4.3. Within the program, one can specify at least 10 stage–storage–discharge relationships for each pond. After a hydrograph is routed through a pond, the detention time and outlet peak discharge are recorded. If the user wishes to change the storage–discharge relationship, it can be done and a new set of detention times and peak discharge are calculated.

Example Problem 4.1 A watershed area of 50 acres with 30% impervious and 50% directly connected impervious acres has a time of concentration of 60 min. The curve number for the pervious area is 61. For design conditions, initial abstraction is assumed equal to zero. The rainfall data are from a standard design storm and are shown in Table 4.4. The hydrograph shape is estimated using the Santa Barbara procedure. The storage–discharge relationship for the detention pond is estimated by a composite linear equation with three segments and follows the data of Table 4.3. Figure 4.3 is an actual screen print from the computer program and illustrates the pond hydrograph plot before and after detention. The printed data relating inflow to outflow and stage are shown in Table 4.5.

4.6 SUMMARY

The number of mathematical models for hydrograph generation and stormwater management in the urban environment are numerous. Ten of these models were reviewed briefly and three of the models were examined further for type and quantity of input and output data. All the models use principles of hydrology and hydraulics and some incorporate loading and concentration data. The material is useful to demonstrate the details neces-

TABLE 4.4

```
-------------------------------------------------SMADA  PAGE 1
Example 4.1     Hydrograph Type :Santa Barbara Method
-----------------------------------------------------------------
```

Time (hr)	Rain (in)	Cumulative (in)	Infiltration (in)	Instantaneous (cfs)	Outflow (cfs)
0.00	0.00	0.00	0.00	0.00	0.00
0.25	0.10	0.10	0.08	3.02	0.34
0.50	0.10	0.20	0.08	3.02	0.93
0.75	0.15	0.35	0.12	4.54	1.57
1.00	0.15	0.50	0.12	4.54	2.23
1.25	0.25	0.75	0.20	7.56	3.08
1.50	0.25	1.00	0.20	7.56	4.07
1.75	0.30	1.30	0.24	10.07	5.13
2.00	0.40	1.70	0.28	20.23	7.35
2.25	0.45	2.15	0.28	30.99	11.41
2.50	0.45	2.60	0.24	38.09	16.55
2.75	0.45	3.05	0.21	43.92	21.98
3.00	0.45	3.50	0.19	48.76	27.40
3.25	0.40	3.90	0.15	46.77	31.92
3.50	0.40	4.30	0.14	49.53	35.53
3.75	0.30	4.60	0.09	38.75	37.44
4.00	0.30	4.90	0.09	39.97	37.87
4.25	0.20	5.10	0.06	27.26	36.92
4.50	0.20	5.30	0.05	27.73	34.83
4.75	0.10	5.40	0.03	14.03	31.73
5.00	0.10	5.50	0.03	14.13	27.81
5.25	0.00	5.50	0.00	0.00	23.20
5.50	0.00	5.50	0.00	0.00	18.04
5.75	0.00	5.50	0.00	0.00	14.03
6.00	0.00	5.50	0.00	0.00	10.91
6.25	0.00	5.50	0.00	0.00	8.49
6.50	0.00	5.50	0.00	0.00	6.60
6.75	0.00	5.50	0.00	0.00	5.14
7.00	0.00	5.50	0.00	0.00	3.99
7.25	0.00	5.50	0.00	0.00	3.11
7.50	0.00	5.50	0.00	0.00	2.42
7.75	0.00	5.50	0.00	0.00	1.88
8.00	0.00	5.50	0.00	0.00	1.46
8.25	0.00	5.50	0.00	0.00	1.14
8.50	0.00	5.50	0.00	0.00	0.88
8.75	0.00	5.50	0.00	0.00	0.69
9.00	0.00	5.50	0.00	0.00	0.53
9.25	0.00	5.50	0.00	0.00	0.42
9.50	0.00	5.50	0.00	0.00	0.32
9.75	0.00	5.50	0.00	0.00	0.25
10.00	0.00	5.50	0.00	0.00	0.20
10.25	0.00	5.50	0.00	0.00	0.15

```
-----------------------------------------------------------------
Total       5.50 in                      432412 cf   431932 cf
Rational Coefficient = 0.433   Peak Flow =  37.87 cfs
Routing Coefficient = 0.111
Watershed File: EX4_1.HYD   Rainfall File: EX4_1.HYD
-----------------------------------------------------------------
```

TABLE 4.5 Pond Hydrographs for Example Problem 4.1

Node: 2—pond node
Pond node: Output by storage–discharge relationship
Maximum volume = 5.90 acre-ft   Maximum stage = 52.35 ft

Time (h)	Inflow (ft^3/s)	Outflow (ft^3/s)	Stage (ft)
0.25	0.34	0.03	50.00
0.50	0.93	0.12	50.02
0.75	1.57	0.27	50.04
1.00	2.23	0.47	50.07
1.25	3.08	0.75	50.11
1.50	4.07	1.09	50.16
1.75	5.13	1.52	50.22
2.00	7.35	2.12	50.30
2.25	11.41	3.08	50.44
2.50	16.55	4.47	50.64
2.75	21.98	6.30	50.90
3.00	27.40	8.12	50.09
3.25	31.92	9.98	51.25
3.50	35.53	12.00	51.42
3.75	37.44	14.01	51.58
4.00	37.87	15.92	51.74
4.25	36.92	17.60	51.88
4.50	34.83	19.00	52.00
4.75	31.73	20.12	52.19
5.00	27.81	20.81	52.30
5.25	23.20	21.06	52.34
5.50	18.04	20.84	52.31
5.75	14.03	20.29	52.21
6.00	10.91	19.50	52.08
6.25	8.49	18.61	52.97
6.50	6.60	17.66	51.89
6.75	5.14	16.67	51.81
7.00	3.99	15.67	51.72
7.25	3.11	14.07	51.64
7.50	2.42	13.70	51.56
7.75	1.88	12.76	51.48
8.00	1.46	11.86	51.40
8.25	1.14	11.00	51.33
8.50	0.88	10.20	51.27
8.75	0.69	9.44	51.20
9.00	0.53	8.73	51.14
9.25	0.42	8.07	51.02
9.50	0.32	7.45	51.04
9.75	0.25	6.83	51.98
10.00	0.20	6.12	50.87
10.25	0.15	5.48	50.78
10.50	0.00	4.89	50.70
10.75	0.00	4.36	50.62
11.00	0.00	3.90	50.56
11.25	0.00	3.48	50.50
11.50	0.00	3.10	50.44
11.75	0.00	2.77	50.40
12.00	0.00	2.47	50.35
12.25	0.00	2.21	50.32
12.50	0.00	1.97	50.28

TABLE 4.5 *(Continued)*

Node: 2—pond node
Pond node: Output by storage–discharge relationship
Maximum volume = 5.90 acre-ft   Maximum stage = 52.35 ft

Time (h)	Inflow (ft^3/s)	Outflow (ft^3/s)	Stage (ft)
12.75	0.00	1.76	50.25
13.00	0.00	1.57	50.22
13.25	0.00	1.40	50.20
13.50	0.00	1.25	50.18
13.75	0.00	1.12	50.16
14.00	0.00	1.00	50.14
14.25	0.00	0.89	50.13
14.50	0.00	0.79	50.11
14.75	0.00	0.71	50.10
15.00	0.00	0.63	50.09
15.25	0.00	0.56	50.08
15.50	0.00	0.50	50.07
15.75	0.00	0.45	50.06
16.00	0.00	0.40	50.06
16.25	0.00	0.36	50.05
16.50	0.00	0.32	50.05
16.75	0.00	0.29	50.04
17.00	0.00	0.25	50.04
17.25	0.00	0.23	50.03
17.50	0.00	0.20	50.03
17.75	0.00	0.18	50.03
18.00	0.00	0.16	50.02
18.25	0.00	0.14	50.02
18.50	0.00	0.13	50.02
18.75	0.00	0.11	50.02
19.00	0.00	0.10	50.01
19.25	0.00	0.09	50.01
19.50	0.00	0.08	50.01
19.75	0.00	0.07	50.01
20.00	0.00	0.07	50.01
20.25	0.00	0.06	50.01
20.50	0.00	0.05	50.01
20.75	0.00	0.05	50.01
21.00	0.00	0.04	50.01
21.25	0.00	0.04	50.01
21.50	0.00	0.03	50.00
21.75	0.00	0.03	50.00
22.00	0.00	0.03	50.00
22.25	0.00	0.02	50.00
22.50	0.00	0.02	50.00
22.75	0.00	0.02	50.00
23.00	0.00	0.02	50.00
23.25	0.00	0.01	50.00
23.50	0.00	0.01	50.00
23.75	0.00	0.01	50.00
24.00	0.00	0.01	50.00
24.25	0.00	0.01	50.00
24.50	0.00	0.01	50.00
24.75	0.00	0.01	50.00
25.00	0.00	0.01	50.00

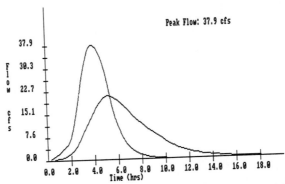

Figure 4.3 Detention pond hydrographs for Example Problem 4.1.

sary to use the models. Most of the programs are resident on personal computers (IBM-PC or compatibles). The following are some "commonsense" notes on the use of mathematical models.

- Always check the input data (garbage in, garbage out). Know the limitations of the model and do not extrapolate results beyond the range of understanding.
- Remember the basic principles of hydrology and water quality and check the results to see if they are reasonable.
- Some urban models do not divide the watershed into directly connected impervious area. Thus the use of the SCS rainfall excess calculation for low rainfall volumes may not produce runoff, when in fact, runoff from the directly connected areas will occur. Again check output quantities for reasonable answers.
- Some programs incorporate optimization analysis, which aids in design, such as pipe sizes with SMADA and hydrograph parameters with HEC-1.
- Breaking an urban watershed into smaller "homogeneous" watersheds usually will produce more accurate hydrographs and loadographs, but this is time consuming.

4.7 PROBLEMS

1. The runoff from an urban area is simulated using the STORM computer program. The urban area is composed of both pervious and impervious areas. The composite runoff coefficient is 0.55 for the particular time of

the simulation. The available depressions storage is 4 mm. What runoff (mm) can be expected from three storm events on the following dates?

April 7, 1978   27 mm
April 14, 1978   12 mm
May 2, 1978   8 mm

What assumptions have been made?

2. What mathematical models could you use if you were interested in predicting:
   a. Total suspended solids discharged for a storm event?
   b. Runoff quantity per year?

3. Describe in your own words the differences among the output of the SWMM, HEC-1, STORM, and the SMADA models.

4.8 COMPUTER-ASSISTED PROBLEMS

1. For the following residential watershed condition, estimate a hydrograph using the SCS method. The peak attenuation factor is available from observed field data and is 340. The rainfall distribution is for a 25-year, 5-h storm:

Time (min)	Rainfall (in.)
30	0.2
60	0.3
90	0.5
120	0.7
150	0.9
180	0.9
210	0.8
240	0.6
270	0.4
300	0.2

The initial abstraction is assumed by the SCS, or 20% of maximum storage. The watershed area is 50 acres, of which 15 acres are impervious and 7.5 acres are directly connected impervious area. The curve number for the pervious area is 61. The total time of concentration is 60 min. Use SMADA or a similar program.

2. Estimate the yearly pollutant loading for the watershed of Problem 1 assuming the loading rates used in the computer program. A retention pond must also be sized with 1:6 side slopes. The rational runoff coefficient is 0.4. Use the runoff hydrograph of Problem 1 and determine the

retention basin size and the final discharge hydrograph (after off-line retention).

3. A reviewing agency wishes you to change your estimate of the time of concentration in Problem 1 from 60 min to 67 min and the total percent impervious from 30 to 25. Estimate the hydrograph shape using the Santa Barbara method. Assume that the resulting hydrograph represents a predevelopment condition.

4. For the same 50 acres of Problem 3, the percent impervious area was increased to 58% with 66% directly connected impervious area. The time of concentration has decreased to 55 min. Using the Santa Barbara method (SMADA), what is the discharge hydrograph values? How much larger is this postdevelopment peak discharge relative to a predevelopment peak discharge?

4.9 REFERENCES

American Public Works Association. 1969. *Water Pollution Aspects of Urban Runoff*, Water Pollution Control Research Series, Report WP-20-15, Federal Water Pollution Control Administration, Washington, D.C.

Crawford, N. H., and Conigian, A. S. 1973. *Pesticide Transport and Runoff Model for Agricultural Lands*, EPA 660/2-74-013, U.S. Environmental Protection Agency, Washington, D.C.

Federal Highway Administration. 1985. *Hydraulic Design of Highway Culverts*, Computer Program HY8, Hydraulic Design Series No. 5, FHWA, McLean, Va.

Huber, W. C., Dickenson, R. E., Roesner, L. A., and Aldrich, J. A. 1988. *Stormwater Management Model User's Manual*, Version 4, EPA 600/33-88-001, U.S. Environmental Protection Agency, Athens, Ga.

Hydrocomp International, Inc. 1972. *Hydrocomp Simulation Programming-Operations Manual*, Hydrocomp. Palo Alto, Calif. (with updates).

Ingram, C. 1975. *An Application of a Computerized Mathematical Model for Estimating the Quantity and Quality of Nonpoint Sources of Pollution from Small Urban and Nonurban Watersheds*, Research Report, College of Engineering, Florida Technological University, Orlando, Fla.

Izzo, J. T. 1975. *An Application of STORM Mathematical Modeling for Evaluation of Nonpoint Source Water Pollution for a Nonurban Watershed*, Research Report at Florida Technological University, Orlando, Fla.

James, E. M., and James, W. 1985. *Proceedings: Conference on Stormwater and Water Quality Management Modelling*, Report R149, McMaster University, Hamilton, Ontario, Canada.

Lager, J. A. 1975. "A Simplified Stormwater Management Model for Planning, Part II," in *Applications of Stormwater Management Models—1975*, EPA Short Course Notes, Amherst, Mass., July 28–Aug. 1.

Lanyon, R. F., and Jackson, J. P. 1974. *A Streamflow Model for Metropolitan Planning and Design*, Urban Water Resources Research Program, Tech Memo 20, American Society of Civil Engineers, New York.

Metcalf and Eddy. 1971. *Storm Water Management Model-SWMM*, Vols. I, II, III, IV, EPA 11024D0C07/71 (with changes), U.S. Environmental Protection Agency, Washington, D.C.

Nix, S. 1990. "Non-point Source Pollution Models: An Overview," in *Urban Non-point Source Pollution and Stormwater Management Symposium*, University of Kentucky, Lexington, Ky.

Shubinski, R. P., and Roesner, L. A. 1973. "Linked Process Routing Models," *AGU Annual Spring Meeting*, Washington, D.C.

Smoot, J. 1977. *Spruce Creek Watershed Non-point Source Loading Model*, Research Report at Florida Technological University, Orlando, Fla.

Stahre, P., and Urbonos, B. 1990. *Stormwater Detention*, Prentice Hall, Englewood Cliffs, N.J.

Terstriep, M. L., and Stall, J. B. 1969. "Urban Runoff by Road Research Laboratory Method," *Journal of the Hydraulics Division*, ASCE, Vol. 95, No. HY4, pp. 1809–1834.

Terstriep, M. L., and Stall, J. B. 1974. *The Illinois Urban Drainage Area Simulator*, Illinois State Water Survey, Urbana, Ill.

U.S. Army Corps of Engineers. 1974. *Urban Stormwater Runoff-Storm*, Computer Program 723-58-L2520, The Hydrologic Engineering Center, U.S. Army Corps of Engineers, Davis, Calif.

U.S. Army Corps of Engineers. 1985. *Flood Hydrograph Package HEC-1*, The Hydrologic Engineering Center, U.S. Army Corps of Engineers, Davis, Calif.

U.S. Department of Agriculture. 1983. *TR20 Computer Programs for Project Formulation—Hydrology*, USDA Soil Conservation Service, Washington, D.C.

U.S. Department of Agriculture. 1986. *TR55 Urban Hydrology for Small Watersheds*, USDA Soil Conservation Service, Washington, D.C.

Wanielista, M. P. 1990. *Hydrology and Water Quantity Control*, Wiley, New York, pp. 221–224.

Wanielista, M. P. 1991. "User Interactive Hydrograph Generation and Design," in *Proceedings: 9th National Conference on Microcomputers in Civil Engineering*, University of Central Florida, Orlando, Fla., pp. 155–159.

Wanielista M. P., and Eaglin R. 1992. *Stormwater Management and Design Aid: Computer Program Manual*. Univ. of Central Florida, Orlando.

Williams, J. R., and LeSeur, W. V. 1976. "Water Yield Model Using SCS Curve Numbers," *Journal of the Hydraulics Division*, ASCE, Vol. 102 No. HY9, pp. 1241–1253.

Wischmeir, W. H., and Smith, D. D. 1965. *Predicting Rainfall-Erosion Losses from Cropland East of the Rocky Mountains*, Agricultural Handbook 282, USDA Agricultural Research Service, Washington D.C., (reprinted, 1972).

Wisner P. E., and Rowney, A. C. 1985. *Proceedings: Conference on Stormwater and Water Quality Management Modelling*, Report R149, McMaster University, Hamilton, Ontario, Canada.

CHAPTER 5

Stormwater Quality

One of the major objectives of stormwater management is the reduction of pollutants within stormwater that cause unwanted physical, chemical, and biological changes in receiving waters. Sources, concentrations, and mass loadings of these materials are discussed in this chapter and their applications are demonstrated in the remaining chapters.

Governments create pollution control laws, such as the U.S. Water Pollution Control Acts, PL 95-270 and PL 92-500, which specify the restoration and maintenance of the chemical, physical, and biological integrity of the nation's waters. Specifics of the U.S. Clean Waters Act require all state programs to meet the act's interim fishable and swimmable numeric goals (Foran, 1990). The integrity of waters can be maintained and restored only after a thorough understanding of the water quality responses from point and nonpoint sources. Point-source effects have been studied for many years, but nonpoint-source effects have only recently become of interest and concern. Changes in the quality of receiving water and its impact on the overall ecology should be investigated to specify the various management alternatives that might be considered to maintain a specified water quality. As such, it is necessary to control pollutant concentrations and mass loadings discharged into receiving water systems.

The water quality responses of pollutants from nonpoint sources are influenced by many complex systems, such as meteorological, hydrological, and geological conditions, plus land-use practices (Figure 5.1). The quantity of flow is characterized by spatial and temporal fluctuations of the precipitation from a storm. Mathematical modeling techniques have been attempted to estimate quantities and qualities resulting from a nonpoint source flows for different land use. Reasonable agreement for predicting mass flows have been observed because mass is a product of concentration and volume (flow rate and time). Volume can be predicted with an accuracy greater than concentrations. Prediction of concentration is much more difficult and complicated.

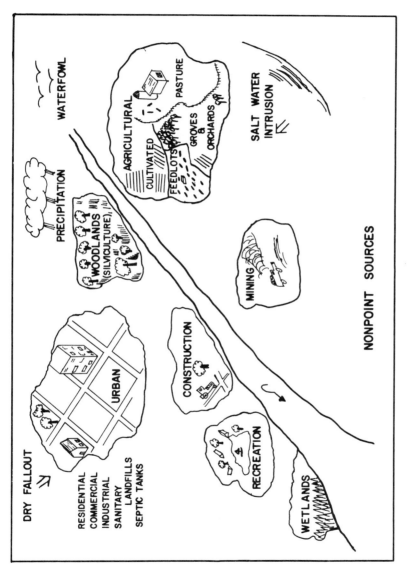

Figure 5.1 Transport mechanisms and land uses.

5.1 WATER QUALITY PARAMETERS

There are many materials in precipitation and in runoff that can cause changes in species and diversity of plant and animal populations. When these changes become socially unfavorable, the material causing the modification is called a pollutant. Frequently, these pollutants are water quality parameters present in large or small quantities relative to the demand for a well-balanced ecosystem.

Water quality parameters for stormwaters comprise a long list and are classified in many ways: (1) sediment, nutrients, and metals; (2) oxygen-demanding and inert material; (3) particulate and dissolved; (4) chemical, biological, and physical; (5) toxic and nontoxic; (6) organic and inorganic; and (7) others. Many specific pollutants are incorporated into one classification if their effects on receiving waters are somewhat the same. Receiving waters can assimilate a limited quantity of each, but there are thresholds beyond which the measured amount becomes a pollutant and results in an undesirable impact. A brief explanation of some impacts caused by a class of pollutants follows.

Sediments. Some sediment is tolerated in a stream; however, too much sediment produces highly turbid waters and adverse consequences have been documented. These impacts are (1) reduced light transmission and thus reduced growth, (2) clogging of fish gills, (3) reduced spawning areas, (4) reduction in organism species and numbers because of food-chain perturbations, (5) rapid sedimentation (filling) in holding ponds, and (6) reduction in aesthetic values. Sediment content in water is measured as suspended solids, volatile and nonvolatile suspended solids, turbidity, and settleable solids. Suspended solids measurements are distinguished from dissolved solids by filtration through a glass fiber or Millipore filter with varying pore size, typically 0.4 μm. Solids passing through the filter are considered dissolved (nonfilterable). Total solids is the sum of dissolved and suspended solids and is the material residue remaining after evaporation. Volatile solids are those that burn off at a temperature of 550°C and are considered to be primarily organic in nature. The non-volatile fraction is that remaining after burning and is considered to be the nonreactive fixed and inert fraction. In stormwater, generally most of the chemicals are associated with the smaller sizes of solids (less than 100 μm) carried in runoff water. Thus the silts and clays that settle slowly may contain a large portion of the pollutant mass.

Oxygen Demand. Free oxygen, also known as dissolved oxygen (DO), is necessary in water to maintain life. In the oxidation of organic matter by biological activities, oxygen from water is used. All oxidizable matter consumes oxygen from the water which is replaced from the atmosphere or produced during algal and plant photosynthesis. A problem from low DO results when the rate of oxygen-demanding material exceeds the rate of replenishment. Fish kills and reduction in aesthetic values have resulted from low-DO conditions. Oxygen demand is estimated by direct measure of DO, and indirect measures such as biochemical oxygen demand (BOD), chemical oxygen demand (COD), oils and greases, and total organic carbon (TOC). Procedures for analyses are standardized by the U.S. Environmental Protection Agency and other organizations.

Bacteria. Common bacteria are coliform, fecal coliform, and specific pathogens, such as *Shigella, Salmonella*, and *Clostridium*. The latter has been related to duck kills in lakes and serious health problems among people. *Salmonella* has been related to gastrointestinal disorders. In some samples of runoff, public health standards on the number of organisms for water contact recreation are violated. During an 18-month study period in Denver, the fecal coliform stream standard was violated 100% of the time (Denver Regional Council of Governments, 1983).

Nutrients. Nutrients are chemicals that stimulate the growth of algae and water plants. They are sometimes called biostimulants. The basic macronutrients are carbon, nitrogen, and phosphorus. Micronutrients are those needed in very small quantities. The orthophosphorus form of phosphorus is readily available for plant growth. Similarly, nitrogen in the form of ammonia and nitrates is available for growth. The increased algal activity will cause increased respiration by plant and animal life, which can cause a decrease in DO at nighttime. The typical nutrient concentrations in runoff are usually more than sufficient to stimulate the growth of algal and plant species. Nutrients in the dissolved (soluble) form are readily available to the plants and animals. Other problems resulting from excess nutrients are (1) surface algal scums, (2) water discoloration, (3) odors, (4) toxic releases, and (5) overgrowth of plants. Common measures for nutrients are total nitrogen, organic nitrogen, total Kjeldahl nitrogen (TKN), nitrate, ammonia, total phosphate, total organic carbon (TOC), and indirectly algamass and chlorophyll *a*.

Metals. A wide variety of metals are present in stormwater. Beyond threshold concentrations most cause toxic effects. Health problems and deaths have been known to result. Specific measures are arsenic, cadmium, chromium, copper, iron, mercury, nickel, lead, selenium, and zinc. A majority of the metals may exist in particulate form and are unavailable for organisms use and bioaccumulation. The most common metals in urban runoff are zinc (Zn), copper (Cu), and lead (Pb). Stream standard concentrations have been violated when stormwater was present (Denver Regional Council of Governments, 1983).

Other Toxic Chemicals. Priority pollutants are generally related to hazardous wastes or toxic chemicals and can sometimes be detected in stormwaters. A listing was developed and published, but others not on the priority pollutants list can be equally harmful, such as chlorides. Measures of priority pollutants in stormwaters include (1) phthalate (plasticizer compound), (2) phenols and cresols (wood preservatives), (3) pesticides and herbicides, (4) oils and greases, (5) metals, and many others. Chlorides are often introduced after and during cold weather conditions. They are used to remove ice and snow from roads, parking lots, and sidewalks. Due to the extreme solubility, almost all chloride applied for snow and ice control ends up in surface or groundwaters. Most organisms adapt to a relatively narrow range of salinity; thus chloride changes are potentially damaging to aquatic life.

5.2 DUSTFALL AND QUALITY OF PRECIPITATION

The term *precipitation* may be used to describe the material deposited by wet (rainfall) and dry (dustfall) processes on land surfaces. The dry dustfall is a continuous process, while wet precipitation (rain, snow, sleet, etc.) is an intermittent process. These processes are effective in removal of airborne pollutants and cleansing the atmosphere. Airborne pollutants are associated with particles and aerosols that accumulate in the atmosphere from industrial stacks, automobiles, planes, and from exposed land. Suspended particles in the atmosphere are deposited by sedimentation, interception processes, Brownian movement, and inertial impaction.

Wet and dry deposition are generally monitored by automatic atmospheric samplers that separate wet from dry deposition by a lid transfer mechanism between collection buckets activated by rainfall. Some investigators may refer to dustfalls or bulk precipitation as the sum of dry and wet precipitation of atmospheric materials and their common means of collection is by dustfall jar samplers, as in Figure 5.2. The jar itself is positioned in a metal bird ring that is 35.5 cm in diameter. The ring is used as an aid in preventing bird droppings. Insects and other nondustfall materials should be removed before dustfall concentrations are determined.

If the bulk precipitation volume and concentration are determined, a simple model is used to estimate dustfall, mainly by material balance.

$$D_w = \frac{\bar{C}V}{ADP + D} \tag{5.1}$$

where D_w = average dustfall, mg/time/collection area
  $\bar{C}$ = average concentration of sample, mg/L
  V = volume of bulk precipitation, L
  ADP = antecedent dry period, time
  D = duration of storm, time

5.2 DUSTFALL AND QUALITY OF PRECIPITATION

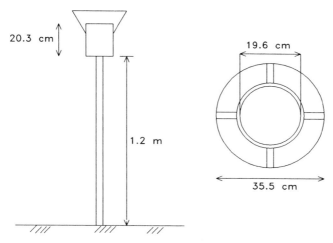

Figure 5.2 Dustfall sampler.

The model above can be used to estimate average (event mean) runoff concentration, when runoff volume (rainfall excess) is substituted for volume of bulk precipitation and the average dustfall concentration is calculated for the area.

$$\overline{C} = \frac{\text{mass}}{\text{volume}} = AD_w \frac{ADP + D}{R} \tag{5.2}$$

where R = rainfall excess for the watershed, L
 A = watershed area

Equation 5.2 is not used to estimate runoff contaminants and does not consider the preceding storm event nor the accumulation or deposition from lower-level sources (cars, pets). D_w is a represented average value of dustfall that is assumed to be removed from the watershed. These are very simplified and limiting assumptions, and thus pollutants should be measured at a discharge point in a watershed. Average annual deposition rates by dustfall are shown in Tables 5.1 and 5.2. Sediment mass rates for various regions in the United States are shown in Table 5.1 and Table 5.2 tabulates specific pollutant rate values for the Washington, D.C. area.

Precipitation (including wet and dry processes) is a major source of impurities in stormwater. Precipitation contributes chemical substances to surface waters and should be considered in analyzing surface water quality. Precipitation cannot be ignored. Around some industrial areas with atmospheric pollution emissions, a lower pH (acid) precipitation has been recorded. Some quality measures of precipitation are shown in Table 5.3. Precipitation impurities will vary from one location to another and one time period to

STORMWATER QUALITY

TABLE 5.1 Average Annual Dustfall Values for Various Water Resources Regions

Area	ton^4/mi^2/month	mt/km^2/month	tons/mi^2/month	mt/km^2/month
New England	8.2	2.87	0.5–152	0.18–53.2
Mid-Atlantic	5.5	1.92	0.3–241	0.1–84.4
Upper Colorado	143.3	50.2	69–281	24–98.4
Pacific Northwest	7.2	2.52	0.3–317	0.1–111
Lower Mississippi	62.0	21.7	18–270	6.3–95
Missouri Basin	34.7	12.2	6–103	2.1–36
Lower Colorado	33.9	12	16–69	5.6–24
South Atlantic	5.0	1.75	1.2–296	0.42–104
Tennessee	4.2	1.47	1.1–17.2	0.38–6.0
Ohio	2.8	0.98	1.7–5.9	0.6–2.1
Upper Mississippi	12.5	4.37	0.3–315.3	0.1–110.4
Great Lakes	32.0	11.2	2–206	0.7–72
Sonris–Red Rainy	23.4	8.2	3.0–73	1.1–26
Rio Grande Region	29.5	10.33	12–269	4.2–94
Texas Gulf Region	32.4	11.34	8–116	2.8–41
Arkansas–White–Red	—	—	—	—
California	16.9	5.9	1–38	0.35–13.3
Great Basin	14.7	5.15	5–56.5	1.8–20

Source: EPA National Aerometric Data Bank, Environmental Monitoring and Support Laboratory, EPA, Research Triangle, Park, N.C., 1971.

TABLE 5.2 Specific Pollutant Deposition Rate for Washington, D.C. Area (lb / acre / yr)[a]

Pollutant	Rural	Suburban	Urban
Total solids	99	155	245
Chemical oxygen demand	199	133	210
Total nitrogen	19.9	12.8	17.0
Nitrate-N	9.4	5.6	6.8
Ammonia-N	5.5	1.1	1.0
Total Kjedldahl N	10.5	7.2	10.2
Total phosphorus	0.71	0.50	0.84
ortho-Phosphorus	0.28	0.26	0.35
Trace metals			
Cadmium	ND[b]	0.09	0.003
Copper	ND	0.21	0.61
Lead	0.006	0.44	0.53
Iron	ND	1.57	5.60
Zinc	0.67	1.35	0.65

Source: Northern Virginia Planning District Commission (1983).
[a]To convert lb/acre/yr to kg/ha/yr multiply by 1.122; to convert kg/ha/yr to mt/km^2/yr, multiply by 0.1 (also see Appendix B).
[b]ND, not detected.

TABLE 5.3 Rainfall Constituents (Concentrations in mg / L)

Constituent	Knoxville Tennessee (Betson, 1978)	Göttingen Germany (Ruppert, 1975)	North Carolina and Virginia (Weibel, 1969)	Melrose, Florida (Brezonik, 1969)	Central Florida (Wanielista, 1976)	Central Florida (Yousef et al., 1985)
Specific conductance (μs/cm)			12	10–30	8–34	
pH	5.1			5.3–6.8	4.7–6.4	5.1
Cl^{-1}	4.0	0.9	0.1–1.1			
SO_4^{2-}	7.1	8.8	1.1–3.2	1.74		1.6
Na^+	1.50	0.8	0.3–1.1	0.8		2.7
K^+	2.6	1.0		0.29–1.85		2.2
Ca^+	3.8	3.0	0.2–1.2	0.13–0.21		1.6
Mg^{2+}	0.74	0.7		0.06–0.35		0.2
Organic N	2.5			0.32	0.01–0.63[a]	0.27
NH_3-N	0.41			0.208		0.11
NO_2-N				0.005		
NO_3-N	0.47[b]		0.045–0.225	0.209	0.02–0.26	0.32[b]
OP-P				0.009	0.01–0.08	
TP-P	0.36	0.36		0.011	0.04–0.20	0.02
TOC					2.2	
BOD_5					1.1	
COD	65					
Suspended solids	16			2.7–14		
Turbidity (NTU)				0.8–2.8		
Alkalinity				5.6–11.1		
Hardness				2–14		
Inorganic carbon				0.5–2.8		

[a] Total organic N + NH_3-N.
[b] NO_2-N + NO_3-N.

another and thus cannot be considered as a constant. For example, chloride ion concentrations are highest along coastal areas. Calcium reaches a peak usually during summer months because of agricultural activities. A survey (Ciaccio, 1971) of ammonium and nitrate concentrations in precipitation showed lower concentrations along the coastline of Florida with increasing concentrations inland.

The constituent loading of nutrients and metals in rainfall can also be compared to runoff into a stormwater pond and to effluent from the pond. In general, particulate matter settles in the pond and dissolved material may

TABLE 5.4 Comparison of Loadings (Pounds)

Constituent	Rainfall	Pond Inflow	Pond Outflow
Organic N	6.0	10.2	7.3
Ammonia-N	5.9	0.8	0.6
NO_x	6.3	2.3	1.0
OP	0.6	4.8	1.5
TP	1.7	5.9	2.1
Zn	0.7	0.4	0.2
Fe	1.7	5.4	3.6

decrease. Rushton (1991) cumulated the pounds of constituents in 22 rainfall events and compared the loads to the input and output loads of a pond. The results are shown in Table 5.4.

5.3 STORMWATER SAMPLING, EVENT MEAN CONCENTRATION, AND LOADING

5.3.1 Stormwater Sampling

Sampling for nonpoint-source response curves should be related to hydrograph and concentration changes. Usually, the most reliable sampling is done by manual means, provided that the labor force understands its tasks and its members are punctual. Manual sampling may not be the best if the corps of sampling personnel are not dependable. Many times storm events are difficult to predict and sampling personnel either miss early parts of hydrographs, do not show up for work, or spend much time doing nothing. Thus instrumentation is often used to collect stormwater runoff and rainfall samples. Minimum instrumentation considered necessary for sampling should provide a record of rainfall, flow rate, concentration, and mass loadings should be recorded. Continuous documentation of precise changes should not be expected because of the many variables related to storm runoff and the cost involved. Thus expensive and time-consuming programs should not be designed. Equipment used for sampling presently operate by suction pump, submersed pump, or other means. Intake line placement, volume of sample, preservation of sample, mobility of instruments, protection of instruments (from vandalism and theft), and cost and power requirements are basic considerations for the use of automatic sampling equipment. Typical diagrams for open channel and sewer line inplacements are shown in Figure 5.3.

Grab sampling is of little use for estimating runoff concentration, flow volume, and mass loadings. One instantaneous value is obtained, which most likely is not representative of the average conditions; even less, the variability of the concentration and mass changes cannot be described. To estimate

5.3 STORMWATER SAMPLING, EVENT MEAN CONCENTRATION, AND LOADING

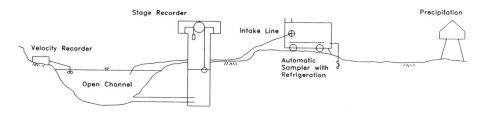

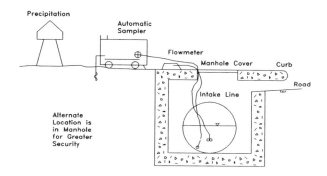

Figure 5.3 Schematic diagrams for sampling.

mass per event and average concentration per event, two methods are generally used: (1) volumes of stormwater are sampled and mixed together to obtain a flow-weighted composite sample, and (2) sequential discrete samples are taken at equal time intervals and each sample concentration is determined. For the flow-weighted method, each sample is mixed with a volume of water directly proportional to the flow at the time of sampling. Sample taken at equal time intervals are weighted relative to the flow at the time of sampling. The event load is determined by multiplying each concentration and flow-weighted volume at the time of collection.

5.3.2 Event Mean Concentration

The event mean concentration (EMC) is defined when the event load for specific contaminant and the event water volume are measured.

$$\overline{C} = \frac{L}{R} \qquad (5.3)$$

where L = loading per event, mg
R = volume per event, L
$\overline{C}$ = event mean concentration (EMC), mg/L

The loading for an event is determined by summing the loadings during each sampling period, provided that flow rate (or volume) data are available for the period. The following equation is used:

$$L = \sum_{i=1}^{n} R_i C_i \qquad (5.4)$$

where R_i = volume proportional to flow rate at time i, L
C_i = concentration at time i, mg/L
n = total number of samples during a single storm event

A plot of concentration, flow rate, and loadings will produce response curves for the nonpoint sources. Typical continuous plots are shown in Figure 5.4.

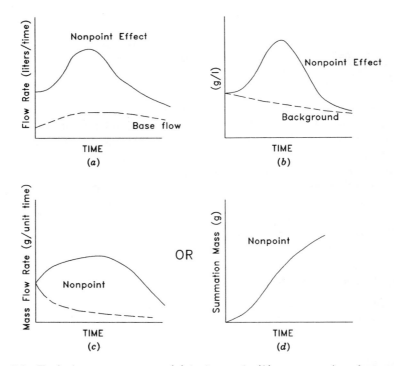

Figure 5.4 Typical response curves: (a) hydrograph; (b) concentration changes with time or pollutograph; (c) loading rate with time; (d) loadograph (cumulative mass).

5.3 STORMWATER SAMPLING, EVENT MEAN CONCENTRATION, AND LOADING

The event mean concentration (EMC) is a statistical parameter used to represent a flow weighted concentration of a desired water quality parameter during a single storm event. If sequential water samples are collected, the EMC can be measured as the concentration in a flow-weighted composite sample with the volume of each fraction directly proportional to flow at time of collection. On the other hand, if sequential discrete samples are collected over the hydrograph, the EMC value can be determined by calculating the cumulative mass of pollutant (loadograph) and dividing it by the volume of runoff (area under the hydrograph).

Example Problem 5.1 Sequential discrete samples of runoff were collected for lead concentration in $\mu g/L$ as follows. Calculate the event mean concentration.

Sampling Time	Flow Rate (L/s)	Lead Concentration ($\mu g/L$)
14:10	0	0
14:20	10	25
14:30	20	40
14:40	50	120
14:50	30	50
15:00	25	35
15:10	20	30
15:20	15	25
15:30	10	20

SOLUTION

Sampling Period Δt (min)	Average Flow Rate (L/s)	Average Runoff Volume ΔR (L)	Average Concentration ($\mu g/L$)	Mass (g)
10	5	3×10^3	12.5	0.038
10	15	9×10^3	32.5	0.292
10	35	21×10^3	80	1.680
10	40	24×10^4	85	2.040
10	27.5	16.5×10^3	42.5	0.701
10	22.5	13.5×10^3	32.5	0.439
10	17.5	10.5×10^3	27.5	0.289
10	12.5	7.5×10^3	22.5	0.169
10	5	3×10^3	10	0.030
		$\Sigma = \overline{108 \times 10^3}$		$\overline{5.678}$

Assumptions are made to determine the first and last sampling periods. The first and last Δt are assumed to be 10 min each, and using equation 5.3, EMC = $\Sigma \text{mass}/\Sigma \Delta R$ = (5.678 g/108 × 10^3 L)(10^6 µg/g) = 52.6 µg/L. If the concentrations are not flow weighted, the average is 38.3 µg/L, which is lower than EMC.

5.3.3 Event Loading

The determination of loading rates can be accomplished for each site by sampling events over a sufficient time period to ensure a representative loading rate. There is a great variability in loadings from one storm to another and from one location to another. Generally, the sampling program guidelines should address the types of weather conditions encountered throughout the year. For example, in the wintertime, runoff from snowmelt would probably produce different loading rates relative to nonsnowmelt conditions. Also, it should be noted that during the leaf-dropping season, loading rates may be higher. In general, one can determine the average mass and event mean storm concentration for a number of storm events sampled per year. By knowing the runoff depth on the drainage basin for these storms, it is possible to calculate a loading rate (mass per unit of runoff) for each storm. When the event mean concentration for measured storms does not change significantly from one storm to the next, a sufficient number of events has been sampled and a mass loading per unit depth of runoff per unit area can be estimated. The greater the number of events sampled representing a wide range of rainfall depths and seasons of the year, the more accurate the estimates of mean concentration and loading rates. The mass loading rates can be expressed as kilogram per square kilometer year or pounds per acre-year.

5.4 RURAL NONPOINT SOURCES

Rural sources are known to contribute major quantities of water pollutants. Where large areas of land are devoted to agricultural activities, high concentrations of pollutants can be measured. These pollutants have their origins in the fertilizers, pesticides, and other agricultural chemicals used in farming practices. Because of the large areas of land involved, the management of rural nonpoint sources proves to be very difficult. Several factors affect the resulting water quality. Among these factors are soil types, climates, land uses, cover crops, management practices, and topography of the area.

Rural land uses are usually classified as agricultural or woodlands. The major portion of nonpoint-source effects in rural areas originate in agricultural areas, although there is considerable variability in the reported literature. Recent studies have begun to document the characteristics and transport mechanisms associated with "rural" effects.

A study conducted by Wallace and Dague (1973) related dissolved oxygen content of Iowa rivers to agricultural sources. Conclusions reached in the study were:

1. Agricultural land runoff causes low dissolved oxygen values during higher river flows each summer.
2. Typical runoffs can cause dissolved oxygen concentrations below 5 mg/L to occur.
3. Land runoff is the only cause of low dissolved oxygen values in the Iowas River.

However, other studies (Wanielista, 1976) have not shown low dissolved oxygen levels in similar situations (possibly due to less intense agricultural activity). Pesticides are found in runoff from rural sources. The quantities of pesticides produced yearly is staggering, the total production of these chemicals in 1971 being 600,000 tons. The use of such large quantities of pesticides necessitated a need to predict runoff quantities. Crawford and Donigan (Crawford, 1973) developed a model for predicting the transport of pesticides, herbicides, fungicides, fumigants, nematocides, algacides, rodenticides, and other biocides. Depending on the types of forms of pesticides used, their effects can be either short or long term.

Various water quality measures, such as biochemical oxygen demand (BOD_5), suspended solids, nitrogen, and phosphorus, have been shown to vary according to agricultural practices. The effects of animal wastes on land runoff have been studied by the North Carolina State University (U.S. Environmental Protection Agency, 1974). Other studies have been summarized and reported by the Soil Conservation Service (SCS) (U.S. Department of Agriculture, 1975).

Reported estimates of manure, solids, organic material, and nutrients produced by various animals and birds exhibit a wide range of values. This is due primarily to feed types, measurement errors, location (climate and production methods), and animal size. Table 5.5 illustrates a comparison among animals. Human daily production rates are shown for comparative purposes. Runoff quality from animal feed areas have a high variability, depending on ground cover, climate, feed type, density, slope, and maturity of animals. In general, these data are available from the Soil Conservation Service for a local area. Although animals vary in weight, average weights can be used for most runoff studies, as shown in Table 5.5.

Rural nonpoint-source effects are generated from several different land uses. Accordingly, the amounts and types of pollutants contained in the runoff will vary widely. Generally, the greater the extent of rural land utilization, the greater the amounts of pollutants introduced, unless management methods for reducing surface runoff are used. Whether the pollutants are in fertilizers, pesticides, or animal wastes, they are dispersed over the

TABLE 5.5 Daily Production and Composition of Livestock Manure: Feces and Urine[a]

	Dairy Cattle	Beef Cattle	Feeder Swine	Poultry	Sheep	Horses	Catfish	People
Weight of Animal (kg)	500–700	360–450	45–100	2–2.5	40–80	450–800	1–3	—
Manure	85	62	69	53	36	50	—	—
	72–90	41–88	50–90	32–67	30–40	40–60	3.1	3.4
Total solids	9.3	8.9	7.2	13.9	9.5	17.5	2.8–3.5	2.4–4.4
	6.8–13.5	6.0–11.1	6.0–9.0	9.0–17.4	8.4–10.7			
Volatile solids	6.9	6.9	5.7	10.8	8.0	—	—	2.0
	5.7–7.9	4.8–8.2	4.0–7.0	8.0–12.9	6.0–9.1			1.1–2.6
BOD$_5$	1.4	1.5	2.3	3.4	0.8	1.4	2.3	1.36
	0.8–1.8	1.0–1.8	2.0–2.8	1.6–5.5	0.7–0.9		1.1–4.9	0.6–2.10
COD	8.4	7.9	5.9	12.5	10.0	—	—	3.12
	4.2–13.3	6.6–9.0	4.7–7.1	9.5–15.8	7.5–12.0			1.0–3.5
Total N	0.37	0.43	0.45	0.86	0.40	0.30	1.6	0.20
	0.29–0.51	0.30–0.58	0.20–0.70	0.45–1.50	0.34–0.45		0.7–2.5	0.14–0.26
Total P	0.069	0.090	0.17	0.40	0.075	0.12	0.25	0.024
	0.026–0.100	0.023–0.170	0.09–0.27	0.20–0.75	0.040–0.120		0.24–0.26	
Total K	0.20	0.23	0.25	0.35	0.32	0.25	1.5	0.064
	0.08–0.35	0.11–0.38	0.10–0.60	0.12–0.50	0.24–0.40		0.7–2.4	

Source: U.S. Department of Agriculture (1975).

[a] Grams/kg wt/day. Upper figure is average; lower figures represent the range given in literature. Dashes indicate data not available or entry not appropriate.

surface of the land, and some quantity eventually enters the aquatic environment. However, some urban nonpoint and point sources of water pollution are perhaps more detrimental than rural nonpoint sources.

Example Problem 5.2 Estimate the average daily loading of nitrogen from a dairy feedlot with 200 cattle if the loadings are not decreased by chemical or biological means during the day.

SOLUTION: The weight of cattle and nitrogen discharged from the feedlot are necessary. Assuming that the average weight of dairy cattle is 600 kg (Table 5.5) and the nitrogen production is 0.37 g/kg for each day (Table 5.5), the load per day is 222 g per head or 44.4 kg per 200 head of cattle.

5.5 URBAN NONPOINT SOURCES

Urban runoff has the potential to contain many polluting materials. The sources of these pollutants vary widely, ranging from the "city" birds, such as pigeons, to vehicle tires and construction activities. Pollutants can consist of solid waste litter, chemicals, air-deposited substances, and vehicle pollutants. In urban areas, some sources of pollution are precipitations, vehicles, animals, buildings, lawns, and humans. These sources result in street and lawn buildup of pollutants. Typical particle-size distribution of street sweeping from the city of Orlando, Florida (Yousef et al., 1990) is shown in Table 5.6 and Figure 5.5. This figure shows the percent weight finer than a specified grain size in millimeters. It appears that the particle-size fraction between 0.25 and 0.5 mm in diameter accounts for about one-half of the total weight. Also, particles less than 0.35 mm in diameter account for one-half of weight and particles less than 0.2 mm in diameter account for 10% of the weight. Particles less than 0.125 mm approximate 2% of the weight on the average, but these particles account for more than 80% of the particles carried in suspension by runoff water (Table 5.7). Most larger particles settle out in pipelines, manholes, and so on, and may be occasionally flushed out

TABLE 5.6   Sieve Analysis of Street Sweepings from the City of Orlando

Sieve	Particle	Percent by Weight						
		1st Run	2nd Run	3rd Run	4th Run	5th Run	Avg.	Std. Dev.
> 18	> 1.0 mm	6.87	7.98	9.70	13.72	9.51	9.55	2.60
18–35	1.0–0.5 mm	32.57	18.09	15.30	14.29	14.99	19.05	7.70
35–60	0.5–0.25 mm	46.21	58.51	50.35	41.62	43.64	48.07	6.69
60–120	250–125 μm	12.27	14.41	23.40	27.47	28.81	21.27	7.55
120–325	124–45 μm	1.80	0.58	0.82	2.42	2.54	1.63	0.90
< 325	< 45 μm	0.28	0.43	0.43	0.49	0.51	0.43	0.09

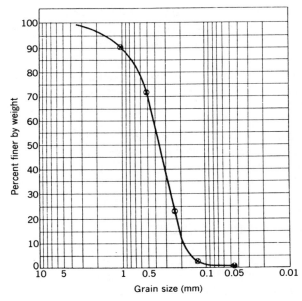

Figure 5.5 Average grain size distribution for street sweepings from Orlando, Florida.

with the first flush at peak flow rates. Smaller particles account for much higher pollutant concentrations per unit mass than larger particles.

As stated before, runoff from urban streets contains contaminants deposited from the atmosphere and accumulations on street and lawn surfaces, containing several pollutants as summarized in Tables 5.8 and 5.9. These materials will vary widely in both quantity and distribution on the street surfaces. Street surface contaminants contain heavy metals, nutrients, pesticides, bacteria, organic, and dust/dirt. Significant amounts of heavy metals are detected in the contaminant materials deposited on street surfaces. The

TABLE 5.7 Wet Sieve Analysis of Highway Runoff Composite Samples

Particle size (μm)	Percent of Suspended Solids				
	Sacramento Hwy. 50	Harrisburg I-81	Milwaukee I-94	Effland I-85	Mean
> 250	1.54	6.10	14.56	3.58	6.45
88–250	9.07	6.70	7.00	1.30	6.02
44–88	10.70	11.70	5.84	8.06	9.08
< 44	78.69	75.50	72.60	87.06	78.45

Source: Rexnord, Inc. (1984).

TABLE 5.8 Quantity and Characteristics of Contaminants Found on Street Surfaces

Measured Constituents	lb/curb mile[a]
TS	1400
Oxygen demand	
$\quad$ BOD$_5$	13.5
$\quad$ COD	95
$\quad$ VS	100
Bacteriological[b]	
$\quad$ Total coliforms	99×10^9 (number/curb mile)
$\quad$ Fecal coliforms	5.6×10^9 (number/curb mile)
Pesticides	
$\quad$ p,p-DDD	67×10^{-6}
$\quad$ p,p-DDDT	61×10^{-6}
$\quad$ Dieldrin	24×10^{-6}
$\quad$ PCB	1000×10^{-6}

Source: Sartor et al. (1974).
[a] lb × 0.454 = kg.
[b] Number of observed organisms/mile.

most prevalent metals found in street sweepings, with the exception of iron, were lead and zinc (Sartor et al., 1974a; Pitt and Amy, 1973; Yousef et al., 1990). Heavy metals appear to be vehicle related and thus are influenced by traffic volume. In addition to heavy metals, vehicles deposit grease and oils and other contaminants on the street surfaces. The quantities deposited vary according to the type and age of vehicles and the kind of fuel used. Pitt and Amy (1973) found grease and oil loadings ranging from 32.8 lb/curb mile/day for industrial areas to 4.90 lb/curb mile/day for commercial areas. Residential areas had a reported loading of 18.6 lb/curb mile/day.

Various nutrients were detected in street sweepings and dustfall on roads and highways (Sartor et al., 1974b; Pitt and Amy, 1973; Wanielista, 1976). Variations in loadings for city streets, rural roads, and highways are presented in Table 5.10. Loading values do vary because of local conditions and the average values are shown in Table 5.10. It appears that loading from highways are greater than loadings from city streets and rural roads, which may indicate that daily traffic volume is important.

Many factors, including climate, surrounding land use, street and highway design, and operation influence the pollutant loads characteristics. The average daily traffic (ADT) volume and type of traffic have an effect on the contaminants present. More contaminants will be present on crowded city streets than on rural roads. Different paving materials impart varied surface characteristics to the street. Sartor et al. (1974a) found that asphalt surfaces had higher solid loadings (kg/km-day) than did concrete surfaces. The

TABLE 5.9 Pollutant Concentration and Loading Data

	Pollutant Concentration (mg/L)		Pollutant Loadings (lb/acre/event)		Pollutant Loadings (lb/acre/in. runoff)	
	Avg.	Range	Avg.	Range	Avg.	Range
pH		6.5–8.1				
TS	1147	145–21,640	51.8	0.04–535.0	260	33–4910
SS	261	4–1156	14.0	0.008–96.0	59	0.9–375
VSS	77	1–837	3.7	0.004–28.2	17	0.2–190
BOD_5	24	2–133	0.88	0.000–4.1	5.4	0.5–30
TOC	41	5–290	2.1	0.002–11.5	9.3	1.1–66
COD	14.7	5–1058	6.9	0.004–34.3	33	1.1–240
TKN	2.99	0.1–14	0.15	0.000–1.04	0.68	0.02–3.17
$NO_2 + NO_3$	1.14	0.01–8.4	0.069	0.000–0.42	0.26	0.002–1.90
TPO_4	0.79	0.05–3.55	0.047	0.000–0.36	0.18	0.011–0.81
Cl	386	5–13,300	13.0	0.008–329	88	1.1–3015
Pb	0.96	0.02–13.1	0.058	0.000–04.8	0.22	0.005–2.97
Zn	0.41	0.01–3.4	0.022	0.000–0.8	0.093	0.002–0.771
Fe	10.3	0.1–45.0	0.50	0.000–3.5	2.34	0.023–10.2
Cu	0.103	0.01–0.88	0.0056	0.000–0.029	0.023	0.002–0.199
Cd	0.040	0.01–0.40	0.00017	0.000–0.014	0.009	0.002–0.091
Cr	0.040	0.01–0.14	0.028	0.000–0.029	0.009	0.002–0.032
$Hg \times 10^{-3}$	3.22	0.13–67.0	0.00059	0.000–0.002	0.730	0.029–15.2
Ni	9.92	0.1–49	0.27	0.007–1.33	2.25	0.023–11.2
TVS	242	26–1522	9.34	0.01–44.0	55	5.98–345

Source: Smith and Lord (1990) using highway runoff data from six sites.

TABLE 5.10 Average Loadings for Common Pollution Parameters and Certain Heavy Metals (kg / curb km-day)

Parameter	City Street	Rural Road	Highway
BOD_5	0.850	0.140	0.900
COD	5.000	4.300	10.000
PO_4	0.060	0.150	0.080
NO_3	0.015	0.025	0.015
N (total)	0.150	0.055	0.200
Cr	0.015	0.019	0.067
Cu	0.007	0.003	0.015
Fe	1.360	2.020	7.620
Mn	0.026	0.076	0.134
Ni	0.002	0.009	0.038
Pb	0.093	0.006	0.178
Sr	0.012	0.004	0.018
Zn	0.023	0.006	0.070

Source: Wanielista (1976) and Pitt and Amy (1973).

physical conditions of the street surfaces are also important. Poor surfaces generally have higher contaminant loadings than good street surfaces. The time of year also plays a role in the amounts of street surface contaminants present. Other factors associated with quantities of contaminants present are the antecedent precipitation, public works department sweeping practices, and quantities of air pollution fallout.

5.6 RUNOFF WATER QUALITY

The impurities in stormwater runoff can cause receiving water impacts that are undesirable. These impacts have been related to the water quality

TABLE 5.11 Rainfall Intensities and Durations for 90% Particle Removal from Impervious Surfaces[a]

0.10 in./h for 300 min
0.33 in./h for 90 min
0.50 in./h for 60 min
1.00 in./h for 30 min

Source: U.S. Environmental Protection Agency (1974).
[a] in./h × 25.4 = mm/h.

TABLE 5.12 Overall Average Dissolved Water Quality Parameters

Parameter	Unit	Rainfall		Runoff				Ratio Rainfall/ Runoff for Maitland
				Maitland		US 17-92		
		$\bar{x}$	σ	$\bar{x}$	σ	$\bar{x}$	σ	
pH (lab)	$-\log[H]$	5.2	0.7	6.9	0.2	6.8	0.2	—
Specific conductance	μS	18	4	123	52	210	64	0.15
Dissolved solids	mg/L	10	2	76	40	160	41	0.13
HCO_3^-	mg/L	1.5	1.3	54	20	49	19	0.03
SO_4	mg/L	2.7	1.4	12	10	23	5	0.23
Cl	mg/L	1.6	0.7	2.9	1.5	22	5	0.55
TN as N	mg/L	0.66	0.50	0.79	0.18	1.37	0.7	0.84
TP as P	mg/L	0.02	0.0	0.05	0.07	0.22	0.13	0.40
Ca^{2+}	mg/L	1.6	1.0	27	28	17	4	0.06
Mg^{2+}	mg/L	0.2	0.1	1.2	1.3	4.2	1.0	0.17
Na^+	mg/L	2.2	2.6	2.9	2.3	17.3	9.9	0.76
K^+	mg/L	0.5	0.4	1.7	1.6	2.6	1.0	0.29
SiO_2	mg/L	0.1	0.1	1.9	1.4	4.6	1.7	0.05
Cd	$\mu g/L$	2.5	0.6	1.6	0.6	18	0.7	1.60
Zn	$\mu g/L$	8.2	1.6	23	4	15	1	0.36
Pb (dissolved)	$\mu g/L$	38	9	41	8	19	1	0.93
Cu	$\mu g/L$	66	58	27	23	9	3	2.40

Source: After Yousef et al. (1985).

parameters of Section 5.1. In approximately 80% of urban areas investigated, the Council on Environmental Quality determined that downstream water quality was controlled by stormwater runoff sources (Council on Environmental Quality, 1972).

Rainfall will cleanse the atmosphere and remove some deposited matter on the impervious surface. The rainfall intensity, runoff rate, and storm duration can affect the amount and type of contaminants removed. Storm duration and intensities considered sufficient to remove 90% of the road surface particles were estimated by the U.S. Environmental Protection Agency (1974) and are shown in Table 5.11. The deposition data presented in tables throughout this chapter are valuable for understanding the magnitude of the problems related to runoff water quality; however, the quantity of pollutants washed from the impervious surfaces have to be determined. Thus mathematical models of both quantity and quality using loading rates, transport functions, degradation effects, and deposition rates are, to a degree, useful in reproducing existing conditions and in predicting future conditions relative to planning, design, or management of surface water quality and quantity.

Recent studies indicate that certain pollutants in stormwater are contributed primarily by rainwater. This is particularly true for various forms of nitrogen and possibly some heavy metals. Various investigators, including Yousef et al. (1985), showed that nitrogen, copper, and cadmium in rainwater

TABLE 5.13 Urban Stormwater Compared to Drinking Water and Ambient Life Standards (mg / L)

Parameter	Stormwater Concentration Greater Than			Drinking Water[a]	Ambient Life Criteria	
	50%	10%	5%		Threshold[b]	Mortality[c]
Total dissolved solids	700	2000	9000	500	—	—
Total phosphorus	0.27	0.65	0.82	—	—	—
Total nitrogen	2.20	4.50	5.60	—	—	—
NO_3-N	0.50	1.10	1.40	10	—	—
Copper	0.02	0.10	0.21	1.3	0.020	0.050
Lead	0.04	0.10	0.50	0.010	0.150	0.350
Zinc	0.10	0.21	0.42	0.030[d]	0.380	0.870
Cadmium	< 0.01	0.01	0.06	0.010	0.003	0.007

Source: Northern Virginia Planning District Commission (1983); Wanielista (1976); U.S. Environmental Protection Agency (1973).

[a] From U.S. Department of Health, Education and Welfare (1962), U.S. Environmental Protection Agency (1974, 1991).
[b] Concentration causing mortality to the most sensitive individual at 50 mg/L CaCO3.
[c] Significantly mortality (50% to the most sensitive plus others) at 50 mg/L CaCO3.
[d] Florida Department of Environmental Regulation Receiving Water Standard 1986 (Chapter 17-3).

may represent a large fraction of the same constituents in runoff water as depicted in Table 5.12.

It should be apparent and expected that stormwater runoff will not have the quality of drinking water. Table 5.13 illustrates this comparison for selected parameters. Also, ambient life standards are compared to stormwater concentration. Data from three different studies on stormwater quality were analyzed to determine concentration equal to or greater than 50%, 10%, and 5% of collected samples, as shown in Table 5.13. However, frequency distributions for all exceedence probabilities would provide insight on water quality issues.

Frequency distribution of average storm concentrations appear to follow log-normal form as shown in Figure 5.6 (Yousef, et al., 1985). A majority of

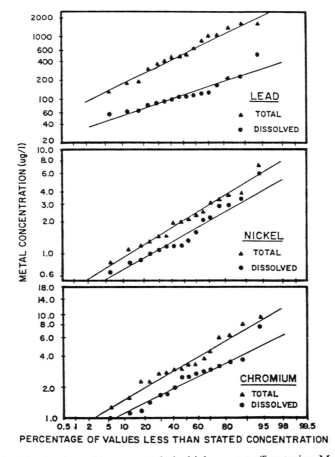

Figure 5.6 Distribution of heavy metals in highway runoff entering Maitland Pond during 1983 and 1984.

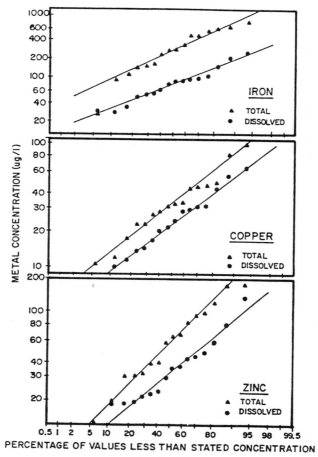

Figure 5.6 (*Continued*)

runoff events with lower average concentration values and higher concentrations are associated with fewer storm events.

Gizzard et al. (1976) reported that about 45% of total phosphorus and 55% of TKN in urban runoff water were measured as dissolved. Tables 5.14 and 5.15 illustrates the magnitude and variability of urban stormwater quality in yearly average percent dissolved estimates. The extreme variability of water quality, as shown in the tables, suggests a need for site-specific studies before design of management systems. Many reasons (land use, sampling times, meteorological) exist for the variability of urban stormwater quality. Impact of land use on the range for event mean concentration for selected water quality parameters are depicted in Figure 5.7.

When comparing the metals found in urban runoff with the metals content of sanitary sewage, loadings of 10 to 100 times the concentration (mg/L) of

5.6 RUNOFF WATER QUALITY

TABLE 5.14 Variability of Urban Stormwater Quality

	Four Sites: USSR, Sweden, and England	Concentration Range, Eight U.S. Sites[a]
BOD_5	36–285	< 1–7700 mg/L
TOC		1–150 mg/L
COD	18–3100	5–3100 mg/L
SS		2–11,300 mg/L
Total solids	30–14,511	200–14,600 mg/L
Volatile solids		12–1600 mg/L
Organic N		0.01–16 mg/L
TKN		0.01–4.5 mg/L
NH_3N		0.1–2.5 mg/L
NO_3N		0.01–1.5 mg/L
Soluble PO_4		0.1–10 mg/L
Total PO_4		0.1–125 mg/L
Oils		0–110 mg/L
Phenols		0–0.2 mg/L
Lead		0–1.9 mg/L
Total coliforms	$40–2 \times 10^5$	$200–146 \times 10^6$/100 mL
Fecal strepococci		$200–1.2 \times 10^6$/100 mL

Sources: Wanielista (1976), U.S. Environmental Protection Agency (1973), American Public Works Association (1969), and Weibel (1969).
[a]Sites are East Bay, California; Cincinnati, Ohio; Los Angeles County; Washington, D.C.; Seattle, Washington; Orlando, Florida; Tampa, Florida; and New York, New York.

TABLE 5.15 Average Dissolved Fractions by Land Use

Element	Percent Dissolved[a]
Single Family	
P	50–67
N	70–80
Zn	43–75
Pb	10–20
Multifamily	
P	50–74
N	65–78
Zn	55–71
Pb	5–10
Commercial	
P	40–68
N	68–90
Zn	45–75
Pb	5–20

Sources: Northern Virginia Planning District Commission (1983) and Wanielista et al. (1982).

[a]Based on event mean concentration.

132 STORMWATER QUALITY

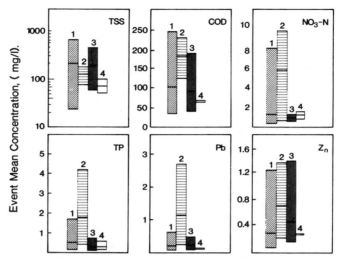

Figure 5.7 Maximum, minimum, and mean values for event mean concentrations from (1) low-density residential, (2) high-density residential, (3) commercial, and (4) industrial land use. (Adapted from Terstriep et al., 1986.)

TABLE 5.16 Some Inorganics and Organics in Urban Runoff

Frequency of Detection (%)	Inorganics	Organics
96	Copper and lead	
95	Zinc	
58	Arsenic	
57	Chromium	
55	Cadmium	
48	Nickel	
20		α-Hexachlorocyclohexane
15		Phenol, pentachloro-
13		Phthalate, bis(2-ethylhexyl)
12		Phenanthrene
11		Naphthalene and pyrene
10		Fluoranthene
5		Benzene
4		Benzene, ethyl- and phenol
2		Toluene

Source: Cole et al. (1984).

TABLE 5.17 Median EMCs for All Sites by Land-Use Category

Pollutant	Residential Median	CV	Mixed Median	CV	Commercial Median	CV	Open/Nonurban Median	CV
BOD (mg/L)	10.0	0.41	7.8	0.52	9.3	0.31	—	—
COD (mg/L)	73	0.55	65	0.58	57	0.39	40	0.78
TSS (mg/L)	101	0.96	67	1.14	69	0.35	70	2.92
Total lead (μg/L)	144	0.75	114	1.35	104	0.68	30	1.52
Total copper (μg/L)	33	0.99	27	1.32	29	0.81	—	—
Total zinc (μg/L)	135	0.84	154	0.78	226	1.07	195	0.66
Total Kjeldahl nitrogen (μg/L)	1900	0.73	1288	0.50	1179	0.43	965	1.00
$NO_2 - N + NO_3 - N$ (μg/L)	736	0.83	558	0.67	572	0.48	543	0.91
Total P (μg/L)	383	0.69	263	0.75	201	0.67	121	1.66
Soluble P (μg/L)	143	0.46	56	0.75	80	0.71	26	2.11

sanitary sewage metals are observed. Then National Urban Runoff Program (1983) studies concluded that lead, zinc, and copper are the most significant metals detected in urban runoff and showed the highest concentrations (Table 5.16). Also, the program results provided median EMC values for four different land uses (Table 5.17). The parameters of Table 5.16, except zinc, showed higher mean concentrations for the residential area.

Decomposition of organic materials can deplete the dissolved oxygen in receiving waters, which can lead to fish kill and deterioration of other aquatic life. Organic matter carried by runoff water into receiving streams can cause immediate and delayed effects to oxygen levels in water. The immediate effect is lower dissolved oxygen concentrations caused by easily biodegradable organic and inorganics in reduced forms. The delayed effect is caused by slowly degradable organic.

Example Problem 5.3 A storm sewer drains a section of city streets 0.2 mi long and 44 ft wide, which has curbing on both sides. What is an estimate of lead concentration in the drainage water following a 30-min storm of 1 in./h intensity? Assume that it has been 3 days since the last storm completely cleaned the road. Also, there is no evaporation, infiltration, or initial abstraction. Adjacent areas contribute runoff water equal to the roadway runoff but do not contain lead. Assume that precipitation carries no or an insignificant amount of lead.

SOLUTION: A typical loading on a city street is 0.093 kg/km-day (Table 5.10). In this section there are 0.2 mi × 2 curbs = 0.4 curb mi × 1.61 km/mi = 0.648 km. The lead loading is

$$Pb = 93(10^3) \text{ mg/km-day} \times 0.648 \text{ km} \times 3 \text{ days} = 1.8 \times 10^5 \text{ mg}$$

A 30-min storm of 1 in./h removes 90% of the particles (Table 5.11) on the roadway, so the estimated amount of lead in the runoff is

$$\text{runoff mass} = Pb_R = 1.8 \times 10^5 \text{ mg} \times 0.9 = 1.62 \times 10^5 \text{ mg}$$

Runoff volume = volume from street + equal volume from other areas.

$$\text{Volume} = \frac{1 \text{ in./h}}{12 \text{ in./ft}} \times 0.5 \text{ h} \times 44 \text{ ft} \times 0.2 \text{ mi} \times 5280 \text{ ft/mi}$$

$$\times \frac{1 \text{ acre}}{43{,}560 \text{ ft}^2} \times 1.235 \times 10^6 \text{ L/acre-ft}$$

$$= 5.5 \times 10^4 \text{ L} \times 2 = 1.1 \times 10^5 \text{ L}$$

$$\text{Lead concentration} = \frac{1.62 \times 10^5 \text{ mg}}{1.10 \times 10^5 \text{ L}}$$

$$= 1.48 \text{ mg/L} \quad \text{(which is the event mean concentration)}$$

However, it must be recognized that several assumptions were made which may not represent the actual deposition loading of lead or uniformity of distribution. Measured concentrations in select storms may be required to verify these assumptions.

Example Problem 5.4 Determine the BOD_5 concentration in the effluent from the wastewater troughs at an egg farm with 10,000 chickens using a total wash rate of 40 gpm of continuous water flow from a poor drinking water source using the following assumptions:

Complete washout of troughs
Poor drinking water source or $BOD_5 = 4$ mg/L
0.0005 lbs of BOD_5/animal/day measured in the discharge.

SOLUTION

1. Determine amount of BOD_5 produced:

$$\text{Number animals} \times \text{rate/day/animal}$$
$$= 10{,}000 \times 0.0005 \text{ lb/day/animal} \times 454{,}000 \text{ mg/lb}$$
$$= 2.27 \times 10^6 \text{ mg/day } BOD_5$$

Note: Loading rate assumed from local data.

2. Determine quantity of water available:

$$40 \text{ gpm} \times 60 \text{ min/h} \times 24 \text{ h/day} \times 3.79 \text{ L/gal} = 2.18 \times 10^5 \text{ L/day}$$

3. Determine concentration:

$$\frac{2.27 \times 10^6 \text{ mg/day}}{2.18 \times 10^5 \text{ L/day}} + 4.0 \ mg/L = 14.4 \ mg/L$$

5.7 MASS LOADINGS

The stochastic nature of rainfall and resulting runoff and infiltration compounded by quality and quantity variations in nonpoint-source and point-source discharges result in nonsteady-state water quality response curves (both in concentration and mass loadings). However, the cumulative pollutant contributions from nonpoint sources can be estimated in terms of mass loadings, regardless of the fact that concentrations are difficult to predict on a time-varying basis.

Loading rates are defined as the washoff quantity of a specific pollutant over a period of time per unit area of total watershed. Average concentrations are generally used where time-related runoff quantities and qualities are not available. Runoff water volumes are then used to convert the concentrations to mass. Using a mixture of metric and English units, the loading rate per acre of watershed can be estimated on a yearly basis using

$$L = \frac{R}{12}\overline{C}(2.716)$$

$$= 0.226 R\overline{C} \qquad (5.5)$$

where L = yearly loading per acre, lb/acre-yr
R = yearly rainfall excess, in.
$\overline{C}$ = average concentration, mg/L
12 = conversion constant, 12 in./ft
2.716 = conversion constant, L-lb/mg-acre-ft

Multiply lb/acre-yr by a factor of 1.1218 to convert to kg/ha-yr.

Using the data from the U.S. National Urban Runoff Program (NURP) (Cole et al., 1984), FHWA (Gupta et al., 1984), and a site-specific study in Orlando, Florida (Wanielista et al., 1981), loading rates were calculated and a comparison of loading rates is shown in Table 5.18. Also shown are average

TABLE 5.18 Highway-Related Metal Average Loading Rate

Study	Loading [kg/ha-day (lb/acre-day)]			
	Lead Loading	Zinc Loading	Iron Loading	Copper Loading
FHWA–Nashville[a]	0.0070	0.0039	0.0766	0.0010
	(0.0062)	(0.0035)	(0.0683)	(0.0009)
FHWA–Hwy. 45[a]	0.0167	0.0076	0.2034	0.0019
(Gupta et al., 1984)	(0.0149)	(0.0068)	(0.1814)	(0.0017)
Orlando	0.0117	0.0101	0.0261	0.0019
(Wanielista et al., 1981)	(0.0104)	(0.0090)	(0.0233)	(0.0017)

[a]Assumes 20 ft of runoff per year.

concentrations. Loadings were calculated from the FHWA data assuming 20 in. of runoff per year.

Preliminary findings of the Nationwide Urban Runoff Program (NURP) (Cole et al., 1984) and the Federal Highway Administration's (FHWA) characteristics of runoff from operating highways (Gupta et al., 1984) indicate concentrations and loading rate data for some *priority pollutants*, a group of toxic chemicals or classes of chemicals listed under Section 307(a)(1) of the Clean Water Act of 1977. There are 10 major groups of 129 specific compound listed. From all the data reviewed, it was evident that metals formed one group of chemicals that were above detection limits and many priority pollutants were below detection limits and were infrequently detected. As an example, 71 priority pollutants were detected in the NURP urban samples. All 13 metals were detected at frequencies greater than 10%; copper, lead, and zinc were found in at least 95% of these samples. The organic pollutants were detected at a much lower frequency. Forty-six of the 57 organics detected were present in only 1 to 9% of the samples. There were 106 possible organic. Some of the aromatic hydrocarbons related to gasoline or oil products were detected at very low concentrations and frequencies. Some are shown in Table 5.16.

To illustrate a comparison of urban to rural nonpoint source loadings, Table 5.19 is constructed. In general, contributions from commercial areas of BOD_5, suspended solids, nutrients, and heavy metals are higher than urban effects from cultivated lands. However, from one location to another for the same land use, extreme variability in the reported data exists. Thus site-specific studies should be given consideration to estimate loading rate data. In addition, the receiving surface-water bodies have varying degrees of assimilative capacities. Therefore, water quality responses will probably differ among locations. Field-reliable instruments that would activate sampling and/or measuring equipment as a result of rainfall and/or flow conditions are used to collect site-specific data and to assess water quality variations with time and location.

TABLE 5.19 Average Yearly Loading Rates and Land Use (Based on Total Area)

Land Use	BOD$_5$	Loading Rate (kg/ha/yr)					
		Suspended Solids	Total Nitrogen	Total Phosphorus	Lead	Copper	Zinc
Urban	50	460	8.5	2.0	0.50	0.20	0.40
1-Acre residential	35	420	6.6	1.8	0.30	0.10	0.25
$\frac{1}{4}$-Acre residential	40	450	7.5	1.9	0.40	0.10	0.40
Commercial	87	840	14.5	2.7	0.85	0.24	1.35
Pasture	11.5	343	6.2	0.50	0.10	0.02	0.08
Cultivated	18	450	26.0	1.05	—	—	—
Citrus	15	25	4.0	0.2	—	—	—
Woodland	5	85	3.0	0.10	0.05	0.01	0.03
Wetlands	14	29	4.9	0.40	—	—	—
Golf course	10	150	4.5	0.78	—	—	—
Highway	87	990	13.8	0.7	0.50	0.08	0.47

Sources: East Central Florida Regional Planning Council (1984); Colston (1974); Kluesener and Lee (1974); U.S. Environmental Protection Agency (1973); Sherwood and Mattra (1975); Wanielista et al. (1981); Harms et al. (1974); Priede-Sedgwick (1983); Driscoll et al. (1990); and Woodward-Clyde (1990).

5.7.1 Mass Loading and Flow Rates

Variabilities in mass loading of pollution carried by a flowing river or stream can be calculated. If flow rates can be predicted or are available (USGS data), pollutant loadings in mass per time period can be estimated or predicted from flow rate data and available concentrations using

$$L = KCQ \qquad (5.6)$$

where L = loadings, kg/day
 K = conversion factor = 0.001
 C = concentration, mg/L
 Q = flow rate, m^3/day

To convert L to pounds per day, multiply by 2.203.

Bivariate regression analysis of loadings on flow rate should produce a strong correlation since loading rates are calculated from flow rates. The pollution concentrations depend on many random factors. The use of regression analysis will minimize errors in the estimate of loading rates using a more exact measure of flow rate. Flow rates can be measured with minimum error and predicted with a greater degree of accuracy in large randomly affected surface-water bodies. Once the mass loading/flow rate relationship is developed, the mass loadings can be cumulated over a period of time, usually a year. Variations in loading rates for adjacent land-use activities during storm events can be estimated in terms of kg/ha-yr, kg/animal-yr, or mass/volume water. The form of the bivariate regression equation relating mass loadings to flow rate is

$$M = a + bQ \tag{5.7}$$

where M = mass loading, kg/time or lb/time
Q = flow rate, m^3/s or ft^3/s
a, b = regression constant

Using USGS flow rate data, a correlation between total nitrogen (lb/day) and flow rate (ft^3/s) was done using 17 sampling points. The physical characteristics of the watershed did not change during the sampling period. As expected, the resulting correlation coefficient was very strong, $r = 0.996$. The line of best fit is shown in Figure 5.8.

Using Figure 5.8, estimates of mass loadings per unit area and time can be made by simulating the daily flow rates. If first flush effects are suspected, more frequent data points on flow rates and concentrations are needed to define the mass discharge during the flushing effect. For very large watersheds (> 10 km^2), concentrations do not change rapidly, and periodic sampling will produce good correlations. However, the sampling should be coordinated with changes in flow rates.

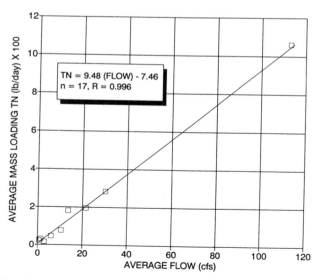

Figure 5.8 Correlation between mass loading–total nitrogen and average flow, Samsula, Florida. (From Smoot, 1977.)

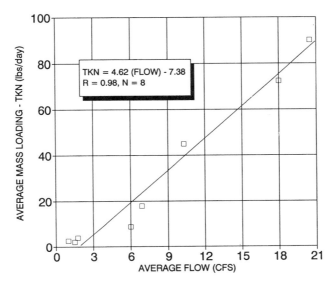

Figure 5.9 Correlation between mass loading, TKN, and average flow.

5.7.2 Other Mass Versus Flow Relationships

Linear relationships for other quality parameters, including suspended solids, total organic carbon, TKN-N, NO_3-N, OP, and TP versus flow, have been constructed. Typical results are shown in Figure 5.9 (Wanielista, 1976). These graphs are used in predicting mass loadings from nonpoint-source effects from large watershed using flow rate data.

Well-defined "pollutographs," which exhibit the first flush pollutants during long-period storms using the same general pattern as a hydrograph, can be constructed for small watersheds. On larger watersheds first flush events are not common. Concentrations do not decrease rapidly with increasing storm volume because distant areas produce high first flush concentrations that mix with flow from areas near the measuring site.

5.7.3 Calculation Steps for Loading Rates from Flow Rates

Mass loadings per unit of area and time from areas that do not have a marked first flush are determined as follows:

1. The data collected from physiochemical analyses are grouped and averaged for selected periods of time when the flow is fairly constant or changing at a constant rate following the rise and/or fall in the hydrograph.

2. The average flow in cubic meters per second or cubic feet per second and mass loading in kilograms per day or pounds per day for various pollutants are calculated.
3. A linear relationship is developed to correlate between the average flows and corresponding mass loadings for various pollutants.
4. Data on yearly average flows and area of drainage basin at the sampling locations can be obtained from the water resources data published by U.S. Geological Survey (USGS) or from specific measuring programs.
5. From straight-line relationships in item 3, mass loadings of pollutants corresponding to average yearly flow are obtained. These mass loadings should be corrected to exclude dry weather conditions.
6. The mass loadings are divided by area to obtain the load per unit area and time period.

5.8 MATHEMATICAL MODELING

Several mathematical models have been attempted to predict water quality parameters from storm events. Simplified assumptions can be made to show that contaminants deposited on street surfaces are washed off by rainfall intensity. Therefore, the amount of pollutants transported from a watershed, dP, in any time interval, dt, is proportional to the amount of pollutants remaining in the watershed, P, which is a first-order reaction:

$$\frac{-dP}{dt} = kP$$

which integrates to

$$P = P_0 e^{-kt}$$

Thus

$$P_0 - P = P_0(1 - e^{-kt}) \qquad (5.8)$$

where P_0 = initial loading, lb
  P = mass remaining after time, t, lb
  k = transport rate constant, per unit time
  t = time
  $P_0 - P$ = mass transported in time, t

The constant, k, is a function of runoff or rainfall excess. To calibrate and verify k, mass of pollutant data relative to the hydrograph must be collected. A rate of 0.5 in./h has been reported to transport 90% of pollutant load in

1 h from a watershed (U.S. Environmental Protection Agency, 1973). For impervious areas, with these data and assumptions, let

$$k = cr \qquad (5.9)$$

where

$$c = \text{constant}$$
$$r = \text{rainfall excess, in./h}$$

and if

$$P/P_0 = 0.1 \text{ for } 90\% \text{ transported in 1 h by 0.5 in./h storm}$$
$$0.1 = e^{-crt} = e^{-c(0.5 \text{ in./h}) (1 \text{ h})}$$
$$-2.3 = -0.5c \text{ (in.)}$$
$$c = 4.6 \text{ in.}^{-1}$$

Substituting into equation 5.8 results in an equation for impervious areas.

$$P_0 - P = P_0(1 - e^{-4.6rt}) \qquad (5.10)$$

Rainfall excess also can be used and is simply the product of rainfall excess rate and time (rt). Other pollutant transport rate coefficients are possible and can be related to transport data. The major limitation of the relationship above is that it can be applied only to situations in which the runoff rate is constant. Unfortunately, this does not hold true for runoff during a particular storm event. The actual phenomenon is graphically portrayed in a hydrograph, where there is a rising and receding limb at the start and finish, respectively, of a rainfall event. After observing a typical hydrograph, it becomes obvious that runoff is not constant at the start or finish of a rain event. Therefore, equation 5.10 may be simplified according to the fundamental definition of numerical differentation:

$$P_{(t+\Delta t)} = P_{(t)} e^{-4.6r \text{ avg} \Delta t} \qquad (5.11)$$

or

$$P_{(t)} - P_{(t+\Delta t)} = \Delta P = P_{(t)}(1 - e^{-4.6r \text{ avg} \Delta t})$$

Then

$$\Delta P = P_0 [4.6 r_{\text{avg}} \Delta t (\text{min})]$$

The term r_{avg} is the average surface runoff within the time interval and can be estimated from the runoff hydrograph of a particular watershed.

TABLE 5.20 Pollutant Loading Factors

Land uses
  $i = 1$ Residential
  $i = 2$ Commercial
  $i = 3$ Industrial
  $i = 4$ Other (assume that $PD_d = 0$)
Pollutants
  $j = 1$ BOD_5, total
  $j = 2$ Suspended solids (SS)
  $j = 3$ Volatile solids, total (VS)
  $j = 4$ Total PO_4 (as PO_4)
  $j = 5$ Total N
Population function
  $i = 1$ $f_i(PD_d) = 0.142 + 0.218 \cdot PD_d^{0.54}$
  $i = 2, 3$ $f_i(PD_d) = 1.0$
  $i = 4$ $f_i(PD_d) = 0.142$
α and β factors for equations
  Storm factors, α and combined factors, β, have units lb/acre-yr-in.

	Pollutant, j				
	1 BOD_5	2 SS	3 VS	4 PO_4	5 N
Storm areas, α					
Residential	0.799	16.3	9.45	0.0336	0.131
Commercial	3.20	22.2	14.0	0.0757	0.296
Industrial	1.21	29.1	14.3	0.0705	0.277
Other	0.113	2.7	2.6	0.00994	0.0605
Combined areas, β					
Residential	3.29	67.2	38.9	0.139	0.54
Commercial	13.2	91.8	57.9	0.312	1.22
Industrial	5.00	120.0	59.2	0.291	1.14
Other	0.467	11.1	10.8	0.0411	0.25

Source: Heaney et al. (1976).

Estimating the initial quantity of surface pollutants, P_0, requires the use of loading data such as that presented in Table 5.19 or estimated by the methods discussed in the following pages (Table 5.20). Initial pollution loadings are a function of many variables, of which population density, precipitation, land use, and type of sewerage system are most important. Other variables are snowmelt, dry weather flow in combined sewer systems, particle size, amounts of additives to residential lawns, directly connected impervious areas, catch basin performance, and so on. There is a definite lack of field data related to deposition and transport of sediment and chemicals as they affect the quality of urban–rural runoff (Sonnen, 1980; Heaney, 1986). However, there exists a need to determine initial loadings and

Heaney et al. (1976) presented the following:

$$P_{0w} = \begin{cases} \beta(i,j) \times P \times f_i(PD_d) & \text{for combined sewer systems} \\ \alpha(i,j) \times P \times f_i(PD_d) & \text{for storm and unsewered areas} \end{cases} \quad (5.12)$$

where P_{0w} = annual wet weather pollutant load, lb/acre-yr
P = annual precipitation, in./yr
$f_i(PD_d)$ = population density for land use, i, people/acre
$\alpha(i,j)$ = coefficient for storm and unsewered areas for pollutant j on land use i, lb/acre-yr-in. (Table 5.20)
$\beta(i,j)$ = coefficient for combined sewered areas for pollutant j on land use i, lb/acre-yr-in. (Table 5.20)

The procedure to calculate runoff pollutographs and hydrographs as a function of time is outlined:

1. Utilizing the impervious area hydrographs for that basin, select an appropriate time interval, Δt, by dividing the time base of the impervious area hydrograph into approximately 20 equal time intervals which are convenient to use, such as 10 min, 20 min, 30 min, and so on. A time interval of 10 min would be appropriate for a hydrograph with a time base of about 200 min.

2. Determine the volume of impervious area runoff within each time interval = average flow (cubic feet per second) multiplied by the time interval (expressed as seconds). The calculated volumes of runoff will be in cubic feet.

$$\text{Volume} = Q_{avg} \times (\Delta t)_{min} \times 60 \quad (5.13)$$

3. Determine the impervious area runoff rate, r_{avg}, by dividing the volume of the impervious runoff in each time interval by the product of Δt (expressed as minutes) times the amount of impervious area within the study area (expressed as acres) times the constant 60.5, which is 43,560 ft²/acre × $\frac{1}{12}$ ft/in.× $\frac{1}{60}$ h/min.

$$r_{avg}(\text{in./h}) = \frac{\text{volume (ft}^3)}{(\Delta t)_{min} \times A_{ac} \times (60.5)} \quad (5.14)$$

4. The value of P_0 for the first time interval can be determined by multipling the contaminant load (expressed as pounds per acre of impervious area per unit time) by the amount of impervious area (expressed as acres) and the antecedent dry period. The remaining

ground pollutant must be calculated at each time step considering the previous loading and the amount removed.
5. Values of ΔP for each time interval can be determined from the pollutant removal equation (equation 5.11).
6. Add each calculated value of ΔP to the summation of loading values.
7. The concentration of a pollutant for each time interval can be determined by dividing the value of ΔP by the volume of impervious area runoff and multiplying the result by the conversion factor 16,019 (mg-ft^3/lb-L).

$$\text{Concentration (mg/L)} = \frac{\Delta P(\text{lb}) \times 16,019}{\text{runoff volume total impervious area (ft}^3)} \quad (5.15)$$

It should be noted that there can be two estimates for volume, one from pervious areas and the other from impervious areas. The loadings can then be separated into two calculations and added together to obtain the final loading. Usually, the total volume from the impervious area hydrograph is greater than the volume from the pervious area.

Using the pollutant concentration procedure, the total solids concentration and the cumulative solids load at corresponding time intervals can be calculated and plotted against time. The graphic plot of concentration versus time represents the resultant pollutograph, while the graphic plot of load versus time represents the resultant loadograph as defined previously.

The development of pollutographs and loadographs is based on the assumption that runoff will remove all size particles at approximately the same rate (i.e., the rate of contaminant removal is independent of particle size). However, each pollutant may not be distributed equally over all particle sizes. Thus solid levels should not be used to represent other pollutants. Pollutographs and loadographs for each pollutant should be developed, thus greatly adding to the workload.

Example Problem 5.5 Using the hydrograph shown in Figure 5.10 for a 1.0-mi^2 impervious area, compute the loadograph and pollutograph for the watershed with the following characteristics:

Total drainage area	2.5 mi^2
Antecedent dry period	60 days
Impervious area (= 40%)	1.0 mi^2(= 640 acres)
Contaminant (solids) loading	2.0 lb/acre-day (impervious area)
BOD composition (within solids)	30,000 mg/kg

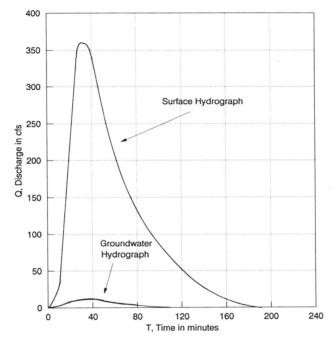

Figure 5.10 Total runoff and groundwater hydrographs.

SOLUTION: The initial solids pollutant loading, P_0, at the start of rainfall considering an antecedent dry period of about 60 days (no significant runoff to remove the solids) is computed as follows:

$$P_0 = 2.0 \text{ lb/acre-day} \times 60 \text{ days} \times 640 \text{ acres} = 76{,}800 \text{ lb (solids)}$$

The actual computations of the loadograph and pollutograph from this basin are shown in Table 5.21 and plotted in Figure 5.11. The maximum flow rate during the first 10 min is read from Figure 5.9 as 50 ft^3/s.

Other transport models for specific pollutants have been developed and are now being used. Of note is a model developed by Novotny et al. (1978) for predicting phosphorus concentration in land runoff. A more recent model was developed by the Federal Highway Administration (Woodward-Clyde Consultants, 1990). Publications associated with this model characterize stormwater pollutant loads from highways and predict water quality impacts caused by the loads. Thirty-one highway sites in 11 states were monitored to provide the data. Median values of concentration are recommended for most loading rate studies. The volume of runoff from the site is then multiplied by the median concentration to calculate the load.

TABLE 5.21 Pollutograph and Loadograph Computation for Example Problem 5.5

Δt (min)	Runoff[a] Volume (ft^3)	r_{avg} (in./h)	P (lb)	ΔP[b] (lb)	$\Sigma \Delta P$ (lb)	Concentration[c] (mg/L) Total Solids	BOD
0–10	15,000	0.039	P_0 = 76,800	2,296	2,296	2,452	73.6
10–20	70,000	0.181	74,504	10,339	12,635	2,366	71.0
20–30	160,000	0.413	64,165	20,317	32,952	2,034	61.0
30–40	187,500	0.484	43,848	16,271	49,223	1,390	41.7
40–50	157,000	0.417	27,577	8,605	57,828	875	26.3
50–60	127,000	0.328	18,972	4,771	62,599	602	18.1
60–70	87,000	0.225	14,201	2,450	65,049	451	13.5
70–80	70,000	0.181	11,751	1,631	66,680	373	11.2
80–90	57,000	0.147	10,120	1,141	67,821	321	9.6
90–100	47,000	0.121	8,979	833	68,654	284	8.5
100–110	42,000	0.108	8,146	674	69,328	257	7.7
110–120	35,000	0.090	7,472	516	69,844	236	7.1
120–130	29,000	0.075	6,956	400	70,244	221	6.6
130–140	22,500	0.058	6,556	292	70,536	208	6.2
140–150	17,500	0.045	6,264	216	70,752	198	5.9
150–160	16,000	0.041	6,048	190	70,942	190	5.7
160–170	12,000	0.031	5,858	139	70,081	186	5.6
170–180	9,000	0.023	5,719	101	71,182	180	5.4
180–190	6,500	0.017	5,618	73	71,255	180	5.4
190–200	4,000	0.010	5,545	43	71,298	172	5.2
200–210	1,500	0.004	5,502	17	71,315	182	5.5

[a] Volume = $Q_{avg} \times (\Delta t)_{min} \times 60$; r_{avg} (in./h) = $\dfrac{\text{volume (ft}^3\text{)}}{(\Delta t)_{min} \times A_{ac} \times 60.5}$

[b] $\Delta P = (4.6 r_{avg} P_0) \times \dfrac{(\Delta t)_{min}}{60}$.

[c] Conc. (mg/L) = $\dfrac{\Delta P \times 16{,}019}{\text{Runoff volume from total imp. area}}$.

Note: Use Impervious area.

5.8.1 Transported Rates Related to Rainfall Excess

The transport rate mass equation 5.8 can be related to rainfall excess. Rewriting equations 5.8 and 5.9 with rainfall excess ($R = rt$) yields

$$\frac{P_0 - P}{P_0} = 1 - e^{-cR} \tag{5.16}$$

A graph (Figure 5.12) illustrating the transport fraction as a function of rainfall excess shows that an exponential equation form can possibly fit the

5.8 MATHEMATICAL MODELING

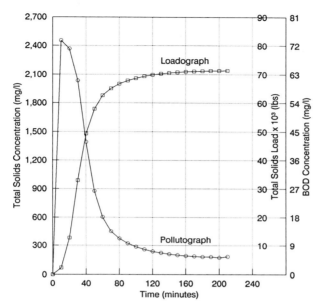

Figure 5.11 Pollutographs and loadographs.

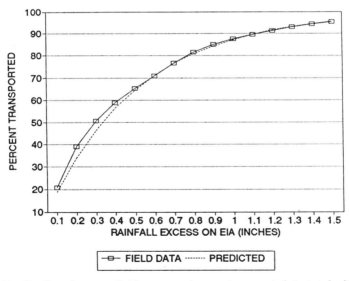

Figure 5.12 Predicted versus field-measured mass transport data total nitrogen for apartment complex.

field data. The data were collected relative to the hydrograph (Miller et al., 1978–1979) and can be related to rainfall excess.

Rewriting equation 5.16 for the fraction remaining gives

$$1 - \frac{P_0 - P}{P_0} = e^{-cR} \qquad (5.17)$$

Use of this equation to estimate the transport coefficient from pollutant mass data as a function of rainfall excess can be done using a best-fit solution procedure. Minimizing the sum of the square of the differences between the predicted and measured fraction remaining after taking the "ln" of both sides of the equation results in the following, letting Y be the fraction remaining.

$$Y = e^{-cR} \qquad (5.18)$$

$$\ln Y = -cR$$

$$F(Y) = \sum_i (\ln Y_i - cR_i)^2$$

$$\frac{\delta F(Y)}{\delta c} = -2 \sum_i (\ln Y_i - cR_i) R_i = 0 \qquad (5.19)$$

and

$$c = \frac{\sum_i \ln(Y_i R_i)}{\sum_i R_i^2} \qquad (5.20)$$

Equation 5.20 was applied to mass transport data for six different pollutant

TABLE 5.22 Transport Rate Coefficient c (per inch) for Six Different Pollutants and Four Different Locations with Land Use and Characteristics

	Land Use			
Pollutant	Residential	Highway	Commercial	Apartment
Total nitrogen	2.83	2.24	2.63	2.07
Total phosphorus	2.64	2.30	2.74	1.66
Total carbon	2.40	2.62	2.44	1.92
Chemical O_2 demand	2.56	2.72	2.82	2.32
Suspended solids	2.61	2.04	2.74	2.97
Total lead	—	2.17	2.95	2.11
Average	2.61	2.35	2.72	2.18
Total area (acres)	40.8	58.3	20.4	14.7
EIA (acres)	2.4	10.5	20.0	6.5
t_c (min)	110.	13.0	7.0	4.0

TABLE 5.23 Rainfall Excess (inches) from the Effective Impervious Area Within Which 80% of the Pollutant Mass Is Transported

Pollutant	Land Use			
	Residential	Highway	Commercial	Apartment
Total nitrogen	0.57	0.72	0.61	0.78
Total phosphorus	0.61	0.70	0.59	0.97
Total carbon	0.67	0.62	0.66	0.84
Chemical O_2 demand	0.63	0.59	0.57	0.69
Suspended solids	0.62	0.79	0.63	0.59
Total lead	—	0.75	0.59	0.76
Average	0.62	0.70	0.60	0.77

parameters from four different watersheds. The data were collected by the U.S. Geological Survey (Miller et al., 1978–1979). The results using equation 5.20 with the mass transport data are shown in Table 5.22. The rainfall excess was calculated as that from the effective impervious area and is shown in Table 5.23. From Table 5.22, the transport rate coefficient is approximately 2.5 per inch for all land uses, but the apartment complex had a lower rate. The rainfall excess over the effective impervious area within which 80% of the pollutants are transported about 0.67 in. (see Table 5.23). The value of R is calculated using equation (5.18) and substituting 0.2 for Y (1 − 0.8). Table 5.24 shows the rainfall excess at which 95% of the pollutant mass is transported. A conservative value of rainfall excess to use is about 1.25 in. over the effective impervious area (for a 95% mass transport).

TABLE 5.24 Rainfall Excess (inches) from the Effective Impervious Area Within Which 95% of the Pollutant Mass Is Transported

Pollutant	Land Use			
	Residential	Highway	Commercial	Apartment
Total nitrogen	1.06	1.23	1.14	1.44
Total phosphorus	1.13	1.30	1.09	1.81
Total carbon	1.25	1.14	1.23	1.56
Chemical O_2 demand	1.17	1.10	1.06	1.29
Suspended solids	1.15	1.46	1.17	1.09
Total lead	—	1.38	1.01	1.42
Average	1.15	1.29	1.12	1.44

Analysis of rainfall excess and mass of pollutants transported from a watershed can yield an estimate of the pollutant mass transport coefficient. This coefficient can be used to establish the rainfall excess, which contains a specified fraction of the runoff pollutants.

5.9 SUMMARY

One of the reasons for stormwater management is to reduce pollutants in runoff waters before discharge to surface and groundwater resources. A material is considered a pollutant when it or its associated effects are detrimental to plant or animal species or is considered socially unacceptable. The pollutants of interest comprise a very long list and have been categorized for ease of use.

- Water quality measures include specific parameters for (1) sediment, (2) oxygen demand, (3) bacteria, (4) nutrients, (5) metals, and (6) toxic chemicals.

- Precipitation includes some stormwater-related pollutants and those in dissolved form usually find their way to a receiving water body. Pollutant deposition and dustfall makes up a significant quantity of stormwater pollutants.

- Concentration is the mass of material per unit volume and loading is the mass of material from an event(s) over time. A loadograph is a plot of cumulative mass versus time.

- Loading rate is the mass of material per unit time and watershed area. In general, the more impervious the watershed area, the higher the loading rate.

- Urban and rural nonpoint sources can be quantified in terms of loading rates and concentration.

- For large river systems, regression analysis can be used to estimate loading rates and concentration.

- Predictions for the removal of pollutants from a watershed are attempted using simplified assumptions, such as those expressed by first-order reactions. Several mathematical models were developed for planning purposes to predict pollutant loadings from a specified watershed.

5.10 PROBLEMS

1. Using Tables 5.18 and 5.19, calculate the average lead and BOD_5 loading rates in kg/yr and tons/yr for a drainage basis with 40 ha of urban highway area and 60 ha of woodland area. Comment and discuss these results.

2. For the initial solids loading in Example Problem 5.5, assume that the suspended solids loading is equivalent to a residential area as expressed by Table 5.19 and compare your answers for initial loading to that of the example.

3. As director of a surface-water hydrologic study of an urban watershed you are asked to develop the average mass loading of organic nitrogen using Table 5.13 in kilograms transported from the basin in 1 year if the average discharge rate is 3 m^3/s (106 ft^3/s). Discuss the results.

4. What are the English equivalents of the answer to Problem 3, in lb/yr?

5. On a rural road of 2-acre size equivalent to 1 mi of curb, what is the average concentration of BOD_5 in the discharge if all stormwater is routed to one discharge point from a storm of 25.4 mm/h (1 in./h) over a 1-h period? Assume that 25% of the deposited BOD_5 particles are transported during the storm to the discharge point; there were 2 days accumulation of particles, no transpiration, 1 mm evaporated on initial impact on the hot roadway surface, 6 mm of depression storage, and 1 mm of infiltration. Assume that BOD_5 loading from the street is approximately 2.4 lb/curb mile/day.

6. Old McDonald has a farm where he raises chickens, pigs, and beef cattle and has a dairy operation. There are 50 head of beef cattle, 80 dairy cows, 250 chickens, and 20 swine on the farm. Wastes from this operation are washed downstream to a nearby lagoon for holding. The lagoon receives drainage from an area of about 10 acres. Assume that a 2-in. rainfall on this site washes 40% of a week's accumulation of animal wastes from this area. Assume that 60% of the rainfall results in runoff and that neither the rainfall nor the receiving channel itself contribute significantly to the total nitrogen loading in the channel. What is the average concentration of total nitrogen in the influent to the lagoon? Assume also that the lagoon is initially empty. Give the answer in mg/L.

7. Describe a field measuring program for determining differences between point-source and nonpoint-source effects. What are some of your assumptions?

152 STORMWATER QUALITY

8. A river draining 432 km² (106,750 acres) of land in Wisconsin is suspected of contributing nonpoint-source mass loadings to a large lake system. You have collected the following data on average flow rates and average loadings over 40 days for three water quality parameters. Water quality was measured relative to the hydrograph using discrete samples.

Period of Time	Number of Days	Average Daily Value (lb/day)			Flow (ft^3/s)
		TKN	NO_3-N	SS	
Apr. 2–Apr. 8	7	675	245	6400	250
May 1–May 6	6	810	250	6500	300
July 5–July 12	8	380	110	4100	160
Sept. 7–Sept. 16	10	210	70	2600	120
Oct. 14–Oct. 22	9	120	22	1212	60

Using linear regression analysis, determine the "best" straight-line relationship between mass loadings (lb/day) and average flow (ft^3/s). Comment on the use of five data points.

9. Using the illustrative example for computing pollutographs and loadographs (Example Problem 5.5); change the impervious area to 2.0 mi², the suspended solids loading to that shown for urban areas in Table 5.19 and assume 10 days of dry conditions before rainfall. Also assume the same BOD_5 to suspended solids concentration. Using the same procedure, calculate a loadograph for BOD_5. What assumptions were made? Discuss your answer.

10. Two rainfall events with different time and rainfall volumes are used to characterize the runoff pollution from a 24.1-acre drainage basin. The impervious portion of the watershed, which is directly connected is 16.2 acres, with a curve number of 98 (found from other work to be accurate) and a rational C value of 0.90. A clay unpaved road covered 1.6 acres of the impervious area. It was considered part of the 16.2 acres. The land use is primarily parking lots and wooded area. The pervious portion has a CN = 65 and a C value of 0.30. The rational coefficient has not been verified. The field quantity and quality data are:

Storm 1: Rainfall = 1.27 in. over a 2-h 5-min period

Time	Flow (ft³/s)	BOD₅ (mg/L)	TN (mg/L)	TP (mg/L)	SS (mg/L)
19:10	(start)				
20:30	9.5	3.1	1.30	0.01	452
20:40	5.6	1.0	1.97	0.82	1424
21:00	3.0	1.0	2.41	0.92	948

Storm 2: Rainfall = 0.25 in. over a 1-h 45-min period

Time	Flow (ft³/s)	BOD₅ (mg/L)	TN (mg/L)	TP (mg/L)	SS (mg/L)
18:30	(start)				
18:45	0.65	2.0	0.38	0.4	2
19:10	2.29	3.0	0.57	0.05	11
19:25	1.77	2.6	0.51	0.05	21.5
19:40	2.10	1.7	0.22	0.04	4
19:55	2.37	1.7	0.16	0.04	4
20:15	2.18	2.7	0.14	0.02	5
20:35	1.04	1.7	0.25	0.04	13

a. Comment on the field sampling program. How would you have done it for average annual loadings rate?

b. Phosphorus is usually related to suspended solids. Why is this not true for this situation?

c. Draw loadographs for BOD$_5$, TN, TP, and SS for the second (0.25-in.) storm.

d. Calculate loadings for BOD$_5$, TN, TP, and SS in kg/ha/yr. Compare these loadings to those presented in Table 5.19. Assume 40 in. of rainfall per year.

e. Calculate the "theoretical" loadograph for a 4-in. storm uniformly distributed over a 2-h period. Use the SCS/SBUH method to generate the hydrograph. Assume 5 days of antecedent dry conditions and a time of concentration equal to 20 min.

f. Compare the hydrograph peak of part (e) with that calculated using the rational formula. Assume that the time of concentration is 1 h. Why are the peaks different?

11. For Problem 8, estimate the average yearly TKN loading rate per acre of land if the time period during which data were collected reflects 40% of the nonpoint-source effects. The remaining loadings per year

can be determined for average flow rate values of:

20 days or 20% of the time 150 ft³/s
30 days or 30% of the time 200 ft³/s
10 days or 10% of the time 350 ft³/s

Note: The TKN load for the first period of time is 675 lb/day × 7 days = 4725 lb.

12. For the data given below, many BOD_5 data are available but only limited TOC measurements.
    a. What is the best estimate of the linear regression equation for TOC as a function of BOD_5?
    b. What is the correlation coefficient?

(Independent) BOD_5 (mg/L)	(Dependent) TOC (mg/L)
3.0	6.0
2.4	6.5
4.0	10.7
7.2	10.9
10.0	27.6
10.1	19.3
6.2	16.7

13. Estimate the annual average loading of lead from a 2-acre highway using event mean concentration data and annual rainfall data for a site of your choice. List all of your assumptions and further assume the runoff volumes 60% of the average yearly rainfall. Express your answer in loading rates of lb/yr. How does this compare to a 2-acre urban area?

14. a. Estimate the initial total phosphorus load (in grams) expected from a watershed with both a 40-acre dairy operation (40 dairy cattle), and a residential 60-acre area (average home site is 1 acre). The last storm event removed previous accumulations; however, there is a 2-day accumulation before a $\frac{1}{2}$-in. rainfall over 60 min occurs. Using the U.S. Environmental Protection Agency (1974) deposition removal estimates, an assumed 40% removal from the dairy operation, and a 0.40 runoff coefficient for the entire area, estimate the average phosphorus concentration (mg/L) in the runoff.
    b. During the first 10 min of runoff the runoff volume was $0.4(10^6)$ L. What is the concentration during this period if the area runoff rate is 0.3 in./h? Assume a first-order removal rate reaction with $c = 4.6$ in.$^{-1}$.

5.11 COMPUTER-ASSISTED PROBLEMS

1. Do Problem 8 of Section 5.10 using a linear regression model.

2. Using the linear regression program, determine a best-fit equation for the loadograph of Problem 9 of Section 5.10. Assume that the summation of mass occurs at the end of each time interval.

3. Do the hydrograph from part (e) of Problem 10 in Section 5.10 using a hydrograph generation computer program and compare the results with your calculations using the SBUH method. Also using a hydrograph generation program, calculate the hydrograph shape assuming an SCS unit hydrograph with shape factor equal to 580. The shape was determined from field-measured hydrographs.

4. Using any spreadsheet computer program (i.e., VP-Planner, Lotus, etc.) or by creating your own program, develop a pollutograph and loadograph for Example Problem 5.5 but increase the impervious area to 80%.

5. Do Problem 12 of Section 5.10 using the hydrology program disk.

6. Calculate using the program RETEN the annual pollutant load from 24 acres of woodland with an average runoff coefficient of 0.1 and an annual rainfall of 40 in.

7. It is planned to make the 24 acres of Problem 6 into a residential development with an average runoff coefficient of 0.5. What is the expected average annual runoff loadings?

8. Compare the results of Problem 7 to the loadings that result from a mixed land use of 12 acres of woodland and 12 acres of residential?

9. The 12 acres of undeveloped land of Problem 8 is developed into a mall area with loadings similar to that of highways with a runoff coefficient of 0.95. What are the annual loadings from this mixed land use?

5.12 REFERENCES

American Public Works Association. 1969. *Water Pollution Aspects of Urban Runoff*, EPA 1103DN501/69, NTIS No. PB215532, APWA, Washington, D.C.

Betson, Roger. 1978. "Bulk Precipitation and Streamflow Quality Relationships in Urban Areas," *Water Resources Research*, Vol. 14, No. 6, pp. 1165–1169.

Brezonik, P.L., et al. 1969. *Eutrophication Factors in North Central Florida Lakes*, Bulletin Series No. 134, University of Florida, Gainesville, Fla.

Ciaccio, L. L., Ed. 1971. *Water and Water Pollution Handbook*, Vol. I, Marcel Dekker, New York, p. 2.

Cole, R. H., Fredrick, R. E., Healy, R. P., and Rolan, R. G. 1984. "Preliminary Findings of the Priority Pollutant Monitoring Project of the Nationwide Urban Runoff Program." *Journal of the Water Pollution Control Federation*, Vol. 56, No. 7, pp. 898–908.

Colston, N. V. 1974. "Characterization of Urban Land Runoff," Reprint 2135, *ASCE National Meeting*, Los Angeles, Jan. 21–25.

Council on Environmental Quality. 1972. *Third Annual Report*, CEQ, Washington, D.C.

Crawford, N. H. 1973. *Pesticide Transport and Runoff Model for Agricultural Lands*, U.S. Environmental Protection Agency, U.S. Government Print Office, Washington, D.C., p. 5.

Denver Regional Council of Governments. 1983. *Urban Runoff Quality in the Denver Region*, DRCG, Denver, Color.

Driscoll, E. D., Shelley, P. E., and Strecker, E. W. 1990. *Pollutant Loadings and Impacts from Highway Stormwater Runoff*, Vol. I, *Design Procedure*. FHWA-RD-88-006, Federal Highway Administration, Washington, D.C.

East Central Florida Regional Planning Council. 1977. *Orlando Area 208*, Winter Park, Fla.

East Central Florida Regional Planning Council. 1984. *Lake Tohopekalega Drainage Area Agricultural Runoff Management Plan*, Winter Park, Fla.

Foran, J. A., 1990. "Toxic Substance in Surface Water: Protecting Human Health—The Great Lakes Experience," *Environmental Science and Technology* Vol. 24, pp. 604–608.

Geological Survey, U.S. Water Resources Division. 1991. "Water Resources Data for 'Your State,'" *Water Year 1991*, published yearly for the water year, U.S. Geological Survey, Washington, D.C.

Gizzard, T. J., et al. 1976. "Assessment of Runoff Pollution Impacts in an Urbanizing Watershed: A Case Study of Northern Virginia's Occoquan Watershed," *EPA Region III Urban Runoff Seminar*, Philadelphia, Nov. 16–17.

Gupta, M. K., Agnew, R. W., Gruber, D., and Kreutzberger, W. 1984. *Constituents of Highway Runoff*, Vol. IV, *Characteristics of Highway Runoff from Operating Highways*, FHWA/RD-81-045, Federal Highway Administration, Washington, D.C.

Harms, L. L., Dornbush, J. N., and Anderson, J. R. 1974. "Physical and Chemical Quality of Agricultural Land Runoff," *Journal of the Water Pollution Control Federation*, Vol. 46, pp. 2460.

Heaney, J. P. 1986. "Research Needs in the Urban Storm-Water Pollution," *Journal of Water Resources Planning and Management*, ASCE, Vol. 112, No. 1, pp. 33–47.

Heaney, J. P., Huber, W. C., and Nix, S. J. 1976. *Stormwater Management Model: Level Preliminary Screening Procedures*, EPA 600/2-76-275, U.S. Environmental Protection Agency, Cincinnati, Ohio.

Kluesener, J. W., and Lee G. F. 1974. "Nutrient Loading from a Separate Storm Sewer in Madison, Wisconsin," *Journal of the Water Pollution Control Federation*, Vol. 46, No. 5, p. 932.

Miller, R. A., Hardee, J., and Mattraw, H. C. 1978–1979. *USGS Open File Reports 79-982, 78-612, 79-1295*, U.S. Geological Survey, Washington, D.C.

Northern Virginia Planning District Commission. 1983. *Washington Metropolitan Area Urban Runoff Demonstration Project*, Metropolitan Washington Council of Governments, Washington, D.C.

5.12 REFERENCES

Novotny, V., Tran, H., Simsimam, G., and Chesters, G. 1978. "Mathematical Modeling of Land Runoff Contaminated by Phosphorus," *Journal of the Water Pollution Control Federation*, Jan., pp. 101–122.

Pitt, R. E., and Amy, G. 1973. *Toxic Materials Analysis of Street Surface Contaminants*, U.S. Environmental Protection Agency, U.S. Government Printing Office, Washington, D.C.

Priede-Sedgwick, Inc. 1983. *Runoff Characterization: Water Quality and Flow*, prepared for the Tampa Nationwide Urban Runoff Program–Phase II.

Rexnord, Inc. 1984. *Source and Migration of Highway Runoff Pollutants*, Vol. 3, Research Report prepared for the FHWA, FHWA/RD-84/059, NTIS PB86-227915.

Ruppert, H. 1975. "Geotechnical Investigations on Atmospheric Precipitation in a Medium-Sized City," *Water, Air and Soil Pollutants*, Vol. 4, pp. 447–763.

Rushton, B. 1991. "Water Quality and Hydrologic Characteristics in a Wet Detention Pond," *Statewide Stormwater Research Workshop*, Southwest Florida Water Management District, Brooksville, Fla., pp. 44–53.

Sartor, J. D., Boyd, G. B., and Acardy, F. J. 1974a, *Water Pollution Aspects of Street Surface Contaminants*, U.S. Environmental Protection Agency, U.S. Government Printing Office, Washington, D.C., p. 77.

Sartor, J. D., Boyd, G. B., and Acardy, F. J. 1974b. "Water Pollution Aspects of Street Surface Contaminants," *Journal of the Water Pollution Control Federation*, Vol. 46, No. 3.

Schuler, T. R. 1987. *Controlling Urban Runoff: A Practical Guide for Planning and Designing Urban BMP's*, Metropolitan Washington Council of Governments, Washington, D.C.

Shaheen, D. G. 1975. *Contributions of Urban Roadway Usage to Water Pollution*, EPA 600/2-75-004, U.S. Environmental Protection Agency, Washington, D.C.

Shannon, E., and Brezonik, P. 1972. "Relationships Between Lake Trophic State and Nitrogen and Phosphorus Loading Rates," *Environmental Science and Technology*, Vol. 6, No. 8, p. 720.

Sherwood, C. B., and Mattraw, H. C. 1975. "Quantity and Quality of Runoff from a Residential Area Near Pompano Beach, Florida," *Proceedings: Stormwater Management Workshop*, University of Central Florida, Orlando, Fla., pp. 147–157.

Smith, D. L., and Lord, B. N. 1990. "Highway Water Quality Control—Summary of 15 Years of Research," *Transportation Research Record*, No. 1279, Washington, D.C.

Smoot, J.L. 1977. *Spruce Creek Watershed Nonpoint Sources Loading Model*, Research Report, University of Central Florida, Orlando, Fla.

Sonnen, M. B. 1980. "Urban Runoff Quality: Information Needs," *Journal of the Technical Councils*, ASCE, Vol. 106, No. TC 1, pp. 29–40.

Terstriep, M. L., Noel, D. C., and Bender, G. M. 1986. "Sources of Urban Pollutants—Do We Know Enough?" *Proceedings of an Engineering Foundation Conference on Urban Runoff Quality—Impact and Quality Enhancement Technology*, New England College, Henniker, N.H., June 23–27.

U.S. Department of Agriculture, 1975. *Agricultural Waste Management Field Manual*, Chapter 4, USDA Soil Conservation Service, Washington, D.C., continuous updates.

U.S. Department of Health, Education and Welfare, 1962. *Drinking Water Standards*, U.S. Public Health Service, Washington, D.C.

U.S. Environmental Protection Agency. 1973. *Methods for Identifying and Evaluating the Nature and Extent of Nonpoint Sources of Pollutants*, EPA 430/9-73/014, U.S. EPA, Washington, D.C.

U.S. Environmental Protection Agency. 1974. U.S. Department of Biological and Agricultural Engineering, North Carolina State University at Raleigh, *Role of Animal Wastes in Agricultural Land Runoff*, U.S. Government Printing Office, Washington, D.C.

U.S. Environmental Protection Agency. 1974. *Water Quality Mangement Planning for Urban Runoff*, EPA 440/9-75-004, U.S. EPA, Washington, D.C.

U.S. Environmental Protection Agency. 1980. *Ambient Water Quality Criteria for Priority Pollutants*, EPA 440/5-80-015, PL 95-217, U.S. EPA, Office of Water Regulation and Standards, Washington, D.C.

U.S. Environmental Protection Agency. 1983. *National Urban Runoff Program*, Vol. I, NTIS PB84-185552, U.S. EPA, Washington, D.C.

Wallace, D. A., and Daque, R. R. 1973. "Modeling of Land Runoff Effects on Dissolved Oxygen," *Journal of the Water Pollution Control Federation*, Vol. 45, No. 8, p. 1796.

Wanielista, M. P. 1976. *Nonpoint Source Effects*, Report ESEI-76-1, University of Central Florida, Orlando, Fla., p. IV-8.

Wanielista, M. P., Yousef, Y. A., and Taylor, J. S. 1981. *Stormwater Management to Improve Lake Water Quality*, EPA 600/12-82-084, U.S. Environmental Protection Agency, Washington, D.C.

Wanielista, M. P., Yousef, Y. A., and Taylor, J. S. 1982. *Stormwater Management to Improve Lake Water Quality*, EPA 600/52-82-048, U.S. Environmental Protection Agency, Washington, D.C.

Water Data Unit, Hydrologic Research Station. 1982. *Surface Water: United Kingdom*, Her Majesty's Stationary Office, Wallingford, Oxon, U.K.

Weibel, S. R. 1969. "Urban Drainage as a Factor in Eutorphication," in *Eutrophication: Causes, Consequences, Correctives*, National Academy of Sciences, Washington, D.C., pp. 390, 449.

Weidner, R. B., et al. 1969. "Rural Runoff as a Factor in Stream Pollution," *Journal of the Pollution Control Federation*, Vol. 41, pp. 377–384.

Woodward-Clyde Consultants. 1990. *Pollutant Loadings and Impacts from Highway Stormwater Runoff*, Vols. I, II, III, and IV, Federal Highway Administration, McLean, Va.

Yousef, Y. A., et al. 1985. *Consequential Species of Heavy Metals*, BMR-85-286 (FL-ER-29-85), Florida Department of Transportation, Tallahassee, Fla.

Yousef, Y. A., et al. 1990. *Efficiency Optimization of Wet Detention Ponds for Urban Stormwater Management (Phase I and II)*, Final Report submitted to Florida Department of Environmental Regulation, DER Contract WM159 (A1 and A2).

Zebuth, H. H., and Wanielista, M. P. 1975. "Some Nonpoint Source Effects and Abatement Measures in Florida," *Proceedings: Stormwater Management Workshop*, University of Central Florida, Orlando, Fla., pp. 30–35.

CHAPTER 6

Receiving Water Quality

The primary intent of water quality laws and regulations is the protection and improvement of surface and groundwater quality. With knowledge of hydrologic and hydraulic principles and water quality measures, runoff volumes and pollutant mass loadings can be estimated for watershed and rainfall conditions. These runoff waters may affect the quality of the receiving waters. In this chapter the primary emphasis is on water quality impacts measured in terms of (1) biotoxicity, (2) dissolved oxygen consumption, (3) sediment accumulation, and (4) eutrophication.

6.1 WATER QUALITY ASSESSMENT

The U.S. Environmental Protection Agency (EPA) frequently publishes an inventory on current water quality trends related to oxygen depletion, harmful substances, eutrophication, health hazards, salinity, and others. From review of these documents (U.S. Environmental Protection Agency, 1984b), it appears that most state agencies report problem areas caused by discharges from point and nonpoint sources. The relative pollution contribution of point sources to nonpoint sources in the United States is shown in Table 6.1. This indicates that nutrients, oxygen-demanding material, bacteria, and some metal impacts are caused primarily by nonpoint sources. Any surface water management must at least address these problem areas.

A general assessment of nonpoint sources will be accomplished in this chapter using conventional mathematical models derived from conservation of mass equations and empirical relationships. Sources and sinks of materials affecting oxygen, sedimentation, and eutrophication are of particular interest. These could limit the beneficial use of receiving waters and cause significant recreation and economic loss. Receiving water quality assessment is usually measured in concentration and mass terms but also includes population numbers to indicate biological assessment, and the presence of chemicals to indicate the toxic or health-related effects. Also, the physical appearance (caused by floating algal and plant species, color, and turbidity) may be important. An impact occurs from a measurable (or statistically significant)

TABLE 6.1 Point and Nonpoint Contributions of Specific Pollutants[a]

Pollutant	Percent from Point Sources	Percent from Nonpoint Sources
Chemical oxygen demand (COD)	30	70
Total phosphorus	34	66
Total Kjeldahl nitrogen	10	90
Oil	30	70
Fecal coliform	10	90
Lead	43	57
Copper	59	41
Cadmium	84	16
Chromium	50	50
Zinc	30	70
Arsenic	95	5
Iron	5	95
Mercury	98	2

Source: U.S. Environmental Protection Agency, (1984b).

[a] The data presented in this table represent the average of individual states' percent contributions, based on average daily loading data for 50 states and the District of Columbia.

increase in pollutant concentration, load, or populations as compared to background values. If the receiving water quality is degraded to a point where the designated use is impaired, the impact is judged as significant and changes in the discharge concentration or mass is needed. There are very visible impacts that impair water uses, such as fish kills, taste and odor problems, and overgrowth of undesirable plant species. Other impairments are more subtle, such as changes in concentration beyond a threshold value. Threshold levels for some pollutants have been set (Table 6.2). If the threshold level of a water quality indicator is violated, evidence exists to manage the source of the violation. Also, biological indicators are becoming of increasing importance as the chemicals that pollute receiving water become more complex and difficult to measure.

The use of Table 6.2 to identify threshold violations does not imply that the violation immediately follows a runoff event. There are both short- and long-term impacts on receiving waters. Short-term impacts may also produce long-term effects. The nature of the effects and the time scale on receiving water is shown in Figure 6.1. In addition to the concentration and mass of pollutant impacts, hydraulic effects due to scouring and resuspension of materials are evident.

Bacterial contaminants are related to public health measures and impair the use of public water supplies, irrigation, fishing, and recreation. Bacterial effects from stormwaters generally occur within a few days to a week

TABLE 6.2 Reference Threshold Level Values of Water Quality Indicators for U.S. Waterways

Parameter	Reference Level	Receptor
Suspended solids (SS)	80 mg/L	Aquatic life
Turbidity	50 NTU	Aquatic life
Nitrate (as N)	0.9 mg/L	Eutrophication
Dissolved solids	500 mg/L	Water supply
Chloride	250 mg/L	Water supply
Sulfate	250 mg/L	Water supply
pH	6.0–9.0	Aquatic life
Dissolved oxygen	4.0 mg/L	Aquatic life[a]
Total coliforms	10,000/100 mL	Recreation

Source: U.S. Environmental Protection Agency (1974).
[a] In some cases, a daily average of 5 mg/L is the threshold.

following a combined sewer system overflow or a storm sewer discharge. However, groundwater inputs high in bacterial populations may persist in receiving water bodies long after a runoff event.

Oxygen-demanding materials are present in most stormwaters. Oxygen depletion is most evident where toxic materials are low in concentration, thus permitting biological activity. When toxic material (e.g., lead, copper) concentrations are high enough, a delayed oxygen consumption is present until the toxic material concentrations are reduced to a level that permits biological activity. Thus oxygen depletion can occur immediately or over a longer period of time. Resuspension of oxygen-consuming material also occurs.

Suspended solids are high in concentration but may settle out of solution very fast, and can be resuspended. High suspended solid levels are related to

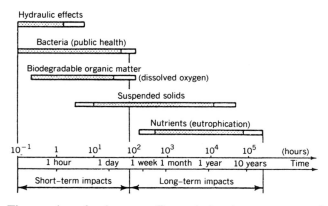

Figure 6.1 Time scales of urban runoff population impacts on receiving waters. (After Hvitved-Jacobsen, 1986.)

longer bacterial survival rates because of an abundance of nutrients and protection from sunlight (Hvitved-Jacobsen, 1986). Solids from discharge pipes can build up mounds of sediment and possibly local pollution and trash. Frequent maintenance of discharge structures (few times a year) are common in areas with erosion problems.

Nutrient mass that enters a receiving water body is both dissolved and in particulate form. The dissolved material is readily available for animal and plant life including algal populations. The particulates that settle to the bottom can be resuspended with high flows. Under anaerobic conditions, nutrients in the sediments can be released back to the water column. The phosphorus mass loadings from urban areas are generally much higher than from natural areas. But nitrogen mass loadings from urban areas approximate those discharged from natural areas. In waters where phosphorus is the limiting nutrient, phosphorus impacts may occur over very long time periods. Removal of a mass of phosphorus rather than concentration reduction is of primary importance.

The impact of toxic compounds were not listed in Figure 6.1; however, their impact is evident as both short-term (hours-days) and long-term (months) effects. The short-term impacts are measured by organism mortality. Mancini and Plummer (1986) found copper to be one of few metals whose concentrations exceed water quality criteria. Their work suggests that copper in urban runoff at almost all locations examined will exceed water quality criteria. Lead and polyvinyl chloride (PVC) concentrations are also important toxicants that have both short- and long-term effects. Their long-term chronic effects are frequently physiological and affect reductions in growth and reproduction (Jenkins and Sanders, 1986). These sublethal effects are more difficult to detect because of long-period delays and other environmental factors. However, the long-term exposure to contaminants make an organism more susceptible to disease, pH changes, temperature changes, and other stresses (Jones, 1986).

Target concentrations for toxic effects from exposure to copper, lead, and zinc have been developed by the U.S. Environmental Protection Agency for the protection of freshwater aquatic life. The acute criteria in addition to the Nationwide Urban Runoff Program (NURP) suggested threshold effect levels listed in Table 6.3. These levels as well as the concentrations for acute criteria increase with increasing water hardness in milligrams per liter (ppm) as calcium carbonate. The acute criteria are based on a continuous-exposure concept and are generally developed from bioassay experiments to determine the toxic concentration to cause 50% mortality of the most sensitive species after 96-h exposure time. Currently, there are no corresponding criteria for short-term intermittent discharges during storm events. However, the NURP studies produced estimates of approximate concentrations that would cause adverse impact from intermittent exposures produced by urban runoff water as depicted in Table 6.3. The acute criteria have substantial safety factors built in, but the suggested threshold effect level values have no safety factor

TABLE 6.3 Target Concentrations for Toxic Effects

Surface Water Total Hardness (mg/L CaCO$_3$)	EPA Acute Criteria (mg/L)			EPA NURP Suggested Threshold Effect Level (mg/L)		
	Copper	Lead	Zinc	Copper	Lead	Zinc
50	0.009	0.034	0.181	0.020	0.150	0.380
60	0.011	0.043	0.210	0.025	0.200	0.440
80	0.014	0.061	0.267	0.030	0.250	0.560
100	0.018	0.082	0.321	0.040	0.350	0.675
120	0.021	0.103	0.374	0.045	0.450	0.785
140	0.024	0.125	0.425	0.055	0.550	0.890
160	0.028	0.149	0.475	0.065	0.650	1.000
180	0.031	0.173	0.523	0.070	0.750	1.100
200	0.034	0.197	0.571	0.080	0.850	1.200
220	0.037	0.223	0.618	0.090	0.950	1.300
240	0.040	0.249	0.664	0.095	1.050	1.400
260	0.044	0.276	0.710	0.100	1.200	1.500
280	0.047	0.303	0.755	0.110	1.300	1.600
300	0.050	0.331	0.800	0.115	1.400	1.700

(Driscoll et al., 1990). It must be realized that impact resulting from intermittent nonpoint discharges into natural streams are extremely difficult to isolate, as it is to determine cause and effect. Also, existing case studies are limited to very few parameters, primarily dissolved oxygen impacts. Biological indicators are considered good tools to evaluate the overall health of streams.

6.2 SEPARATING POINT FROM NONPOINT SOURCES

On large watersheds where first flush effects are not obvious, it may be difficult to separate the nonpoint-source effects from sewage treatment plant discharges. Mass loadings are useful to differentiate between point- and nonpoint-source effects. It has been indicated previously that concentrations in mg/L for certain constituents in water from nonpoint sources may decrease during rainfall events. However, the decrease in concentration mixed with the increase in flow rate can produce mass load increases. The mass pollutant contribution during wet weather conditions is usually several times greater than that contribution during dry weather conditions.

Example results of hydrograph-related measurements for TOC are shown in Figure 6.2 for a specific stream location. The streamflow during the dry weather period was less than the flow from the two point sources in the watershed (municipal secondary biological facilities) minus evaporation. This figure reflects conditions on the stream during dry conditions (point effects)

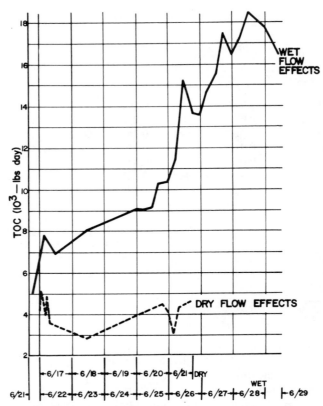

Figure 6.2 Shingle Creek, SC-3, total organic carbon loadings: runoff and infiltration conditions. (From Wanielista et al., 1977; copyright 1977 WPCF and reprinted with permission.)

and wet conditions (nonpoint plus point effects). The dry flow condition provides a baseline indicative of point-source effects and movement of subsurface water.

Two measuring sites were chosen to document point and nonpoint effects. One site was named SC3, the other SC1. SC3 received drainage water from an estimated area of 21,670 ha (83.6 mi^2) and SC1 receives drainage water from an estimated area of 46,000 ha (180 mi^2) (U.S. Geological Survey, 1974). The watershed is mainly a nonurban area, which constitutes 74%, while 17% of the area is urban and the remaining 9% of the area is surface water providing depression storage and, in effect, reducing the topographic drainage area.

Water samples were collected during the months of April through June 1975. A summary of the analysis is presented in Table 6.4. The data collected show a general decrease in pH values, specific conductance, inorganic car-

TABLE 6.4 Changes in Water Quality Characteristics Due to Nonpoint Sources on a Specific Stream

	Range at SC3		Range at SC1	
Parameter	Dry Period (Point Source)	Wet Period (Nonpoint)	Dry Period (Point Source)	Wet Period (Nonpoint)
Flow (m^3/s)	0.25–0.62	1.24–6.03	0.57–1.38	1.78–8.57
pH	7.26–7.86	7.54–6.84	7.47–7.92	7.62–6.73
SPCOND (μS/cm)	385–435	282–135	278–365	270–153
Turbulence (JTU)	0.6–1.3	0.4–2.4	1.2–12.5	0.9–8.8
SS (mg/L)	1.0–6.0	2.0–35.3	2.0–18.0	2.0–34.7
TDS (mg/L)		192–160		226–152
Alkalinity (mg/L as CaCO$_3$)	82–92	83.3–33.3	59.1–100.6	69.4–33.3
Hardness (mg/L as CaCO$_3$)	82–94	73.2–3.0	58–96	81.3–42
BOD (mg/L)	2.0–4.8	0.8–9.2	3.5	1.4–8.2
COD (mg/L)	32–104	24–68.8	28–64.8	25.6–70.4
TOC (mg/L)	9.0–12.1	10.5–19.8	7.2–19.1	11.5–25.1
IC (mg/L)	16.5–17.7	12.0–3.7	12.4–19.6	10–5.6
TKN (mg/L N)	0.63–0.97	0.01–1.98	0.1–0.88	0.06–0.88
NO$_3$ (mg/L N)	0.05–0.06	0.02–0.14	0.1–0.15	0.03–0.20
OP (mg/L P)	2.2–2.4	1.8–1.35	2.3–1.7	1.95–1.65
TP (mg/L P)	2.6–2.9	1.9–1.4	2.0–2.7	1.7–1.0

Source: Courtesy of Water Pollution Control Federation.

bon, alkalinity, hardness, and phosphorus concentration in water as the flow in the stream increases. However, as concentrations decrease, mass generally increases during a runoff event. During the same event, an increase in BOD, COD, TOC, and nitrogen concentration is observed when the stream flow increases. These trends are similar to those observed for separate storm sewer–affected sites.

Point-source effects can be separated from nonpoint-source effects if the sampling and analysis methods consider flow rates. The mass loading curves are valuable in this respect.

Another example of receiving water impacts is the depressed dissolved oxygen concentrations during rainy months relative to drier months (Yousef et al., 1976; Wanielista et al., 1981). The mass of organic materials entering the study streams increased significantly during the heavy rainfall periods. Actual field-collected dissolved oxygen data were used for the comparison (see Figure 6.5, page 179).

Contaminants found in nonpoint-source flows may decrease in concentration, but the total mass continues to increase as long as there is flow. Therefore, special attention should be given to change in mass as well as concentration during runoff events. Long-term situations may exist where an increase in mass with lower concentration may be more detrimental than lower flows and higher concentrations. Thus mass loading standards are usually initiated, in addition to concentration, to define the extent of pollutant effects in receiving waters when nonpoint-sources produce quality effects.

6.3 COMBINED SEWERS AND COMPARISON TO SEPARATE SEWERS

Separate sewers are so named because they are designed to carry separately only stormwater or only sewage. The water quality impacts from separate systems are only from nonpoint sources. Sewage impacts are similar but are not stressed in this work or covered in any detail.

Investigators have reported various degrees of impacts caused by stormwaters from separate sewer systems (U.S. Department of Transportation, 1986). Everything from no effect to extensive fish kills and nutrient enrichment have been reported. Table 6.5 is presented to illustrate some of the observed impacts. Some were caused by specific highway-related activities, which occur very infrequently, such as bridge painting. Others are due to deicing agents and vegetation control (herbicides). Still others are reported from the daily operation of highways. Some impacts have been documented and Maryland, Texas, and Florida have instituted laws for control of separate sewer systems. The laws were developed for both biological and chemical controls and related to eutrophication, dissolved oxygen, and sedimentation.

Combined sewers are constructed to transport both sewage and stormwater. During wet weather flow (stormwater runoff), stormwater mixes with sewage. When the holding and treatment capacity of the water pollution control facilities are exceeded or the overflow within the combined sewer is exceeded, wastewater will overflow into a receiving body of water. Comparison of these overflows for antecedent dry conditions less than or equal to and greater than 4 days are shown in Table 6.6.

Hvitved-Jacobsen (1986) also compared separate storm sewer and combined sewer overflow concentrations. The concentration of total nitrogen, total phosphorus, fecal coliforms, BOD, and COD was higher in the combined sewer than in the storm sewer.

The *first flush* of stormwater is defined as waters with higher concentration, but if runoff continues, the concentration will decrease as shown in Table 6.7. If the first flush can be treated or otherwise removed from direct discharge, usually a majority of the pollutants could be prevented from directly entering the surface waters. More extensive comparative data on a nationwide level are available in an Environmental Protection Agency publication (Manning et al., 1977).

TABLE 6.5 Summary of Highway Runoff Water Quality Impact Studies from Separate Sewer Systems

Author(s)/Affiliation	Location	General Comments
Smith and Kaster (1983)/ University of Wisconsin	Wisconsin, river	Change in benthos community
Shutes (1984)/ Middlesex Polytechnic	England, river	Spatial, temporal change in benthic populations
Portele (1982)/ Washington, DOT	Washington	Algal and fish bioassay with soluble fraction runoff indicate effects
Gjessing (1984a)/ Norwegian Institute of Water Research	Norway	Bioassay showed stimulated bacteria and algae growth
Winters and Gidley (1980)/ California DOT	California, streams	Range from stimulation to inhibition of growth
Dupuis et al. (1984)/ Rexnord, Inc.	Wisconsin and North Carolina	Bioassays showed varied toxic effects in benthos Chronic sublethal stress in fish was evident Higher metal concentration in cattails near drain Algae growth inhibition Higher metal levels near highway
Wang et al. (1981)/ Washington DOT	Washington, marsh	Marsh sediments background and surface levels high in chemicals
Wanielista (1980)/ Florida DOT	Florida, lake	Higher levels of hydrocarbons and metals Recommends *Tubifex* sp. as highway runoff indicator Higher sediment concentrations near highway Contaminants immediately settled to sediments

TABLE 6.5 *(Continued)*

Author(s)/Affiliation	Location	General Comments
Hvitved-Jacobson et al. (1984)/University of Aalburg, Denmark	Florida, pond	Nutrient input of P goes to sediment, N removed
Gjessing (1984b)/Norwegian Institute of Water Research	Norway, lake	Effects by metals and salt observed
Irwin and Losey (1979)/USGS	Florida, stream	Contribution not significant
Lenat and Moody (1983)/North Carolina DOT	North Carolina, lake	Higher Al and Zn in benthos lake after sandblasting
Nakao et al. (1977)/California DOT	California, river	High chemical loadings
Hunt (1984)/California DOT	California	Bioassay LC_{50} range from 100 to 4000 mg/L (fish)
		Bioassay LC_{50} was about 12,000 mg/L (zooplankton)
Parks and Winters (1982)/California DOT	California, river near bridge	Higher percentage of sediment
		Water quality apparently not affected
Patenaude (1979)/Wisconsin Division of Highways and Transportation	Wisconsin, pond, ditch	Variation in salt residue
Hutchinson (1981)	Maine, pond, streams	Water near highways by salt affected
Hoffman et al. (1981)/California DOT	California, lakes	Temporary stratification was established
Frost et al. (1981)/USGS	Massachusetts	No effects
Peters and Turk (1981)/USGS	New York, river	Higher sodium and chloride levels
Kramme et al. (1985)/URS Dalton Inc.	Ohio	Bioassay of runoff showed inhibition of hypocotyl growth and 2,4-D in runoff
Morre (1976)/Indiana State Highway Commission	Indiana	Bioassays conducted with minor effects
Kramme et al. (1985)/URS Dalton Inc.	Ohio	Low mortality of *Daphnia magna*

Source: U.S. Department of Transportation (1986).

TABLE 6.6 Comparison of Combined Sewer Overflow Quality for Two Antecedent Dry Conditions and Secondary Treated Effluent with Highway Runoff

Water Quality Parameter:	COD (mg/L)	BOD (mg/L)	SS (mg/L)	VSS (mg/L)	Total P (mg/L)	Total N (mg/L)
	Combined Sewer—Antecedent Dry Period Shorter Than 4 Days					
Mean	150	40	150	80	10	40
	Combined Sewer—Antecedent Dry Interval Longer Than 4 Days					
Mean	400	132	390	228	15	50
	Secondary Effluent					
Range	40–80	20–30	30–60	20–40	2–3	30–40
	Highway Runoff					
Mean	147	24	261	77	0.79[a]	4.13

Sources: Bastian (1986), Rex Chainbelt, Inc. (1972), Hvitved-Jacobsen (1986), and Gupta et al. (1981).
[a]As TPO_4.

TABLE 6.7 Comparison of Quality Characteristics from First Flush and Extended, Combined-Overflow Data

Analysis	Concentration During First Flush[a]	Concentration of Extended Overflow[b]
COD	581 ± 92	161 ± 19
BOD	186 ± 40	49 ± 10
Total solids	861 ± 117	378 ± 46
Total volatile solids	489 ± 83	185 ± 23
Suspended solids	522 ± 150	166 ± 26
Volatile suspended solids	308 ± 83	90 ± 14
Total nitrogen	17.6 ± 3.1	5.5 ± 0.8
Orthophosphate	2.7 ± 1.0	—
PH	7.7 ± 0.1	7.2 ± 0.1
Coliform density per mL	$142 \pm 108 \times 10^3$	$62.5 \pm 27 \times 10^3$

Source: Rex Chainbelt, Inc. (1972).
[a]Data represent 12 overflows at 95% confidence level range.
[b]Data represent 44 overflows at 95% confidence level range.

6.4 TOXICITY

A toxic compound or element is one that has the potential to be a poison to a living organism. Some toxicants in stormwater include metals (lead, chromium, copper, zinc, nickel, and cadmium), petroleum hydrocarbons (gasoline, oil, herbicides, pesticides), and deicing chemicals. The relative human impact and metal bioaccumulation tendency of edible portions of fish is shown in Table 6.8. Note that a relative health hazard to humans is possible, but threshold levels are not specified.

RECEIVING WATER QUALITY

TABLE 6.8 Relative Hazard to Humans Presented by the Usual Occurrence of Metals in the Edible Portions of Fish and Shellfish[a]

Metal	Toxicity to Humans from Oral Ingestion		Freshwater Fish Muscle		Marine Fish Muscle		Marine Shellfish or Crustaceans		Human Hazard Rating		
	Low	High	Low	High	Low	High	Low	High	Low	Med.	High
Aluminum	×				×	×	b	b	×		
Arsenic		×	×	×		×		×			×
Cadmium		×	×	×	×			×			×
Chromium	×		×		×		×		×		
Copper	×		×		×			×		×	
Iron	×			×		×		×		×	
Lead		×	×	×	×			×			×
Nickel	×		×		×		×		×		
Zinc	×		×		×			×		×	

Source: Phillips and Russo (1978).
[a] × Entry indicates the bioaccumulative tendency.
[b] Insufficient information available.

The threshold level is an indication of the toxic effects and is usually based on laboratory procedures using the dissolved form of the toxicant. The tests are identified as bioassays. The tests must use the actual stormwater and receiving water to reflect accurately the possible synergistic or antagonistic effects of all constituents. Synergistic effects result when after mixing substances the overall effect is greater than the sum of the individual effects taken separately. Antagonistic effects reduce the expected additive toxicity from various elements, but synergistic effects result in an excess of additive toxicity.

6.4.1 Bioassay

A bioassay is used to identify toxicant concentrations that are lethal to organisms during a short- or long-term period. Acute mortality results if a percent of the population dies (usually 50%) in a short period of an organisms' life cycle. For fish, the time period is generally 3 to 7 days. Chronic mortality tests are performed to expose organisms to lower concentrations over their entire life cycle. Over this time period, bioaccumulation of toxicants are noted as they affect reproduction and mortality rates.

Considerable effort has been expended toward laboratory determination of acute and chronic toxicity of a wide range of individual pollutants, including most stormwater-generated toxicants, especially metals. It is important that toxic threshold concentrations be developed and established using aquatic organisms associated with receiving waters. Such criteria are derived from scientific facts obtained from laboratory experiments or in situ observations that depict organisms' responses to a defined pollutant under controlled environmental conditions for a specified period. From such toxic criteria, limits on concentrations may be imposed which are considered safe for intended usage of a water body. Of course, toxicity is affected by many water quality parameters in natural or man-made streams which can singularly or collectively reduce or enhance the toxic levels of contaminants. For example, parameters affecting heavy metal toxicity include: pH, temperature, dissolved oxygen, hardness, alkalinity, and organic complexing agents. Moreover, antagonistic and synergistic effects resulting from combinations of more than one metal in water have been demonstrated.

Numerous methods have been used to assess toxicity in the aquatic environment, but all the methods assess either sublethal or lethal effects of the toxicant. Short-term lethal effects or acute toxicity can be expressed as the lethal concentration for a stated percentage death of organisms tested which are exposed to the toxicant for a specified period of time, or the reciprocal, which is the tolerance limit of surviving organisms. Acute toxicity for aquatic organisms generally has been expressed for 24 to 96 h of exposure time to a toxicant. The lethal concentration (LC) is generally associated with a numeral indicating the percentage of test organisms killed at a given toxicant concentration. However, results of aquatic bioassay are often expressed as the median lethal concentration for a 96-h exposure, which are referred to as 96-h LC_{50}.

Sublethal effects over a long-term exposure may cause physical and biochemical changes on test organisms such as changes in growth rate and swimming speed of fish, or impaired respiration and reduced activity. Reproduction seems to be one of the most sensitive of sublethal or chronic responses. The U.S. Environmental Protection Agency has accumulated bioassay data and issued water quality criteria documents published by EPA in 1976, 1980, and 1984. These criteria for several heavy metals are summarized in Table 6.9. The data in the table show the relative toxicity of seven different metals and indicate that the tolerance limit for metals is increased as the water hardness increases.

Example Problem 6.1 Mosquitofish (*Gambusia affinis*), typical of the local water environment, are collected from highway stormwater ditches, transported to a laboratory, and acclimated. The fish are to be tested for their response to metal toxicity. Dilution water is obtained from a detention pond receiving highway runoff and dosed with copper to a known concentration. Bioassays are conducted in twelve 16-L all-glass aquaria that provided duplicates of five different concentrations and the control. Known number of fish (50) are added randomly to each of the duplicate test concentration and control tanks along with dilution water 24 h prior to the addition of the metal toxicant. Water in the aquaria is aerated constantly and the number of fish that died for each metal concentration are observed and recorded at time

TABLE 6.9 EPA Water Quality Criteria for Metals for Protection of Freshwater Aquatic Life

Parameter	Metal Concentration (μg/L)						
	Hg	Pb	Zn	Cu	Cr[a]	Cd	Ni
Acute criteria at water hardness:							
50 mg/L as CaCO$_3$	1.1	25	18	8.4	870	2	1100
100 mg/L as CaCO$_3$	1.1	64	32	16	1500	4.5	1800
200 mg/L as CaCO$_3$	1.1	160	57	29	2700	10	3100
Issuance date	1984	1984	1980	1984	1984	1984	1980
Chronic criteria (30-day exposure) at water hardness:							
50 mg/L as CaCO$_3$	0.2	1	47	5.8	42	2	56[b]
100 mg/L as CaCO$_3$	0.2	25	47	11	74	4.5	96
200 mg/L as CaCO$_3$	0.2	64	47	20	130	10	160
Issuance date	1984	1984	1980	1984	1984	1984	1980

[a] Trivalent chromium.
[b] 24-h averaging exposure period.

intervals of 12, 24, 30, 48, 60, 72, 84, and 96 h of exposure as shown. Dead fish are removed daily. Determine the 96-h LC$_{50}$ value.

Copper Concentration (mg/L)	Average Percent Survival at Exposure Time (h)							
	12	24	36	48	60	72	84	96
0.1	100	100	100	95	90	80	70	65
0.3	100	100	95	85	72	60	50	45
0.5	100	95	80	55	40	30	20	15
0.7	100	80	55	30	15	10	5	0

SOLUTION: LC$_{50}$ values are determined graphically by a method in which percent of fish surviving are plotted on a probability scale versus exposure plotted on a logarithmic scale, for various concentrations of metal tested (Figure 6.3). Straight lines are drawn through points for each of the concentrations used in a particular bioassay. The result is several lines representing metal concentrations used. The 96-h LC$_{50}$ values are determined by fitting an interpolated line made through the intersection of 50% survival and 96-h coordinate.

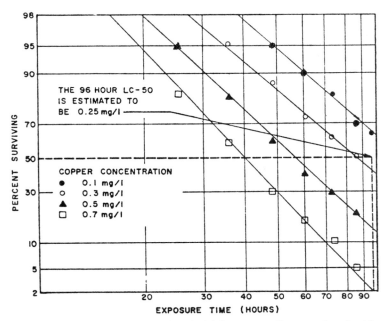

Figure 6.3 Survival of *Gambusia affinis* in highway runoff water dosed with copper.

6.4.2 Algal Assays

Algal assays are laboratory bottle tests with either specific species or in situ algae to determine growth rates, toxic effects, limiting nutrients, or inhibitory effects by measuring growth responses. A typical presentation of data from an algal assay is a plot of mass changes versus incubation time (Figure 6.4). Figure 6.4 illustrates a stimulation of growth when stormwaters high in nutrients are added to lake water. When the runoff waters are coagulated with alum, growth rates are reduced but are still evident for this experiment.

A simplified growth rate is defined by an exponential process, or

$$\frac{dN}{dt} = +KN \qquad (6.1)$$

where N = mass of algae, mg/L
t = time, usually hours or days
K = growth rate, h^{-1} or day^{-1}

Integration produces an estimate of the growth rate, or

$$N_t = N_0 e^{Kt} \quad \text{or} \quad K = \frac{\ln(N_t/N_0)}{t} \qquad (6.2)$$

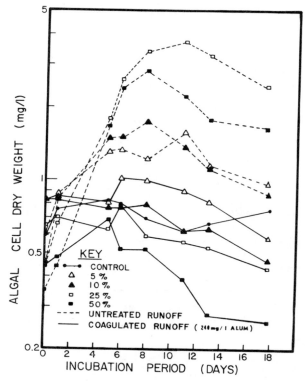

Figure 6.4 Responses of indigenous algal species in Lake Eola to various concentrations of stormwater runoff and coagulated runoff. (After Harper, 1979.)

where t = elapsed time
N_t, N_0 = time and beginning algal mass

Growth rates can be compared by plotting the mass as a function of time and then calculating the line of best fit between concentration and time, which would then be an estimate of the growth rate. If the growth rate is negative or concentrations decrease after the addition of stormwater, this is an indication of toxic or other inhibitory activity.

6.4.3 Toxicity of Heavy Metals in Stormwater

Stormwater runoff may contain significantly higher concentrations of trace metals, particularly lead (Pb), zinc (Zn), iron (Fe), cadmium (Cd), chromium (Cr), nickel (Ni), and copper (Cu) than the adjacent water environment (Yousef et al., 1984). As these metals reach the ecosystem, they will undergo physical, chemical, and biological transformations. They may be adsorbed on clay particles, taken up by plant and animal life, or remain in solution. The particulate metals will settle to the bottom sediments and heavy metals may resuspend or redissolve back into solution if environmental conditions, such as changes in DO concentration, pH values, temperature, and water chemical composition, permit. Fate and transformations of trace elements in natural environments follow complex processes and much information is needed before one can predict their impact.

The total concentration of a particular metal in natural waters can be very misleading. A water with high total metal concentration may be, in fact, less toxic than another water with a lower concentration of different species of that metal. For example, ionic copper is far more toxic toward aquatic organisms than is organically bound copper. Also, biotoxicity of copper species decreases as their stability increases. Therefore, it is of prime importance to fully understand different dissolved metal species in an aqueous environment. It is equally important that toxic threshold concentration for metals be developed and established from experimental or in situ observations that depict responses of selected organisms to a defined metal under controlled environmental conditions for a specified exposure time.

Toxicity is affected by complex interactions associated with many parameters in natural surface waters which can singularly or collectively reduce or enhance the toxic action of metals. These parameters include complexing agents, hardness, alkalinity, pH, temperature, and dissolved oxygen. Also, combinations of heavy metals in water can cause antagonistic or synergistic effects. The precise definition of such effects on aquatic life has not been fully developed.

Heavy metals may accumulate in the sediments, phytoplankton, benthic organisms, and fish. Toxicity of metals may reduce diversity and abundance of the sensitive aquatic organisms and replace them with pollution-tolerant organisms. A study of highway runoff entering Lake Ivanhoe in Orlando, Florida, through a series of scupper drains running along the entire length of the bridge crossing over the lake had been completed by Yousef et al. (1984). It was concluded that metals in sediment core samples collected from areas beneath scupper drains were at least twice as high as those under a bridge without scupper drains. Similarly, phytoplankton (*Spirogyra* and *Hydrilla*) exposed to highway runoff contained twice as much Zn, Pb, Ni, Cd, Cr, and Fe concentrations. Benthic organisms such as the crustaceans and *Tubifex* worms exhibited high metal concentrations.

Other algal assays indicate severe inhibition of mixed algal populations when exposed to low percentages of runoff (1, 5, and 10%) from a high-traffic urban freeway (185,000 ADT) in California after 2 weeks of antecedent dry period (Winters and Gidley, 1980). However, runoff from lower ADT rural highways, or from urban freeways with short antecedent dry days, is generally stimulatory to the algal populations. Also, the amphipod *Gammarus* did exhibit acute toxicity to undiluted runoff from urban highways (120,000 ADT) and rural highways (7400 ADT) in Milwaukee, Wisconsin (Rexnord, Inc., 1984). Of course, the response of an ecosystem to increasing toxic stress shows no effect at low toxic concentrations with the gradual elimination of susceptible species and survival of the most resistant biological forms as the toxic stress increases. Eventually, the entire community may be eliminated or altered at fairly high toxic stress. It is interesting to notice that runoff water coagulated with alum has stimulatory effects as shown in Figure 6.4.

6.4.4 Deicing

Deicing agents are salts of sodium and chloride but may also contain nickel, chromium, and cyanide (Kobriger and Geinopolos, 1984). Salt compounds applied to roadways in the United States in 1976–1977 measured over 10 million tons/year (9 million metric tons per year) (Kobriger and Geinopolos, 1984). When removed from the street, metal and salt concentrations in the receiving waters may be toxic or at least out of balance for particular organisms and vegetation. Chronic effects in vegetation are the loss of foliage and reduced growth. Salt concentrations exceeding 1 g in 100 g of water can affect reproduction and the health of all living species (Adams, 1972). A specific sodium salt used in Wisconsin produced cyanide in adjacent drainage ditches 30 min after application. The salt contained sodium ferrocyanide as an anticaking additive. The loading of cyanide and other metals from a highway surface with rock salt is toxic to plant and animal species.

6.5 DISSOLVED OXYGEN IMPACTS

The potential dissolved oxygen depletion in receiving streams and lakes resulting from discharges of stormwater and combined sewer overflows have been investigated by many researchers. They detected oxygen sags related to urban runoff discharges, particularly from combined sewer overflow systems. The average stormwater 5-day biochemical oxygen demand (BOD_5) may equal or exceed the concentrations measured in wastewater treatment plant effluents, but the receiving water impacts are not comparable. Stormwater BOD loadings are acute, intermittent, and associated mainly with particulate matter. These organic deposits may accumulate near the inlet to receiving water bodies and may result in delayed effects caused by increased bottom sediment oxygen demand (SOD).

It is extremely difficult, if not impossible, to determine the impacts of long-term accumulation of organic matter form stormwater runoff into lakes and streams and to separate those impacts from other sources of pollution. The rate and extent of bacterial degradation of stormwater organic material is considerably different from the kinetics for other point sources, such as secondary effluent discharges. The nature of stormwater discharge as well as the kinetics involved tend to complicate the modeling process. However, simplified assumptions can be made to predict dissolved oxygen concentrations after selected storm events. Many investigators over the years have attempted to develop computer programs for modeling of stream water quality to evaluate waste load allocations, discharge permit determinations, and other pollutant evaluations in the United States. The U.S. Environmental Protection Agency has funded research program to develop the stream water quality model QUAL 2E and various modifications (U.S. Environmental Protection Agency, 1987). These models consider eight areas: (1) algal, nitrogen, phosphorus, and dissolved oxygen interactions; (2) algal growth rate; (3) temperature; (4) dissolved oxygen; (5) arbitrary nonconservative constituents; (6) hydraulics; (7) downstream boundary concentrations; and (8) input/output modifications. These models can be used to study the magnitude and instream quality characteristics of nonpoint waste loads. For example, the user can model diurnal variations in meteorological data and their impact on instream water quality, particularly dissolved oxygen.

Dissolved oxygen (DO) is used very frequently to indicate the "general health" and the impacts of biodegradable organic discharges. Decline in dissolved oxygen concentration at a particular location generally reflects a discharge of pollution load of oxygen-consuming matter. Predictions of changes in DO concentrations, however, may be more difficult if the sources, sinks, and environmental changes are not fully assessed. The dissolved oxygen concentration in surface water is affected by temperature, salinity, and altitude. Table 6.10 presents the solubility of dissolved oxygen in water under various temperatures ranging from 0 to 30°C and chloride concentrations from 0 to 20,000 mg/L. The values indicated in the table are saturation levels of dissolved oxygen in surface water which decrease by increasing temperature and increasing chloride concentration. Altitude influence can also be approximated by a 7% decrease in DO saturation per 2000-ft increase in elevation. An example of dissolved oxygen changes during a year in a stream as shown in Figure 6.5. Decrease in oxygen during the summer months is caused by increased temperature and increased nonpoint-source loadings.

Warwick and Edgmon (1988) simulated dissolved oxygen over a 3-year period using a combined sewer overflow in Altoona, Pennsylvania. They used six storm events to calibrate the model and then simulated 208 storm events, of which 42% of those caused instream DO levels to fall below 4.0 mg/L.

6.5.1 DO Mass Balance

Dissolved oxygen concentration in water bodies is considered critical in determining their suitability for the support of specific water use classification, such as recreation, fishing, and others. Therefore, mathematical models have been used to simulate ecological and water quality interactions and to predict DO levels in the water body. Water bodies have the capacity to assimilate some pollutional load due to natural self-purification processes. DO models require good understanding of transport processes into flowing waters, physiocochemical and biological interactions in the aquatic environ-

TABLE 6.10 Solubility of Oxygen in Water

Temp. (°C)	Chloride Concentration in Water (mg/L)					Difference per 100 mg Chloride
	0	5000	10,000	15,000	20,000	
	Dissolved Oxygen (mg/L)					
0	14.6	13.8	13.0	12.1	11.3	0.017
1	14.2	13.4	12.6	11.8	11.0	0.016
2	13.8	13.1	12.3	11.5	10.8	0.015
3	13.5	12.7	12.0	11.2	10.5	0.015
4	13.1	12.4	11.7	11.0	10.3	0.014
5	12.8	12.1	11.4	10.7	10.0	0.014
6	12.5	11.8	11.1	10.5	9.8	0.014
7	12.2	11.5	10.9	10.2	9.6	0.013
8	11.9	11.2	10.6	10.0	9.4	0.013
9	11.6	11.0	10.4	9.8	9.2	0.012
10	11.3	10.7	10.1	9.6	9.0	0.012
11	11.1	10.5	9.9	9.4	8.8	0.011
12	10.8	10.3	9.7	9.2	8.6	0.011
13	10.6	10.1	9.5	9.0	8.5	0.011
14	10.4	9.9	9.3	8.8	8.3	0.010
15	10.2	9.7	9.1	8.6	8.1	0.010
16	10.0	9.5	9.0	8.5	8.0	0.010
17	9.7	9.3	8.8	8.3	7.8	0.010
18	9.5	9.1	8.6	8.2	7.7	0.009
19	9.4	8.9	8.5	8.0	7.6	0.009
20	9.2	8.7	8.3	7.9	7.4	0.009
21	9.0	8.6	8.1	7.7	7.3	0.009
22	8.8	8.4	8.0	7.6	7.1	0.008
23	8.7	8.3	7.9	7.4	7.0	0.008
24	8.5	8.1	7.7	7.3	6.9	0.008
25	8.4	8.0	7.6	7.2	6.7	0.008
26	8.2	7.8	7.4	7.0	6.6	0.008
27	8.1	7.7	7.3	6.9	6.5	0.008
28	7.9	7.5	7.1	6.8	6.4	0.008
29	7.8	7.4	7.0	6.6	6.3	0.008
30	7.6	7.3	6.9	6.5	6.1	0.008

Source: Adapted from *Standard Methods*, latest edition.

ment, and movement of material across air–water and sediment–water interfaces. Models can give good results only if the physical mechanisms in the stream are reflected precisely and the rate constants and coefficients are accurately estimated from sufficient data. After all the parameters have been identified and assessed, the DO model can be constructed from a material balance about an elemental volume in the stream. The net rate of dissolved

6.5 DISSOLVED OXYGEN IMPACTS

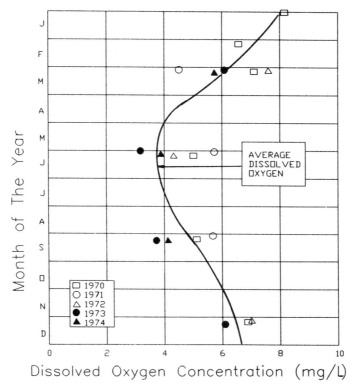

Figure 6.5 Changes in dissolved oxygen concentration at station WN 14, Bowless Creek, Florida.

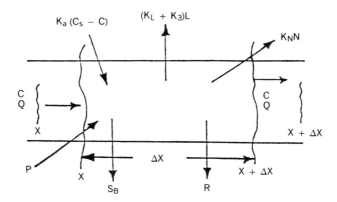

Figure 6.6 DO mass balance.

oxygen mass in and out of the volume plus the time rate of change of mass within the volume must equal that produced by the sources and reduced by the sinks as shown in Figure 6.6.

Figure 6.6 is a schematic of a stream section of volume V with all sources and sinks affecting dissolved oxygen. The nomenclature used in Figure 6.6 is defined as follows:

Q = advective flow rate, L^3/day

$x, \Delta x$ = distance, m

C_s = saturated dissolved oxygen, mg/L

C = average dissolved oxygen within the elemental volume ΔV, mg/L

K_a = reaeration coefficient, day^{-1}

K_L = carbonaceous BOD decay coefficient, day^{-1}

L = ultimate CBOD remaining, mg/L

K_N = nitrogenous BOD decay coefficient, day^{-1}

N = NBOD remaining, mg/L

K_3 = BOD sedimentation coefficient, day^{-1}

P = photosynthetic oxygen production rate, mg/L/day

R = algal respiration oxygen consumption, mg/L/day

S_B = benthic (bottom sediment organisms) oxygen demand, mg/L/day

The mass balance may now be written:

$$\text{input} - \text{output} \pm \text{production} = \text{change}$$

where

$$\text{Input} = QC \bigg\}_x \Delta t$$

$$\text{Output} = -QC \bigg\}_{x+\Delta x} \Delta t$$

$$\begin{aligned}\text{Production} = &+K_a(C_s - C)V\Delta t \\ &-(K_L + K_3)LV\Delta t \\ &-K_N N V \Delta t \\ &-S_B V \Delta t \\ &-RV\Delta t \\ &+PV\Delta t\end{aligned}$$

$$\text{Change} = V\Delta C \tag{6.3}$$

where V is the volume in liters. The mass balance expresses the dynamics of oxygen change in a freshwater flow situation due to the addition of oxygen by reaeration and photosynthesis and the depletion of oxygen from carbonaceous oxidation, nitrogenous oxidation, benthic demands, and algal respiration.

Dividing equation 6.3 by volume and time, allowing Δx to approach zero and substituting oxygen deficit (D) for $C_S - C$, the following equation is obtained for steady-state conditions.

$$D = D_0 \exp\left(\frac{-K_a X}{U}\right) + \frac{L_0 K_L}{K_a - K_L}\left[\exp\left(\frac{-K_L X}{U}\right) - \exp\left(\frac{-K_a X}{U}\right)\right]$$

$$+ \frac{N_0 K_N}{K_a - K_N}\left[\exp\left(\frac{-K_N X}{U}\right) - \exp\left(\frac{-K_a X}{U}\right)\right]$$

$$+ \frac{R + S_B - P}{K_a}\left[1 - \exp\left(\frac{-K_a X}{U}\right)\right] \quad (6.4)$$

where U = advective velocity, m/day
X = distance from discharge
D = DO deficit, mg/L
D_0 = initial DO deficit, mg/L at the point of discharge, $X = 0$
L_0 = initial ultimate CBOD, mg/L at $X = 0$
N_0 = initial NBOD, mg/L at $X = 0$

The initial ultimate CBOD concentration is estimated from

$$L_0 = \frac{W + L_u Q_u}{Q_W + Q_U} \quad (6.5)$$

where W = CBOD discharge, g/day or kg/day
L_U = concentration upstream of discharge, g/L or kg/m^3
Q_U = advective flow upstream, L/day
Q_W = stormwater discharge, L/day or m^3/day

The nitrogenous BOD concentrations can be estimated by

$$\text{NBOD} = N_0 = 4.57(\text{TON}) \quad (6.6)$$

where TON is the total oxidizable nitrogen, mg/L, which is the sum of the organic plus ammonia nitrogen.

The ultimate CBOD can also be estimated from the 5-day CBOD and the decay rate constant, which is determined from field or laboratory experimentation. If $K = 0.1$ day^{-1} (base 10), then

$$L_0 = \frac{CBOD_5}{0.68} \tag{6.7}$$

Since the decay coefficients K_L and K_N are temperature dependent, decay coefficients for other than standard 20°C conditions can be estimated using

$$K_T = K_{20} \times 1.047^{T-20} \tag{6.8}$$

where K_{20} = decay coefficient at 20°C/day
K_T = decay coefficient at T°C/day

Methods are available to estimate K_L and K_N (Eckenfelder and O'Connor, 1961). There are several equations available in the literature to calculate K_a for various streams and receiving water bodies; however, two equations are presented here, one by O'Conner for stream depths of 1 to 30 ft and the other by Owens for shallow streams (0.4 to 2.4 ft). Using English units, the formulas are, respectively,

$$\text{at 20°C, } K_a = \frac{12.9U^{0.5}}{H^{1.5}} \quad \text{for } 1 \leq D \leq 30 \text{ ft} \tag{6.9}$$

and

$$\text{at 20°C, } K_a = \frac{21.6U^{0.67}}{H^{1.85}} \quad \text{for } 0.4 \leq D \leq 2.4 \text{ ft} \tag{6.10}$$

where K_a = reaeration coefficient, day^{-1}, base e
U = stream velocity, ft/s
H = stream depth, ft

The temperature effects on the relationship is

$$(K_a)_T = (K_a)_{20} \times 1.024^{T-20} \tag{6.11}$$

Covar (1976) plotted the data used by three groups of researchers—O'Conner and Dobbins (1958), Churchhill et al. (1962), and Owens et al. (1964)—for the reaeration expression (K_a) they developed. These expressions were based on stream velocity and depth. Analysis of these data

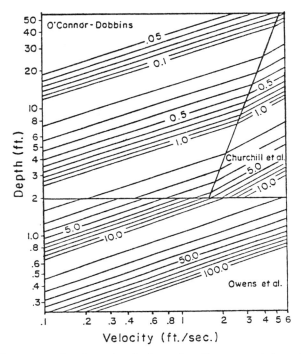

Figure 6.7 K_a versus depth and velocity using the suggested method of Covar (1976). (From U.S. Environmental Protection Agency, 1978.)

assisted him in the development of families of curves (Figure 6.7) to predict stream aeration as a function of velocity and depth for streams with velocities up to 6 ft/s and depths up to 50 ft (U.S. Environmental Protection Agency, 1978).

Example Problem 6.2 A stream in Pennsylvania has been reported to be oxygen deficient (below 4 mg/L) when a paper plant and stormwater discharge occur at the same time. Field data on average flow rate, velocity, and decay rates are available. Sedimentation, photosynthesis, and other discharges are considered negligible. Data for estimating the CBOD decay rates are (1) average velocity of 4 km/day (2.5 mi/day or 0.15 ft/s); and (2) four CBOD readings with distance as depicted in Table 6.11. From a semilog plot, the K_L, base 10, is determined as 0.062 day^{-1}, or K_L, base e, as 0.15 day^{-1}. The calculation of K_L assumes a first-order decay reaction rate. By similar methods, K_N is calculated as 0.12 day^{-1}. The stream depth was 6 m (19.7 ft) in the section of interest. Thus the K_a at 20°C was calculated using equation 6.9 as 0.06 day^{-1}. Similar results are obtained from Figure 6.7. If the ultimate

TABLE 6.11 Measured CBOD Concentration

CBOD (mg/L)	X Distance
9	Point of discharge at zero distance
6.5	8.2 km downstream
5.2	12.1 km downstream
2.5	28.9 km downstream

concentrations after dilution of CBOD and NBOD are 9 and 2 mg/L, respectively, what is the deficit concentration at 20°C 8 km (5 mi) downstream? The initial deficit is 2 mg/L.

Using the first three expressions of equation 6.4, the calculation for downstream deficit at the 8-km marker is

$$D = 2\exp\left(\frac{-0.06 \times 8}{4}\right) + \frac{9(0.15)}{0.06 - 0.15}\left[\exp\left(\frac{-0.15 \times 8}{4}\right) - \exp\left(\frac{-0.06 \times 8}{4}\right)\right]$$

$$+ \frac{2(0.12)}{0.06 - 0.12}\left[\exp\left(\frac{-0.12 \times 8}{4}\right) - \exp\left(\frac{-0.06 \times 8}{4}\right)\right] = 4.4 \text{ ppm}$$

At 20°C, the saturation dissolved oxygen content is 9.2 ppm; therefore, the instream concentration at 8 km is 9.2 − 4.4 or 4.8 ppm. This calculation illustrates procedures for calculating DO deficit at a point along a stream. Critical conditions must now be calculated.

6.5.2 Critical DO Conditions

A typical dissolved oxygen profile is shown in Figure 6.8. Frequently, the investigator is interested in the minimum DO (critical) and the time (distance) for the occurrence of the minimum DO. Neglecting nitrogen demand, the critical equations are

$$t_c = \frac{1}{K_a - K_L} \ln\left\{\frac{K_a}{K_L}\left[1 - \frac{D_0(K_a - K_L)}{K_L L_0}\right]\right\} \quad (6.12)$$

and

$$X_c = U \times t_c$$

Critical dissolved oxygen can be found by substituting X_c into equation 6.4 with other appropriate terms.

6.5 DISSOLVED OXYGEN IMPACTS

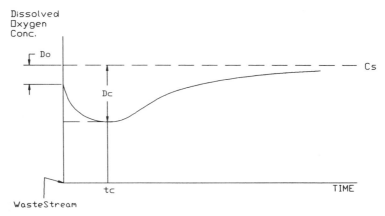

Figure 6.8 DO sag curve.

Example Problem 6.3 For the discharges of Example Problem 6.2, calculate the distance downstream in kilometers where the minimum dissolved oxygen occurs using equation 6.12.

SOLUTION: Assuming the data of Example Problem 6.2 and substituting into equation 6.12, the solution illustrates the calculations.

$$t_c = \frac{1}{0.06 - 0.15} \ln\left\{\frac{0.06}{0.15}\left[1 - \frac{2(0.06 - 0.15)}{(0.15)(9)}\right]\right\}$$

$$= \frac{1}{0.09} \ln[0.4(1.13)]$$

$$= 8.8 \text{ days}$$

and

$$X_c = (4 \text{ km/day})(8.8 \text{ days})$$
$$= 35.2 \text{ km}$$

Also, the critical deficit D_c can be calculated:

$$D_c = 2 \exp\left[\frac{(-0.06)(35.2)}{4}\right]$$

$$+ \frac{9(0.15)}{0.06 - 0.15}\left\{\exp\left[\frac{(-0.15)(35.2)}{4}\right] - \exp\left[\frac{(-0.06)(35.2)}{4}\right]\right\}$$

$$+ \frac{2(0.12)}{0.06 - 0.12}\left\{\exp\left[\frac{(-0.12)(35.2)}{4}\right] - \exp\left[\frac{(-0.06)(35.2)}{4}\right]\right\}$$

$$= 1.1796 + 4.8397 + 0.9536 = 7.0 \text{ mg/L}$$

Therefore, the instream concentration at 35.2 km is $9.2 - 7.0 = 2.2$ mg/L or ppm dissolved oxygen.

TABLE 6.12 Average Values of Gross Photosynthetic DO Production

Water Type	Average Gross Production (g/m²-day)	Average Respiration (g/m²-day)
Truckee River—bottom-attached algae	9	11.4
Tidal Creek—diatom bloom (62–109 (10⁶) diatoms/L)	6	
Delaware estuary—summer	3.7	
Duwamish River estuary, Seattle, Washington	0.5–2.0	
Neuse River System, North Carolina	0.3–2.4	
River Ivel	3.2–17.6	6.7–15.4
North Carolina streams	9.8	21.5
Laboratory streams	3.4–4.0	2.4–2.9

Source: Zison et al. (1977).

6.5.3 Photosynthesis, Respiration, and Benthic Demand

If evidence exists from chlorophyll tests, algal counts, diurnal DO readings and/or observation indicating that photosynthetic activity is prevalent, these must be incorporated into the DO analysis. Photosynthetic rates are not constant. A sine curve function can be used to estimate the diurnal variability. Selected photosynthesis production rates were presented by the EPA (Zison et al., 1977) and are shown in Table 6.12. Also variable are the respiration values for use in equation 6.4. The photosynthesis and respiration rates must be divided by average depth in meters, or

$$\text{rate} = (g/m^2 - \text{day})/m = mg/L\text{-day} \qquad (6.13)$$

Benthic demands are also variable and temperature dependent. Sandy bottom streams exhibit average rates of 0.5 g O_2/m^2-day compared to municipal outfall sludges of 2 to 10 g O_2/m^2-day. Thomann (1972) presents other data illustrating the range of values for different situations. Generally, if benthic and photosynthesis data are not available, mathematical models will not include the terms or gross average values will be assumed. Physical observation of the area may be sufficient to determine the need for this level of analysis.

6.5.4 Impoundment Dissolved Oxygen

Within a water column of an impoundment (long-detention water body), similar sources and sinks for DO exist as for a stream, but the advective forces are minimal and are assumed nonexistent. Consider a mass balance for

6.5 DISSOLVED OXYGEN IMPACTS

oxygen-consuming materials with no inflow term, or

$$-QL + W - K_S LV - K_1 LV = V\frac{dL}{dt} \tag{6.14}$$

where Q = export flow rate, L/day
L = oxygen-consuming materials concentration, mg/L/day
W = waste loading from all sources, mg/day
V = impoundment volume, L
K_S = settling rate for materials, day^{-1}
K_1 = deoxygenation rate in the water column, day^{-1}

Integration of the first-order linear differential equation 6.14 yields.

$$L_t = \frac{W}{(K_S + K_1 + Q/V)V}\left[1 - \exp\left(K_S + K_1 + \frac{Q}{V}\right)t\right]$$
$$+ L_0\left[\exp\left(K_S + K_1 + \frac{Q}{V}\right)t\right] \tag{6.15}$$

where L_0 is the initial or starting concentration. The equilibrium or steady-sate concentration, L_e, is obtained when time approaches infinity, or

$$L_e = \frac{W}{V(K_S + K_1 + Q/V)} \tag{6.16}$$

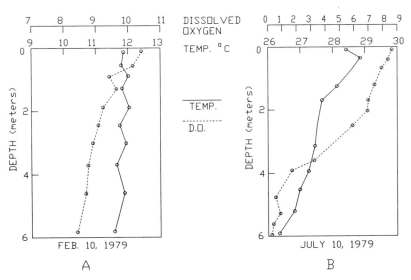

Figure 6.9 Typical temperature and dissolved oxygen profiles for Lake Eola, Orlando, Florida.

The stormwater impacts on Lake Eola in downtown Orlando, Florida can be seen in Figure 6.9. The lake is landlocked and is fed by stormwater during the rainy season, April through October. The dissolved oxygen is depressed in the deep areas of the lake during the summer months.

6.6 SEDIMENT ACCUMULATION

Sediment discharges can add contaminants to the water and bottom layers but also cause physical changes near the discharge area. This may affect stream fauna because of erosion and deposition having an effect on food sources and reproduction areas. Also, sediment probably carries oxygen-demanding materials. The spatial location of these sediments as it affects fauna can be documented as having a regional impact (Osborne and Herricks, 1988). Nutrients are also adsorbed to sediment and can have a direct effect on the rate of plant growth in the local and regional area. However, in some cases, the biodegradable materials and nutrients are low relative to the quantity of nonbiodegradable fractions and nutrients.

Deposited sediments in streams, lakes, and ponds receiving stormwater runoff reduce capacity, fill shallow areas near inlets, and cover areas of habitat for aquatic organisms (Striegl, 1987). Particulate matter from various sources are transported by washout, traffic volume, air transport, and other mechanisms. Also, biomass die-off and the deposition of debris on the bottom of the ponds leads to an increased accumulation of bottom sediments and a reduction in storage volume. The rates of sediment accumulation depend on a number of factors. These factors include the surrounding land use, intensity–duration–interevent dry period between rainfall events, volume of traffic, season of the year, type of street surface, condition of street surface, and public works maintenance practice.

Pollutants in stormwater have a strong affinity to suspended solids transported during storm events. Subsequently, the removal of suspended solids in receiving water will often result in removal of the pollutants (Lager et al., 1977; Horner and Mar, 1982). It had been indicated that large fractions of heavy metals, organic pollutants, and nutrients attach to solid particles, and most of the pollutants are associated with the smaller particle fractions, less than 100 μm in diameter. Settling of pollutant-saturated particles occurs as discrete individual particles and clusters of smaller particles fuse into large ones, thus accelerating their settling rates. Whipple and Hunter (1981) concluded that settling rates of pollutants in runoff vary greatly and particle-size distributions cannot be transposed into settling rates for absorbed constituents. Runoff water contains a wide variety of sediment particles that are different with respect to their specific gravities and how they absorb metals and other pollutants.

Estimates for runoff quantities of solids were presented in Chapter 5. If the degradable fraction can be estimated or determined; the remaining quantities can be treated as conservative substances. Thus a mass balance of inputs and outputs with no decaying terms can be written assuming that no settling occurs.

$$\text{Input} - \text{output} = 0$$

$$S_u Q_u + S_w Q_w - S_0(Q_u + Q_w) = 0$$

$$S_0 = \frac{S_u Q_u + S_w Q_w}{Q_u + Q_w} \quad (6.17)$$

where S_u = upstream or before discharge concentration, g/L
Q_u = before discharge flow volume, L
S_w = source discharge concentration, g/L
Q_w = source discharge volume, L
S_0 = instream at the discharge concentration, g/L

If a uniformly distributed source characterized by a loading per unit length along a river is present, the resulting concentration, assuming no settling is

$$S_0 Q_0 + W_x Q_d X - S_x(Q_0 + Q_d) = 0 \quad (6.18)$$

For nonpoint surface discharges

$$S_x = \frac{S_0 Q_0 + W_x Q_d X}{Q_0 + Q_d} \quad (6.19)$$

where S_x = downstream mixed concentration at distance X, g/L
X = distance downstream, km
$Q_0 = Q_u + Q_w$, L
W_x = distributed load, g/L-km
Q_d = flow associated with distributed nonpoint load, L

and for only scour effects, w = g/km, $Q_d = 0$,

$$S_x = S_0 + \frac{wx}{Q_0} \quad (6.20)$$

where w is the scour load per unit length of stream, g/km.

Estimation of the bedload (scour material) is difficult to determine. Many methods are useful and a special manual (Task Committee on Preparation of Sedimentation Manual, 1971) has been developed. If data on specific size particles, shear stress, and bed stress are not available, statistical estimators can often be useful.

TABLE 6.13 Eutrophication Potential Using Algal Cells and Nutrients

Potential	Units	Dry Algal Cells	TP-P	TN-N
Threshold	mg/L	1.45	0.013	0.092
Likely problem	mg/L	14.50	0.130	0.920

6.7 EUTROPHICATION AND TROPHIC ANALYSIS

Trophic analysis is an examination of a water body for biological productivity. The elements believed most responsible for or highly related to the productivity of surface waters that may accelerate the natural decay of surface waters are nitrogen and phosphorus. Algal growths may or may not be present as a result of increasing nutrients. In very general terms eutrophication potential can be related to dry algal cells, phosphorus levels, and nitrogen levels as shown in Table 6.13 (Zison et al., 1977). Eutrophic lakes are ones that have high nutrient concentrations, resulting in overgrowth of algal and plant species while oligotrophic lakes have low values.

If the quantity of nutrients discharged from point sources and nonpoint sources can be estimated or measured, the mass and concentrations can be developed for certain physical conditions of surface water bodies. If the nutrients do not settle, resuspend, and/or otherwise change, concentrations can be estimated if flow rates are known. However, the mass of nitrogen and phosphorus will rarely remain in solution. Nitrate, ammonium, and orthophosphate ions are generally available to waterborne plants and are needed for optimal plant growth.

If nitrogen and/or phosphorus are limited, many impoundments will not have excessive plant growths. However, impoundments with long residence times have accumulated nutrients resulting in excessive plant growth.

The hydraulic residence time, t_w, is the average time a water particle is in an impoundment and is computed the same way as detention time:

$$t_w = \frac{V}{Q} \qquad (6.21)$$

where V = average flow volume, m^3
Q = average exchange flow rate, m^3/day

The flow volume is that area receiving or discharging flows which does not include stagnant waters. Stagnant waters may be caused by a density gradient or physical separation. The exchange flow rate is the average or steady-state value. It may be difficult to estimate the flow volume for large and/or irregular-shaped impoundments. Depth profiles, influent locations, effluent locations, and thermal gradients are important for estimating flow volume and exchange flow rate.

Chlorophyll is frequently used to indicate and quantify the level of productivity. A lake with a high overall chlorophyll level (about 50 to

100 mg/m^3) is frequently referred to as *eutrophic*, while a low chlorophyll level (less than 10 mg/m^3) is frequently referred to as *oligotrophic* (underenriched). The exchange rate is indirectly related to chlorophyll levels. Many other factors have also been related to trophic conditions. Several methods have been developed for evaluating trophic conditions, and some are discussed here.

6.7.1 Shannon–Brezonik Trophic State Index

The trophic state index (TSI) was developed by Shannon and Brezonik (1972) after investigating 55 central Florida lakes. These are shallow lakes with average depths generally less than 10 ft and surface areas varying from a few acres to several thousand acres. The TSI was established as a function of several critical parameters.

$$TSI = 0.18T + 0.008CD + 1.1TN + 4.2TP \\ + 0.01PP + 0.44CL + 0.39CR + 0.26 \quad (6.22)$$

where T = turbidity, JTU
CD = conductivity, mS/cm
TN = total organic nitrogen, mg/L
TP = total phosphorus, mg/L
PP = primary productivity, mg-C/m^2-h
CL = chlorophyll *a*, mg/m^3
CR = [Ca] + [Mg]/[Na] + [K]

Shannon and Brezonik (1972) assigned the following values of each trophic classification: 0–3, oligotrophic; 3–7, mesotrophic; 7–10, eutrophic, and greater than 10, hypereutrophic.

Example Problem 6.4 Consider a lake with the following water quality data: T = 5.2 JTU; CD = 226 µS/cm; TN = 0.39 mg/L; TP = 0.034 mg/L; CL = 4.3 mg/m^3; CR = 1.25; and PP = 30 mg C/m^2-h. What is the trophic state index? Interpret the results.

SOLUTION: The TSI = 6.3, which indicates a mesotrophic condition.

In a further analysis, Shannon and Brezonik (1972) developed through multivariate analysis an equation that relates the TSI to nitrogen and phosphorus loading.

$$TSI = 26.10(P) - 242.36(P)^2 + 1.12(N)^2 + 28.68(N)(P) + 2.37(N) \quad (6.23)$$

where P = phosphorus loading, g/m^3/yr
N = nitrogen loading, g/m^3/yr

Example Problem 6.5 Using the loadings mentioned earlier in Example Problem 6.4 and a lake volume of 4.67 million m^3, then for $\bar{P} = 0.143$ g/m^3/yr and $\bar{N} = 1.247$ g/m^3/yr, the TSI is 8.48, which implies a lag in lake response to nutrient loading, which means that further degradation of water quality is likely to occur.

6.7.2 Vollenweider Model

Vollenweider (1969) expresses the trophic status of a lake as a function of the areal phosphorus loading rate, the mean depth of the lake, and the flushing rate. He classified several lakes according to this relationship and from these investigations developed a trophic analysis graph, as shown in Figure 6.9. The y axis—phosphorus leading—is merely the grams of phosphorus contributed annually to each square meter of lake surface. The x axis—$\bar{Z}/t_w$—is the mean depth, $\bar{Z}$, in meters, divided by the hydraulic residence time, in years.

Example Problem 6.6 For a lake, phosphorus loading is 0.69 g/m^2/yr, $\bar{Z}$ is 5.2 m, and t_w is 8.2 years; therefore, $\bar{Z}/t_w$ is 0.60. Plotting on the Vollenweider graph, it may be seen that this lake is classified within the eutrophic range (see point 1 on Figure 6.10). The ratio of watershed to lake area is 8:1.

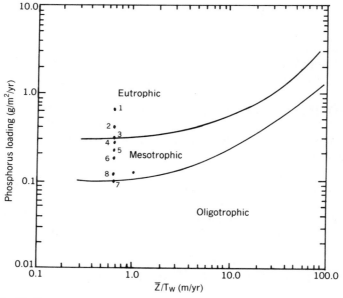

Figure 6.10 Vollenweider model. Reductions in urban phosphorus load: 1, 0%; 2, 50%; 3, 60%; 4, 70%; 5, 80%; 6, 90%; 7, 100% (hypothetical natural conditions); 8, 97%.

SOLUTION: What is convenient about these loading graphs is that the impacts of reducing phosphorus loading can be seen. For example, suppose that the lake watershed were completely natural. Using the woodland loading rates (Table 5.13), the phosphorus load would be reduced to 0.08 g/m^2/yr. Plotted on the graph, it may be seen that the lake at this time was oligotrophic. This point may serve as a convenient goal. Other percent reductions of a hypothetical urban reduction are shown in Figure 6.10.

6.7.3 Dillon Model

After review of Vollenweider's work, Dillon (1975) indicated that one of the Vollenweider weaknesses was inability to consider the nutrient retention capabilities of the lake sediments. Further, he developed a model that took this retention into consideration.

Basically, Dillon used the input–output concept used by Vollenweider. He expanded the model, however, by developing a nutrient retention coefficient, R, that is defined as being equal to the ratio of the phosphorus mass retained to the incoming load. For the preceding example, assume that this value is 450 lb/1560 lb or 0.29. Dillon used this retention coefficient as part of a loading parameter, $L(1 - R)/\rho$, which he utilized as the y axis of the loading graph (Figure 6.11). In this, L is the phosphorus loading (g/m^2/yr) and ρ is the flushing per year. The x axis on his graph is the mean depth.

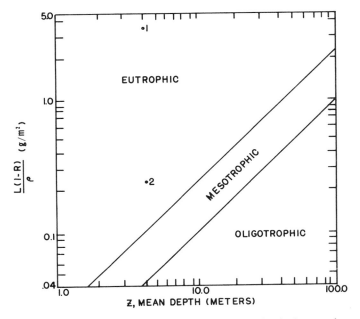

Figure 6.11 Dillon model. 1, Present condition; 2, hypothetical natural condition.

Example Problem 6.7 Applying this model to the lake of Example Problem 6.6 and using $\rho = 0.12$ per year, $L = 0.69$ g/m^2/yr, and $R = 0.29$, $L(1 - R)/\rho$ is 4.1. This places the lake in a hypereutrophic ranking as shown in Figure 6.11. However, under natural conditions, $L = 0.08$ g/m^2/yr and $\rho = 0.24$ per year, the lake is still classified eutrophic if it is assumed that R remains constant. Therefore, according to Dillon, the lake would be classified as a naturally eutrophic lake. Such a classification would probably be accurate for many closed-system lakes. Indeed, it must be understood that because of the poor flushing, these lakes are particularly vulnerable to excessive nutrient loading. As would be expected, they depend heavily on their sediments to detain and immobilize many of the incoming nutrients.

6.7.4 Larsen–Mercier Model

In line with the Dillon model, Larsen–Mercier (Larsen et al., 1974) developed a eutrophication model that considered nutrient retention. Their graph (Figure 6.12) utilizes as the y axis the mean phosphorus concentration of the incoming runoff. To calculate this for the example problems, consider that the total load, 1560 lb/yr, is divided by the hydraulic input, which is 25% of

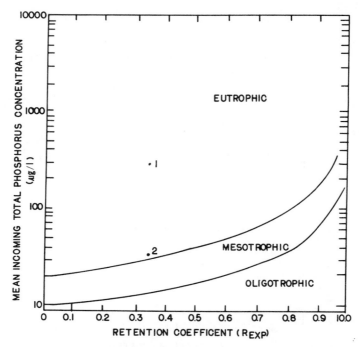

Figure 6.12 Larsen–Mercier Model. 1, Present condition; 2, hypothetical natural condition.

the watershed rainfall and 100% of the direct rainfall to the lake, or 667 million gal/yr. This division results in a value of 297 μg/L. As shown in Figure 6.12, this implies a eutrophic condition. Under natural loadings, the incoming concentration is estimated at 34 μg/L. As can be seen, the Larsen–Mercier model also indicates a natural eutrophic status for a lake.

6.8 MATHEMATICAL MODELING FOR RECEIVING WATERS

A mathematical model is an abstraction in an equation form of a real-world situation. Mass balance models are frequently used to calculate concentration and mass changes in receiving waters and can also be used for the performance of stormwater management practices. Receiving water quality models require knowledge of sedimentation mechanisms and kinetics for removal of dissolved fractions. Removal rate constants and physical, chemical, and biological interactions must be defined under various environmental conditions. These are complex processes and simplification of the models may be appropriate.

Additional factors that affect water quality are dispersion and density stratification. Dispersion is the transport of material due to turbulence and molecular diffusion. It is usually represented as a function of concentration gradient.

Density stratification occurs when flows of different density or flow characteristics meet. Examples are a small flow (tributary) entering a larger water body, a stormwater discharge into a large river or lake, or fresh water meeting salt water. Fresh water has a different density than salt water and the separation is most distinctive. However, other very slight differences in density result when untreated stormwaters are discharged into receiving waters. These differences are due to temperature and sediment load. These nonmixing conditions affect the volume used in the assumptions of water quality modeling. The volume of mixing may be significantly less than the volume of the receiving water body. The receiving water body may have to be divided into separate models and interconnected.

6.8.1 Reaction Kinetics

Concentrations of water quality parameters exiting a detention pond change with residence time. The rate constants for these changes can be determined from pilot or in situ experimentation and the change in the concentration can be described by the *reaction rate*, which is a term used to define the formation or removal of the material from a specified volume of water. When writing a mass balance for the material, it forms part of the generation or utilization terms. At constant temperatures, empirical data on the formation or removal have been observed to be a function of the concentration of the material, or the rate of reaction can be described as

$$r = kC^\alpha \tag{6.24}$$

where r = reaction rate, mg/L-day
k = reaction rate constant, day^{-1}
C = concentration, mg/L
α = order of the reaction

Other time and concentration units are also used. If growth of material results, as in biological organisms, the reaction rate constant is positive. Chemical changes for pollutants are usually decreasing with time; thus the reaction rate is negative.

The usual method for determining the order of the reaction is to use a batch reactor (no inflow or outflow) in which the material is placed in an environment (oxygen, pH, temperature, etc.) that is expected in the stormwater pond. The amount of material remaining with time is tabulated and usually plotted to determine the reaction order.

Zero-Order Reaction. A zero-order reaction is one in which concentration change is a function of time and not concentration. Assuming a decreasing concentration, the reaction equation is

$$r = \frac{d(C)}{dt} = -k \tag{6.25}$$

Integration between the limits of $C = C_0$ to C when $t = 0$ to t yields

$$C = C_0 - kt \tag{6.26}$$

The reaction rate can be determined by plotting concentration remaining versus time on an arithmetic scale, as shown in Figure 6.13a.

First-Order Reaction. A first-order reaction that is common in water quality and hydrologic analysis results in an exponential relationship. The reaction equation states that concentration change is a function of remaining concentration:

$$r = \frac{dC}{dt} = -kC \tag{6.27}$$

Integration between the limits of $C = C_0$ to C when $t = 0$ to t yields

$$\ln \frac{C}{C_0} = -kt \quad \text{or} \quad C = C_0 e^{-kt} \tag{6.28}$$

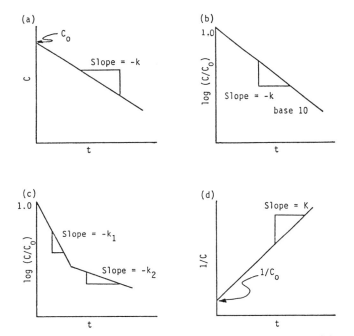

Figure 6.13 Typical reaction kinetics curves: (*a*) zero-order; (*b*) first-order; (*c*) composite exponential; (*d*) second-order.

or

$$\log \frac{C}{C_0} = \frac{-kt}{2.3} \tag{6.29}$$

The reaction rate is determined by plotting $\log C/C_0$ versus t with the slope equal to k as shown in Figure 6.13*b*.

Composite Exponential. A composite exponential (Figure 6.13*c*) may result when on a plot of $\log C/C_0$ versus time, multiple straight-line equations result. The analysis procedure is similar to hydrograph recession limb analysis.

Second-Order Reaction. A second-order reaction equation states that concentration change is a function of remaining concentration squared:

$$r = \frac{dC}{dt} = -kC^2 \tag{6.30}$$

and integration from $C = C_0$ and $t = 0$ to t yields

$$\frac{1}{C} - \frac{1}{C_0} = -kt \tag{6.31}$$

Figure 6.13*d* illustrates the procedure for determining a second-order equation.

If one has no knowledge of the order of the reaction, all three graphical presentations are completed (Figure 6.13*a*, *b*, and *d*). The plot that provides the best straight-line fit then determines the order of the reaction and the slope of the straight line is the reaction rate constant.

Example Problem 6.8 For a 1-acre impervious area, 1 in. of rainfall removes 206 g of iron and places the iron in a landlocked pond. The initial concentration of iron and volume of water in the pond were both near zero. After 3 days, the iron concentration was 1.00 mg/L, and after 6 days, the iron concentration was 0.61 mg/L. The concentration was 0.27 mg/L after 10 days. What is the reaction order, zero or first, for the iron concentration?

SOLUTION: First calculate the initial concentration of iron:

$$C = \frac{\text{mass}}{\text{volume}} = \frac{206 \text{ g} \times 10^3 \text{ mg/g}}{(1/12 \text{ ft})(43{,}560 \text{ ft}^2/\text{acre})(2835 \text{ L/acre ft})} = 2 \text{ mg/L}$$

By plotting the concentration remaining versus time as shown in Figure 6.14, a first-order equation is the best fit with $k = 0.20 \text{ day}^{-1}$ and $C = 2e^{-0.2t}$.

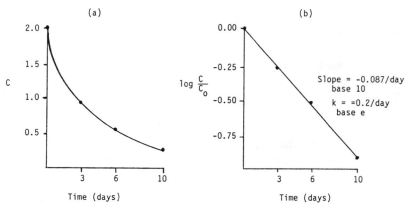

Figure 6.14 Determination of the reaction rate order for Example Problem 6.8.

6.8.2 Types of Models

There are five frequently used models for water quality control: (1) batch, (2) complete-mix continuous-flow, (3) plug-flow, (4) intermediately mixed, and (5) packed-bed. Each can assume certain reaction kinetics that describes the transformation of a pollutant.

The batch reactor assumes that the kinetics of the process takes place with neither flow entering or leaving the system. It generally is used for modeling landlocked shallow lakes and stormwater ponds designed to retain water for a specific period. Receiving waters are assumed to undergo concentration changes according to specified reaction kinetics.

The complete-mix continuous-flow model assumes that the flow and pollutant material are dispersed immediately in the volume of the system. It can be used to approximate water quality conditions in rivers, shallow lakes, and detention systems, with the assumption that over the time period of analysis there are no changes in kinetics, flow rates, and volumes. Thus worst conditions of high flow rates (minimum detention times) can be evaluated, or conditions can be averaged over short time periods. The complete-mix continuous-flow reactor is considered in greater detail in this chapter.

The plug-flow reactor assumes that the flow and pollutant material pass through the system and are unmixed in the longitudinal direction. Thus each particle remains in the system for a time interval equal to the detention time. This model is used for narrow rivers and detention systems in which the reaction kinetics gradually change in the system. A series of complete-mix continuous-flow sections of a system with varying reaction rates do produce results similar to those of the plug-flow model. However, plug-flow models may be more reasonable to approximate the existing system conditions.

An intermediately mixed pond has a detention-time frequency distribution that lies between a plug-flow and a completely mixed reactor. Most stormwater ponds are classified as intermediately mixed (Nix, 1985; Martin, 1988). The mean value of detention time exceeds the median value, indicating a right-skewed distribution. To calculate water quality concentrations, a numerical integration technique is used (Ferrara and Hildick-Smith, 1982).

The packed-bed model is named because of flow through a packing medium, such as soils, carbon, gravel, or other media for water quality enhancement. Their use is to describe water movement and quality improvement in stormwater filtration/infiltration ponds and groundwater movement.

Complete-Mix Continuous-Flow Model. A complete-mix continuous-flow model is one that has no concentration gradients within the system. A concentration entering the system is uniformly dispersed such that the concentration leaving is the same as that in the reactor volume. A reach (short distance) of a river system may be modeled as performing in this way, as well as stormwater detention ponds also modeled (Ferrara and Hildick-Smith, 1982). The differential equations for the mass balance must be solved

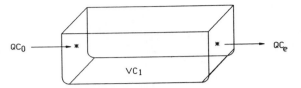

Figure 6.15 Complete-mix continuous-flow system.

numerically. A schematic of the system is shown in Figure 6.15. A mass balance for the one-dimensional state of the complete-mix continuous-flow system for a time period is

mass inflow − mass outflow + mass generation = accumulation

$$QC_0 - QC_e + kV = \frac{dC}{dt}V \qquad (6.32)$$

Nonreactive Complete-Mix Continuous-Flow Model. For a conservation material (i.e., nonreactive) the generation term is zero ($k = 0$). Using a conservative substance, the hydraulic detention time of a system can be determined by tracing the appearance of the substance in the effluent after injecting the material upstream. For this case, integration of equation 6.32 is done from $C = 0$ to C_t when $t = 0$ to t_t:

$$\frac{dC}{dt} = \frac{Q(C_0 - C)}{V} \qquad (6.33)$$

Assuming that Q and V are constants,

$$\int_0^{C_t} \frac{dC}{C_0 - C} = \frac{Q}{V} \int_0^t dt \qquad (6.34)$$

$$\frac{\ln(C_0 - C_t)}{C_t} = -t\frac{Q}{V}$$

$$C_t = C_0[1 - e^{-(Q/V)t}] \qquad (6.35)$$

As the conservative substance is introduced into the system, it builds to its initial value and approximates it when t approaches $5(V/Q)$, or

$$1 - e^{-5} \approx 0.993 \qquad (6.36)$$

The hydraulic detention time of a system is defined as $t_d = V/Q$. When $t = 5t_d$, an approximation of steady-state conditions occurs. This is important when modeling a system because a change in input mass is considered to reach steady state after five detention times. (In some cases three detention times are considered a reasonable approximation of steady state.)

First-Order Complete-Mix Continuous-Flow Model. Chemicals and sediment that enter a river or stormwater detention pond will settle and otherwise undergo chemical and biological transformation. Sometimes, this change can be represented by a first-order reaction. For a reach of a river or detention pond, assume a complete-mix continuous-flow situation with the generation term represented by a first-order reaction showing a removal, determine an equation for concentration at the end of the reach (effluent of a detention pond). Using equation 6.34 with a generation term $-kC$, integrate the expression to solve for C_t with limits of integration $C = 0$ to C_t when $t = 0$ to t.

$$\int_0^{C_t} \frac{dC}{C_0 - C[1 + k(V/A)]} = \frac{Q}{V}\int_0^t dt \quad (6.37)$$

Let $t_d = V/Q$; integration results in

$$\frac{-1}{kt_d + 1} \ln \frac{C_0 - (1 + kt_d)C}{C} = \frac{t}{t_d}$$

and

$$C = \frac{C_0[1 - e^{-(1+kt_d)(t/t_d)}]}{1 + kt_d} \quad (6.38)$$

As the rate of removal approaches zero or that of a conservative substance, equation 6.38 approximates the response curve of equation 6.35.

The greater the reaction rate, the faster the system approaches the assumptions of steady state. As time approaches infinity, equation 6.38 is reduced to

$$C \simeq \frac{C_0}{1 + kt_d} \quad (6.39)$$

Relative equilibrium conditions can be determined as a function of time and the product of reaction rate and detention time as shown in Figure 6.16. The greater the product of reaction rate and detention time, the faster the equilibrium. As shown in Figure 6.16, the greater the reaction rate for a fixed detention time, the faster the stormwater system approaches the approximations for a steady-state complete-mix system. On a river system with slowly responding hydrographs, this may be a reasonable assumption and the solution can be determined for discrete flow rates and storage volumes within

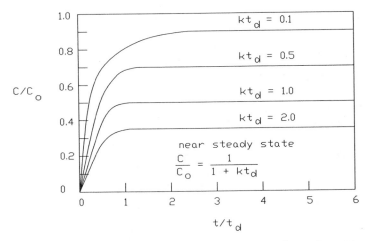

Figure 6.16 Effluent response curves for a first-order complete-mix continuous-flow river or detention system.

a reach. The flow and volume assumptions may not be as critical as those of the reaction rate. Other reaction rates may be more appropriate for use in the model and should be tested.

Series of First-Order Complete-Mix Continuous-Flow Systems. A river reach or detention pond may be separated into a series of reactors because of the natural physical divides that exist. A schematic representation of the model is shown in Figure 6.17. Assuming steady-state conditions exist and removal takes place in each reach (section), the mass balance equations for

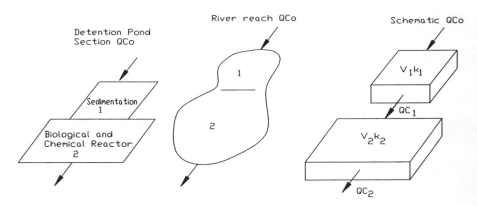

Figure 6.17 Series model schematic.

each reach (section) can be written as follows:

$$\text{First: } QC_0 - QC_1 - k_1 C_1 V_1 = 0$$
$$\text{Second: } QC_1 - QC_2 - k_2 C_2 V_2 = 0$$

Solving for concentration with $t_{d1} = V_1/Q$ and $t_{d2} = V_2/Q$ yields

$$C_1 = \frac{C_0}{1 + k_1 t_{d1}} \tag{6.40}$$

$$C_2 = \frac{C_0}{(1 + k_2 t_{d2})(1 + k_1 t_{d1})} \tag{6.41}$$

Every time cross-sectional area, volume, or reaction rate changes, a mass balance can be written and the equations developed to solve for concentration. In addition, when flow rate changes, another set of equations can be developed.

Example Problem 6.9 A wet-detention pond performance is described by two complete-mix continuous-flow compartments in series for a time period when flow and volume is relatively constant. The removal of solids follows a first-order reaction and the reaction rate constant is 0.5 and $0.05/h^{-1}$ for compartments 1 and 2, respectively. If the initial solids concentration into the pond is 600 mg/L with hydraulic residence times of 10 h and 48 h for compartments 1 and 2, respectively, calculate the effluent concentration for the pond

SOLUTION: Using the complete-mix continuous-flow equation 6.41 and the assumptions of long-term periods relative to detention time and a first-order reaction, the effluent concentration can be estimated:

$$C_2 = \frac{C_0}{(1 + k_1 t_{d1})(1 + k_2 t_{d2})} = \frac{600}{[1 + 0.5(10)][1 + 0.05(48)]} = 29.4 \text{ mg/L}$$

For other flow rate conditions, the effluent concentrations can be estimated given the influent concentrations. Then a frequency distribution relating influent to effluent conditions can be developed.

6.8.3 Dead Volume

Because of density differential or physical restrictions of systems, volumes of water will not react with an input. This occurs in the corners of detention ponds, below inlets and outlet structures, and on the inside bank along a river bend. The result is that short circuiting of flows occurs, which means that not all of the receiving water volume in used for assimilation. This is represented schematically as shown in Figure 6.18.

The dead volume in a pond (receiving water) will decrease the reaction time and the resulting effluent concentrations will be higher than expected. To illustrate this, consider a conservative substance in a single complete-mix continuous-flow model. The resulting concentration is estimated using equation 6.35.

$$C_t = C_0(1 - e^{-Q/V(t)})$$

If the volume of mixing decreases to V_n, the effluent concentration change is given by C_t^n.

$$C_t^n = C_0(1 - e^{-Q/V_n(t)}) \tag{6.42}$$

and the increase in concentration because of the reduced volume is

$$C_t^n - C_t = C_0(e^{-Q/V(t)} - e^{-Q/V_n(t)}) \tag{6.43}$$

Also consider a conservative substance as a pulse input to a system over a time period that has a constant flow. Using a pulse input makes the system function according to the following mass balance after the pulse:

$$\text{input} - \text{output} + \text{generation} = \text{accumulation}$$

$$O - CQ + \text{zero} = V\frac{dC}{dt} \tag{6.44}$$

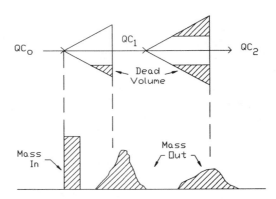

Figure 6.18 Two complete-mix continuous-flow model with dead volume.

6.8 MATHEMATICAL MODELING FOR RECEIVING WATERS

Integration occurs between C = pulse concentration (C_p) to C when $t = 0$ to t:

$$\int_{C_p}^{C} \frac{dC}{C} = \frac{-Q}{V} \int_0^t dt$$

and

$$\frac{V}{Q} = t_d$$

$$\ln C - \ln C_p = \frac{-t}{t_d}$$

$$t_d = \frac{-t}{\ln(C/C_p)}$$

or

$$C = C_p\, e^{-t/t_d} \qquad (6.45)$$

where C_p = pulse concentration
t = time from peak concentration

Example Problem 6.10 Short circuiting and reduced volume performance can be approximated for a detention pond or receiving water body by injecting as a pulse a conservative substance into the influent of the water body. Determine the dead volume for an initial concentration of 100 mg/L and the following conditions assuming relatively constant volume and flow rate:

Volume 50,400 ft^3
Flow rate 2 ft^3/s

Concentration data at the effluent:

Time (h)	Concentration (mg/L)
0.0	100
0.2	95
1.0	78
2.0	60
3.0	46
4.0	35
6.0	20
8.0	12

206 RECEIVING WATER QUALITY

For stormwater ponds, the pond volume and rate vary with time, and thus the reduced volume should be estimated for other different volume and flow rate conditions.

SOLUTION:

1. If the concentration of conservative substance is removed from the pond faster than that estimated using the detention time based on total volume, a lesser volume of pond is being used as the mixing zone. Thus calculated detention time based on total volume, or

$$t_d = \frac{V}{Q} = \frac{50{,}400(\text{ft}^3)}{2(\text{ft}^3/\text{s}) \times 3600(\text{s/h})} = 7\text{ h}$$

2. Using this detention time as if we were using the total volume, compute the expected effluent concentration and detention time using equation 6.45.

Time (h)	t/t_d	C
0.0	0.000	100
0.2	0.029	97
1.0	0.143	87
2.0	0.286	75
3.0	0.429	65
4.0	0.571	56
6.0	0.857	42
8.0	1.143	32

Comparing the total volume concentration to the actual, the actual is less, which means that the material was removed faster in the actual situation because less volume was available.

3. Thus compute the actual detention time using $t_d = -t/[\ln(C/C_p)]$ (equation 6.45).

Time (h)	C/C_p	$\ln(C/C_p)$	t_d (h)
0.0	1.00	–	–
0.2	0.95	−0.05	3.9
1.0	0.78	−0.25	4.0
2.0	0.60	−0.51	3.9
3.0	0.46	−0.78	3.9
4.0	0.35	−1.05	3.8
6.0	0.20	−1.61	3.7
8.0	0.12	−2.12	3.8

average $t_d = 3.9$ h

4. Thus the active volume is 3.9 h × 2 ft³/s × 3600 s/h or 28,080 ft³.
5. The dead volume is the difference between the total volume and the active volume or 50,400 ft³ − 28,080 ft³ = 22,320 ft³.

At the stated flow rate of 2 ft³/s, the unused volume of receiving water is 22,320 ft³ which indicates short circuiting. It should be noted that there is a time lapse after the pulse is injected and some loss of initial concentration. Thus adjustments must be made to reduce initial concentration and time for sampling.

6.9 SUMMARY

Receiving water impacts can be determined by an examination of water quality data. The mass loadings of chemicals in stormwater and other stormwater-related pollutants can be compared to those from point sources. In many cases, mass loadings are greater from stormwaters. Next, receiving water concentrations can be compared to determine relative impacts. Toxic impacts, dissolved oxygen concentrations, nutrient loadings, and sediment loads are related to specific threshold levels. Some important concepts are:

- Toxic compounds may exist in stormwaters.
- Algal assay tests using stormwater and receiving water are indications of impacts due to stormwater.
- Deicing agents are salts that contain toxic contaminants.
- The dissolved oxygen model may incorporate more than one generation term. Threshold levels for the survival of game fish are usually 4 or 5 mg/L. Mathematical models are available to predict DO impacts from stormwater discharges.
- Trophic state analysis is usually done using many equation forms: Vollenweider model, Shannon and Brezonik index, Dillon model, Larsen–Mercier model, and others. The results are then compared and conclusions on trophic state are made.
- Kinetic reaction rates can be determined using batch (no inflow, no outflow) vessels.
- The order of the reaction can be obtained by graphical comparisons. A first-order reaction is commonly used.
- A widely used water quality model is the complete-mix continuous-flow one or some combination of a series of plug-flow units with the complete-mix model.
- Conservative substances are used to determine hydraulic detention times and dead volumes in mixing zones.
- The mass balance models of this chapter require the assumption of constant flow rate over a period of time. This is not impossible for large rivers because many of the data used for the analysis are changing for a

Figure 6.19 Bioassay shaker table with flasks and controlled light intensity and duration.

fixed time period. For wet-detention ponds, short time intervals may have relatively constant values.

6.10 PROBLEMS

1. a. For two of the following water quality areas, discuss potential impacts on receiving waters: (1) bacteria, (2) nutrients, (3) metals, and (4) organics.

   b. For one other area [do not use the same two as in part (a)], list at least five water quality parameters as a measure for this area.

2. What is a combined sewer, and how does it compare to a storm sewer in terms of solids and organic concentrations? Discuss your answer considering dry and wet weather flow, first-flush conditions, and concentration of pollutants.

3. Discuss the algal assay test with reference to Figure 6.19 and state reasons for doing this test. How do you measure algal estimates in these tests and state the problems associated with laboratory bioassays and their applications in field situations.

4. If algal mass increased from 10 μg/L chlorophyll a to 60 μg/L chlorophyll a in 2 days and follows an exponential growth pattern, what is the rate of growth?

5. For Example Problem 6.2 develop the complete DO sag curve using at least six points, of which one is the minimum DO point. Neglect the nitrogen demand. Also draw the DO profile for $K_a = 0.30$ day^{-1} (base e).

6. Develop a sediment rating curve that best fits the data given below. Discuss which form is the best and why.

Dependent Variable Sediment Discharge (mt/day)	Independent Variable Discharge Rate (m^3/s)
9.5	0.15
22.0	0.21
85.0	0.29
180.0	0.35
650.0	0.72

7. If the flow volume of a lake is 650(10^9) m^3, the exchange rate is 112(10^9) m^3/yr, the average depth is 17 m, and a phosphorus loading of 1.0 g/m^2/yr, how is the lake classified?

8. Comment on management methods to restore the lake of Problem 7 to oligotrophic conditions. List all assumptions.

9. A point-source BOD$_5$ discharge into a small lake with a volume of 500,000 m^3 occurs at a rate of 0.56 m^3/s and a concentration of 20 mg/L. If the combined removal rate (settling plus deoxygenation) is 0.69 day^{-1} (base e), determine the concentration at equilibrium.

10. What is the trophic state using the Vollenweider model for a 440-acre, $\frac{1}{4}$-acre-lot-size residential area discharging into a 100-acre lake that has an average depth of 10 ft. The average lake residence time is 0.5 year.

11. Present a mass balance with a resulting equation for a chemical substance in a large detention pond which undergoes degradation as a first-order reaction. Assume steady-state conditions over a long time period.

12. For the following concentration data in a detention pond, determine the order of the reaction and the value of the reaction constant.

Time (h)	Concentration (μg/L)
0	500
1	360
2	260
3	180
4	120
5	90
6	70

13. For a segment of a drainage canal with a detention time of 10 days and an initial pulse of stormwater with 0.6 kg of phosphorus in 300,000 L of runoff water, what is the concentration after 10 days if the phosphorus decays according to a first-order reaction with $k = 0.2$ day^{-1}.

14. Assuming a first-order complete-mix continuous-flow condition for a small detention pond with volume equal to 70,000 ft^3 and a flow rate of 2 ft^3/s, calculate the effluent concentration after one detention time if the kinetic rate term is 2.4 day^{-1}. The initial concentration is 20 mg/L.

15. If a detention pond volume used for waste assimilation (water quality control) decreased by 50%, how much would the concentration of conservative substance in a complete-mix continuous-flow model change? Express your answer as a function of initial concentration, time, and detention time. Hold flow rate as a constant value.

16. For Problem 15, assume that the detention time is 8 h at a flow rate of 6 ft^3/s. The upstream concentration is 50 mg/L. It is proposed to reduce the detention time to 4 h; what is the expected effluent concentration at both the 8- and 4-h detention time after 5 h?

17. A lake can be modeled by two complete-mix continuous-flow compartments in series for each time frame when flows and volumes are constant. For the two compartments, the hydraulic detention times are 2 and 24 h, respectively, with first-order reaction rates of 0.4 and 0.03 h^{-1}. If the initial concentration is 800 μg/L, what is the concentration leaving the second compartment?

18. A river receives stormwater discharge at mile marker zero. The concentration of lead averages 600 μg/L. The river can be divided into three segments and the worst flow condition produces lead detention times of 6, 18, and 24 h for river segments 1, 2, and 3. If the first segment can be approximated by a batch reactor, detention time of 6 h, and reaction rate $= \frac{1}{6}$ h^{-1}, while the second and third segments are complete mix with first-order reaction rates of 0.1 and 0.05 h^{-1}, and travel times of 18 and 24 h, what is the expected concentration after segment 3? Assume constant flow.

6.11 REFERENCES

Adams, F. J. 1972. "Highway Salts: Social and Environmental Concerns," paper presented at the *Highway Research Board, 5th Summer Meeting*.

Bastian, R. K. 1986. "Potential Impacts on Receiving Waters," in *Urban Runoff Quality*, B. Urbonas and L. Roesner, Eds., American Society of Civil Engineers, New York.

Churchill, M. A., Elmore, H. L., and Buckingham. 1962. "The Prediction of Stream Reaeration Rates," ASCE, *Journal Sanitary Engineering*, Vol. 88, No. SA4, pp. 1–46.

Covar, A. P. 1976. "Selecting the Proper Reaeration Coefficient for Use in Water Quality Models," U.S. Environmental Protection Agency Conference on Environmental Simulation and Modeling, April 19–22, Cincinnati.

Dillon, P. J. 1975. "The Phosphorus Budget of Cameron Lake, Ontario: The Importance of Flushing Rate to the Degree of Autrophy of Lakes," *Limnology and Oceanography*, Vol. 20.

Driscoll, E. D., Shelley, P. E., and Strecker, E. W. 1990. *Pollutant Loadings and Impacts from Highways Stormwater Runoff*. FHWA-RD-88-006, Federal Highway Administration, Washington, D.C.

Dupuis, T. V., et al. 1984. *Effects of Highway Runoff on Receiving Waters*, Vol. II, *Results of Field Monitoring Program* (Draft Report), prepared by Rexnord Inc. for the Federal Highway Administration, Washington, D.C.

Eckenfelder, W. W., and O'Connor, D. J. 1961. *Biological Waste Treatment*, Pergamon Press, Elmsford, N.Y.

Ferrara, R. A., and Hildick-Smith, A. 1982. "A Modeling Approach for Storm Water Quantity and Quality Control via Detention Basins," *Water Resources Bulletin*, Vol. 18, No. 6, pp. 975–981.

Frost, L. R., Jr., Pollock, S. J., and Wakelee, R. F. 1981. *Hydrologic Effects of Highway-Deicing Chemicals in Massachusetts—Executive Summary*, Open-File Report 81-210, U.S. Geological Survey, Washington, D.C.

Gjessing, E. 1984a. "Acute Toxicity and Chemical Characteristics of Moderately Polluted Runoff from Highways," *Science of the Total Environment*, Vol. 33, pp. 225–232.

Gjessing, E. 1984b. "Effect of Highway Runoff on Lake Water Quality," *Science of the Total Environment*, Vol. 33, pp. 245–257.

Gupta, M. K., et al. 1981. *Federal Highway Administration, Constituents of Highway Runoff*, Vols. I, II, III, FHWA/RD-81/042, 043, 044, Federal Highway Administration, Washington, D.C.

Harper, Harvey. 1979. "Ecological Responses of Lake Eola to Urban Runoff," M.S. Thesis, University of Central Florida, Orlando, Fla.

Hoffman, R. W., Goldman, C. R., Paulson, S., and Winters, G. R. 1981. "Aquatic Impacts of Deicing Salts in the Central Sierra Nevada Mountains, California," *Water Resources Bulletin*, Vol. 71, No. 2, pp. 280–285.

Horner, R. B., and Mar, B. W. 1982. *Guide for Water Quality Assessment of Highway Operations and Management*, Report to the Washington State Department of Transportation by the Department of Civil and Environmental Engineering, University of Washington, Seattle, Wash.

Hunt, H. 1984. *Effects of Bridge Repainting Operations on the Environment: Random Samples*, California Department of Transportation, Division of Engineering Services, Transportation Laboratory, Sacramento, Calif.

Hutchinson, F. E. 1981. "The Influence of Salts Applied to Highways on the Levels of Sodium and Chloride Ions Present in Water and Soil Sample," *Proceedings and Transactions of the Research Board of Maine, Landscape and Environmental Design*.

Hvitved-Jacobsen, T. 1986. "Conventional Pollutant Impacts on Receiving Water," *Proceedings of the NATO Workshop on Urban Runoff Pollution*, Springer-Verlag, Heidelberg, pp. 345–378.

Hvitved-Jacobsen, T., Yousef, Y. A., Wanielista, M. P., and Pearce, D. B. 1984. "Fate of Phosphorus and Nitrogen in Ponds Receiving Highway Runoff," *Science of the Total Environment*, Vol. 33, pp. 259–270.

Irwin, G. A., and Losey, G. T. 1979. *Water Quality Assessment of Runoff from a Rural Highway Bridge near Tallahassee, Florida*, USGS/WRI 79-1, U.S. Geological Survey, Water Research Division, Tallahassee, Fla.

Jenkins, K. D., and Sanders, B. M. 1986. "Assessing Biological Effects of Contaminants in Situ," in *Urban Runoff Quality*, B. Urbonas and L. Roesner Eds., American Society of Civil Engineers, New York.

Jones, J. 1986. "Urban Runoff Impacts on Receiving Waters," in *Urban Runoff Quality*, B. Urbonas and L. Roesner, Eds., American Society of Civil Engineers, New York.

Kobriger, N. P., and Geinopolos, A. 1984. *Sources and Migration of Highway Runoff Pollutants—Research Report*, Vol. III, FHWA/RD-84/059, Federal Highway Administration, Washington, D.C.

Kramme, A., et al. 1985. *Reference Manual for Assessing Water Quality Impacts from Highway Maintenance Practices*, FHWA/RD-85/060, prepared for the Federal Highway Administration, Office of Engineering and Highway Operations Research and Development, Washington, D.C.

Lager, J. A., Smith, W. G., Lynard, W. G., Finn, R. M., and Finnemore, J. E. 1977. *Urban Stormwater Management and Technology: Update and User's Guide*, EPA 600/8.77-014, U.S. Environmental Protection Agency, Washington, D.C.

Larsen, D. P., et al. 1974. "Modeling Algal Growth Dynamics in Shagawa Lake, Minnesota," in *Modeling the Eutrophication Process*, E. J. Middlebrooks, et al., Eds., Ann Arbor Science Publishers, Ann Arbor, Mich.

Lenat, D., and Moody, D. 1983. *Effects of Bridge Maintenance on Benthic Macroinvertebrates in Salem Lake, North Carolina*, Dec. 8, Memorandum to Steve Tedder, North Carolina Department of Transportation, Division of Environmental Management.

Linsley, R. K., Kohler, M. A., and Paulhus, J. H. 1958. *Hydrology for Engineers*, McGraw-Hill, New York.

Mancini, J. L., and Plummer, A. H. 1986. "Urban Runoff and Water Quality Criteria," in *Urban Runoff Quality*, B. Urbonas and L. Roesner, Eds., American Society of Civil Engineers, New York.

6.11 REFERENCES

Manning, M. J., et al. 1977. *Nationwide Evaluation of Combined Sewer Overflows and Urban Stormwater Discharges*, EPA 600/2-77-064c, U.S. Environmental Protection Agency, Washington, D.C.

Martin, E. 1988. "Effectiveness of an Urban Runoff Detention Pond-Wetland System," *Journal of Environmental Engineering, ASCE*, Vol. 11, No. 4, pp. 810–827.

Morre, D. J. 1976. *Chemical Control of Brush and Environmental Safety of Roadside Vegetation Management Chemicals*, Joint Highway Research Project Report JHRP-76-32, Indiana State Highway Commission.

Nakao, D. I., et al. 1977. *The Effects on the Aquatic Environment Due to the Cleaning and Repainting of the Middle River Bridge*, CA-TL-7108-77-29, California Department of Transportation, Office of the Transportation Laboratory, Sacramento, Calif.

Nix, S. J. 1985. "Residence Time in Stormwater Detention Basins," *Journal of Environmental Engineering, ASCE*, Vol. 111, No. 1, pp. 95–100.

O'Connor, D. J. and Dobbins, W. E. 1958, "Mechanism of Reaeration in Natural Streams," *ASCE Transactions*, Paper No. 2934, pp. 641–684.

Osborne, L. L., and Herrick, E. E. 1988. "Habitat and Water Quality Considerations in Receiving Waters," in *Urban Runoff Quality Controls*, L. Roesner, B. Urbonas, and M. Sonnen, Eds., American Society of Civil Engineers, New York, pp. 29–47.

Owens, M., Edwards, R. N., and Gibbs, J. W. 1964. "Some Reaeration Studies in Streams," *J. Air and Water Poll.*, Vol. 8, pp. 469–486.

Parks, D. M., and Winters, G. R. 1982. *Long-Term Environmental Evaluation of Point Residue and Blast Cleaning Abrasives from the Middle River Bridge Repainting Project*, FHWA/CA/TL-82/09, California Department of Transportation, Office of the Transportation Laboratory, Sacramento, Calif.

Patenaude, R. 1979. *Investigation of Road Salt Content of Soil, Water and Vegetation Adjacent to Highways in Wisconsin*, Progress Report III, Wisconsin Division of Highway Transportation Facilities, Materials Section, Soils Unit.

Pedersen, E. R., and Perkins, M. A. 1986. "The Use of Benthic Invertibrate Data for Evaluating Impacts of Urban Runoff," *Hydrobiology*, Vol. 139, pp. 3–22.

Peters, N. E., and Turk, J. T. 1981. "Increases in Sodium and Chloride in the Mohawk River, New York, from the 1950's to the 1970's Attributed to Road Salt," *Water Resources Bulletin*, Vol. 17, No. 4, pp. 586–598.

Phillips, G. R., and Russo, R. C. 1978. *Metal Bioaccumulation in Fishes and Aquatic Invertebrates: A Literature Review*, EPA 600/3-78-103, U.S. Environmental Protection Agency, Washington, D.C.

Portele, G. J. 1982. *Effects of Seattle Area Highway Stormwater Runoff on Aquatic Biota*, Report 11 prepared for the State Department of Transportation, University of Washington, Seattle, Wash.

Rex Chainbelt, Inc. 1972. *Screen/Flotation Treatment of Combined Sewer Overflows*, EPA 11020FDCO 1/72, NTIS PB 215695, U.S. Environmental Protection Agency, Washington, D.C.

Rexnord, Inc. 1984. "Source and Migration of Highway Runoff Pollutants," Vol. 3, Research Report prepared for the FHWA, FHWA/RD-84/059, NTIS PB 86-227915.

Shannon, E. E., and Brezonik, P. L. 1972. "Eutrophication Analysis: A Multivariate Approach," *Journal of the Sanitary Engineering Division, ASCE*, Vol. 89, No. SA1, pp. 350–57.

Shutes, R. B. E. 1984. "The Influence of Surface Runoff on the Macro-Invertebrate Fauna of an Urban Stream," *Science of the Total Environment*, Vol. 33, pp. 271–282.

Smith, M. E., and Kaster, J. L. 1983. "Effect of Rural Highway Runoff on Stream Benthic Macroinvertebrates," *Environmental Pollution, Series A*, Vol. 32, pp. 157–170.

Standard Methods (1985 or latest edition). Published jointly by American Public Health Association, American Water Works Association and Water Pollution Control Federation, 1015 Fifteenth Street, N.W., Washington, DC 20005.

Striegl, R. G. 1987. "Suspended Sediment and Metal Removal from Urban Runoff by a Small Lake," *Water Resources Bulletin*, Vol. 23, No. 6.

Task Committee on Preparation of Sedimentation Manual. 1971, "Sedimentation Discharge Formulas," *Journal of the Hydraulics Division, ASCE*, Vol. 97, No. HY4, Proc. Paper 7786.

Thomann, R. V. 1972. *Systems Analysis and Water Quality Management*, McGraw-Hill, New York.

U.S. Department of Transportation. 1986. *Highway Runoff Water Quality Training Course*, Federal Highway Administration, Washington, D.C.

U.S. Environmental Protection Agency. 1974. *National Water Quality Inventory*, Report to Congress, EPA 440/9-74-001, U.S. EPA, Washington, DC.

U.S. Environmental Protection Agency. 1975. *National Water Quality Inventory*, Report to Congress, EPA 440/9-75-014, U.S. EPA, Washington, DC.

U.S. Environmental Protection Agency. 1978. *Rates, Constants, and Kinetics Formulations in Surface Water Quality Modeling*, EPA/600-3-78-105, Center for Water Quality Modeling, Environmental Research Laboratory, Athens, Ga.

U.S. Environmental Protection Agency. 1979. *Quality Criteria for Water*, EPA-440/9-76-023, U.S. EPA, Washington, D.C.

U.S. Environmental Protection Agency. 1980. "Water Quality Criteria Documents; Availability," *Federal Register*, Vol. 45, No. 231, pp. 97318–97379.

U.S. Environmental Protection Agency. 1984a. "EPA Request for Comments on Nine Documents Containing Proposed Ambient Water Quality Criteria," Vol. 49, Federal Register, 4551, Feb. 7.

U.S. Environmental Protection Agency. 1984b. *Nonpoint Source Pollution in the U.S.*, Report to Congress, EPA 440/9-84-001, Office of Water Programs, Washington, D.C.

U.S. Environmental Protection Agency. 1987. *The Enhanced Stream Water Quality Models QUAL 2E and QUAL 2E-UNCAS: Documentation and User Model*, EPA/600/3-87-007, Environmental Research Laboratory, Athens, Ga.

U.S. Geological Survey. 1974. *1973 Water Resources Data for Florida*, Part 1, Vol. 2, USGS, Tallahassee, Fla., pp. 25–30.

Vollenweider, R. A. 1969. "Moglichkeiten and Grenzen elementarischen Modelle der Stoffbilanz von Seen," *Archiv fuer Hydrobiologie*, Vol. 66, No. 1, pp. 1–36.

Wang, T. S., et al. 1981. *Transport, Deposition, and Control of Heavy Metals in Highway Runoff*, Report 10 prepared for the Washington State Department of Transportation by the Universty of Washington, Department of Civil Engineering, Seattle, Wash.

Wanielista, M. P. 1976. *Nonpoint Source Effects*, University of Central Florida, Orlando, FLA.

Wanielista, M. P. 1980. *Management of Runoff from Highway Bridges*, Contract 99700-7198, Florida Department of Transportation, Tallahassee, Fla.

Wanielista, M. P. 1983. *Stormwater Management*, Ann Arbor Science Publishers, Ann Arbor, Mich.

Wanielista, M. P., Yousef, Y. A., and McLellon, W. M. 1977. "Nonpoint Source Effects on Water Quality," *Journal of the Water Pollution Control Federation*, Part 3, Mar., pp. 441-451.

Wainelista, M. P., Yousef, Y. A., Golding, B. L., and Cassagnol, C. L. 1981. *Stormwater Management Manual*, University of Central Florida, Orlando, Fla.

Warwick, J. J., and Edgmon, J. D. 1988. "Wet Weather 'Quality Modeling,'" *Journal of Water Resources Planning and Management*, Vol. 114, No. 3, pp. 313-325.

Whipple, W., Jr., and Hunter, J. V. 1981. "Settleability of Urban Runoff Pollution," *Journal of the Water Pollution Control Federation*, Vol. 53, No. 12, pp. 1726-1731.

Winters, G. R., and Gidley, J. L. 1980. *Effects of Roadway Runoff on Algae*, FHWA/CA/TL-80/24, California Department of Transportation, Office of Transportation Laboratory, Sacramento, Calif.

Yousef, Y. A., Wanielista, M. P., Carroll, W. E., Fagan, R., and Elmi, H. 1976. *Waste Load Allocation for Tampa Bay Tributaries*, Technical Report ESEI-5, University of Central Florida, Orlando, Fla.

Yousef, Y. A., Wanielista, M. P., Hvitved-Jacobsen, T., and Harper, H. H. 1984. *The Science of the Total Environment*, Vol. 33, Elsevier, Amsterdam.

Zison, S. W., et al. 1977. *Water Quality Assessment*, EPA 600/9-77-023, U.S. Environmental Protection Agency, Athens, Ga.

CHAPTER 7

Stormwater Management Alternatives for Water Quality Improvement

Given a number of stormwater management methods that are socially and politically acceptable, the ones that are chosen for use should be those that achieve the desired technical and economic objective, such as the removal of pollutants at least cost. Presented are some of the technical considerations that lead to the design and operation of some management practices: off-line retention (infiltration) ponds, exfiltration trenches, sedimentation ponds, coagulation with alum, filtration, and swales. In Chapter 8, wet-detention ponds, on-line infiltration, and reuse ponds are discussed. Other methods for management are listed in Appendix G with a brief description of their use and expected mass removal effectiveness. Regulation and other efforts are being made to determine water volume or flow rate criteria for each design based on a specified pollutant removal effectiveness.

7.1 PHILOSOPHIES FOR STORMWATER QUALITY MANAGEMENT

Chemicals, debris, organisms, and other materials in runoff waters may be of a concentration or mass loading that cause an unwanted condition within the receiving waters. These conditions may produce algal blooms, low oxygen levels, violation of receiving water standards, fish kill, depressed property values, aesthetic changes, floating debris, and other alterations of social activities. Philosophies for the management of these stormwater pollutants depend on many factors, some of which are stormwater quantity and quality impacts, mass loadings, desired removal efficiencies, political limitations, local laws, technical data, and cost.

When abundant data on stormwater quality characteristics and receiving water conditions are available, a waste load allocation based on a mass balance can be developed. The quantity and accuracy of the data help determine the level of detail by which the treatment efficiencies can be specified. Since stormwater quality characteristics and receiving water responses to chemicals, organisms, and others are all extremely variable, average conditions or probability distributions are generally specified. The average values for treatment efficiencies are expressed as a mass or concentration removal over a season or a 1-year period of time. More precise descriptions for treatment efficiencies are available in the form of frequency distributions for percent removal per storm event. Stormwater treatment systems for the removal of water-quality-related impurities can be developed based on average yearly or seasonal mass removal or a risk of exceeding a specific effluent concentration or mass discharge.

The specified level of removal also depends on the philosophy of the regulatory and regulated communities. There are many philosophies, although in some respects many of them are similar to those used for point-source water quality control. Some of the frequently used philosophies for stormwater quality control apply to hydrologic changes from new developments, which are summarized as follows:

1. Equate post-land-use mass loadings to pre-land-use mass loadings.
2. Specify a level of treatment that is equitable to other treatment systems (i.e., nonpoint mass removal approximates those of point-source removals).
3. Develop a cost-effective removal quantity beyond which the marginal cost (monetary value per percent pollutant removal or per mass removed) is equal to that resulting from sewage and industrial treatment systems.
4. Specify a level of treatment based on least cost to achieve a desired water quality goal, such as a concentration level or mass removal.

In some locations, the water quality may need to be improved by retrofitting existing land uses. Some philosophies of retrofitting are:

1. Equate current land-use mass loadings and frequency distributions on concentrations to baseline (i.e., woodland, prairie) conditions.
2. Specify a level of treatment based on a least cost allocation of treatment capacity to achieve a stated water quality goal.

The specification and implementation of the philosophy is complicated and depends on the resources of the regulated community along with the expected benefits from the stormwater regulation. Some of the benefits are intangible and must be represented by surrogate measures, such as community preferences. No matter what philosophies are used to regulate stormwater, the efficiency of each stormwater management practice with respect to design parameters and removal mechanisms must be available.

Hydrologic, chemical, biological, and physical processes affect the reduction of pollutant mass and concentration when using stormwater quality management alternatives. All of these processes affect the design and operation of the stormwater systems. Because of the anticipated variability in removal efficiency, average annual efficiencies or probability distributions can be used for effectiveness measures.

7.2 ALTERNATIVE STORMWATER QUALITY MANAGEMENT PRACTICES

A stormwater management practice for quality control is one that reduces the mass or concentration of pollutants in stormwater before its ultimate discharge. The management practice will then improve the quality of the receiving groundwaters and surface waters. Many management practices cited in the references for this chapter indicate the diversity of options adopted by the planning, engineering, and regulatory agencies. In particular, the Delaware (Tourbier and Westmacott, 1974) and Metropolitan Washington (Schueler, 1987) publications are very helpful and aid in further expanding knowledge of some of the management practices reported in this chapter.

Detailed evaluations of selected stormwater management practices have been completed by many, and include those by the EPA on urban planning and control (Kaiser et al., 1974), control of erosion (Thronson, 1971, 1972), combined sewer microstraining and disinfection (Glover and Herbert, 1973), swirl concentrator for combined sewers (Sullivan et al., 1973), and urban controls (Schueler, 1987).

There are three general classifications for stormwater management methods: (1) permit, (2) structural, and (3) nonstructural. Conversion of nonpoint-source runoff to a point source implies control and management through an issuance of a permit to discharge. This first classification is preferred for those cases in which the water quality is obviously degraded resulting from stormwater flows. Because of the tremendous number of stormwater sources and the nondegrading nature of many small flows, the permit process is used for large urban and concentrated rural sources. The second management classification requires hydrologic modification and is frequently accomplished using some structural changes in the flow transport system (i.e., diversion for infiltration, wet-detention ponds). If the structural change is done after the water enters the sewer system, the change is referred to as off-site. If the pollutants can be reduced before transport by sewers to receiving water bodies, the stormwater is said to be managed on-site. The third management classification requires changes that reduce the quantity of pollutants on-site without any structural change. A most successful nonstructural practice is the reduction of lead in gasoline. Listed in Figure 7.1 are some of the structural and nonstructural methods, with a description of each given in Appendix G.

7.2 ALTERNATIVE STORMWATER QUALITY MANAGEMENT PRACTICES 219

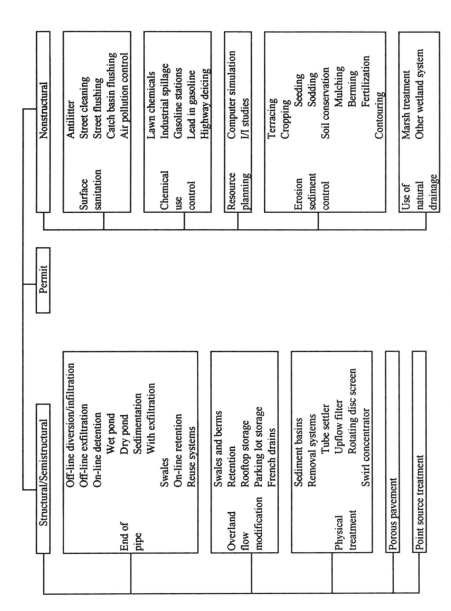

Figure 7.1 Water Quality Management practices (modified from Field et. al., 1977).

7.3 FUNDAMENTAL EFFICIENCY CONSIDERATIONS

The pollutant removal effectiveness of any stormwater system depends on the mass removed or concentration decrease relative to the runoff input mass and concentration over a period of time. Stormwater flows vary with time, and thus the efficiency of a system will vary between events. However, the fundamental considerations of mass balances and the probability distributions for rainfall volume are useful for assessing performance related to size.

7.3.1 Mass Balances

Stormwater management methods used for water quality enhancement are developed to remove a fraction of the mass of a particular pollutant species. Thus the fundamental concepts of a mass balance are applicable. The general mass balance is

$$\text{mass in} - \text{mass out} \pm \text{generation} = \text{accumulation} \quad (7.1)$$

For a stormwater practice, the mass out of the system should be less than the mass into the system. The generation term can be negative, positive, or zero, depending on particular pollutant species.

The mass removal efficiency for a pollutant is calculated relative to the mass input from stormwater, or as presented in previous chapters,

$$\% \text{ removed} = \left[\frac{\text{mass in} - \text{mass out}}{\text{mass in}}\right] \times 100 \quad (7.2)$$

Concentration changes are also used to describe the efficiencies, especially when flow rates and mass of water into and out of a process are equal, thus:

$$\% \text{ removal} = \frac{\text{concentration in} - \text{concentration out}}{\text{concentration in}} \times 100 \quad (7.3)$$

For a pond without an impermeable liner, there may be additional mass from groundwater that enters the pond, resulting in a greater mass out of the pond than mass into the pond from stormwater. The pond is producing more mass, even though the concentration may be reduced. Thus it is important to design the ponds to limit groundwater inputs by increasing storage volume and placing effluent weir controls above the groundwater table levels.

7.3.2 Rainfall Processes and Runoff Diversion

Probability histograms and empirical frequency distributions are used to describe rainfall and runoff events provided that the events are independent of each other. The distributions will aid in determining return periods and exceedence probabilities associated with the independent events, namely rainfall volumes and runoff volumes. A minimum interevent dry period between rainfalls of 4 to 5 h has been used for rainfall independence (DiToro, 1980; Wenzel and Voorhees, 1981; Arnell, 1982; Schilling, 1983). When using rainfall data for runoff analyses, the rainfall interevent dry periods should be long enough to ensure that runoff events are independent of one another.

7.3.3 Diversion Based on the Number of Rainfall Events per Year

The treatment of the runoff from a percentage of all storm events given a minimum interevent dry period can be specified. As an example, consider controlling the runoff pollution associated with 90% of the storm events separated by a minimum interevent dry period. The depth of rainfall associated with the 90% frequency will vary from one region to another as shown in Table 7.1, which assumes a 4-h minimum interevent dry period between storm events. The frequency of storm events distribution as a function of storm volume (see Figure 2.1a) assumes that no rain occurred for a specified period of time. In general, inland regions with relatively low annual rainfalls have storm volumes of about $\frac{1}{2}$ in. at the 90% event frequency level. Mountain regions with high annual rainfalls have greater volumes per storm event. Coastal regions in the mid-Atlantic and south Atlantic states have frequency distributions for which about 90% of the storm events have a rainfall volume of 1 in. or less. These statistics will vary among locations.

The importance of the rainfall event frequency distribution is that it is related to annual mass of pollutants in the runoff. A portion of the rainfall will appear as runoff and can be assumed to follow the event frequency distribution of rainfall. Both appear to have a distribution approximated by an exponential or gamma distribution. If all the runoff water from 90% of the storm events can be diverted, the mass associated with 90% of the storm

TABLE 7.1 Approximate Storm Event Depth (inches) for the 90% Event Frequency of Storms and Yearly Volumes (with Minimum Interevent Dry Periods of 4 Hours)

Location	Approximate Volumes (in.)	
	≥ 90% Frequency	Annual Volume
Florida	1.0	52
Boston, Massachusetts	0.9	45
Baltimore, Maryland	1.0	42
Denver, Colorado	0.5	15
Washington, D.C.	1.0	44
Austin, Texas	1.0	32
Cardiff, Wales	1.0	45
Treherbert, Wales	2.0	92

events will be removed. Thus a design parameter for the runoff volume associated with any event frequency can be specified to establish the size of the treatment. If this volume is diverted for treatment and never discharged directly (off-line treatment), the mass of pollutants associated with the volume will be diverted from direct surface discharge. For the state of Florida, the runoff from all storm events with 1 in. or less of rainfall were specified to be diverted in Florida (Livingston, 1985). However, for the Colorado area (Douglas County, 1986) the runoff from $\frac{1}{2}$ in. over the impervious area was specified. If the runoff waters from 90% of the rainfall events can be diverted (off-line) or held on-line for complete mass removal before discharge to a ground or surface water, a pollutant mass removal efficiency as a frequency distribution and as an annual average can be estimated.

7.3.4 Diversion Based on the Volume of Runoff per Year

Cumulative volume of rainfall and runoff can be used to specify performance standards in addition to the frequency of storm events. The volume of runoff water can be calculated from a frequency distribution of event rainfall volumes if all the rainfall events result in runoff. The volume of rainfall diverted as a percentage of the total rainfall for each stipulated diversion rainfall event depth is shown for a specific central Florida location in Figure 7.2. Fifteen years of rainfall data were used to calculate the yearly average shown in Figure 7.2. One or a few years should not be used because the short record may not represent an average. The rainfall volume is calculated from a frequency distribution of rainfall events. The equation used to calculate the rainfall volume associated with the events in the range 0 to 0.10 in. given a histogram is calculated using an expected value equation and is written as

$$\text{volume}_{0.1} = \Pr(v_{0.1}) N \bar{v}_{0.1}$$

where $\text{volume}_{0.1}$ = volume of rainfall in the histogram interval 0 to 0.1 in.
$\Pr(v_{0.1})$ = empirical probability calculated from the number of events in the interval (n_i) and the total number
N = total number of events
$\bar{v}_{0.1}$ = average volume of rainfall in the interval

The stated diversion depth will be diverted no matter how large the storm. As an example, consider the diversion event depth equal to $\frac{1}{2}$ in. This specifies that events up to $\frac{1}{2}$ in. will be diverted and those events over $\frac{1}{2}$ in. will have the first $\frac{1}{2}$ in. diverted and the remaining amount will be directly discharged. If an off-line infiltration pond was specified, the cumulative volume of runoff for a stated diversion event depth can be calculated knowing the time for infiltration. The minimum interevent dry period for calculating the percent diversion volume is set equal to or greater than the

7.3 FUNDAMENTAL EFFICIENCY CONSIDERATIONS   223

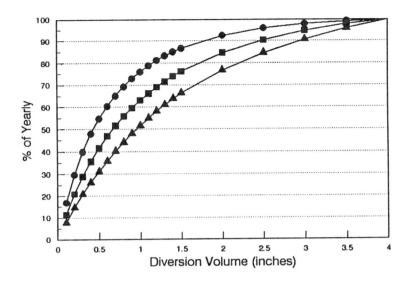

Figure 7.2 Percentage of yearly rainfall volume diverted (in inches) for fixed diversion volumes and minimum interevent dry periods of 4, 24, and 72 h for one location. Formula for calculation of diversion volume given a histogram of rainfall events with average rainfall for each interval:

$$\text{volume} = \sum \text{Pr}(v) N \bar{v}_i + \sum [(1 - \text{Pr}(v_{dv}))](dv) N$$

where volume = volume of rainfall per year for diversion dv
 $\text{Pr}(v_i)$ = probability of event volume in the interval i
 N = number of total events
 dv = diversion volume
 $\bar{v}_i$ = average volume in interval i

time for infiltration. Then, before the next storm event starts, the infiltration pond will be empty. The percent volume diversion increases by decreasing the interevent dry period from 72 h to 4 h (Figure 7.2). Of course, increasing the interevent dry period would reduce the number of events and increase the runoff volume per event. It appears that volume diversion may be overestimated when infiltration occurs during the storm event; however, the interevent dry period time can be reduced by the infiltration time during the storm event. If concentrations of pollutants were relatively constant with time, the mass of pollutants removed would be equal to the fraction of yearly runoff diverted. On very large watersheds, it was shown that mass was related

224 STORMWATER MANAGEMENT ALTERNATIVES FOR WATER QUALITY IMPROVEMENT

to flow by a linear relationship, which implies that the concentrations can be represented by a constant value (Wanielista, 1977). This is equivalent to saying that on large watersheds, the first-flush effect is nonexistent. Figure 7.2 can therefore represent the average yearly efficiency (percent removal) given a stated event design level, provided that first-flush events do not exist. Detailed calculations are shown in Table 7.2.

TABLE 7.2 Diversion Volume Analysis for Orlando, Florida: 15 Years of Data, 4-Hour Minimum Interevent Dry Period

Prec.[a]	n_i^b	$\bar{v}_i$	$\sum n_i$	$n_i \bar{v}_i$	$\sum n_i \bar{v}_i$	Col.[c] (in.)	Vol. (%)
0.10	396	0.05	396	19.8	19.8	125.0	18.0
0.20	284	0.15	680	42.6	62.4	216.0	31.1
0.30	133	0.25	813	33.3	95.7	286.2	41.2
0.40	107	0.35	920	37.5	133.1	344.3	49.5
0.50	93	0.45	1013	41.9	175.0	392.5	45.4
0.60	70	0.55	1083	38.5	213.5	432.5	62.2
0.70	46	0.65	1129	29.9	243.4	466.7	67.1
0.80	43	0.75	1172	32.3	275.6	496.4	71.4
0.90	36	0.85	1208	30.6	306.2	522.2	75.1
1.00	36	0.95	1244	34.2	340.4	544.4	78.3
1.10	35	1.05	1279	36.8	377.2	563.1	81.0
1.20	23	1.15	1302	26.5	403.6	578.8	83.2
1.30	18	1.25	1320	22.5	426.1	592.5	85.2
1.40	13	1.35	1333	17.6	443.7	604.7	87.0
1.50	17	1.45	1350	24.7	468.3	615.3	88.5
2.00	47	1.75	1397	82.3	550.6	652.6	93.9
2.50	22	2.25	1419	49.5	600.1	672.6	96.7
3.00	11	2.75	1430	30.3	630.3	684.3	98.4
3.50	8	3.25	1438	26.0	656.3	691.3	99.4
4.00	4	3.75	1442	15.0	671.3	695.3	100.0
> 4.00	6	4	1448	24.0	695.3	695.3	100.0
	1448						

[a]Precipitation interval from histogram of rainfall volumes (0.04–0.10 in., 0.11–0.20 in., etc.)
[b]Number of rainfall events in each precipitation of interval.
[c]Vol., diversion volume for the 15 years of records associated with a diversion depth of 0.10 in. and is calculated as

$$\text{vol. at } 0.10 \text{ in.} = \sum_{i=0}^{0.10} \frac{n_i}{N} \bar{v}_i N + \left(1 - \sum_{i=0}^{0.10} \frac{n_i}{N}\right)(dv)N$$

$$= \sum_{i=0}^{0.10} n_i v_i + \left(N - \sum_{i=0}^{0.10} n_{0.10}\right)(dv)$$

$$= (396)(0.05) + (1448 - 396)(0.10)$$

$$= 19.8 + 105.2 = 125.0 \text{ in.}$$

If first-flush events (higher concentrations and mass in the early time of runoff rather than latter times) are present, the average yearly efficiency should increase because the concentrations or quantities of pollutants are treated during the early part of a runoff event. Thus, for small watersheds, either the efficiencies expected from diversion systems will be higher or the stated volume of treatment can be reduced.

7.4 OFF-LINE RETENTION (DIVERSION) SYSTEMS

Once runoff water is diverted for off-line stormwater management, it is treated and either returned for surface discharge or exfiltrated into the ground and possibly evaporated. Off-line treatment can be either sedimentation basins, chemical coagulation tanks, exfiltration (frequently called infiltration) ponds, exfiltration pipe, or large storage ponds which provide a complex system for settling, infiltration, evaporation, and chemical and biological treatment.

In urban areas with a large impervious area and very high land cost, the off-line system is placed underground in the form of perforated or slotted pipe. The pipe is placed in exfiltration gravel (washed $\frac{3}{4}$-in. size) and the gravel wrapped in permeable filter cloth. Any off-line system that depends on the use of exfiltration and percolation can be designed so that the bottom of the pond or exfiltration pipe is above the water table most of the time. Generally, this can be achieved by placing the pond bottom approximately 1 ft above the seasonal high water table.

During a storm used for flood control design (i.e., 25-year, 24-h), the off-line system treats the runoff from the first flush, and as such, the peak discharge may not be attenuated because the peak occurs later in the runoff event. However, lower-volume storms will have an attenuated peak discharge.

The quantity of runoff diverted can be expressed in terms of the depth of runoff over the entire watershed or the diversion of the runoff from the first quantity (depth) of precipitation. Examples of both criteria are: (1) divert the first $\frac{1}{2}$ in. of runoff (over the total watershed) and (2) divert the runoff from the first inch of rainfall. If the runoff coefficient for a watershed is 0.5, the runoff from the first inch of rainfall is the same as the first $\frac{1}{2}$ in. of runoff. Thus the criteria may be consistent with one another.

A percent annual diversion volume control can be determined from an annual volume diversion graph (Figure 7.2). Using Figure 7.2 as an example, 80% of yearly volume is diverted if the diversion depth is set to capture the first 1.1 in. of runoff (runoff = rainfall) for each and every storm, and the treatment system recovers full capacity in 4 h. An annual diversion volume curve can be developed for specific areas for minimum interevent dry periods to determine precipitation volume (inches). The runoff coefficient used to convert rainfall to runoff is usually a conservative estimate (large) because the complete infiltration volume may not be recovered before the next rainfall event.

The timing and volume of rainfall with different treatment rates affect the size of the treatment volume. The treatment volume may be estimated using a pond mass balance by simulating the pond water level over a long period of time. The volume of runoff in the pond is affected by both the addition of new runoff water and the removal of water already in the pond in a given period of time.

A general mass balance for an off-line exfiltration pond, trench, sump, or rock well can be written as a special form of equation 7.1 using volume units:

rainfall excess − exfiltration = change in storage

$$R - F = \Delta S \qquad (7.4)$$

A computer simulation for this mass balance equation can be accomplished using a commercially available spreadsheet program or a user-defined computer code. Following are the variable definitions and equations that can be used in a simulation for an equivalent impervious area of one acre (EIA in ft^2) and volume units of ft^3:

EIA   Equivalent impervious area = 43,560 ft^2
f Exfiltration rate, in./h through an area
S_{min} Storage volume within the off-line structure that is always below the groundwater table and thus is not exfiltrating, ft^3
S_{max} Storage volume of the structure (ft^3) at full capacity
P Precipitation depth, in.
R Runoff as rainfall excess that equals precipitation if watershed is 100% contributing, ft^3
A_c Area of structure available for exfiltration, ft^2
F Volume of water exfiltrated during hour, ft^3

Some of the simulation equations are:

Rainfall excess: $R = (\text{EIA})P/(12 \text{ in./ft})$

Filtration volume: $F = fA_c(\Delta t, \text{h})/(12 \text{ in./ft})$

Storage: $\Delta S = R - F$ with $S_{min} < S < S_{max}$

At the beginning of the simulation, the structure is considered to be at the initial design condition. For each hour simulated, the mass balance is calculated as follows:

1. $S_{initial} = S_{final}$ from the previous hour.
2. R is calculated from rainfall and equivalent impervious area and converted to ft^3.
3. $S_{average} = \frac{1}{2}[S_{initial} + (S_{initial} + R)]$.
4. From $S_{average}$, depth in structure is calculated so that depth is less than or equal to design maximum depth (note: a volume–depth relationship must be available).
5. A_c is calculated from depth of the structure and storage volume.
6. F is computed as $fA_c(\Delta t)/(12 \text{ in.}/\text{ft})$, where $\Delta t = 1$ h.
7. During any period, if $S_{init} + R - F > S_{max}$, discharge ($D$) is calculated as $D = S_{init} + R - F - S_{max}$.
8. $S_{final} = S_{init} + R - F - D$.
9. S_{final} is carried forward as S_{init} for the next time period.

This balance is repeated for every hour during the simulation record. At the end of the period efficiency is calculated as (total F)/(total R). Exfiltration rate (f) can be varied until efficiency values of 50, 60, 70, 80, 85, 90, and 95% are determined. This was repeated by Wanielista et al. (1991) for structure volumes of 0.2, 0.4, 0.6, 0.8, 1, and 1.2 in. over the equivalent impervious watershed. Curves of equal diversion efficiency can be plotted, relating exfiltration rate [(ft^3/s)/EIA] to design volume (inches over the equivalent impervious area). Results of this simulation are shown in Figure 7.3. The exfiltration rates are multiplied by the total exfiltration area (ft^2) above the GWT to convert the exfiltration rates (in./h) through the structure exfiltration area to a more general removal rate (ft^3/s).

Another simulation for the mass balance expressed by equation 7.4 can be constructed. This time, define the variables exfiltration rate (ft^3/s per equivalent impervious area) and design volume (inches over the equivalent impervious area). Also, include the transport rate for pollutant mass expressed as a first-order transport rate (see Section 5.8.1). The mass balance can be developed using volume units of ft^3/EIA:

1. Structure design volume (V_{design}, ft^3), transport rate (K, /in.), and removal rate [R_Q, (ft^3/s)/EIA] are fixed for each simulation.
2. For the first time step the storage volume is assumed to be empty, $S_{initial} = 0$.
3. Hourly runoff volume (R) is input to the structure, $S_1 = S_{initial} + R$.

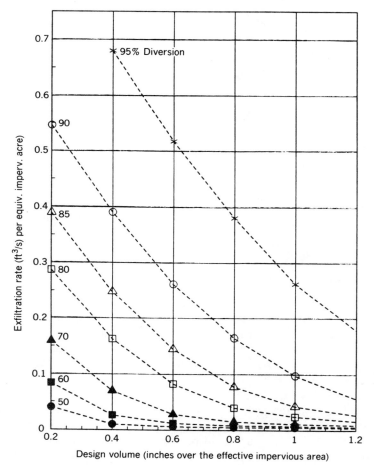

Figure 7.3 Water removal rate as a function of design volume for seven different annual diversion efficiencies using rainfall data from Orlando, Florida.

4. Volume removed is calculated for one acre of EIA as:

$$F = R_Q(1 \text{ h})\left(\frac{3600 \text{ s}}{\text{h}}\right)$$

$$= 0.9917 R_Q \qquad (7.5)$$

where F = volume exfiltrated in 1 h (ft^3)
  R_Q = exfiltration rate = $fA_c/(12)(3600)$ [(ft^3/s)/EIA]

The maximum volume that can be exfiltrated is initial storage + input, or if $F >= S_1$, then $F = S_1$.

7.4 OFF-LINE RETENTION (DIVERSION) SYSTEMS

5. $S_{final} = S_{initial} + R - F$, with a maximum equal to the trench design volume.
6. If applicable, discharge is calculated as $D = S_{final} - V_{design}$, and $S_{final} = V_{design}$.
7. S_{final} is carried forward as $S_{initial}$ for the next hour.
8. Cumulative runoff and exfiltration volume are calculated, and annual average diversion efficiency is computed as diversion $= (\Sigma F)/(\Sigma R)$.

This mass balance is continued each hour in the period under consideration. For the purposes of developing generalized design curves that are, however, specific to a location, the following apply:

1. Watershed area is 1 acre, 100% directly connected impervious or the equivalent impervious area is 1 acre (43,560 ft^2).
2. Hourly precipitation data are used to calculate runoff, all runoff is assumed to arrive at the off-line structure at the beginning of the hour.

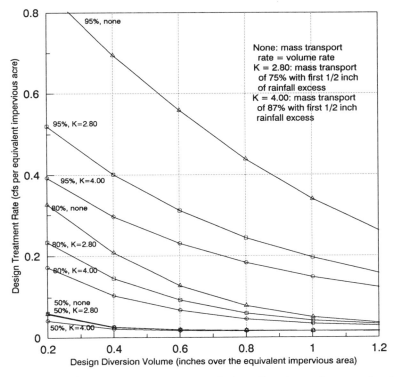

Figure 7.4 Diversion mass percentages at design diversion volume and rate for transport rates at West Palm Beach, Florida.

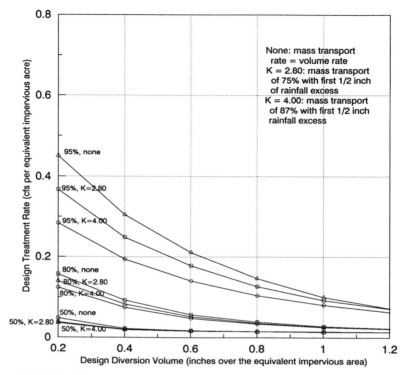

Figure 7.5 Diversion mass percentages at design diversion volume and rate for transport rates at Atlanta, Georgia.

Data should be available for the period of analysis (15 years has been used before; Wanielista et al., 1991).

3. Diversion efficiency is calculated for 1510 combinations of V_{design} and R_Q, with V_{design} varying from 0.2 to 2.0 in. in 0.2 in. steps, and R_Q varying from 0.005 to 0.750 ft^3/s in steps of 0.005 ft^3/s.

4. First-flush transport of pollutant mass is assumed to be associated with runoff (rainfall excess): expressed by the transport rate factors $K = 2.8$ and 4.0 per inch, developed from equations 5-16 through 5-20.

To study the effects of geographic location (i.e., precipitation volume, intensity, and interevent dry periods), the mass balance for pollution transport can be repeated for other areas (Wanielista et al., 1991). Typical results are shown in Figures 7.4, 7.5, and 7.6. The curves are different for the three areas because of the rainfall volumes and interevent dry periods, and the pollutant transport rates affect the results at higher annual average diversion percentages.

7.5 SEDIMENTATION

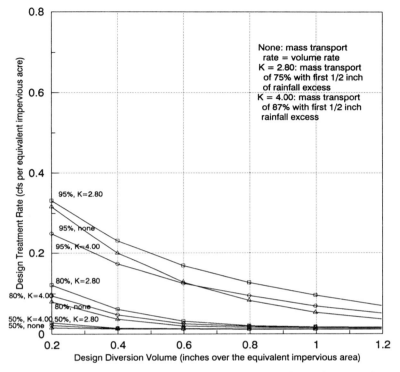

Figure 7.6 Diversion mass percentages at design diversion volume and rate for transport rates at Roanoke, Virginia.

Example Problem 7.1 An area in West Palm Beach has 100 acres of EIA, from which 80% of the annual average runoff volume must be diverted for surficial aquifer recharge. The regulatory agency requires an off-line exfiltration pond to store the runoff of $\frac{1}{2}$ in. from the EIA. What is the area of the pond assuming an exfiltration rate of 4 in./h?

SOLUTION: Using the results of the mass balance simulation for West Palm Beach (Figure 7.4) and with no first flush (mass transport cure), the design treatment rate is 0.16 (ft^3/s)/EIA, and for 100 acres of EIA, it is 16 ft^3/s. The exfiltration area is

$$A_c = \frac{16 \text{ ft}^3/\text{s}(3600 \text{ s/h})(12 \text{ in./ft})}{(4 \text{ in./h})(43{,}560 \text{ ft}^2/\text{acre})}$$

$$= 4 \text{ acres}$$

7.5 SEDIMENTATION

Sediment will settle to the bottom of stormwater ponds (Yousef et al., 1986*b*; McCuen, 1980; Whipple, 1979). These ponds can be designed to remove sediment and are then called sedimentation ponds. These should not be confused with wet-detention ponds, which are designed to obtain higher pollution control efficiencies, including the removal of smaller particles and dissolved materials.

7.5.1 Sedimentation Ponds: Physical Configuration

Sedimentation ponds were first used to control sediment that was eroded from agricultural lands. More recently, they are being used in urban environments to remove sediment before wetlands and before wet-detention ponds, so that the capacity of the wet-detention ponds are not reduced. The physical configuration of a pond consists of an inlet control, settling volume, debris and sedimentation volume, and an outlet (Figure 7.7). Dual-purpose sediment ponds are used to manage both peak discharge and remove sediment.

The inlet control system to the pond may be constructed to spread the influent and reduces the velocity. Velocity dissipation devices such as berms and baffles placed at the bottom of the incoming channel or in the pond are used to maximize sediment accumulation near the inlet.

The settling volume is greatest when velocities are minimum. The influent peak discharge, particle size, and desired efficiency determine the surface area and depth. The influent and effluent flow rates must produce a detention time long enough to settle certain-size particles. The debris and solids carried in runoff water accumulate in the sediment and debris volumes. The volume is determined by the annual sediment load, the yearly efficiency of removal, and the density.

The outlet controls the rate of discharge and holds velocities low to improve settling and prevent the carryover of solids during high flow conditions. In Figure 7.7 a perforated riser pipe is shown; however, weirs are also used. Floating-debris (paper, balls, etc.) traps are also needed to prevent these from leaving the pond. The overflow spillway is designed for the rare event (10-year, 25-year, or greater return period). The outlet control for the settling volume is designed for a volume of runoff and an interevent dry period (time of no rainfall). The dry period is selected equal to the maximum time for settling. The time for settling is based on the particle settling velocity.

7.5.2 Settling Velocities

Approximate settling velocities for various-size sediment particles are listed in Tables 7.3 and 7.4. Note that stormwaters have a wide range of particle sizes as expressed by general soil classifications of clay, silt, and sand. Thus, from Tables 7.3 and 7.4, the effectiveness of any settling pond for the removal of solids is a direct function of the frequency distribution on the size of particles in stormwater. Using stormwater inputs into sedimentation ponds and other stormwater tests for distributions on particle sizes, the percentage

7.5 SEDIMENTATION 233

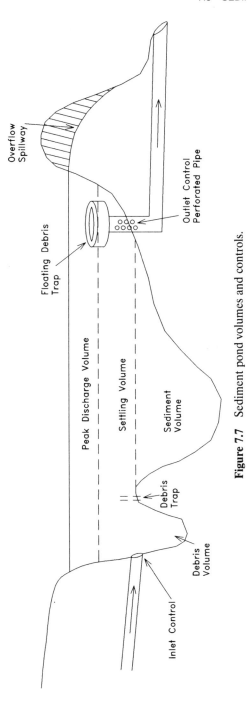

Figure 7.7 Sediment pond volumes and controls.

TABLE 7.3 Approximate Settling Velocities for Various Sediment Particles

Diameter		Type of Particle	Settling Velocity
mm	μm^a		
8–10	8000–10,000	Gravel	100–200 ft/min
0.5–1	500–1000	Coarse sand	15–20 ft/min
0.25–0.5	250–500	Medium sand	2–10 ft/min
0.08–0.1	80–100	Fine sand	30–80 ft/h
0.002–0.05	2–50	Silt	1–2 ft/h
< 0.0001	< 2	Clay	0.005–0.01 ft/day

Source: U.S. Department of Agriculture 1981.
$^a\mu m$ = 0.001 mm or about 1/25,000 of an inch.

TABLE 7.4 Spherical Particle Diameters and Settling Velocities Calculated Using Stokes Law with Specific Gravity of 2.65 and 20° C

Diameter (μm)	Velocity (ft/s)	Diameter (μm)	Velocity (ft/s)
8	0.00019	30	0.0027
10	0.00029	40	0.0047
12	0.00042	50	0.0074
15	0.00066	60	0.011
20	0.0012	80	0.019
25	0.0018	100	0.029

Sources: Malcom and New (1975).

TABLE 7.5 Percentage of Particle Mass in Stormwater as Related to Average Settling Velocity

Percent of Mass in Urban Runoff	Average Settling Velocity (ft/h)
0–20	0.03
20–40	0.33
40–60	1.5
60–80	7.0
80–100	70.0

Source: Adapted from U.S. Environmental Protection Agency (1983).

of particles in urban runoff with average settling velocity can be estimated using Table 7.5.

7.5.3 Design Storm Sizing

For some reservoirs, engineers have formulated empirical equations from sediment removal (Brune, 1953). These are based on assumed particle sizes for a specific location and specific design flow. Design conditions are specified and sediment removal estimate for the design condition.

Stormwater particles may be discrete or flocculent. Classical equations for settling mechanics can be used to estimate discrete particle removal efficiencies. For discrete nonflocculating particles without hindrance or help from each other, Stokes law can be used to define settling. There are many simplifying assumptions using Stokes law which cannot be duplicated in stormwater ponds, such as discrete nonflocculant particles, equally distributed velocity vectors, plug flow, and complete removal once the particle strikes the bottom. Nevertheless, the settling mechanisms for discrete particles are important for some design and for an understanding of the removal mechanisms.

Most particles that settle with velocities greater than that required by the design should be effectively removed, where the design velocity is defined as

$$v_0 = \frac{h_0}{t_d} \quad (7.6)$$

where v_0 = design settling velocity, ft/h
h_0 = depth of settling, ft
t_d = pond detention time, h

and the pond detention time is

$$t_d = \frac{V_p}{3600 Q} \quad (7.7)$$

where t_d = detention time, h
V_p = volume of pond, ft^3
Q = outflow rate based on a design storm, ft^3/s
3600 = seconds per hour

with a pond overflow rate of

$$\text{OR} = \frac{Q \times 3600}{A} \quad (7.8)$$

where OR = overflow rate, ft/h
A = surface area, ft^2

Thus the surface area as a function of design settling velocity, which is equal to the overflow rate, is calculated from

$$A = \frac{Q \times 3600}{v_0} \tag{7.9}$$

The outflow rate depends on a storage–discharge relationship for the control device. The change in storage depends on the difference between the inflow and outflow rates, and the inflow rate is based on a specific rainfall volume, rainfall distribution, and watershed characteristics. If average precipitation volume and intensity are used, the overflow rate and detention time will be available over 50% of the time (because the mean is greater than the median for these type of distributions).

Example Problem 7.2 Environmental impact on fishing bed areas requires that particles greater than 8 µm in diameter be removed from stormwater 90% of the time. The watershed area is 240 acres with a runoff coefficient of 0.3. The settling pond volume is 4.0 ft deep. The rainfall intensity associated with the 90% or less occurrence is 0.4 in./h. (Data for intensity are given in Figure 2.11, Chapter 2 for large geographic areas in the United States.)

SOLUTION: From Table 7.4, the settling velocity is 0.00019 ft/s. The design runoff rate is calculated as

$Q = (0.4 \text{ in.}/\text{h})(0.3)(240 \text{ acres})(43{,}560 \text{ ft}^2/\text{acres})(1 \text{ ft}/12 \text{ in.})(1 \text{ h}/3600 \text{ s})$

$= 29 \text{ ft}^3/\text{s}$

Using equation 7.9 and assuming outflow rate equals runoff rate:

$$A = \frac{29 \times 3600}{(0.00019 \text{ ft/s})(3600 \text{ s/h})} = 152{,}632 \text{ ft}^2 \text{ or } 3.5 \text{ acres}$$

Since depth is 4.0 ft, volume = 14.0 acre-ft. With this volume, check the design settling velocity using equations 7.6 and 7.7.

$$t_d = \frac{V_p}{3600 Q} = \frac{609{,}840 \text{ ft}^3}{3600(29)} = 5.84 \text{ h}$$

and

$$v_0 = \frac{h_0}{t_d} = \frac{4.0}{5.84} = 0.68 \text{ ft/h} = 0.00019 \text{ ft/s}$$

Thus the design will settle particles with faster rates (0.00019 ft/s) 90% of the time.

7.6 REMOVAL OF DISSOLVED CONTAMINANTS IN STORMWATER

The flow rate through a pond is rarely constant and usually follows a hydrograph form. Also, the size/weight composition of particles vary. Thus detention time, storm efficiencies, and overflow rate vary.

7.6 REMOVAL OF DISSOLVED CONTAMINANTS IN STORMWATER

Some organics and metals are dissolved in solution but can be removed by adsorption. Adsorption is the process whereby an adsorbate (some organics and metals) moves from the solution phase to the surface of an adsorbent, where it is held by attractive forces. These forces of attraction may be physical, chemical, electrical, or a combination of the three. Some soils and carbon are excellent for adsorption. Soils high in silt content may remove some impurities. Heavy metals from highway runoff were greatly attenuated in the soils and most heavy metals in stormwater ponds were associated with the bottom sediments (Yousef, 1984). Maximum contact with soils was suggested, which leads to a design using swales, selected soil filtration material, or activated carbon. Other alternatives for removing dissolved chemicals in stormwater are the addition of a chemical to induce coagulation, and biological uptake.

7.6.1 Activated Carbon

Commercial carbons can be prepared from a variety of raw materials, including wood, lignite, coal, bone, petroleum, and nut shells. After preparation, the carbons are termed activated carbon and in a granular form are called granular activated carbon. Activated carbons are carbonaceous materials that are subjected to selective oxidation to produce a highly porous structure. The high porosity and surface area gives activated carbon its adsorptive properties.

In water treatment activated carbon is used to remove compounds that cause objectionable taste, odor, or color. Recently, suspect carcinogens have been partly removed. In industrial wastewater and stormwater treatment it is mainly used to adsorb toxic, organic compounds, and metals.

Three of the major considerations in the application of activated carbon are the form of the activated carbon, the carbon capacity, and the rate of adsorption. The capacity and rate of adsorption are influenced by the nature and concentration of the activated carbon, and nature and concentration of the impurities present, the nature of the solvent, and the environmental and operating conditions.

The equilibrium adsorptive capacity may be determined from simple isothermal batch tests. Various dosages of activated carbon are added to different containers of the stormwater to be treated, and the amount of organics removed calculated by the expression

$$X = (C_0 - C_e)V \qquad (7.10)$$

where X = mass of organics adsorbed, mg
C_0 = initial organic concentration, mg/L
C_e = equilibrium concentration achieved, mg/L
V = test volume treated, L

The specific adsorption capacity, X/M or q, is then calculated by dividing both sides of equation 7.10 by the mass of carbon added, M (mg).

$$\frac{X}{M} = (C_0 - C_e)\frac{V}{M} \qquad (7.11)$$

If we call the mass of carbon added per unit volume the dosage, D (mg/L), equation 7.11 can be expressed by

$$\frac{X}{M} = \frac{C_0 - C_e}{D} \qquad (7.12)$$

Langmuir and Freundlich Isotherms. The Langmuir and Freundlich isotherms are used to relate the amount of solute adsorbed to the per unit mass of adsorbent. The Langmuir equation may be written as

$$q = \frac{X}{M} = \frac{b(X/M)_0 C_e}{1 + bC_e} \qquad (7.13)$$

where b = constant related to energy of adsorption
$(X/M)_0$ = adsorbate required to produce a complete monolayer or ultimate adsorption capacity

The Freundlich equation for isothermal adsorption was developed from empirical relations and is written

$$\frac{X}{M} = K_f C_e^{1/n} \qquad (7.14)$$

where K_f and n are constants. The Freundlich equation may also take the form

$$q = \frac{C_0 - C_e}{D} = K_f C_e^{1/n} \qquad (7.15)$$

A linear form of the Freundlich equation may also be obtained by taking the logarithm of the equation such that $\log X/M = \log K_f + 1/n \log C_e$. This linear form may be used to evaluate the system constants K_f and $1/n$ from plotting experimental data either on log-log paper or by taking the log of the values and plotting on arithmetic graph paper.

When using columns of activated carbon, complete exhaustion of the carbon can occur and the entire adsorption column is in equilibrium with the influent and effluent flows (Reynolds, 1982). From laboratory column studies using granular activated carbon (Filtrasorb 300, Calgon Corporation) and highway runoff waters, about 3 to 5 lb of granular activated carbon were needed per acre-inch of runoff. Ten thousand pounds will probably provide a 1-year operation before breakthrough for an area of 40 acres and 42 in. of runoff per year (Wanielista et al., 1988). The use of granular activated carbon is a very expensive alternative and should be considered only when all other alternatives are either not available or not economical.

7.6.2 Coagulation

Coagulation is the process for the removal of colloidal or fine suspended matter by agglomeration. Coagulation results from the addition of an iron or aluminum salt. Alum, $Al_2(SO_4)_3 \cdot 14H_2O$, is the coagulant most commonly used and aluminum hydroxide is the precipitate that forms. However, the pH of the liquid is usually reduced which can produce a problem in receiving waters.

$$Al_2(SO_4)_3 \cdot 14H_2O + 6HCO_3 \rightarrow 2Al(OH)_3 + 3SO_4 + 6CO_2 + 14H_2O$$

Laboratory experimentation using alum injection into stormwaters indicates that a dose as low as 6 mg/L, but as high as 25 mg/L, may be necessary for coagulation. It is important to maintain a minimum alkalinity of 25 mg/L and a pH of 6.0 after the application of alum treatment (Harper and Wanielista, 1984). A concentration of 10 mg/L applied to stormwater was sufficient to reduce total nitrogen by 60%, dissolved orthophosphorus by 90%, and suspended solids by 80%. The concentration of alum added to the stormwater should reflect the chemical concentrations in the stormwater.

The injection system consists of a pump, a chemical holding tank, a flow measurement device, a feed hose, and various fittings. Chemicals are pumped out of the holding tank through a connecting hose. The system is designed to be installed inside a modified manhole or in a separate housing unit. The chemical injection rate is controlled by two factors: flow rate and concentration of alum. A constant chemical injection rate can be obtained by controlling the overflow from the system in a linear fashion. A flow proportional dosage pump may be used. Also, this can be accomplished through the use of a linear proportional weir. The stormwater and chemical are then mixed rapidly by passing through a pipe before discharge.

A concern is the accumulation of sludge produced by this system. For a 100-acre watershed and a dose of 10 mg/L, alum sludge is projected to accumulate to a depth of approximately 16.0 cm/yr (6.3 in./yr) in a 1-acre detention pond. This floc would slowly become incorporated into the sediments, as has been observed for inactivants applied to sediments. This gradual slow deposition onto the sediments would also inactivate existing nutrients in the sediments, preventing their release into the water column.

Construction cost of four automatic alum injection systems to treat a yearly runoff of about 150 acre-ft is estimated at $100,000. Annual alum use would be approximately 5000 gal/yr and total nitrogen removal between 50 and 80%. If sand filters are used, the rate of filtration has been reported to be approximately 8 to 10 in./h (0.90 gal/ft^2-min). The city of Austin, Texas has experience with these filters and specifies treatment of the first $\frac{1}{2}$ in. of runoff (Oswald, 1986).

The applications of these filters are for areas with high sinkhole potential, as pretreatment systems for exfiltration pipes (underground pipes), and in areas where surface detention ponds cannot be used. In high-sinkhole-potential areas, excessive waters must be prevented from infiltrating into the ground and thus creating a sinkhole. In exfiltration pipes sediment is more difficult to remove than from a filter bed.

7.7 SWALES

Swales are vegetated open channels that are designed to transport, treat, and store runoff. By constructing a swale with low velocities and high infiltration capacity, impurities in the runoff waters can be removed by settling, filtration, physical, chemical, and biological processes. It was shown by Wanielista et al. (1978) that shoulder areas of highways were very effective for the removal of hydrocarbons, metals, and solids. By incorporating the hydrologic processes of runoff and infiltration with those processes for the removal of impurities, a swale design based on quantity and quality is possible.

Erosion control also results when swales are used. However, strict limits on water velocities are important to prevent transport and loss of soil. The vegetation within the swales have been reported to be very effective for the removal of solids and retention on soil (Lord, 1986).

Contaminants in surface runoff, such as nutrients (nitrogen and phosphorus) and heavy metals (lead, zinc, copper, nickel, iron, cadmium, and chromium), can be reduced and retained on the site. However, regeneration and relocation of loosely bound contaminants may occur at intermediate locations throughout the length of the swale. Particulate contaminants are filtered out by the grassy cover on swales and settle and settle to the bottom sediments.

7.7.1 Swale Hydraulics and Hydrology

Flow of water over swales is characterized by a thin depth in comparison with the width of flow. The flow may become a wide-open-channel flow shown as sheet flow or overland flow. Uniform flow over swales may be turbulent or laminar, depending upon such factors as discharge, slope, viscosity, and

7.7 SWALES 241

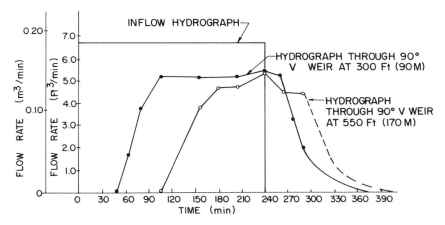

Figure 7.8 Flow hydrographs for swale experiments conducted adjacent to ramp A, northwest of S-41 exit at Epcot interchange on May 16, 1983.

surface roughness. The flow is laminar if velocities and depths of flow are relatively small. Uniform surface flow becomes turbulent if the surface is rough and the depth of flow is sufficiently large to produce persisting eddies. It must be realized that the natural ground surface is not even and uniform in slope; therefore, the flow over swales can change from laminar to turbulent over a short distance.

Eight different swale areas along highways were instrumented for water quality and quantity studies in central Florida (Yousef et al., 1985; Wanielista et al., 1988). The sites were different in terms of the volume of runoff and groundwater conditions. Controlled stormwater flow rates from adjacent retention ponds were discharged over selected sections of swale areas at Maitland interchange and EPCOT interchange near Orlando, Florida. A water and contaminant mass balance were performed for each experiment to study retention of water by swales and their effectiveness to remove nutrients and heavy metals.

Flow hydrographs were developed as shown in Figure 7.8. The hydrographs reflect clearly the water retention and excess runoff from swale areas under various inflow rates. Representative water flow cross sections through swales adjacent to Maitland interchange are shown in Figure 7.9. The hydraulic water depth, which is defined as the cross-sectional area divided by the top width of flow, did not exceed 0.041 m (1.6 in.), and the calculated water velocities varied from 0.90 to 2.98 m/min during the swale experiments. Hydraulic and hydrologic characteristics of swales are presented in Table 7.6.

The data presented describe the swale characteristics and experimental conditions. The inflow rates for the Maitland area varied between 0.026 and 0.227 m^3/min (7 to 60 gal/min) and averaged 0.189 m^3/min (50 gal/min) for

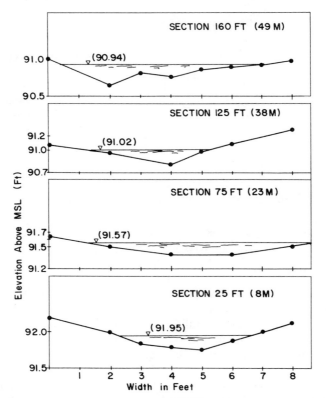

Figure 7.9 Cross sections for water flow through swale adjacent to Maitland interchange and I-4.

the EPCOT area. The inflow rate was calibrated periodically during the experiments and adjustments of the values were made to ensure constant flow.

The hydrographs reflect clearly the water retention and the excess runoff from swale areas under various inflow rates. The average loading rates varied from 0.036 to 0.154 m^3/m^2-h (1.42 to 6.06 in./h) on the Maitland swale area. These rates resulted in excess runoff averaging 0.0 to 0.066 m^3/m^2-h (0 to 2.6 in./h). The EPCOT site loading rates averaged 0.053 to 0.105 m^3/m^2-h (2.08 to 4.13 in./h) and the excess runoff averaged 0.039 to 0.071 m^3/m^2-h (1.52 to 2.8 in./h). The flow rates were calculated from the area under the hydrograph divided by the area of the swale covered with water and duration time of flow. Under the experimental conditions, there was no excess runoff for flows less than 0.036 m^3/m^2-h (1.42 in./h). Excess runoff reached more than 90% of average flow at areas of the EPCOT site when the soil was saturated with moisture.

TABLE 7.6 Hydraulic and Hydrologic Characteristics of Swale Experiments

No.	Location	Swale (m) Length	Swale (m) Hydraulic Depth	Average Velocity (m/min)	Flow In (m³)	Flow Out (m³)	Rainfall Excess Loading Rates $\frac{m^3}{m^2\text{-h}}$	Rainfall Excess Loading Rates $\frac{in.}{h}$	Infiltration $\frac{m^3}{m^2-h}$	Infiltration $\frac{in.}{h}$	Excess Runoff $\frac{m^3}{m^2\text{-h}}$	Excess Runoff $\frac{in.}{h}$
1	Maitland	53	0.038	2.58	40.9	17.6	0.154	6.06	0.088	3.46	0.066	2.60
2	Maitland	53	0.033	1.37	20.7	8.3	0.072	2.83	0.043	1.69	0.029	1.14
3	Maitland	49	0.017	0.90	8.14	0	0.036	1.42	0.036	1.42	0	0
4	EPCOT	90	0.040	2.76	57.7	39.2	0.105	4.13	0.034	1.33	0.071	2.80
		180	0.040	2.08	39.2	35.8	0.079	3.10	0.007	0.17	0.072	2.83
5	EPCOT	90	0.041	2.98	46.3	30.8	0.094	3.69	0.032	1.26	0.062	2.43
		180	0.039	1.94	30.8	23.0	0.053	2.08	0.014	0.55	0.039	1.52
6	Maitland	53	0.031	2.35	35.1	26.0	0.092	3.62	0.024	0.94	0.068	2.68

The double-ring infiltrometer was used to estimate infiltration in the swale area at Maitland. The limiting infiltration rate using this method was approximately 0.125 m³/m²-h (5 in./h) with higher values of 0.305 to 0.375 m³/m²-h (12 to 15 in./h) reported during time of infiltration. For most of the hydrograph input/output studies these infiltration rates were never realized. Thus the double-ring infiltrometer overestimates the swale infiltration rate. The calculated infiltration rate from the measured hydrographs at the Maitland site did not exceed 0.088 m³/m²-h (3.46 in./h).

Since rainfall excess, slope, average time of concentration, and length are measured directly, a relatively accurate estimate of the channel roughness measured by Manning's overland coefficient N should result. These results are shown in Table 7.7 (Yousef et al., 1985). Also, there are many formulas

TABLE 7.7 Manning's Roughness Coefficient (N) and Calculated Versus Field Measured Peak Discharge Values

Location	Date	Runoff Coefficient C	Manning's N	Peak Discharge (m³/min) Measured	Peak Discharge (m³/min) Calculated
Maitland	1/24/83	0.43	0.058	0.098	0.098
	2/07/83	0.41	0.096	0.038	0.035
	5/31/83	0.75	0.055	0.118	0.110
EPCOT	3/23/83	0.68	0.044	0.131	0.156
	3/23/83	0.91	0.059	0.118	0.143
	5/16/83	0.66	0.035	0.145	0.127
	5/16/83	0.75	0.046	0.131	0.097
			$\overline{N} = 0.056$		

for estimating peak discharge. With the well-defined areas and relatively short time of concentration, the rational formula should provide an accurate estimate. Since the runoff coefficient, flow area, and rainfall excess rate were available, it would be easy to compare flow measured with calculated peak discharge (equation 7.16). The shape of the discharge hydrograph also gives an indication that the rational formula may be used.

The rational formula with rainfall excess rate substituted for rainfall intensity is written as

$$Q_p = CrA \qquad (7.16)$$

where Q_p = peak discharge (triangular hydrograph), m^3/min
C = runoff coefficient
r = rainfall excess to swale, m/min
A = flow area (wetted perimeter times length), m^2

The average value for Manning's roughness coefficient was 0.056 for both areas. During one heavy growth period, the roughness coefficient was as high as 0.096. If this value was discarded when calculating an average value for roughness, the coefficient would be 0.05. This value may help other designers of grassy swale systems.

When using the rational formula, peak discharge for the shorter swale (53 m) was accurately predicted. However, for the longer swale (90 m), the calculated value was not as accurate as the measured value. The average watershed area may be the variable, which was difficult to measure accurately.

7.7.2 Swale Design

The design must incorporate both quality and quantity considerations. Since the primary mechanisms for management are infiltration and flow rates, quantitative estimates must be made. Infiltration rates are usually estimated from a knowledge of the soil permeability characteristics and the depth of the water table. Flow rates are estimated knowing the slope, cross-section flow areas, rainfall excess, and the roughness characteristics of the swale.

Wanielista et al. (1986) reported on the hydrologic and hydraulic design of swales that had a seasonal high water table at least 1-ft below the bottom of a swale. They conducted 20 field tests doing a mass balance on runoff input to a swale and its output. The difference in input and output was the infiltration volume, from which an infiltration rate was calculated. Limiting infiltration rates during actual swale operation were estimated at 5 to 7.5 cm/h (2 to 3 in./h). For the same swales, the double-ring infiltrometer was used to estimate the limiting infiltration rates. They were recorded at 12.5 to 50 cm/h (5 to 20 in./h).

Using a mass balance of input and output waters in a swale system, equation 7.17 was developed to estimate the length of a swale necessary to

7.7 SWALES

TABLE 7.8 Maximum Permissible Design Velocities to Prevent Erosion and Manning's N for Swales

Cover	Manning's N for vR:[a]			Slope Range (%)	Maximum Permissible Velocity (ft/s)
	0.1	1.0	10		
Tufcote, Midland, and coastal Bermuda grass	0.25	0.150	0.045	0.0–5.0 5.1–10.0 Over 10.0	6.0 5.0 4.0
Reed canary grass	0.40	0.250	0.070	0.0–5.0	5.0
Kentucky 31 tall fescue	0.40	0.250	0.070	5.1–10.0	4.0
Kentucky bluegrass (mowed)	0.10	0.055	0.030	Over 10.0	3.0
Red fescue and Argentine Bahia	0.10	0.055	0.030	0.0–5.0	2.5
Annuals[b] and ryegrass	0.10	0.050	0.030	0.0–5.0	2.5

Source: After Ree and Palmer (1949) and Wanielista et al. (1988).
[a] Product of velocity and hydraulic radius (ft^2/s).
[b] Annuals—use only as temporary protection until permanent vegetation is established.

infiltrate all the input and rainfall excess from a specific storm event using a triangular cross-sectional area (Wanielista et al., 1988). Another equation was developed for a trapezoidal cross-sectional shape:

$$L = \frac{K\overline{Q}^{5/8}S^{3/16}}{N^{3/8}f} \qquad (7.17)$$

where L = length of swale, m or ft
$\overline{Q}$ = average runoff flow rate, m^3/s or ft^3/s
S = longitudinal slope, m/m or ft/ft
N = Manning's roughness coefficient (for overland flow) (Table 7.8 and 7.9)
f = infiltration rate, cm/h or in./h
K = constant that is a function of side slope parameter Z (1 vertical/Z horizontal) and is defined in Table 7.10

The slope of a swale should be as flat as possible to promote infiltration and reduce resuspension caused by high flow velocities. Table 7.8 indicates maximum or permissible velocities to reduce erosion or resuspension.

For most watersheds, the length of a swale necessary to infiltrate the runoff from 3 in. of rainfall waters was found to be excessive or at least twice the distance available. A 3-in. rainfall was chosen because it has a frequency of occurrence equal to once per year for many Gulf and east coast areas of the United States. This assumes independent storm events defined by an interevent dry period of 4 h. Thus some type of swale block (berm) or on-line

TABLE 7.9 Overland Flow Manning's N Values for Shallow Flow $(vR < 1.0)^a$

	Recommended Value	Range of Values
Concrete	0.011	0.01–0.013
Asphalt	0.012	0.01–0.015
Bare sand	0.010	0.010–0.016
Graveled surface	0.012	0.012–0.030
Bare clay-loam (eroded)	0.012	0.012–0.033
Fallow (no residue)	0.05	0.006–0.16
Plow	0.06	0.02–0.10
Range (natural)	0.13	0.01–0.32
Range (clipped)	0.08	0.02–0.24
Grass (bluegrass sod)	0.45	0.39–0.63
Short grass prairie	0.15	0.10–0.20
Dense grass	0.24	0.17–0.30
Bermuda grass	0.41	0.30–0.48

Source: Engman (1983), with additions from the Florida Department of Transportation (1986).

[a] These values were determined specifically for overland flow conditions and are not appropriate for conventional open-channel flow calculations.

detention/retention may be more helpful. Pitt (1986) indicated that the most cost-effective solution for the reduction of runoff volume, residual solids, and bacteria was infiltration. This result is similar to that of Yousef et al. (1985), who recommend that swale blocks should be considered to reduce further the chemical constituents and runoff volumes.

Using as a design criteria, the runoff volume from 7.5 cm (3 in.) of rainfall and storage of noninfiltrated runoff, Wanielista et al. (1988) has developed swale block designs for highway applications. Basically, the swale block

TABLE 7.10 Swale Length Formula Constant

Z (Side Slope) (1 vertical/Z horizontal)	K (SI Units) (i = cm/h, Q = m^3/s)	K (U.S. Units) (i = in./h, Q = ft^3/2)
1	98,100	13,650
2	85,400	11,900
3	71,200	9,900
4	61,200	8,500
5	54,000	7,500
6	48,500	6,750
7	44,300	6,150
8	40,850	5,680
9	38,000	5,255
10	35,760	4,955

volume can be calculated for a fixed length of swale and a triangular cross section using

volume of runoff − volume of infiltrate = swale block volume

$$Q(\Delta t) - Q_f(\Delta t) = \text{volume of swale}$$

$$Q(\Delta t) - \left(\frac{LN^{3/8}f}{KS^{3/16}}\right)^{8/5}(\Delta t) = \text{swale volume} \qquad (7.18)$$

where Q_f = average infiltration rate, m³/s
Δt = runoff hydrograph time, s

Example Problem 7.3 Consider as an example a swale section with the parameters for equation 7.18 and a length of 76 m:

$N = 0.05 \qquad f = 7.5$ cm/h (3.0 in./h)
$S = 0.01 \qquad Z = 7$
$Q = 0.0023$ m³/s (0.08 ft³/s) for $\Delta t = 100$ min

(a) What swale length would be necessary to infiltrate all the waters?
(b) Only 76 m (250 ft) was available; thus how much storage volume is necessary?

SOLUTION: (a) Using equation 7.17, we have

$$L = \frac{44{,}300(0.0023)^{5/8}(0.01)^{3/16}}{(0.05)^{3/8}(7.5)} = 172 \text{ m}$$

(b) Using equation 7.18, we obtain

$$0.0023(60)(100) - \left[\frac{(76)(0.05)^{3/8}(7.5)}{44{,}300(0.01)^{3/16}}\right]^{8/5} 60(100) = \text{volume of swale}$$

and volume of storage = 10.0 m³.

Volume in the swale must be available to contain the runoff waters. In highway designs for high-speed situations, safety must be considered; thus a maximum depth of water equal to 0.5 m (about 1.5 ft) and flow line slopes on the berms of 1 vertical/20 horizontal are recommended. Along lower-speed highways or in some residential/commercial urban settings, steeper flow line berm slopes (1:6) are acceptable.

7.7.3 Berms (Swale Blocks)

The volume of runoff water stored in a swale (swale block runoff) with a downstream berm is calculated using the formula for the volume of a pyramid. With reference to the nomenclature of Figure 7.10 the volume of swale block water stored is the volume of runoff upstream of the berm minus the volume of berm from the crest upstream, or

volume of swale block = volume of runoff − volume of berm

$$V_s = \frac{L}{3}\frac{WH}{2} - \frac{L_B}{3}\frac{WH}{2}$$

$$= \frac{WH}{6}(L - L_B) \qquad (7.19)$$

where L = length upstream from the berm to a point where the maximum ponded water ends (same elevation as the crest of the berm), ft
  W = width of swale water line at the crest of the berm, ft
  H = crest depth at center line of swale, ft
  L_B = length of berm upstream of the crest, ft

For swales with longitudinal slopes less than 1%, the swale block length can be dropped from equations 7.19 if it does significantly affect the resulting calculations. In practical applications, the height of swale is fixed at 1 to 3 ft and the upstream length is calculated for a specified swale block volume. The calculated upstream length may not be sufficient for the specified swale block volume and maximum height; thus multiple on-line berms must be used. The

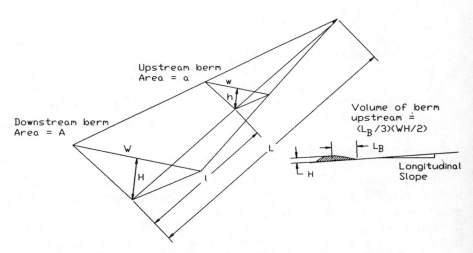

Figure 7.10 Berm (swale block) geometry for volume calculations.

volume of storage between berms can be calculated using equation 7.20.

$$V_B = \frac{L}{3}(A + a + \sqrt{Aa}) \qquad (7.20)$$

where V_B = volume at end of berm with an upstream berm, ft^3
A = cross-sectional area at berm, ft^3
  = $WH/2$
W = width of swale water line at the crest of the berm, ft
a = cross-sectional area upstream of berm if obstructed upstream, ft^2
  = $wh/2$
w = width of swale water line upstream, ft
h = height of swale water line upstream, ft

For a 6:1 side slope and various longitudinal slopes, Wanielista et al. (1986) developed a berm spacing chart as shown in Figure 7.11. Similar charts can be developed for other sets of assumptions on rainfall storms and watershed characteristics.

7.7.4 Water Quality Considerations

From the results obtained in swale experiments, it appears that the chemistry of heavy metals in natural waters is a fairly complex and site-specific phenomenon. In the studies conducted at Maitland in which only inorganic species were measured, the solubility and removal efficiencies obtained for dissolved species appeared to be related to the dominant inorganic complex present. Those metal species that were present as a charged ion, such as zinc and iron, were removed to a significant degree. Those that were complexed with inorganic species and carried a diffuse charge or zero charge were not removed.

The importance of organic complexing in regulating solubility was demonstrated in the two EPCOT experiments. Of the metal ions present, copper and iron are known to form significant metal–organic complexes, and as a result, no removal was found to occur. Other metals that formed no important organic complexes were regulated by their inorganic species. It also appears that the removal of dissolved metal species occurs more rapidly in bare earth swales than in grass-lined channels. Concentrations of iron, copper, and chromium were actually increased by passage through a grassed swale.

Data collected over an 8-month period, from both highway and swale areas, indicate lower removal efficiencies than were obtained in the controlled experimental spiked runoff. It is possible that certain metals may change forms between storm events and become solubilized. This is particularily likely in species that have a change in species from a charged free ion to a neutral ion in the pH range 6 to 7.5.

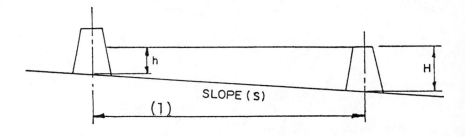

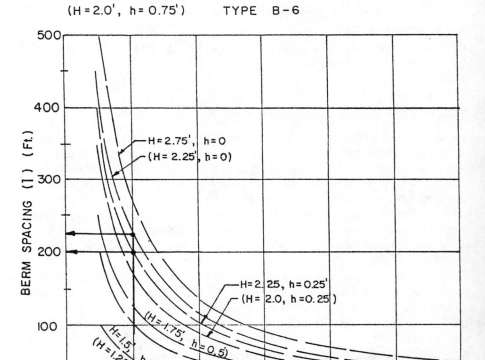

Figure 7.11 Berm spacing (1) as a function of berm heights (H, h) and slope (s) for swale of a side slope of 6:1 for four-lane interstate highways using a 3-in. rainfall design storm. (From Wanielista et al., 1986.)

Similarly, ionic nitrogen species (NH_4^+, NO_2^-, NO_3^-) and phosphorus species ($H_2PO_4^-$, HOP_4^{2-}, PO_4^{3-}) may be retained on the swale site by sorption, precipitation, coprecipitation, and biological uptake processes. These processes can reduce the nutrient concentrations in highway runoff flowing over swales. Also, it appears that the removal of dissolved heavy metal and nitrogen and phosphorus species occurs more rapidly in bare earth swales than in grass-lined channels. Additionally, thinner grass seems to be more efficient than a thicker grass cover in decreasing contaminants. It is believed that thick grass cover may affect available sorption sites and increase organic debris (grass clippings, mower debris, litter). The organic debris is then subjected to decay processes and relocation. This was evident by the decline in the removal efficiency of soluble NO_3 and NH_4 forms of nitrogen and organic nitrogen in thick grassy swales (EPCOT study 5-16-83, Maitland study 5-31-83). Also, the decrease in the removal of organic N concentration may be attributed to an increase in organic deposition in the swale due to organic debris that exists during periods of rapid grass growth.

Occasional increases in highway contaminants were observed at intermediate stations during swale experiments, particularly close to inflow areas. This appears possible due to the initial flow effects on resuspension and resolubilization of loosely bound contaminants. The swale experiments showed better removal efficiencies at slow flow rates than at high velocities. The removal of nitrogen in swales on a concentration basis (measured in this study as $\mu g/l$), is inversely related to the velocity of the runoff through the swale. There seems to be very little removal of nitrogen concentrations when the excess runoff is above 3.00 in./h. Therefore, it is apparent that if swales are designed to produce low inflow rates and velocities, some nitrogen concentration removal could be expected, with the amount of removal being a function of site conditions, such as swale cover and soil characteristics.

The removal of heavy metals and nitrogen and phosphorus species on a mass basis is directly related to infiltration losses through swales. Therefore, retention of as much water as possible on the swale area will reduce the pollutant loadings to adjacent receiving waters.

The results from swale experiments concluded that heavy metals and nutrients may be removed from runoff water flowing over swales. Table 7.11 shows average percent removal of dissolved heavy metals, phosphorus, and nitrogen in runoff water flowing over low- and high-water-table swales. The infiltration rates were calculated from a mass balance of flow into and out of the swales. The infiltration rates varied from a low of 1.3 to a high of 8.6 cm/h. For all eight sites, the average infiltration was about 6.2 cm/h (2.5 in./h). The low rates were associated with the higher (wet soil) water table.

The removal of nitrogen in swales was inversely proportional to the velocity of runoff or directly related to residence time. When the rainfall excess rate exceeded 3 in./h, there was no removal of nitrogen and, in fact, slight increases. This could be the result of resuspension and resolubilization of loosely bound contaminants.

TABLE 7.11 Average Percent Removal of Dissolved Heavy Metals and Nutrients in Runoff Water Flowing over Swales[a]

Parameter	Reduction in Total Mass (%)		Reduction in Concentration (%)	
	Low Water Table	High Water Table	Low Water Table	High Water Table
Pb	−56	−76	—	−57
Zn	−93	−77	−86	−62
Cu	−70	−49	−17	−8
Fe	−89	−35	−69	+9
Ni	—	−71	—	−51
Cd	—	−63	—	−43
Cr	−61	−50	−11	0.0
OP-P	−61	−46	−24	−9
TP-P	−63	−42	−25	−3
Inorganic N	−54	−43	−28	+5
Organic N	−28	−41	+13	+8
TN	−51	−41	−11	+7

Source: Yousef et al. (1985).
[a] Negative values indicate a reduction.

Comparison of the effectiveness of the high-groundwater site to that of the low-groundwater site showed better removal efficiencies for the low groundwater (better infiltration) site. The high groundwater site remained wet longer. Both swale types appear to be more effective than nitrogen and phosphorus for the removal of metals. These efficiencies may be governed by the predominant ionic species and complexes. Charged species are retained by sorption processes. Some heavy metals, in particular lead, have a small fraction of their total concentration in dissolved form. Thus sedimentation and filtration remove a large fraction of the total. Slow velocities and large contact areas should be promoted to increase the sedimentation and infiltration processes. Wherever possible, the side slopes of the swales should be as flat as possible. An additional design feature to provide time for additional infiltration is the use of berms (swale blocks) across the longitudinal flow path.

Average mass removal rates for the high- and low-water-table swales were related to contact area and time of flow (Yousef et al., 1985) and are tabulated in Table 7.12. The results are site specific but indicate that longer residence time in a swale and greater contact area (product of wetted perimeter and swale length) results in higher removals.

Example Problem 7.4 A swale for a commercial watershed in a low-water-table area is designed to have an average wetted perimeter of 3.5 m and is 180 m long. If Table 7.12 is representative of the lead removal rates and

TABLE 7.12 Swale Average Mass Removal Rates as a Function of Wetted Area and Residence Time

	Mass Removal Rate (mg/m^2/h)			
	Low Water Table[a]		High Water Table[b]	
Parameter	Range	Average	Range	Average
Pb	0.2–2.8	1.14	1.3–3.9	2.61
Zn	0.7–3.1	1.85	0.8–10.7	5.76
Cu	0.1–0.8	0.42	0.5–0.6	0.60
Fe	4.3–30.0	15.23	1.5–9.2	5.33
Ni	—	—	0.9–4.4	2.63
Cd	—	—	0.2–0.4	0.26
Cr	0.3–3.4	1.85	0.3–1.5	0.90
OP-P	5.1–56.0	23.80	1.4–25.8	13.60
TP-P	6.9–61	27.10	1.4–26.4	13.96
Inorganic N	32–104	62.50	4.6–12.9	8.75
Organic N	9–51	22.80	33.9–38.7	36.30
TN	41–122	85.30	42.8–46.7	44.80

Source: Yousef et al. (1985).
[a]Average velocity = 0.03 m/s.
[b]Average velocity = 0.04 m/s.

Table 5.16 is the EMC value for influent to the swales, what is the removal efficiency for this one event? The runoff volume from a design storm event is 25,727 L ($\frac{1}{4}$ in. over 1 acre) with a swale flow velocity of 0.025 m/s.

SOLUTION

1. From Table 5.17, the EMC for lead from a commercial area is 0.104 mg/L and the influent loading is (0.104)(25,727) = 2676 mg.
2. The swale detention time is

$$t_d = \left(\frac{180/0.025}{3600} \right) = 2 \text{ h}$$

3. The mass reduction rate from Table 7.12 is 1.14 mg/m^2-h:

$$\text{removal} = 1.14(180 \times 3.5)(2) = 1436 \text{ mg}$$

4. The percent removal = (1436/2676)100 = 54%.

7.8 SUMMARY

Stormwater quality management practices are those used to reduce both the concentration of and mass of stormwater pollutants. The practices are classified as structural, nonstructural, or permit types. Within each class, there exists a number of practices. A brief description of some practices was presented.

- Efficiency of operation is measured by concentration change and mass reduction.
- A mass balance on any stormwater system is fundamental for analyzing efficiencies.
- Off-line retention (exfiltration or other treatment) system efficiencies can be estimated from a frequency distribution for rainfall volumes. A simulation using a mass balance equation for pond volume is needed to size off-line system to treat a fraction of runoff from each and every event. Other equations that approximate the more mathematically complex first flush can be developed.
- Sedimentation pond sizes can be estimated using settling velocity distributions, rainfall intensity, and watershed characteristics.
- Average detention time can be calculated by dividing the average pond volume by the average effluent flow rate. Also, overflow rates can be calculated.
- Filtration treatment can be designed to remove sediment and some particulate forms of nutrients, metals, and organics from stormwaters.
- Swales are open channels that both transport and infiltrate stormwaters. Berms in swales can be used to enhance the volume of stormwater infiltrated into the ground. The water quality effectiveness can be estimated by the mass of water infiltrated with respect to the mass input to the swale.

7.9 PROBLEMS

1. What is the yearly volume of rainfall and the average number of storms per year in central Florida associated with the diversion of the runoff from the first inch of rainfall? Compare results using general data of this chapter with site-specific data for your area.
2. Settling volume and overflow rate for a sedimentation pond must be specified at average conditions for the removal of solids found in stormwater from a 10-acre watershed that has area for only an 1800-ft^2

pond with a 4-ft storage depth. The runoff coefficient is 0.7, and rainfall statistics are:

Precipitation volume: $\bar{P} = 0.40$ in.; $C_{V_p} = 1.50$

Duration: $\bar{D} = 6.0$ h; $C_{V_D} = 1.20$

Intensity: $\bar{i} = 0.07$ in./h; $C_{V_i} = 1.20$

Interval between: $\bar{\delta} = 72$ h; $C_{V_\delta} = 1.10$

3. What is detention time based on the average conditions of Problem 2?
4. A stormwater detention pond is gaged to determine the removal quality characteristics for each storm event. For the following data on performance for one storm, determine the percent removal effectiveness using a mass balance approach. Percent removed is defined as the mass removed divided by mass input × 100. The average conditions are:

Input: $Q = 0.3$ m³/s, $\bar{C} = 100$ mg/L, $t = 1$ h

Output: $Q = 0.1$ m³/s, $\bar{C} = 20$ mg/L, $t = 3$ h

5. Calculate the annual quantity (pounds) of alum to coagulate stormwater runoff from a 100-acre residential watershed that has a runoff coefficient of 0.30, annual rainfall of 30 in., and effluent using an alum dosage of 10 mg/L. Also, what is the annual mass loading and average concentration for total nitrogen if the average influent concentration is 1.6 mg/L?
6. For the watershed of Problem 5, calculate the quantity (pounds) of activated carbon needed to treat detained runoff water that exfiltrates through a side bank underdrain into perforated pipes which go into an activated carbon chamber.
7. An off-line exfiltration trench with a unit cross-sectional area for exfiltration of 20 ft² per 1-ft length of trench and a unit volume of 14 ft³/per 1-ft length is being designed to recharge a surficial aquifer with 80% of the yearly runoff from a 5-acre watershed with a 60% directly connected impervious area. Compare the length of trench for the Orlando area to that necessary for the Atlanta area considering that each is designed for a storage volume of ½ in. over the EIA and the exfiltration rate is 2 in./h. If the construction cost of the trench is $250 per 1-ft length in Atlanta, would it cost less to run a water quality sampling program at $10,000 to prove first flush with a transport

coefficient of 4.0 per inch of rainfall excess? Assume that you have to make a decision based on the total cost of the trench plus the sampling cost.

8. Determine a length of swale of Bermuda grass with an infiltration rate of 5 cm/h and a triangular cross section with 1:4 side slopes, and a longitudinal slope of 0.005 to effectively percolate all the runoff water averaging 0.028 m^3/s. If only 400 m of swale area is available and the runoff hydrograph lasts for 200 min, how much additional storage is necessary, if any?

9. Calculate the height of a swale block necessary to store 4032 ft^3 of runoff water remaining after infiltration if the side slope on the swale is 1:6 and the maximum length of swale is 300 ft. Assume that there is no need for an additional swale block upstream. The longitudinal slope on the berm is 1:6. Do the calculation assuming that the swale block volume is significant and then do the calculation assuming that it is not significant. Comment on the results. Also, what is the slope of the swale?

10. What is the slope of a swale with argentine bahia cover, triangular cross section, and side slope of 1:10 to transport 3 ft^3/s without causing erosion? The maximum width of flow cannot exceed 10 ft. How does the slope change or show the sensitivity of the solution with respect to the choice of N (pick another N value and solve for slope)?

11. For Example Problem 7.5, charge only the EMC to 0.06 mg/L and calculate the removal effectiveness.

12. A residential area has available only a 150-m-long area for a swale with low water table that can be used to infiltrate the runoff from a 2-acre parcel of land with a runoff coefficient of 0.15 and a design rainfall of 1 in. The slope of the land will produce a flow velocity of 0.02 m/s. What is the average wetted perimeter for the swale to remove over 60% of zinc in the influent waters?

7.10 COMPUTER-ASSISTED PROBLEMS

The following residential watershed characteristics are used for the first six problems of this section:

Watershed area	48 acres
Impervious area	24 acres
Percentage of impervious area directly connected	50
Pervious area curve number	50
Time of concentration	30 min
SCS hydrologic soil class	A

The rainfall data are specified by regulation to be a 6-h event with return frequency of 25 years. The data in 15-min increments expressed as inches is (or you may pick your own);

Hour: 1st	2nd	3rd	4th	5th	6th
0.10	0.16	1.08	0.24	0.14	0.12
0.11	0.17	1.14	0.24	0.13	0.12
0.12	0.27	0.32	0.18	0.13	0.11
0.15	0.30	0.28	0.16	0.12	0.11

1. Using the hydrology program diskette and the SMADA program, develop a stormwater management system by specifying the volume for water quality treatment using off-line retention ponds with side slopes of 1:4. Use the Santa Barbara urban hydrograph generation routine and the stochastic equation for off-line retention design. You must achieve 80% removal. The depth to seasonal high water table is 4 ft.

2. For Problem 1, change the concentration of suspended solids to 140 mg/L. What is the approximate diversion volume to match the pollution loading of suspended solids of Problem 1? Compare the size of this retention pond to that of Problem 1. Also present a graphic display of the hydrograph after off-line retention. Is the peak discharge attenuated?

3. A swale is used on-site that produces an initial abstraction of 1 in. over the impervious area. Does this have an effect on the peak discharge using the same conditions of Problem 2?

4. For the conditions of Problem 3, a more realistic estimate of time of concentration is 75 min because of the swale attenuation. How does the peak discharge change before off-line attenuation?

5. Using the original watershed characteristics, the developer decides to reduce the impervious watershed area to 12 acres. Also, the time of concentration increases to 60 min. How much is the hydrograph attenuated when:
   a. The impervious area is reduced to 12 acres?
   b. The impervious area is reduced to 12 acres and the time of concentration is equal to 60 min?
   Use the Santa Barbara urban hydrograph method.

6. Solve Problem 5 except when calculating off-line volume use the rational formula with a runoff coefficient of 0.60.

7. Calculate the length of swale and if necessary the swale block volume to store and infiltrate an average runoff of 0.4 ft^3/s over 1 h. The swale

limiting infiltration rate is 3 in./h with a Manning roughness coefficient of 0.10, a side slope of 1:6, and a longitudinal slope of 0.005. The available swale length is 500 ft. Use the RETEN computer program.

8. An off-line exfiltration trench constructed of a 3-ft-diameter perforated pipe and 12 in. of 0.4 porosity rock surrounded by a textile fabric is planned for a 1-acre commercial watershed near Roanoke, Virginia. The watershed has a runoff coefficient of 0.8. The trench has a limiting exfiltration rate of 2 in./h through the fabric. Three feet of cover is required and the seasonal high water table is 8 ft below the surface. What is the trench length to exfiltrate on the average 80% of the yearly runoff volume? If the average annual removal were increased to 95%, what is the length of trench? Use the RETEN computer program.

9. The trench of Problem 8 is planned for Orlando, Florida. Compare the trench lengths between the two locations. All site conditions are the same.

10. For a 200-acre residential area with a runoff coefficient of 0.5, an off-line surface retention pond is planned to store the first $\frac{1}{2}$ in. of runoff from the equivalent impervious area. If the pond has side slopes of 1:4, what is the volume and area of the pond water for pond water depths of 4 and 5 ft?

7.11 REFERENCES

Arnell, V. 1982. *Rainfall Data for the Design of Sewer Pipe Systems*, Report Series A: 8, Department of Hydraulics, Chalmers University of Technology, Gothenburg, Sweden.

Arnell, V., Harremoes, P., Johansen, N. B., and Niemcyznowicz, J. 1983. "Review of Rainfall Data Application for Design and Analysis," *Proceedings of a Specialized Seminar on Rainfall—The Basis for Urban Runoff Design and Analysis*, Copenhagen.

Brune, M. 1953. "Trap Efficiency of Reservoirs," *Transactions of the American Geophysical Union*, Vol. 34, No. 3.

DiToro, D. M. 1980. "Statistics of Receiving Water Response to Runoff," *Proceedings of a National Conference on Urban Stormwater and Combined Sewer Overflow Impact on Receiving Water Bodies*, EPA 600/9-80-056, U.S. Environmental Protection Agency, Orlando, Fl.

Douglas County, Colorado, 1986. Chapter 15, "Water Quality Enhancement," in *Douglas County Storm Drainage Design and Technical Criteria*, Douglas County, Colo.

Driscoll, Eugene D. 1983. "Performance of Detention Basins for Control of Urban Runoff Quality," *Proceedings of the International Symposium on Urban Hydrology, Hydraulics, and Sediment Control*, University of Kentucky, Lexington, Ky.

7.11 REFERENCES

Driscoll, Eugene D. 1988. "Long Term Performance of Water Quality Ponds," in *Design of Urban Runoff Quality Controls*, L. Roesner, B. Urbonas, and M. Sonnen, Eds., American Society of Civil Engineers, New York, pp. 145–163.

Engman, E. T. 1983. "Roughness Coefficients for Routing Surface Runoff," in *Proceedings of the Conference on Hydraulic Engineering*, ASCE, New York, pp. 560–565.

Field, R., Tafuri, A., and Masters, H. 1977. *Urban Runoff Pollution Control Technology Overview*, EPA 600/2-77-047, U.S. Environmental Protection Agency, Cincinnati, Ohio.

Florida Department of Environmental Regulation. 1990. *Florida Administrative Code*, Chapter 17.25, Tallahassee, Fla.

Florida Department of Transportation. 1986. *Drainage Manual*, Florida DOT, Tallahassee, Fla.

Gizzard, T. J. Randall, C. W., Weand, B. L., and Ellis, K. L. 1986. "Effectiveness of Extended Retention Ponds," *ASCE Proceedings of an Engineering Foundation Conference on Urban Runoff Quality—Impact and Quality Enhancement Technology*, New England College, Henniker, N.H., June 23–27.

Glover, G. E., and Herbert. G. R. 1973. *Microstraining and Disinfection of Combined Sewer Overflows*. EPA-R2-73-124, Environmental Protection Agency, Washington, D.C.

Harper, H. H., and Wannielista, M. P. 1984. *An Investigation into Alum Application for Sediment Nutrient Inactivation in Megginnes Arm, Lake Jackson*. Florida Department of Environmental Regulation, Tallahassee, 194 p.

Kaiser, E. J., et al. 1974. *Promoting Environmental Quality Through Urban Planning and Controls*, Socioeconomic Environmental Studies Series, EPA 600/55-73/015, U.S. Environmental Protection Agency, Washington, D.C.

Lager, J., et al. 1977. *Urban Stormwater Management and Technology: Update and User's Guide*, EPA 600/8-77-014, U.S. Environmental Protection Agency, Edison, N.J.

Livingston, E. 1985. "The Stormwater Rule: Past, Present, and Future," in *Stormwater Management: An Update*, M. Wanielista, ed., University of Central Florida, Orlando, Fla.

Lord, B. N. 1986. "Effectiveness of Erosion Control," *Urban Runoff Technology, Engineering Foundation Conference*, New England College, Henniken, N.H., June.

Malcom, H., and New, V. 1975. *Design Approaches for Stormwater Management in Urban Areas*. North Carolina State University, Raleigh, N.C.

McCuen, R. H. 1980. "Water Quality Trap Efficiency of Storm Water Management Basins," *Water Resources Bulletin*, Vol. 16, No. 1, pp. 15–21.

Miller, Robert A. 1985. *Percentage Entrainment of Constituent Loads in Urban Runoff, South Florida*, Report 84-4329, U.S. Geological Survey, Tallahassee, Fla.

Miller, R. A., Hardee, J., and Mattraw, H. C. 1978–1979. *USGS Open Fil Reports* 79-982, 78-612, 79-1295, U.S. Geological Survey, Washington, D.C.

Oswald, George E. 1986. *The Evolution of Erosion and Sedimentation Control in the City of Austin, Texas*, City of Austin, Texas.

Pitt, R. 1986. "The Incorporation of Urban Source Area Controls in Wisconsin's Priority Watershed Projects," *Urban Runoff Technology, Engineering Foundation Conference*, New England College, Henniker, N.H., June.

Ree, W. O., and Palmer, V. J. 1949. *Flow of Water in Channels Protected by Vegetative Lining*, Soil Conservative Bulletin 967, U.S. Department of Agriculture, Washington, D.C.

Reynolds, T. D. 1982. *Unit Operations and Processes in Environmental Engineering*, PWS Publishers, Boston, p. 568.

Schilling, W. 1983. "Univariate Versus Multivariate Rainfall Statistics: Problems and Potentials," *Proceedings of a Specialized Seminar on Rainfall—The Basis for Urban Runoff Design and Analysis*, Copenhagen.

Schueler, Thomas R. 1987. *Controlling Urban Runoff*, Metropolitan Washington Council of Governments, Washington, D.C.

Sullivan, R. H. et al. 1973. *The Swirl Concentrator as a Grit Separator Device*, EPA-670/2-74-026, Environmental Protection Agency, Washington, D.C.

Thronson, R. E. 1971. *Control of Sediments Resulting from Highway Construction and Land Development*, Report, Office of Water Programs, U.S. Environmental Protection Agency, Washington, D.C.

Thronson, R. E. 1972. *Guidelines for Erosion and Sediment Control—Planning and Implementation*, Environmental Protection Technology Series, EPA R2-72-015, U.S. Environmental Protection Agency, Washington, D.C.

Tourbier, J., and Westmacott, R. 1974. *Water Resources Protection Measures in Land Development—A Handbook*, Water Resources Center, University of Delaware, Newark, Del.

U.S. Department of Agriculture. 1981. *Soil Survey of Porter County, Indiana*, USDA Soil Conservation Service, Washington, D.C.

U.S. Environmental Protection Agency. 1972. *Screening/Flotation Treatment of Combined Sewer Overflows*, Report, Water Pollution Control Research Series 11020 FDC 01.72, U.S. EPA, Washington, D.C.

U.S. Environmental Protection Agency. 1983. *Final Report of the Nationwide Urban Runoff Program*, Water Planning Division, U.S. EPA, Washington, D.C.

Wanielista, M. P. 1977. "Off-line Retention Pond Design," in *Proceedings of the Stormwater Retention/Detention Basins Seminar*, Y. A. Yousef, Ed., University of Central Florida, Orlando, Fla., pp. 48–71.

Wanielista, M. P., and Shannon, E. E., 1977. *Stormwater Management Practices Evaluation*, East Central Florida Regional Planning Council, Orlando Metropolitan 208 Study.

Wanielista, M. P., Yousef, Y. A., and Bell, J. 1978. *Shallow-Water Roadside Ditches for Stormwater Purification*, Report ESEI-78-11, Florida Department of Transportation, Tallahassee, Fla.

Wanielista, M. P., Yousef, Y. A., Van DeGraaff, L., and Rehmann-Kuo, S. 1986. *Best Management Practices for Highway Runoff Erosion and Sediment Control*, Florida Department of Transportation, Tallahassee, Fla.

Wanielista, M. P., Yousef, Y. A., and Avellaneda, E. 1988. *Alternatives for the Treatment of Groundwater Contaminants: Infiltration Capacity of Roadside Swales*, FL-ER-38-88, Florida Department of Transportation, Tallahassee, Fla.

7.11 REFERENCES

Wanielista, M. P., Gauthier, M. J., and Evans, D. L. 1991. *Design and Performance of Exfiltration Systems*, Florida Department of Transportation, Tallahassee, Fla.

Wenzel, H. G., and Voorhees, M. L. 1981. *An Evaluation of the Urban Design Storm Concept*, Research Report UILU-WRC-81-0164, Water Resources Center, University of Illinois at Urbana-Champaign.

Whipple, W. 1979. "Dual-Purpose Detention Basins," *Journal of Water Resources and Planning, ASCE*, Vol. 105, No. WR2.

Yousef, Y. A. 1984. "Fate of Heavy Metals in Stormwater Runoff From Highway Bridges," *Science of the Total Environment*, Vol. 33, pp. 233–244.

Yousef, Y. A., and Wanielista, M. P. 1990. *Efficiency Optimization of Wet Detention Ponds for Urban Stormwater Management*, Florida Department of Environmental Regulation, Tallahassee, Fla. Contract WM 159.

Yousef, Y. A., Wanielista, M. P., Harper, H. H., Pearce, D. B., and Tolbert, R. D. 1986a. "Nutrient Transformation in Retention/Detention Ponds Receiving Highway Runoff," *Journal of the Water Pollution Control Federation*, Vol. 58, No. 8, pp. 838–844.

Yousef, Y. A., Hvitved-Jacobsen, T., Wanielista, M. P., and Tolbert, R. D. 1986a. "Nutrient Transformation in Retention/Detention Ponds Receiving Highway Runoff," *Journal of the Water Pollution Control Federation*, Vol. 58, No. 8, pp. 838–844.

Yousef, Y. A., Wanielista, M. P., and Harper, H. H. 1986b. "Design and Effectiveness of Urban Retention Basins," *ASCE Proceedings of an Engineering Foundation Conference on Urban Runoff Quality-Impact and Quality-Enhancement Technology*, New England College, Henniker, N.H., June 23–27.

CHAPTER 8

Wet-Retention/Detention Ponds

Retention and detention ponds are generally open excavated or natural depressions on the ground surface of varying size and depth. They are generally located in excess land areas, green space of developments, and/or recreational sites. Wet-detention ponds maintain a permanent pool of water and they are among the most common stormwater management practices. *Retention* is long-term storage that results in residence times averaging weeks or months and most of the time there is no surface discharge. Short-term storage time of hours up to a few days with discharge to adjacent surface waters is known as *detention*. Storage areas that are occasionally dry and hold water only during heavy storm events are known as dry ponds. In highly urbanized areas, they may be underground storage containers such as tunnels and concrete vaults.

Ponds are usually designed to store a selected volume of runoff from a specified rainfall event with a predetermined interevent period during which there is a controlled surface discharge, infiltration, or irrigation of the stored water. Ponded water may be released to surface waters, or reused for irrigation periodically, or otherwise reused until the elevation reaches its design dry weather stage. The maximum surface water discharge rates are regulated to conform to the capacities of a downstream drainage system. A portion of the retained volume (temporary storage) is also removed by evapotranspiration and infiltration into the ground.

As shown in previous chapters, stormwater contains heavy metals, algal nutrients, and other contaminants that are transported in solution and/or particulate forms. Direct discharge of urban runoff to adjacent water bodies may cause adverse effects on water quality; however, the receiving water impacts are often site specific and the extent of the problem directly depends on runoff quantities, qualities, additional point sources, land use, and the biological and chemical properties of the water body. The use of wet-retention/detention ponds for storage and attenuation of flow peaks from urban areas has been established, but studies on their effectiveness in removal of contaminants are continuing. Stormwater retention/detention facilities can cause a distinct reduction in runoff flow peaks and much of the suspended solids transported can settle out before entering adjacent waterways. Also,

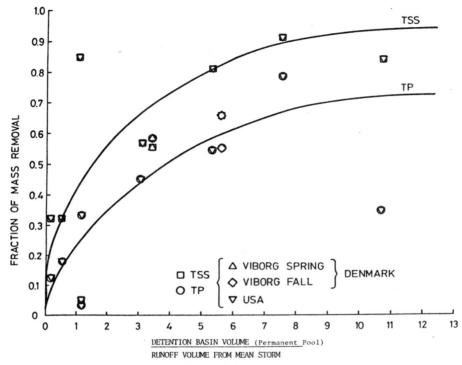

Figure 8.1 Removal of suspended solids and total phosphorus in detention ponds.

physical, chemical, and biological processes reduce the concentrations and mass flow for soluble contaminants.

Water quality data collected during various storm events (Driscoll, 1983; Yousef and Wanielista, 1989) indicate that the wet-detention pond permanent volume (excluding sediment storage) divided by mean runoff volume (V_B/V_R) is related to water quality improvement and the relationship is shown in Figure 8.1. The mean runoff volume assumes a minimum interevent dry period of 6 h. The higher the ratio, the greater the average annual percent mass removal. For example, phosphorus and suspended solids removal in wet-detention ponds are near a maximum when the ratio of V_B/V_R approximates 6 to 7.

Many similarities exist between contaminant removal processes in surface-water reservoirs, lakes, and streams and in wet-retention/detention ponds. For example, bottom sediments in these systems act as a source or sink for phosphorus, nitrogen, and heavy metals (Mortimer, 1971; Hutchinson, 1957). Therefore, the exchange processes at the sediment–water interface and their impact on the overlying water quality are important parameters. Also, vertical and horizontal distribution of contaminants into bottom sediments deserves further evaluation.

8.1 DESIGN CONSIDERATIONS

Temporary storage in a pond is designed for a specific volume of rainfall excess from a design storm to improve water quality. Also, pond input hydrographs are attenuated. The temporary storage can be removed by infiltration, reuse by mechanical means (irrigation), or by gradual surface discharge.

8.1.1 Temporary Storage Removal by Infiltration and Irrigation

In some designs, volume reduction by infiltration and irrigation may be possible. The soil infiltration or irrigation rates must have the capacity to remove the temporary stored water between storm events. This is particularly suitable in areas with deep permeable soil. Infiltration rates are based on water storage capacity with example minimum infiltration rates for surrounding soils shown in Table 8.1. The effective water storage capacity of a soil is the void space available for water storage. The minimum vertical infiltration rates during saturated conditions can be simulated in the field using double-ring infiltrometers. Relationships between infiltration rates estimated using standard double-ring infiltrometers and infiltration rates for on-line swale systems with sandy, sandy loam, and loamy sand soils indicate that the infiltration rate in the swale is about one-third of that rate from the double-ring test. The estimates were based on a materials balance of swale input and swale output waters. Minimum rates of about 2 to 3 in./h were noted for the swales, while the double-ring infiltrometer results were about 5 to 10 in./h for the same sites. The soil texture specified in Table 8.1 is based on the U.S. Department of Agriculture soil texture triangle. It should be realized that minimum infiltration rates measured in the field may be reduced by continuous deposition of fine and colloidal particles and/or oil and grease films transported with runoff waters. Therefore, maintenance procedures are essential and should not be ignored.

TABLE 8.1 Hydrologic Soil Properties Suitable for Retention Basins

Soil Texture Class	Effective Water Capacity (in. water/ in. soil)	Minimum Infiltration Rate	
		in./h	cm/h
Sand	0.35	8.27	21
Loamy sand	0.31	2.41	6.1
Sandy loam	0.25	1.02	2.6
Loam	0.19	0.52	1.2

Source: Data from Rawls et al. (1982).

Irrigation rates from ponds will vary with the use of the water, local weather conditions, vegetation, and soil water availability. Average rates of irrigation for a year are generally between 0.5 and 2.0 in./week (1.39 to 5.1 mm/week) in the temperature zone, which can result in near complete removal of the temporary storage during average rainfall conditions.

8.1.2 Temporary Storage Removal by Direct Surface Discharge

On-line wet ponds are designed with a permanent pool of water depth varying between less than 1 ft (0.3 m) to about 12 ft (3.66 m). For systems that maintain a permanent pool of water, shallow ponds with depth less than 4 to 6 ft (1.22 to 1.83 m) are desirable to promote a more aerobic environment and enhance water quality improvement. The temporary storage is above the permanent pool. Also, the invert of the exit structure from the pond should be above the average water table elevation to avoid continuous flow of groundwater seepage through the pond. The average detention time depends on the permanent plus temporary storage volume and the rate at which detained water is discharged from the pond. An increase in detention time occurs when permanent plus temporary storage volume increases or discharge rate decreases. Also, average yearly detention time depends on the interevent dry period. The elevation between the exit invert and the ground-

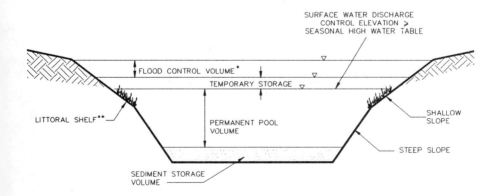

*Can be measured above permanent pool, however some regulatory agencies measure above the temporary storage.

**The reader should consult local water management districts and other regulatory agencies to determine specific geometric and littoral zone design requirements.

Figure 8.2 Schematic of a wet-detention pond illustrating pond volumes.

water depends on backwater effects in the sewer system, but typically range from 1 to 3 ft (0.30 to 0.91 m). An example pond profile detail is shown in Figure 8.2.

8.2 MODELING CONCEPTS FOR WET-DETENTION PONDS

The operation of detention/retention ponds can be described by mathematical models that can be used to predict the performance of stormwater treatment systems. Design of detention ponds for water quality improvement using mathematical modeling can be achieved by two conceptual approaches. These approaches are based on (a) settling of incoming suspended solids and removal of particulate fractions of contaminants and (b) removal of fractions from dissolved contaminants by physical, chemical and biological processes. These models may be referred to as (1) sedimentation models and (2) pond water quality models. Sedimentation models require knowledge of a particle-size distribution for suspended solids in runoff water entering the pond and particulate fraction of various contaminants. However, the pond water quality models require knowledge of sedimentation mechanisms and kinetics for removal of dissolved fractions. Removal rate constants and physical, chemical, and biological interactions must be defined under various environmental conditions. These are complex processes and simplification of the models may be appropriate. Design of wet-detention ponds based on sedimentation models has been discussed in Chapter 7. This chapter stresses the pond water quality models.

Design of wet detention ponds based on either sedimentation or pond water quality improvements suggest that pollutant removal efficiency could be directly proportional to hydraulic detention time. Therefore, it is essential to understand the variations and distribution of detention times within the contained water volume in wet-detention ponds. In general, the mechanisms of contaminant removal within wet-detention ponds are expected to be the same as mechanisms studied in many shallow small lakes and ponds (see Chapter 6). In addition to sedimentation, there are also mechanisms to remove dissolved nutrients, heavy metals, and other contaminants, such as binding nutrients in algal and plant growth, sorption of phosphorus and heavy metals by sediments, nitrification/denitrification processes, chemical precipitation, and others. It is extremely difficult to evaluate precisely the role and extent of various mechanisms in removal of contaminants separately. Therefore, simplified relationships similar to that shown in Figure 8.1 may have to be developed.

Eutrophication models can be used to describe the relationships between algal growth and incoming available nutrients. Algal growth may be desirable to tie up nutrients (N and P) in algal cells, but excessive growth may be undesirable. Excessive algal growth may lead to oxygen depletion, odor problems, and unsightly conditions. The algal growth depends on both nutrient loadings and mean residence times. Residence time also affects the dominant algal species present in the pond. The residence time is directly related to the pond volume, the rainfall-runoff pattern, and flow regime in the pond, such as completely mixed, plug-flow, or a combination of two or more flow regimes.

8.2.1 Mass Balance for Concentration Changes

Mass balances for various suspended and dissolved fractions of nutrients, heavy metals, and other contaminants entering detention ponds can be developed. Accumulation, release, and transport of contaminants between various compartments of the pond system can be described by the following general mass balance equation. The uptake or release processes by sediments, biota, and other mechanisms should be adequately evaluated to provide the proper numbers for input to the mass balance equations.

Rate of accumulation = rate of flow − rate of flow + rate of
of reactant within   of reactant   of reactant   generation of
the pond boundary   into the pond   out of the   reactant within
                                              pond   the pond

− rate of utilization
(disappearance) of
reactant within the pond

or

$$\frac{d(VC)}{dt} = QC_0 - QC + V(r_g) - V(r_u) \qquad (8.1)$$

where V = pond volume, L^3
C = concentration of component, ML^{-3}
t = time (t)
QC_0 = incoming load of component, Mt^{-1}
QC = outlet load of component, Mt^{-1}
r_g = generation rate, $ML^{-3}t^{-1}$
r_u = utilization rate, $ML^{-3}t^{-1}$

A numerical method solution can be used to solve the system described by several linked differential equations. The volume in the pond may not be constant. Therefore, $d(VC)/dt = V(dC/dt) + C(dV/dt)$, or

$$\frac{dC}{dt} = \frac{1}{V}\frac{d(VC)}{dt} - \frac{C}{V}\frac{dV}{dt} \qquad (8.2)$$

If dV/dt is considered negligible during a specified time period,

$$\frac{dC}{dt} = \frac{1}{V}\frac{d(VC)}{dt} \qquad (8.3)$$

The time period can range from a few minutes during a storm runoff event to more than 1 day in between events. The change in volume with time can be calculated from the mass balance of a pond water budget. Also, the rates of generation and/or disappearance for various parameters such as suspended solids, nitrogen, phosphorus, heavy metals, and so on, should be fully understood. Reaction rates should be determined as defined in Chapter 6.

During the operation of a wet detention pond, the runoff water is stored and either infiltrated into the ground, reused, or discharged directly to adjacent surface waters. During infiltration, there is removal of solids from the infiltrated water and the detained stormwater remaining for longer periods of time in surface storage will undergo the removal of solids and dissolved pollutants. Dissolved pollutants may be removed during the infiltration process by physical, chemical, or biological processes, although special soils and monitoring may have to be recommended (Dornbush, 1981; Bell and Wanielista, 1981).

8.2.2 Temporary Pond Volume in a Wet-Detention Pond

A wet-detention pond is constructed to provide a storage volume for water quality treatment and a runoff volume to attenuate hydrograph peaks. The water quality treatment volume is composed of a permanent pool and a temporary pool of water that have the primary purpose of providing sufficient holding (detention) time for the pollutants to settle and undergo chemical and biological changes. The peak attenuation runoff volume is commonly designed to be added above the temporary pool volume.

Typical calculations for temporary pond volume (above the permanent pool) are based on a detention (holding) time after runoff events. The detention time is equated to a minimum interevent dry period and is necessary to achieve water quality improvement. A minimum interevent dry period (i.e., 24 h, 48 h, 72 h, or greater) can be specified and the cumulative volume of rainfall between the consecutive dry periods is calculated. The volumes are rank ordered and a cumulative conditional probability frequency distribution is calculated. The distribution is for the volume of rainfall in the interval of time between the minimum dry periods. In mathematical terms,

$$F(P \mid \Delta) = \sum \Pr[P \mid \Delta] \qquad (8.4)$$

where $F(P|\Delta)$ = cumulative conditional probability distribution for volume
P = precipitation between a specified minimum interevent dry time, in.
Pr = probability symbol
Δ = minimum interevent dry period, h

The cumulative conditional probability is developed from hourly (or more frequent) rainfall records. The empirical conditional probability distribution is calculated using the Weibull plot position, or

$$F(P|\Delta) = \frac{m}{N+1} \tag{8.5}$$

where m = less than or equal to ranking
N = number of precipitation events

The exceedence probability for any values of rainfall (p) that exceed a specific value (P) is defined as

$$\Pr\{p \geq P\} = 1 - F(P|\Delta) = 1 - \frac{m}{N+1} \tag{8.6}$$

and the return period in years is

$$T_r(\text{years}) = \frac{1}{\Pr\{p \geq P\}n}$$

or

$$T_r = \frac{1}{[1 - m/(N+1)]n} \tag{8.7}$$

where n is the average number of conditional events per year.

A graphical expression for the conditional probability distribution is a plot of precipitation volume versus interevent dry period for various return periods (frequencies). Figure 8.3 was developed for the Orlando, Florida area and is called a PIF curve for "*p*recipitation, *i*nterevent, *f*requencies." It is similar to FID curves for frequency, intensity of precipitation, and duration of average intensity. By specifying a frequency of violation (once per 4 months) and a treatment time (interevent period), the volume of rainfall can be read from a PIF curve. Since no direct human life-threatening situation exists in specifying a pond volume for water quality enhancement, a 5% or greater exceedence probability level is probably technically sufficient for design (Maass et al., 1966). However, cost/benefit analysis could be done (see Chapter 9) and a probability of exceedence determined.

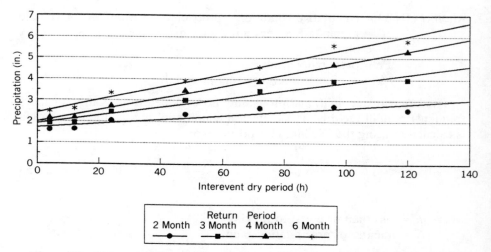

Figure 8.3 PIF curve for Orlando, Florida; precipitation ≥ 0.04 in., 15-year data.

Example Problem 8.1 The volume of rainfall conditional upon a 72-h interevent dry period and associated with an exceedence probability of 0.05 is 3 in. If there are on the average 60 of these conditional rainfalls per year, what is the return period such that the volume of pond is not exceeded?

SOLUTION: The empirical conditional probability distribution was determined to specify the volume and exceedence probability. Thus the return period for these conditions needs to be calculated. Using equation 8.7 yields

$$T_r = \frac{1}{\Pr\{p \geq P\}n} = \frac{1}{(0.05)(60)} = \tfrac{1}{3} \text{ year (4 months)}$$

Thus over a long time period, we can expect the volume to be exceeded once every 4 months.

8.2.3 Permanent Pool Volume in a Wet-Detention Pond

There are two very different operating systems for wet-detention ponds: (1) a flow-through intermediately mixed reactor pond (Nix, 1985; Martin, 1988) and (2) a batch reactor. The pond holds up to a predetermined volume of runoff for a specific treatment time period and then releases or reuses the stored water. Runoff water into a flow-through pond is resident for a variable detention time, with detention time being a function of reactor volume and outflow rate. Both the pond volume and output flow rate varies with time.

Flow-through wet-detention pond volume will fluctuate over time and longer detention times are needed to remove additional pollutant concentrations by sedimentation, chemical interaction, and biological uptake. Deten-

8.2 MODELING CONCEPTS FOR WET-DETENTION PONDS

tion of 24 h or more has been shown to remove 90% or more of the suspended solids (Gizzard et al., 1986). However, soluble and colloidal fractions may not be significantly removed. In addition, Driscoll (1983) noted that approximately 20% of the particles have very slow settling velocities (0.3 ft/h or less).

Yousef et al. (1986b), observed a decline in orthophosphorus concentration discharged into a flow-through intermediate mixed retention pond receiving highway runoff to background level within 3 days following the rainfall event. Nutrients in stormwater are both in the dissolved and particulate form. Much of those in particulate form will settle to the bottom of a wet detention pond and attach to the sediment. Also, some other dissolved forms may be removed by incorporation into plant form or attached to other solid forms and then removed from the water column. Bottom sediments in detention ponds may act as sources or sinks for phosphorus and nitrogen. A source releases nutrients while sinks hold nutrients. The sediments act as a sink during aerobic environments (in the presence of free oxygen) and as a source during anaerobic environments (in the absence of free oxygen) (Yousef et al., 1986b). Under anaerobic conditions near the bottom sediment, phosphorus and ammonia nitrogen can be released. Nitrogen gas can also be released from the bottom sediments due to nitrification/dentrification processes.

Shallow ponds (1 to 2 m deep) with a bottom sediment layer of primarily inorganic matter will probably provide the environment for a permanent oxidized sediment layer. However, excessive growth of certain plants that produce organic materials that settle and decay in the bottom of ponds should be avoided or at least controlled. The organic materials will lower the oxygen content and will cause the release of soluble phosphorus and nitrogen into the water column.

In a recent study using model detention ponds (Yousef and Wanielista, 1989), it was concluded that a minimum detention time of at least 72 h is needed to remove more than 95% of suspended solids and 30 to 70% of nutrients and heavy metals. This is consistent with data reported in the field. Therefore, it may be desirable to design these ponds based on statistical analysis of rainfall records using storm events separated by a minimum dry period consistent with desired pollutant removal effectiveness in the pond. A minimum dry period between two successive events will allow a slow bleed-off rate, maximize detention periods, and enhance removal efficiencies of pollutants. This dry period should be selected on the basis of time required to minimize cumulative effects. If design storm events are based on a minimum of 72 to 120 h of dry period, sufficient time may be available to achieve near-maximum removal of pollutants by settling and biological processes.

Detention time is the time a particle of runoff water is resident in a wet pond. It will vary with pond volume and outflow rate changes. Based on 1 year of pond operating data, a log-Pearson type III distribution (see Chapter 2) was fit to event average detention times calculated for four different permanent pool designs (see Figure 8.4). The runoff into the pond was from a

272 WET-RETENTION / DETENTION PONDS

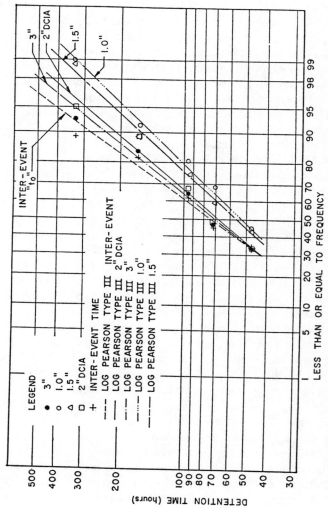

Figure 8.4 Comparison of detention-time frequency distributions using a log-Pearson type III theoretical distribution.

8.2 MODELING CONCEPTS FOR WET-DETENTION PONDS

TABLE 8.2 Comparison of Permanent Pool Designs Using Detention Time (Hours)[a]

Percent Less Than	1.0 in. over Watershed	1.5 in. over Watershed	2 in. DCIA	3 in. over Watershed	t_0
50	52	56	73	74	75
59	65	70	84	88	99
90	150	165	202	230	280

[a]Approximate average.

19.2-acre residential area with about 5 acres of directly connected impervious area. The runoff from 3 in. of rainfall (3-in. curve) was found to produce the highest detention times relative to pool volumes calculated using 2 in. over the directly connected impervious area and the runoff from 1.5 in. over the total area. Table 8.2 shows a comparison of detention times for the four designs and the time between runoff events, t_0.

As noted from Figure 8.4, the "higher" the frequency distribution, the greater the detention time, and the greater the permanent pool storage, the higher the detention time. However, detention time can never exceed interevent runoff time, which may be assumed to approximate the rainfall interevent time.

For a flow-through wet-detention pond designed for the runoff from 3 in. of rainfall on the watershed, about 50% of the interevent dry periods are less than 70 h. Most ponds that function to remove nutrients have long detention times (Yousef et al., 1986b; Camp et al., 1985). Thus, if 72 to 120 h is required for high (possibly maximum) removal effectiveness, a flow-through wet-detention pond will not achieve the desired level of removal for selected water quality parameters about 50% of the time. For greater removal, a detention pond that stores a specific volume above the permanent pool without release for a specified period of time can be designed and operated (Section 8.3).

8.2.4 Wet-Pond Construction Details

A wet-detention pond consists of water quality (permanent pool) and peak-discharge attenuation storage. The water quality storage is based on a volume of runoff from the watershed. If PIF curves used to specify rainfall volumes for 72-h interevent dry periods and a 2-, 3-, or 4-month return, a rainfall volume of between 2.5 and 4.5 in. appears reasonable for the Orlando, Florida area (Figure 8.3). Most ponds that function to remove nutrients have long detention times (Yousef et al., 1986b; Camp et al., 1985). The permanent pool of water should be deep enough to allow sedimentation in some areas, but on the average the depth of the permanent pool should not exceed 5 to 6 ft (Yousef et al., 1986). Based on a plug-flow regime operation, the length-to-width ratio should be greater than 4:1, which can be achieved with baffles (Yousef and Wanielista, 1989). The Southwest Florida Water Management District (1987) and other management entities in Texas and Maryland have specified some construction details to encourage littoral zone planting and growth and discourage the recylcing of nutrients and other impurities from the bottom sediment to the water column. Other details are available for many different types of stormwater detention ponds (Stahre and Urbonas, 1990). Figure 8.2 illustrates some construction details. The littoral zone plants should be diversified. Some of the more common types used in warm (temperate) climates are shown in Table 8.3. The detention time is longer than required for sediment removal. Pond storage for water quality control with average detention times of about 3 to 5 days (72 to 120 h) may be appropriate. The fluctuations in depth of pond during runoff should not cause the littoral zone to die. Recommended is a 6- to 18-in. pond fluctuation from a storm, which does not cause the littoral zone depth to exceed 36 to 42 in. (Southwest Florida Water Management District, 1987). Whether the littoral zone enhances or degrades the water quality depends on appropriate maintenance measures and control of vegetation growth. This may be more desirable for multiuse ponds such as those used for recreation and stormwater management.

The runoff storage volume for peak discharge control using a wet-detention pond is the volume of runoff water calculated from the specified design storm (e.g., 25-year, 24-h). This storage is in addition to the permanent pool.

8.2.5 On-Line Infiltration Pond Volume

An on-line infiltration pond both infiltrates and infrequently discharges stored water. The infiltration area must be defined with an average infiltration rate. The pond may or may not contain a permanent pool. If the exponential distribution can be used for runoff volumes, runoff durations, and interevent times, another exponential distribution can probably be developed for overflow volume from an on-line infiltration pond. Thus the resulting probability distribution can be used to estimate storage as a function of yearly overflow percentage and infiltration rate. The major advantage of using this probability approach is the simplicity of the equation used to specify pond size and infiltration rate. The simplicity is relative to a simulation of the rainfall, runoff, infiltration pond operation.

Roesner (1974) and others (Goforth et al., 1983; Padmanabhan and Delleur, 1978) also developed deterministic and stochastic relationships among overflow, infiltration rate, and size of facilities. From their work it is also expected that the exponential distribution will reasonably fit the empirical distribution on rainfall volume, runoff duration, and interevent times. Also see Figure 2.2a, which illustrates a reasonable fit of the exponential distribution to a histogram of rainfall volumes.

Using rainfall data from 25 one-hour rainfall recording stations in Florida, the following average statistics were developed for storms with at least

TABLE 8.3 Plant Species Suitable and Sometimes Available from Nurseries for Littoral Zone Plantings of Detention Ponds[a]

Common Name	Scientific Name
American elm[b]	*Ulmus americana*
Arrowhead	*Saggiteria lancifolia*
Bananna water lily[c]	*Nymphaea mexicana*
Black gum[b]	*Nyssa biflora*
Bulrush	*Scirpus* spp.
Button-weed	*Diodia virginiana*
Buttonwood[b]	*Cephalanthus occidentalis*
Carolina willow[b]	*Salix caroliniana*
Coinwort	*Hydrocotyle umbellata*
Cypress[b]	*Taxodium* spp.
Cinnamon fern	*Osmunda cinnamomea*
Day flower	*Commelina diffusa*
Floating hearts[c]	*Nymphoides aquatica*
Frog's-bit	*Limnobium spongia*
Hat pins	*Eriocaulon decangulare*
Iris	*Iris hexagona*
Knotweed (Smartweed)	*Polygonum* spp.
Lizard tail	*Sarurus cernuus*
Loblolly bay[b]	*Gordonia lasianthus*
Maidencane	*Panicum hemitomon*
Pickerelweed	*Pontederia cordata*
Popash[b]	*Fraxinus caroliniana*
Redbay[b]	*Persia palustris*
Red maple[b]	*Acer rubrum*
Rushes	*Eleocharis* spp.
Sawgrass	*Cladium jamaicense*
Sedges	*Cyperus* spp.
Southern cutgrass	*Leersia hexandra*
Soft rush	*Juncus effusus*
St. John's wort	*Hypericum fasciculatum*
Swamp bay[b]	*Magnolia virginiana*
Swamp lily[c]	*Crinum americanum*
Virginia willow[b]	*Itea virginica*
Wax myrtle[b]	*Myrica cerifera*
White water lily	*Nymphaea odorata*
Yellow-eyed grass	*Xyris* spp.
Yellow pond lily	*Nuphar luteum*

[a] Other species may be suitable and may be used with approval.
[b] Tree or shrub; more suitable for higher elevations.
[c] More suitable for deeper water (i.e., greater than 1 to 2 ft).

0.04 in. of rainfall and separated by at least 12 h of no rain (interevent dry period). Twelve hours is the approximate time for infiltrating the full storage volume.

Mean rainfall volume: $\bar{P} = 0.72$ in.

Mean interevent time: $\bar{\Delta} = 92$ h

Mean storm duration: $\bar{D} = 14$ h

A comparison of mean volumes and durations for two minimum interevent dry periods are shown on a statewide basis in Table 8.4. The mean rainfall volume had the least variability among the sites when considering the statistics for seven minimum interevent dry periods.

Loganathan et al. (1985) reported a mean interevent time of 124 h for the Atlanta, Georgia area. In addition, Loganathan also reported an average runoff volume of 0.22 in. with an average duration of 6.9 h. The runoff

TABLE 8.4 Comparison of Rainfall Statistics for 4- and 72-Hour Minimum Interevent Dry Periods by Geographic Region

Region	4-h Minimum Interevent Dry Period		72-h Minimum Interevent Dry Period	
	Mean Volume (in.)	Mean Duration (h)	Mean Volume (in.)	Mean Duration (h)
Apalachicola	0.63	5.5	1.54	49.9
Daytona Beach	0.53	4.8	1.34	51.0
Fort Myers	0.54	2.6	1.82	70.3
Gainesville	0.50	2.9	1.34	51.8
Inglis	0.55	3.0	1.45	47.2
Jacksonville	0.52	5.3	1.34	56.7
Key West	0.45	3.8	1.13	45.8
Lakeland	0.50	3.1	1.52	61.3
Melbourne	0.46	2.7	1.25	46.5
Miami	0.46	3.9	1.71	83.1
Moore Haven	0.49	4.1	1.35	58.4
Niceville	0.62	3.8	1.70	51.7
Orlando	0.49	4.1	1.41	64.0
Parrish	0.52	2.6	1.49	55.6
Tallahassee	0.67	5.2	1.74	56.5
Tampa	0.54	4.4	1.32	54.7
West Palm Beach	0.53	4.3	1.71	68.9
Average	0.53	3.9	1.48	57.3

8.2 MODELING CONCEPTS FOR WET-DETENTION PONDS

volumes and duration are site specific and depend on watershed-related factors.

Given exponential distributions for the independent variables (1) runoff volume, (2) runoff duration, and (3) interevent dry time, the expected values (averages) for the distributions can be calculated (Loganathan et al., 1985). These can be used to estimate the size of an infiltration on-line pond given a flow capture efficiency and infiltration rate. The flow capture efficiency is related to risk of no infiltration by

$$\varepsilon = 1 - \text{flow capture efficiency (infiltration)} \tag{8.8}$$

where ε is the risk (fraction). It is required to infiltrate 80% of the runoff, the risk of overflow is 20% or $\varepsilon = 0.20$.

The storage capacity over the watershed V_E is computed from the following formula assuming exponential distributions (Loganathan et al., 1985):

$$V_E = \frac{1}{\alpha} \ln \frac{\beta}{(\alpha a + \beta)\varepsilon} \tag{8.9}$$

where a = infiltration (treatment) rate, in. over watershed/h
 $\alpha = 1/\overline{R}$ such that $\overline{R}$ is the average runoff volume, in.$^{-1}$
 $\beta = 1/t_b$ such that t_b is the average runoff duration, h^{-1}

Example Problem 8.2 For a flow capture efficiency of 95% and a pond infiltration rate of 0.5 in./h, calculate the storage size assuming exponential distributions for runoff volume and runoff duration with the respective averages as 0.45 in. and 4 h. The pond area is about 1% of the total watershed area.

SOLUTION: First express the pond infiltration rate as an infiltration rate over the entire watershed or (0.5 in./h)/100 = 0.005 in./h. Then use equation 8.9 with

$$a = 0.005 \text{ in./h} \qquad \varepsilon = 0.05$$

$$\alpha = \frac{1}{0.45} = 2.22 \text{ in.}^{-1} \qquad \beta = \tfrac{1}{4} = 0.25 \text{ h}^{-1}$$

$$V_E = \frac{1}{2.22} \ln \frac{0.25}{[2.22(0.005) + 0.25]0.05}$$

$$= 1.33 \text{ in. over watershed}$$

For design purposes with the stated assumptions, one would design the size of the pond for 1.33 in. of runoff over the watershed.

8.2.6 Probability Models

Since the particles in stormwater are variable in density and composition from one storm to another, it is difficult to predict accurately a level of sediment removal for a year using design storm data. It may, however, be more accurate to estimate long-term efficiencies given an assumed frequency distribution on solids density and size. The distribution of stormwater settling velocities is usually well defined by a log-normal distribution (U.S. Environmental Protection Agency, 1986).

Using frequency distribution for precipitation intensity, volume, duration, and interevent times, long-term performance during runoff events (R_L) was estimated by the U.S. Environmental Protection Agency (1986). The efficiencies were shown to be a function of removal at mean runoff (R_M) and the coefficient of variation of runoff volume, C_{VR}. The results are shown in Figure 8.5. The variable (Z) is defined as maximum removal at very low flow, expressed as a fraction and is frequently assumed equal to 1.0 for stormwater

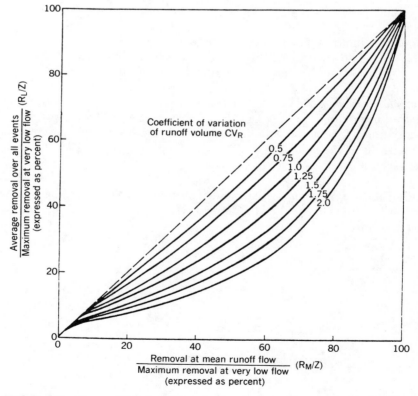

Figure 8.5 Long-term performance of a device where removal mechanism is sensitive to flow rate during runoff. (From U.S. Environmental Protection Agency, 1986.)

planning studies. The average removal at mean runoff flow is estimated assuming an exponential relationship as a function of overflow rate or detention time, or

$$R_M = 1 - e^{(v_s/\text{OR})} \tag{8.10}$$

$$R_M = 1 - e^{-kt_d} \tag{8.11}$$

where R_M = efficiency at mean runoff conditions (fraction)
OR = overflow rate, ft/h
t_d = detention time, h
k = rate coefficient = v_s/h_0; h_0 = settling depth
v_s = sedimentation rate, ft/h

The two equations give equivalent results. Using them requires data for the physical dimensions of a pond with estimates of settling velocities.

Captured stormwaters, after runoff ceases, also undergo removal. These ponded volumes also follow a gamma distribution whose integration resuls are shown in Figure 8.6. The average long-term performance of a sedimentation pond is estimated based on "effective" pond volume (V_E). DiToro and

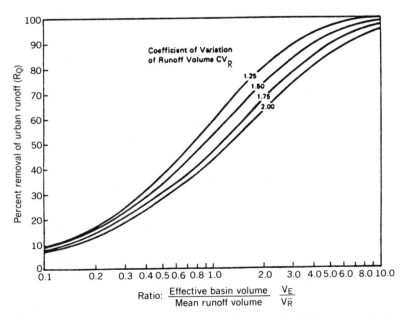

Figure 8.6 Average long-term performance during nonrunoff. (From U.S. Environmental Protection Agency, 1986.)

Small (1979) presented an empirical equation for approximating effective pond volume based on an emptying rate ratio (E):

$$E = \frac{\delta \Omega}{V_R} \qquad (8.12)$$

where δ = average interval between the midpoints of two storms, h
Ω = rate at which pond empties, ft^3/h
V_R = runoff volume from mean storm, ft^3

and

$$\Omega = v_s A \qquad (8.13)$$

where A is the surface area (ft^2).

The volume of water removed between storms on the average is the product $\delta \Omega$. In cases where the volume that can be removed between storms is small on the average relative to the storm volume that enters much of the volume may be occupied with carryover from prior storms. Thus effective volume may be smaller. Figure 8.7 indicates the effect that emptying rate has on effective storage volume. The fraction removals during runoff and during nonrunoff are assumed independent; thus the removals can be multiplied by each other or the combined percent removals can be estimated by

$$\text{overall percent removal} = 100 - \frac{F_P F_Q}{100} \qquad (8.14)$$

where F_P = percentage not removed during runoff conditions
F_Q = percentage not removed during quiescent conditions

Driscoll (1989) estimated the overall percent solids removal using a mean rainfall volume of 0.40 in. ($C_v = 1.50$) for all storm events and compared the results to solids removal for a mean rainfall volume of 0.52 in. (and no runoff for rainfall less than 0.1 in.) with a $C_v = 1.28$ and found that the overall (long-term) performance efficiency was about 75% for both calculations. For this efficiency, the pond surface area was about 2% of watershed.

Example Problem 8.3 A sedimentation pond with a storage volume below the outlet invert elevation is used for a 5-acre commercial site with a runoff coefficient of 0.5. The distribution of solids is given by Table 7.5. The physical

8.2 MODELING CONCEPTS FOR WET-DETENTION PONDS 281

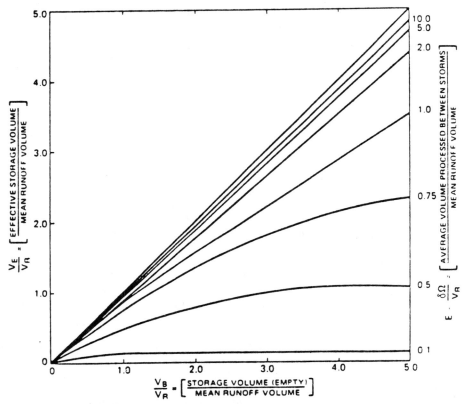

Figure 8.7 Effect of previous storms on long-term effective storage capacity. (From U.S. Environmental Protection Agency, 1986.)

storage volume is 5000 ft^3 with an average surface area of 1260 ft^2. The rainfall statistics for the area were calculated for a local area in the United States.

Average volume of precipitation: $\bar{P} = 0.53$ in., $C_{VP} = 1.44$

Average rainfall intensity: $\bar{i} = 0.086$ in./h, $C_{Vi} = 1.31$

Average duration: $\bar{D} = 7.2$ h, $C_{VD} = 1.09$

Average interval between: $\bar{\delta} = 85$ h, $C_{VS} = 1.00$

What is an estimate of solids removal on a long-term (average annual) basis?

WET-RETENTION / DETENTION PONDS

SOLUTION: Since the pond has a control invert above the bottom (wet rather than dry) removal will occur both during runoff and quiescent conditions. Average overflow rate is calculated using equation 8.15.

$$\text{OR} = \frac{\overline{Q}}{\overline{A}_s} \qquad (8.15)$$

$$= \frac{0.086 \text{ in./h}}{1260 \text{ ft}^2} \times 0.50 \times 5 \text{ acres} \times 43{,}560 \text{ ft}^2/\text{acre} \times 1 \text{ ft}/12 \text{ in.}$$

$$= 0.62 \text{ ft/h}$$

and average volume of runoff as

$$V_R = \overline{P}C(\text{area})\left(\frac{43{,}560}{12}\right)$$

$$= 0.53 \times 0.50 \times 5 \times 3630$$

$$= 4810 \text{ ft}^3$$

Assume that the variability of runoff is the same as the corresponding rainfall variability and rainfall excess is the same as rainfall volume:

$$C_{VQ} = 1.31 \quad \text{and} \quad C_{VR} = 1.44$$

where C_{VQ} = coefficient of variability of runoff rate
C_{VR} = coefficient of variability on rainfall excess (volume)

During Runoff Conditions: Using equation 8.10 gives

$$R_M = 1 - e^{-(v_S/\text{OR})}$$

and R_L from Figure 8.5, or as shown in the following table ($z = 1.0$).

Average Settling (v_s) Velocity (ft/h)	Percent of Time	R_M (%) (Equation 8.10)	R_L (%) (Figure 8.5)
0.03	20	5	5
0.33	20	41	20
1.5	20	91	77
7.0	20	100	100
70	20	100	100

$$\text{Overall average removal} = 60\% = \frac{\sum R_L}{S}$$

F_D or percentage not removed = 40%

During Quiescent Conditions: (*Note*: Assume that larger particles have settled during the storm event, thus the particle size distribution has changed.)

$$\frac{\text{pond volume}}{\text{runoff volume}} = \frac{V_B}{V_R}$$

$$\frac{V_B}{V_R} = \frac{5000}{4810} = 1.04$$

and

$$E = \frac{\delta\Omega}{V_R} = \frac{85(v_s)1260}{4810}$$

$$= 22.28 v_s$$

Average Settling (v_s) Velocity (ft/h)	Percent of Time	E	V_E/V_R (Figure 8.7)	R_Q (%) (Figure 8.6)
0.03	25	0.7	0.60	40
0.33	25	7.4	1.00	54
1.50	25	33.4	1.04	55
7.0	25	156.0	1.04	55
70	0			

Overall average removal = 51%

F_Q or percentage not removed = 49%

$$\text{Overall percent removal} = 100 - \frac{F_D F_Q}{100} = 100 - \frac{49(40)}{100} = 80\%$$

Note: During quiescent conditions, the efficiencies are less than during the runoff conditions. This is because some portion of the solids are not available for quiescent settling or are removed from the pond.

8.2.7 Fate of Pollutants

Nutrient and metal interactions with bottom sediments in wet-detention ponds have not been fully investigated. Only limited data from literature are available to describe the interaction, but the processes studied in small lakes may apply to stormwater wet-detention ponds. To determine uptake and release rates from bottom sediments, aquatic polyethylene isolation chambers (IC) are constructed and submerged with the open top down on the bottom sediments of wet-detention ponds in Orange County, Florida (Yousef et al., 1986b). The exterior of the chambers was painted black with epoxy to eliminate light penetration inside the chambers. The effective volume of water retained by each chamber varies depending on the size of the chamber and the bottom sediments area isolated varies with the area of the chamber.

Aerobic conditions could be maintained in each IC by compressed air supply from cylinders connected by tygon tubing to stone diffusers located near the bottom of the chambers. Three chambers could be placed with water–sediment contact, but a polyethylene cover is placed on the pond bottom of the third one, IC 3, to prevent exchange between sediments and water inside this chamber. From a study completed by Yousef et al. (1986b), a known solution of ammonia nitrogen and phosphorus was injected into both IC 2 and IC 3, but IC 1 was left as a control chamber during each experiment. The experiments were run under aerobic environment for selected periods of time, then anaerobic conditions were established by cessation of the airflow to the chambers. Decline in dissolved oxygen concentrations and nutrient analyses were conducted during the anaerobic phase in order to evaluate oxygen uptake rates and nutrient transformations between water and sediments inside the isolation chambers.

Oxygen Uptake. The condition of the bottom sediments was reflected by the low sediment oxygen demand (SOD) values. SOD values were based on measurement of dissolved oxygen (DO) in isolation chambers after cessation of the air supply. DO depletion was best approximated by first-order reactions as presented in Table 8.5. Average overall removal rates were calculated arithmetically by dividing the total oxygen removal by the period of incubation. The average rates in IC 1, IC 2, and IC 3 placed in the EPCOT pond were 0.51, 0.61, and 0.23 g O_2/day. The difference in O_2 depletion rates between IC 2 and IC 3 is 0.38 g O_2/m^2-day, which was representative of the sediment oxygen uptake at an average temperature of 25°C. The SOD measurements in another pond (Maitland) averaged 0.9 g O_2/m^2-day, which

TABLE 8.5 Uptake of Dissolved Oxygen in Isolated Chambers During Transition from Aerobic to Anaerobic in EPCOT Pond

Experiment	Relationships		Number of Observations	Average Oxygen Removal (g O_2/m^2-day)
First	$O_2 = 9.40e^{-0.28t}$	1	5	0.33
	$O_2 = 7.81e^{-0.34t}$	2	5	0.43
	$O_2 = 6.93e^{-0.13t}$	3	5	0.31
Second	$O_2 = 10.1e^{-0.11t}$	3	5	0.15
	b	2	—	0.78
	b	1	—	0.68

[a] O_2 in g dissolved oxygen per m^2 sediment, and t is incubation time in days.
[b] Not enough to produce relationships.

was higher than in EPCOT pond, but are fairly low compared with eutrophic lakes. These values correspond with the high redox potentials measured in water and sediment samples. The stormwater ponds are less than 10 years old and did not accumulate much sediment deposits.

Phosphorus Transformation. Uptake and release of phosphorus under aerobic and anaerobic conditions inside isolation chambers were tested. Figure 8.8 shows the decline in phosphorus of the water column under aerobic environment and Figure 8.9 shows the increase in phosphorus under anaerobic conditions. Maintaining aerobic conditions at the water–sediment interface usually results in the decline of soluble phosphorus due to absorption by the sediments and the control of its release under aerobic environment (Fillos, 1976). In Maitland pond it was observed that OP concentration increased to 130 μg P/L following two heavy rain events, but was reduced to a low level (10 μg/L) in approximately 3 days.

The uptake and release of phosphorus (Figures 8.8 and 8.9) appeared to follow relationships presented in Table 8.6. The control chamber IC 1, with no added phosphorus, did not show a decline in concentration with time under aerobic conditions. Also, the phosphorus uptake inside IC 3 occurred at a much slower rate in the absence of sediment contact with the overlying water column. The uptake rate constant in IC 2 was 0.15 day^{-1} and in IC 3 was 0.02 day^{-1}. Bottom sediments contact was responsible for the much higher removal rate in IC 2 than IC 3. These data suggested that two-thirds of the incoming OP would be removed within 1 week and sediments may be responsible for 80% of the removal rate. The calculated average removal rate was estimated at 18.9 mg/m^2-day for TP.

Release of phosphorus under anaerobic environment was observed in this study and modeled as presented in Table 8.6. As expected, the slope of the

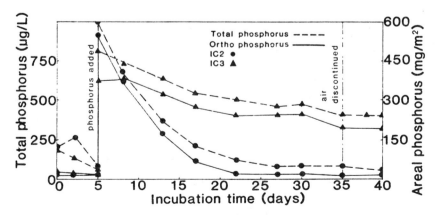

Figure 8.8 Phosphorus removal inside isolation chambers under aerobic conditions in Epcot Pond receiving highway runoff.

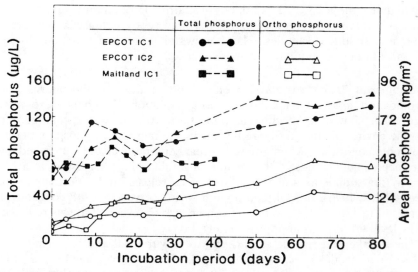

Figure 8.9 Release of phosphorus inside isolation chambers after establishing anaerobic environment in retention ponds receiving highway runoff.

regression line between phosphorus concentration and contact time inside IC 1 in Maitland pond was more than three times the slope of IC 1 in EPCOT ponds. The release of phosphorous is a function of its concentration in the top sediment layer and the interstitial water. Sediment from the Maitland pond contained more than three times the phosphorus in the EPCOT pond. This was consistent with results obtained from IC 2 inside EPCOT pond. IC 2 was dosed with phosphorus prior to anaerobic experimentation and showed a higher release rate of OP than IC 1. Release rates were averaged as shown in Table 8.7. They were very low relative to values recorded from lake bottom sediments, which varied between 1 and 30 mg P/m^2-day (Theis and McCabe, 1978).

TABLE 8.6 Uptake and Release of Orthophosphorus Inside Isolation Chambers

Environment	Location	Chamber	Relationship	Number of Observations	Average (mg P/m^2-day)
Aerobic	EPCOT	IC 2	$OP = 723e^{-0.15t}$	8	18.9
		IC 3	$OP = 640e^{-0.02t}$	8	8.9
Anaerobic	Maitland	IC 1	$OP = 5 + 1.34t$	12	0.9
	EPCOT	IC 1	$OP = 11.4 + 0.38t$	10	0.3
		IC 2	$OP = 16.7 + 0.79t$	10	0.5

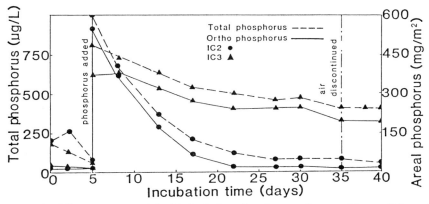

Figure 8.10 Nitrogen removal inside isolation chambers under aerobic conditions in Epcot Pond receiving highway runoff.

It has been shown that nitrate might prevent release of phosphorus from anaerobic sediments (Riple and Lindmark, 1978). It is assumed that denitrification processes maintain a high redox potential at the upper sediment layers, thus increasing the absorption properties of the top layer of sediments and preventing the release of phosphorus from lower anaerobic sediments. This effect was also demonstrated in these studies. Nitrate added to isolation chambers in Maitland pond was depleted at a rate between 50 and 500 mg N/m²-day, which confirms the importance of sediments in nitrogen removal by denitrification.

Changes in concentrations of inorganic nitrogen forms (NH_4-N, NO_2-N, and NO_3-N) and organic nitrogen (ON) concentrations inside the isolation chambers were studied under aerobic and anaerobic environments. Under aerobic environment, NH_4-N declined gradually while NO_3-N increased. However, a noticeable overall decline in TN concentrations was observed as shown in Figure 8.10. Relationships between TN and incubation time (t) were developed as shown in Table 8.7. The decline in total nitrogen in an aerobic environment followed an exponential function with rate constant of

TABLE 8.7 Changes in Nitrogen with Incubation Time Inside Isolation Chambers

Environment	Location	Chamber	Relationship	Number of Observations n	Average (mg N/m²-day)
Aerobic	EPCOT	IC 2	TN = $6.36e^{-0.06t}$	9	139
		IC 3	TN = $3.7e^{-0.02t}$	9	35
	Maitland	IC 2	TN = $8.46e^{-0.054t}$	13	79
Anaerobic	EPCOT	IC 2	NH_4-N = $0.05t + 0.305$	12	31
		IC 2	NH_4-N = $0.044t - 0.015$	6	36

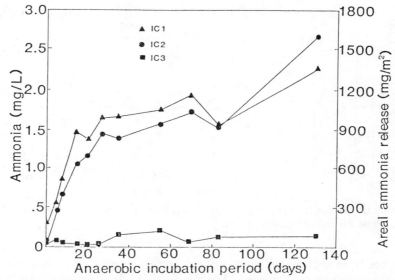

Figure 8.11 Ammonia release inside isolation chambers at EPCOT Pond under anaerobic environment.

0.06 and 0.054 day^{-1} for IC 2 in EPCOT and Maitland ponds, respectively. The role of the sediments in nitrogen removal was obvious by comparing rate constants for IC 2 in EPCOT and Maitland ponds, respectively. The role of the sediments in nitrogen removal was obvious by comparing rate constants for IC 2 (0.06/day^{-1}) and IC 3 (0.02 day^{-1}). The removal rate constant, 0.06 day^{-1}, corresponds to a half-life of 12 days, showing efficient soluble nitrogen removal by the sediments.

Under anaerobic environment, the release of ammonia from the sediments took place as depicted in Figure 8.11. IC 1 and IC 2 showed gradual increase, but IC 3 (without sediment contact) showed very little or no increase in ammonia nitrogen in an anaerobic environment. The increase in NH_4-N with time inside IC 2 in EPCOT and Maitland ponds followed linear relationships as presented in Table 8.7. The slope of lines is almost identical. Ammonia released from the bottom sediments may be oxidized by the top aerobic layer to nitrate nitrogen or taken up by plants. The nitrate is eventually reduced to nitrogen or taken up by plants. The nitrate is eventually reduced to nitrogen gas by the lower layers of the bottom sediments, which may explain the overall reduction in total nitrogen.

Metal Transformation. The removal of dissolved metal species is rapid, with as much as 90% removal occurring in 4 days, as shown in Table 8.8. Soluble concentrations of copper, zinc, iron, and lead were added to two test chambers in concentrations between 0.5 and 1 mg/L. However, 4 days later

TABLE 8.8 Uptake and Release of Heavy Metals Inside Isolation Chambers at EPCOT Pond

Sample Location	Metal	Total Metals Concentration (μg/L) in Water									
		4–1[a]	4–1[a]	4–4	4–18	4–21	4–25	5–5	5–9	5–12	5–19	5–24
IC 1	Cu	23	—	15	17	27	8	11	9	13	15	22
	Zn	7	—	9	5	6	4	4	3	4	4	4
	Fe	596	—	614	455	596	743	781	916	904	1118	1267
	Pb	23	—	32	26	23	27	24	28	26	23	23
IC 2	Cu	21	683| 71	17	19	24	24	26	20	19	46
	Zn	14	857| 82	10	10	11	10	5	12	5	10
	Fe	744	790	648	499	772	1059	1282	1703	2008	1654	1666
	Pb	24	904| 93	23	29	41	27	39	37	29	22
IC 3	Cu	23	590| 61	19	17	22	14	64	28	28	41
	Zn	13	749| 50	3	10	4	5	7	7	7	10
	Fe	401	468	720	617	788	612	360	341	535	269	300
	Pb	27	724| 56	30	32	45	30	48	38	25	33
Pond	Cu	22	—	35	16	25	26	26	27	24	28	29
	Zn	12	—	4	0	6	4	3	2	3	4	3
	Fe	603	—	855	454	820	848	423	404	371	421	174
	Pb	28	—	52	21	36	48	46	49	51	44	45

Diffused air   Diffused air was shut-off
| ← was supplied → | ←   → |

[a] After addition of nutrient and heavy metal solution into IC 2 and IC 3.

TABLE 8.9 Changes in Aqueous Metal Concentration with Incubation Time Inside Isolation Chambers at EPCOT Site

Environment	Chamber	Metal	Relationship	Number of Observations n	Average (mg/m^2-day)	Remarks
Anaerobic	IC 1	Fe	Fe(μg/L) = 507 + 19.6t	8	+13.5	Release
	IC 2	Fe	Fe(μg/L) = 728 + 34.7t	8	+19.5	Release
	IC 3	Fe	Fe(μg/L) = 687 − 12t	8	−5.8[a]	Uptake

[a] Average Fe removal rate during period of air shutoff.

when the next sample was collected, concentrations of copper, zinc, and lead had been substantially reduced by an average of 90%. By the next sample collection after 18 days, concentrations in the dosed chamber were indistinguishable from the control, which received no metal additions. No change was noted either with or without sediment contact in these metal concentrations throughout the test period, even when anaerobic conditions were established. During aerobic environment when diffused air was supplied, the metals Cu, Zn, and Pb declined rapidly inside IC 2 with sediment contact and IC 3 without sediment contact. One-tenth or less of the added Cu, Zn, and Pb remained in the water phase within the first 3 days. Iron did not reflect the same removal efficiencies. The average uptake of Cu, Zn, Fe, and Pb in IC 2 during aerobic environment are 23.5, 29.9, 10.3, and 31.1 mg M/m^2-day, respectively.

The release of heavy metals inside the isolation chambers after cessation of the air supply source in order to promote an anaerobic environment was studied. It is obvious from Table 8.9 that Fe is the only metal to show sizable release inside the control chamber IC 1 and IC 2. IC 3 did not show Fe release since dissolved oxygen stayed high and an anaerobic environment did not develop during the entire experimental period. It took IC 1 and IC 2 from 3 to 7 days to go anaerobic after shutting off the air source, but it took IC 3 over 1 month. The release of Fe can be modeled as shown in Table 8.9. IC 2 showed the highest release rate because of excess iron added initially. IC 3 did not show any release during the experimental period, and on the contrary, it showed uptake of iron. It never became anaerobic inside IC 3.

Example Problem 8.4 Soil samples were incubated under aerobic and anaerobic conditions, with soluble phosphorus being measured in the solution and solid phase after the addition of soluble phosphorus to the solid phase. Each 300-g soil sample was kept in suspension by use of a magnetic stirrer in 1500 mL of water in a sealed 2-L flask. Slow streams of air for the aerobic treatments and argon for the anaerobic treatments were continuously bubbled through the suspensions. The samples were incubated for 17 days at 30°C before additions of phosphorus as $Ca(H_2PO_4)_2$. Samples of the suspension were removed 24 h after the addition of phosphorus and filtered. The resulting chemical analysis at equilibrium is shown in Table 8.10. What effects do anaerobic conditions have on the release of phosphorus from the soil? Do this by developing a mass balance for soluble phosphorus that is used to calculate the phosphorus concentration in the soil phase.

SOLUTION: A material balance for phosphorus mass in the solid and water phases is developed and applied to the first set of data:

$$P_1 V_1 + P_s M_s = PM_s$$

TABLE 8.10 Soil–Water Equilibrium for Phosphate in Soil

Added P (μg P/g soil)	Aerobic		Anaerobic	
	Water P_1 (μg P/mL)	Calculated Soil P_s (μg P/g soil)	Water P_1 (μg P/mL)	Calculated Soil P_s (μg P/g soil)
420	10.6	367.0[a]	30.0	270.0
780	30.0	630.0	90.0	330.0
1080	40.0	880.0	100.2	579.0

[a] Example calculation:

$$(10.6 \ \mu g\ P/mL)(1500\ mL) + (P_s)(300\ g) = (420\ \mu g\ P/g\ soil)(300\ g)$$

$$P_s = 367\ \mu g\ P/g\ soil$$

where P = phosphorus added, μg/g soil
 M = mass of the soil, g
 P_1 = phosphorus concentration in water, μg/L water
 P_s = phosphorus concentration in soil, μg/L soil

From the data for this particular soil, the release of phosphorus under anaerobic conditions is greater than the release under aerobic conditions. The example problem also illustrates the use of a mass balance. The ratio of phosphorus in the soil to phosphorus in the water also can be developed for specific initial concentrations and soil conditions.

8.3 WET-DETENTION POND AS A HOLDING TANK AND A REUSE POND

A wet-detention pond can also operate to hold runoff water for a time period equal to a minimum interevent dry time before discharging, and then release the treated water without exceeding downstream flow rate and volume constraints. Thus pond volume for the storage of the next runoff event is available. Runoff water stored in the pond may also be reused for irrigation, car washing, cooling water makeup, or other beneficial uses. This type of pond can be called a "smart" pond because it uses the principles of stormwater management and reuse technologies. As shown in Figure 8.12, there are four storage volumes, one for sediment and debris, another called a permanent pool for water quality enhancement, a reuse or temporary storage volume, and a fourth for peak flow rate attenuation.

The reuse or temporary storage volume depends on the meteorological conditions, which produce precipitation volume and interevent dry periods. These will vary from one location to another and are expressed as PIF curves. A comparison of three regions are shown in Figure 8.13 (Wanielista

8.3 WET-DETENTION POND AS A HOLDING TANK AND A REUSE POND

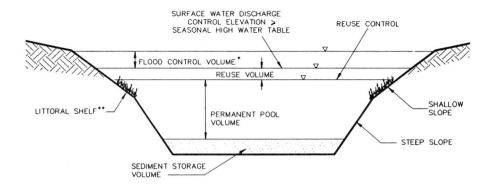

*Can be measured above permanent pool, however some regulatory agencies measure above the reuse volume.

**The reader should consult local water management districts and other regulatory agencies to determine specific geometric and littoral zone design requirements.

Figure 8.12 Schematic of a stormwater reuse pond.

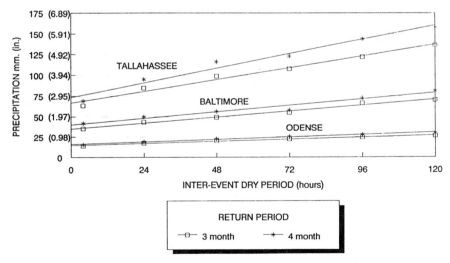

Figure 8.13 Comparison of PIF curves for three geographic areas.

et al., 1990). The Odense Denmark data were developed for interevent dry period ranging from 2 to 96 h. The Tallahassee and Baltimore interevent dry period ranged from 4 to 120 h. For a 72-h dry period and a 4-month return frequency, the design rainfall volume is 125 mm for the Tallahassee area, 65 mm for Baltimore, and 33 mm for the Odense, Denmark area. The comparison illustrates that the design volumes and PIF curves can be expected to be different for different geographic areas. When runoff water in temporary storage is removed for reuse, treated, or infiltrated into the ground, the mass discharge of pollutants will be reduced.

8.3.1 Holding Pond with No Discharge for a Minimum Interevent Dry Period

The interevent time for treatment is measured by a depth sensor connected to a timing device and could be wired to a data logger to record changes in pond elevation. As pond elevation increases above a set interval (i.e., 1 in.) that represents a runoff volume to the pond, the timing device is set (or reset) to zero. If no pond elevation change is noted within the interevent dry period, the water quality control valve or gate is opened and the treated water is discharged as long as there is no rainfall. When the pond elevation returns to the elevation of the permanent pool, no more ponded water will discharge.

The holding pond can also be used in combination with chemical coagulation using alum. When this practice is used, the bottom debris and sediment area is made larger but the water quality treatment volume is reduced because the detention time necessary for coagulation is between 12 and 24 h. The volume of precipitation from the PIF curves is lower at 24 h than 72 or 96 h.

Example Problem 8.5 Size the water quality volume (ft^3) for a holding pond with a specification of a minimum detention time of 72 h and a return period of once every 4 months in the Baltimore area. The watershed is 120 acres with a runoff coefficient of 0.50.

SOLUTION: Using Figure 8.13 with a return period of 4 months and a detention time of 72 h, the precipitation volume is 2 in. The runoff volume is calculated as

$$R = 2 \text{ in.} \times 120 \text{ acres} \times 0.50 \times 43{,}560 \text{ ft}^2/\text{acres}$$

$$= 5{,}227{,}200 \text{ ft}^2\text{-in.} \times \tfrac{1}{12} \text{ ft/in.} = 435{,}600 \text{ ft}^3$$

8.3.2 Reuse Ponds Operating on a Schedule

Reuse of ponded water is an economical alternative relative to the use of potable water. A stormwater or irrigation utility can produce a reliable source of water at about $0.20 to $0.50 per 1000 gal compared to potable water at $1.00 or more per 1000 gal.

The reuse of stormwater is not a new concept. For years, systems have been built which use stormwater in some capacity. However, the reuse components have generally been additions to existing detention pond designs with no consideration for the quantity of discharge. Thus a design procedure that considers both reuse and discharge would be beneficial.

Ecological Benefits. No-impact stormwater management practice has become an increasingly important professional goal and is more often publicly demanded. Although the term *reuse* alone implies environmental benefits, there are tangible arguments supporting the benefits of reusing stormwater. Foremost is the reduction of volume and pollution discharges to surface waters: Water reused is water that is not discharged. Typical detention ponds draw down the temporary storage volume using control devices that discharge into adjacent natural water bodies. However, reuse systems deliver the reuse volume back over the watershed or use the water in some other productive way. Because a reuse pond discharges only waters above the temporary storage and reuses the temporary storage volume, it will discharge less than will a similarly sized detention pond.

Reuse is a good practice for the conservative management of groundwater resources. As urbanization increases there is a change in the hydrologic balance of the region. An increase in watershed discharges will decrease the amount of water that had previously infiltrated into the ground and evaporated from the watershed. However, as the stormwater is reapplied to the watershed, there is greater potential for groundwater replenishment and evapotranspiration. Also, when reuse systems replace irrigation systems dependent on groundwater, there is decreased use of groundwater, whether the original source for irrigation was potable water or pumped on site. Stormwater ponds usually receive nutrients (nitrogen, phosphorus, etc.) from surrounding watershed areas. Dissolved nutrients can be recycled back to the landscape by reuse systems that irrigate the stormwater.

Economic Benefits. The concept of reuse may be ecologically sound, but unless the inclusion of reuse is monetarily profitable, it would not become widely implemented into design. There are several economic advantages to reusing stormwater. A significant monetary savings will result from not using and paying for potable water. This fact is exaggerated for large land users such as golf courses. The annual cost of potable water for a 100-acre golf course irrigating at a rate of 2 in./per week can be estimated given an average cost of potable water equal to $1.00 per 1000 gal.

$$\$/\text{yr} = \frac{2 \text{ in.}}{\text{week}} \times 100 \text{ acres} \times \frac{1 \text{ ft}}{12 \text{ in.}} \times \frac{325{,}828 \text{ gal}}{\text{acre-ft}}$$

$$\times \frac{52 \text{ weeks}}{\text{yr}} \times \frac{\$1.00}{1000 \text{ gal}}$$

$$= \$282{,}385/\text{yr} \tag{8.16}$$

The golf course would pay close to $300,000 per year for water.

This figure could be substantially reduced with a reuse system. A reuse system would use the same irrigation network but would require a pumping system to deliver the water. The initial cost of the pump system is estimated to be between $25,000 and $35,000, with an electrical and maintenance cost of $15,000 to $30,000 per year. An annual cost considering amortization of the equipment over 20 years at 10% is

$$\$/\text{yr} = P(\$35{,}000, 10\%, 20 \text{ yr}) + \$30{,}000$$

$$= (\$35{,}000 \times 0.1175) + \$30{,}000$$

$$= \$34{,}112.50/\text{yr} \tag{8.17}$$

This cost does not include the cost of supplemental irrigation. A reuse pond, having land-use requirements comparable to those of a wet-detention pond, will save significant money in water costs.

Different water restrictions generally apply to operators of stormwater reuse systems. Most of the water used would come from the storage of stormwater runoff.

Iron and other minerals contained in groundwater will have an opportunity to oxidize and settle in the reuse pond before being distributed by the reuse system. Traditional groundwater irrigation systems may cause rust stains on walls and sidewalks, thus causing an economic impact.

Behavior of a Reuse Pond. The response of a typical reuse pond to a rainfall event is summarized. During and following a rainfall event, there is runoff into the pond and the water level rises to some depth above the permanent pool. If this new water level exceeds the level of the surface discharge control, there will be discharge at some rate until the water level drops back below the control structure. The pond water level is incrementally (daily) removing or adding an amount of water from or to the pond. If the reuse volume is expended, supplemental water, such as groundwater, is used to maintain the permanent pool volume. This could occur as seepage through the sides of the pond or by mechanical pumping using a controller. This scenario can be simulated by creating a mass balance of inputs and outputs.

8.3 WET-DETENTION POND AS A HOLDING TANK AND A REUSE POND

A Model. A model is based on the continuity equation

$$\text{inputs} - \text{outputs} = \Delta S \qquad (8.18)$$

By considering all potential water movements, a complete hydrologic balance may be expressed in volume units as

$$R + G + P \pm F - RU - D - ET = \Delta S \qquad (8.19)$$

where R = rainfall excess or runoff volume
G = supplemental water (groundwater)
P = precipitation directly on the pond
F = water movement through the sides of the pond
RU = reuse volume on the EIA
D = discharge
ET = evapotranspiration
S = storage in pond

The average evapotranspiration rate is generally negligible and the precipitation on the pond is accounted for in the equivalent impervious area. Additionally, evaporation data are generally only available in mean monthly values compared to the daily time step of the model making the estimate of evaporation potentially inaccurate. These parameters were dropped from the mass balance. Also, because of the difficulty in estimating groundwater movement into and out of a pond with permeable sides, and the practice of constructing lined ponds, the flow of groundwater through the sides of the pond was assumed to equal zero. Equation 8.19 was further simplified to

$$R + G - RU - D = \Delta S \qquad (8.20)$$

Assumptions have been made on variables that are not easily quantified. These assumptions were addressed using sensitivity analysis and the evapotranspiration daily data assumed for model runs was not considered to affect the design procedure (Harper, 1991). For modeling purposes, there were two inputs, runoff and supplement, and two outputs, reuse and discharge (Figure 8.14). Runoff was established from known precipitation and watershed data. The reuse rate was a controlled variable. Both supplemental water and discharge were functions of the water level of the pond, or the storage volume. Since groundwater movement was assumed to equal zero, supplemental water will be considered as that which is pumped into the pond mechanically. Supplement will occur at a rate that is necessary to maintain the permanent pool; the maximum required rate would equal that of reuse. Because storage capacity below the permanent pool is eliminated by supplement, the design may be considered conservative with respect to discharge

298 WET-RETENTION / DETENTION PONDS

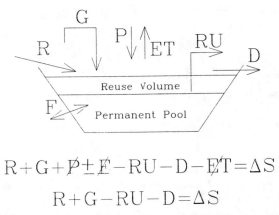

$$R+G+\cancel{P}\pm\cancel{F}-RU-D-\cancel{ET}=\Delta S$$
$$R+G-RU-D=\Delta S$$

Figure 8.14 Summary of mass balance for a reuse pond.

from the pond. With the previous simplifications, the pond operation is simulated by the model.

Pond Reuse Effectiveness. The effectiveness of a reuse pond used for irrigation can be measured by the percentage of yearly runoff water volume that is not discharged. Using daily rainfall records over a 15-year period (from January 1974 through December 1988) as collected at rainfall measuring stations along the east coast of the United States, reuse pond operation was simulated using the following operational and design criteria.

1. Temporary storage volume based on rainfall volumes of 0.25 to 7.00 in. over the effective impervious area.
2. Irrigation rates per day of 0.06 to 0.30 in. over an area equal to the equivalent impervious area.
3. Direct surface discharge if pond volume exceeds the temporary storage volume.
4. Stored water interchange with the surrounding groundwater is negligible because of a relatively impermeable layer between the pond and the ground.
5. Runoff occurs when precipitation exceeds or equals 0.04 in.

The results of the simulation are shown in the charts of Figure 8.15. There are three variables that require explanation to use the charts: *r*euse rate, *e*fficiency of reduced discharge, and *v*olume of temporary storage—thus the name REV charts. By knowing or specifying two of the three variables, the other can be found from a REV chart. In general, as the yearly average

8.3 WET-DETENTION POND AS A HOLDING TANK AND A REUSE POND

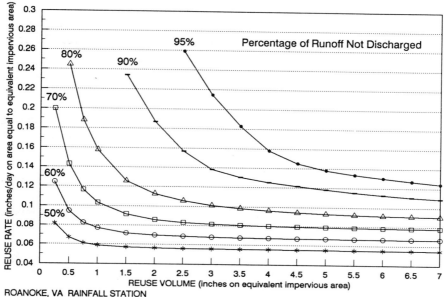

ROANOKE, VA RAINFALL STATION
MAY 1974 - DEC. 1988
MEAN ANNUAL RAINFALL ≈ 40.9 in.

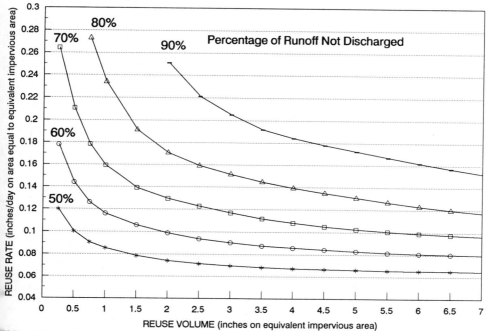

ORLANDO RAINFALL STATION
MAY 1974 - DEC. 1988
MEAN ANNUAL RAINFALL = 48.2 in.

Figure 8.15 Some examples of REV charts.

300 WET-RETENTION / DETENTION PONDS

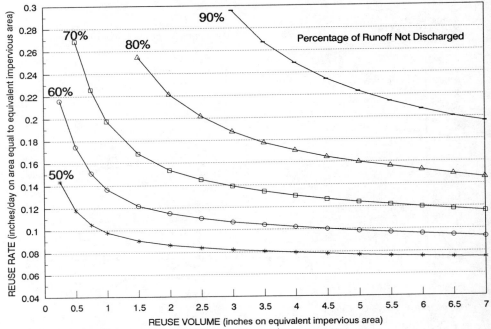

MIAMI RAINFALL STATION
MAY 1974 - DEC. 1988
MEAN ANNUAL RAINFALL = 54.5 in.

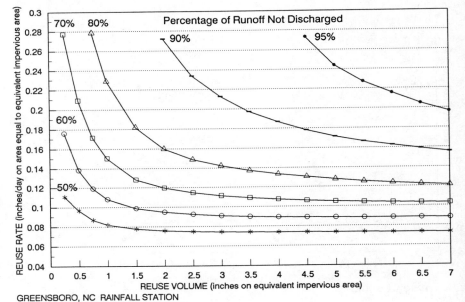

GREENSBORO, NC RAINFALL STATION
MAY 1974 - DEC. 1988
MEAN ANNUAL RAINFALL = 53.9 in.

Figure 8.15 *(Continued)*

8.3 WET-DETENTION POND AS A HOLDING TANK AND A REUSE POND

rainfall increases, the temporary storage volume increases to maintain a specified efficiency and reuse rate. However, the rate of reuse and the interevent dry periods also affect the temporary storage, and REV charts can change among geographic areas with similar yearly rainfall volumes. The pond mass balance considers reuse water to be available every day and above the permanent pool. This operational limitation is generally what is done to maintain a relatively high and attractive shoreline for the reuse pond while calculating a conservative estimate of the pond discharge volume. Simulating 15 years of pond operation, Harper (1991) allowed water reuse from the permanent pool and found that the pond would operate at or below the permanent pool elevation a higher percent of the time during a year relative to maintaining an elevation at the permanent pool. Therefore, the probability of flooding or the exceedence of the flood control elevation would be lowered when pond elevation was permitted to go below the permanent pool, and is relative to either a wet-detention pond design or the operation of the reuse pond when the lowest pond elevation was controlled at the permanent pool elevation.

Example Problem 8.6 A detention pond must be designed for a new community of multifamily condominiums in the Orlando, Florida area. The watershed information is as follows:

40 acres total area
24% directly connected impervious area
No runoff contribution from pervious area for rainfall up to 5.5 in.
80% of runoff not discharged
Maximum reuse storage is 4 in. on equivalent impervious area
1.5 in./week of reuse by irrigation

How much land area is needed for irrigation, and what is the volume of the reuse storage?

SOLUTION: Since the intensity of local irrigation is given, the area to be used for irrigation can be found by knowing the reuse rate. This is a function of the reuse volume and the discharge efficiency and can be obtained from the REV chart. From the Orlando chart, Figure 8.15, for a volume of 4 in. on the EIA and an efficiency of 80%, a reuse rate of 0.14 in./day on the EIA is obtained.

$$RU = f(4 \text{ in.}, 80\%)$$
$$= 0.14 \text{ in.}/\text{day} \qquad (8.21)$$

The EIA is equal to 40 acres times 0.24, or 9.6 acres. Therefore, the irrigation (reuse) area (A_R) can be calculated using

$$A_R = 9.6 \text{ acres} \times \frac{0.14 \text{ in./day}}{1.5 \text{ in./week}} \times 7 \frac{\text{days}}{\text{week}}$$

$$\approx 6.3 \text{ acres} \qquad (8.22)$$

The reuse storage volume is stated in the problem as being 4 in. on the EIA, or

$$V_R = 9.6 \text{ acres} \times 4 \text{ in.} \times \frac{\text{ft}}{12 \text{ in.}}$$

$$= 3.2 \text{ acre-ft} \qquad (8.23)$$

Therefore, the area needed for irrigation is approximately 6.3 acres and the reuse storage volume is 3.2 acre-ft.

Runoff from Both Impervious and Pervious Areas. If the pervious area of a watershed contributes runoff when rainfall is equal to the reuse volume, the size of the EIA will be influenced by the pervious portion as well as the impervious portion of the watershed.

Example Problem 8.7 Solve Example Problem 8.6 except with the pervious area having an SCS curve number (CN) of 80.

SOLUTION: The same procedure is followed. To find the reuse rate, the efficiency and the reuse volume must be known. The efficiency has remained at 80%, and although the actual reuse volume has changed due to the increase in EIA, it is still calculated using 4 in. on the EIA. Therefore, the REV chart can be used in the same way and the reuse rate remains at 0.14 in./day on the EIA. To find the EIA we must determine the effect of the pervious area in response to a 4-in. rainfall event. The curve number method will be used to calculate the resulting runoff. Using the SCS saturation equation, the potential soil/cover storage at saturation, S', is calculated as

$$S' = \frac{1000}{CN} - 10$$

$$= \frac{1000}{80} - 10$$

$$= 2.5 \text{ in.} \qquad (8.24)$$

The resulting rainfall excess from the pervious area for a rainfall of 4 in. is calculated using the SCS rainfall excess equation:

$$R_p = \frac{[4 - 0.2(S')]^2}{4 + 0.8(S')}$$

$$= \frac{[4 - 0.2(2.5)]^2}{4 + 0.8(2.5)}$$

$$= 2.0 \text{ in.} \tag{8.25}$$

and the runoff coefficient is

$$C_p = \frac{R_p}{P} = \frac{2 \text{ in.}}{4 \text{ in.}}$$

$$= 0.50 \tag{8.26}$$

The effective runoff coefficient can then be calculated using

$$C = \frac{1.0(40 \text{ acres} \times 0.24) + 0.50(40 \text{ acres} \times 0.76)}{40 \text{ acres}}$$

$$= 0.62 \tag{8.27}$$

The EIA is then computed as 24.8 acres.

$$\text{EIA} = 0.62 \times 40 \text{ acres}$$

$$= 24.8 \text{ acres} \tag{8.28}$$

We can now calculate the local irrigation area:

$$A_R = \frac{24.8 \text{ acres} \times 0.14 \text{ in./day}}{1.5 \text{ in./week}} \times 7 \frac{\text{days}}{\text{week}}$$

$$\approx 16.2 \text{ acres} \tag{8.29}$$

The land area needed for irrigation is 16.2 acres. As expected, this is considerably more area than what was needed in Example Problem 8.6, due to the significant runoff from the pervious area. The reuse volume can be calculated in the same way as before:

$$V_R = 24.8 \text{ acres} \times 4 \text{ in.} \times \frac{\text{ft}}{12 \text{ in.}}$$

$$= 8.3 \text{ acre-ft} \tag{8.30}$$

Similarly, the reuse volume is greater because the overall imperviousness (EIA) of the watershed has been increased.

304 WET-RETENTION / DETENTION PONDS

Calculation of Supplement Rate. One of the benefits of reusing stormwater is the cost savings of not having to buy water. A reuse pond sometimes requires the use of groundwater or other sources to supplement the reuse storage volume. Since reuse rates may average as high as 2.0 in. per week over 52 weeks, a reuse volume of 104 in. per year would be required, but runoff may only supply a portion of the reuse needed. The designer should be interested, for economic and conservation considerations, in how much supplemental water is necessary for a certain design. This can easily be done using a REV chart.

The procedure is based on a mass balance of the reuse pond. By equating the inputs and outputs to the pond using units of inches/year, we obtain

$$\text{inputs} = \text{outputs}$$
$$\text{rainfall excess} + \text{supplement} = \text{reuse} + \text{discharge}$$
$$R + G = RU + D \tag{8.31}$$

and

$$G = RU + D - R \tag{8.32}$$

However, assuming runoff not discharged is equal to reuse,

$$D = (1 - E)R \tag{8.33}$$

where E is the annual efficiency fraction of the pond. Therefore,

$$G = RU + (1 - E)R - R$$
$$= RU - [R - (1 - E)R]$$
$$= RU - [R - (R - RE)] \tag{8.34}$$

or

$$G = RU - (R \times E) \tag{8.35}$$

Equation 8.35 indicates that supplement use is a function of the reuse rate, the amount of rainfall excess, and efficiency of the system. The rate of supplement use in equation 8.35 has units of inches per year on the EIA.

Example Problem 8.8 An economic feasibility study is being done for a reuse pond in a small apartment development in Orlando, Florida. It is necessary to know the quantity of groundwater supplement to a pond per year. The reuse storage is 1.5 in. on the EIA, and the reuse rate is estimated at 0.2 in./day on a previous area equal to the EIA. The average rainfall is 50 in. per year.

8.3 WET-DETENTION POND AS A HOLDING TANK AND A REUSE POND

SOLUTION: From the Orlando REV chart we find that the efficiency of the system will be 81%. We can now use equation 8.35.

$$G = RU - (R \times E)$$

$$= \left(0.2 \frac{\text{in.}}{\text{day}}\right)\left(365 \frac{\text{days}}{\text{year}}\right) - \left(50 \frac{\text{in.}}{\text{year}}\right)(0.81)$$

$$= 32.5 \frac{\text{in.}}{\text{year}}$$

This system will use approximately 32.5 in. of groundwater or a volume equal to 32.5 in. times the EIA per year.

Example Problem 8.9 The owners of a golf course are considering a water reuse system to decrease the use of potable water. They will be using an existing water hazard for reuse storage, the volume of which is 6.5 inches on the EIA. They plan to irrigate at a daily rate of 0.12 inches on an area equivalent to the EIA. How much water do they need to supplement the reuse volume over one year if the annual rainfall volume is 50 in?

SOLUTION: A no discharge efficiency of 80% is obtained from the Orlando REV chart. Again from equation 8.35,

$$G = RU - (R \times E)$$

$$= \left(0.12 \frac{\text{in.}}{\text{day}}\right)\left(365 \frac{\text{days}}{\text{year}}\right) - \left(50 \frac{\text{in.}}{\text{year}}\right)(0.80)$$

$$= 3.8 \frac{\text{in.}}{\text{year}}$$

we see that only 3.8 in./yr of supplemental groundwater is needed.

The necessary rate of supplemental water depends on the design of the system. Referring to the mass balance, since runoff will not change, a higher reuse rate (irrigation in this case) will require more groundwater supplement and a larger reuse volume will decrease the intensity of fluctuations in the level of the pond, which will decrease the demand for supplement. Thus there is an economic trade-off between the cost of land for reuse storage and the cost of supplemental water. If a pond is to be designed in an area where the cost of supplemental water is high and a certain efficiency must be maintained, the designer would probably prefer a larger storage volume and a lower reuse rate. On the other hand, in an area where land is relatively expensive, a higher quantity of supplemental water might be desirable.

306 WET-RETENTION / DETENTION PONDS

Groundwater has been used exclusively as the supplement in previous examples, but sources such as graywater and other surface-water bodies may also be used.

Reuse Rates. Usually, less than 50% of a watershed can be used for irrigation. However, area outside the watershed may be available and not cause any hydrologic inbalance. Recommended rates during irrigation months are 1.0 to 2.0 in. (38 to 51 mm) per week for most turfs and ornamentals. Some agricultural interests may need more. Car wash water makeup is usually estimated at 3 to 5 gal per car. Cooling-water makeup in large buildings may be very substantial and variable from one location to the next.

8.4 SUMMARY

Probability distributions were used to help understand hydrologic processes and wet-detention pond treatment efficiencies. Reaction kinetics were then used with these probability distributions to aid in water quality management and the design of detention ponds.

- On-line wet-retention/detention ponds can be designed as flow-through detention ponds, reuse ponds, and infiltration ponds.
- The pollutant removal efficiencies of wet-detention ponds have been related to the volume of the pond.
- An on-line detention pond consists of four pools of water: sediment, permanent, temporary, and flood control.
- Detention time is a function of the pond volume and the outflow rate. Short circuiting and dead volume should be minimized.
- Interevent times longer than 4 h must be used to ensure independence between runoff events or holding pond elevation when probability distributions are used to specify design criteria.
- Reuse ponds can be designed (using REV charts) for water conservation and water quality treatment.
- The development of PIF curves was illustrated and the curves were shown to be useful for the design of wet-detention ponds.

8.5 PROBLEMS

1. Comment on the specification of an interevent dry period if the design of a pipe for peak flow and detention pond for irrigation and peak flow reduction is needed.

2. Comment on construction details for a wet detention pond that will minimize recycling of bottom pollutants and maximize detention time.

3. Using a REV chart of your choice, calculate the land area in acres for irrigation if the average weekly irrigation rate is 0.9 in. and the temporary storage is 3 in. over a 20-acre equivalent impervious area. If the yearly average rainfall is 45 in., how much supplement should be added?

4. For the Baltimore, Maryland region on the east coast of the United States, a PIF curve has been developed using on the average 40 storm events per year with a minimum interevent dry period of 72 h. If the design exceedence probability set by risk analysis is 0.10, what is the return period (months) that is used for design and the cumulative rainfall volume if the PIF curve of Figure 8.13 is used as representative of this east coast region?

5. What is the water quality volume (cubic feet) for a detention pond to hold the runoff water resulting from the cumulative rainfall of Problem 4 on a watershed of 10 acres with a runoff coefficient of 0.40? If this water quality volume occupies an average depth of 4 ft, what is the average surface area? If the pond width can be only an average of 50 ft, what is the length of pond? Comment on the length-to-width ratio.

6. Assuming runoff volume, interevent times for rainfall, and that runoff duration follows exponential distributions, change the flow capture efficiency of Example Problem 8.2 to 90% and the treatment rate to 0.01 in./h and calculate the storage. Assume that all other data given in the example problem are correct.

7. The travel time of water in the ground from a stormwater retention pond to a lake is 5.79 days. Filtration material has been placed in the ground between the pond and the lake. Determine the concentration of a chemical discharged to the lake if the initial concentration is 5000 mg/L and the rate of chemical change in the soil follows a first-order reaction with decay constant of 0.5 day^{-1} (base e).

8. Comment on the construction details on a wet-detention pond to effectively remove pollutants. Use a rainfall volume (from a PIF curve) of 3 in. from 10 acres with a runoff coefficient of 0.70. Can this pond occupy a 5000 = ft^2 area? For a sediment plus quality volume of 80,000 ft^3, calculate average detention time based on an average outflow rate of 1.5 ft^3/s.

9. Soils from a particular location were incubated under aerobic and anaerobic conditions with the addition of soluble phosphorus, and the resulting soluble phosphorus in the water was determined (Table 8.11). Each sample (300 g) was kept in suspension by use of a magnetic stirrer in 1500 mL of water in a sealed 2.0-L flask. Slow streams of air for the aerobic treatments and argon for the anaerobic treatments were continuously bubbled through the suspensions. The samples were incu-

TABLE 8.11 Soil–Water Equilibrium for Phosphate in Soils

Aerobic		Anaerobic	
Added μg P/g soil	Equil. Solution μg P/mL	Added μg P/g soil	Equil. Solution μg P/mL
15	0.025	15	0.025
30	0.05	30	0.04
45	0.12	45	0.095
60	0.8	60	0.4
180	4.0	180	5.0
420	10.6	420	30
780	30	780	90
1080	40	1080	100.2

bated for 17 days at 30°C before incremental additions of phosphorus as $Ca(H_2PO_4)_2$. Samples of the suspension were removed 24 h after each addition of phosphorus and filtered prior to analyzing for phosphorus. The chemical analysis at equilibrium is shown in the table. Is phosphorus partitioned more strongly in the soil or in the water? What effect do anaerobic conditions have on the release of phosphorus from the soil? Show the range of partion coefficients.

10. Using the EPA statistical model, rainfall, watershed, and pond data of your choice, develop a removal efficiency and sedimentation rate (as a overall yearly average). Example data for eight locations are given in Appendix H. The results are compared to core samples at each of the locations as shown in Appendix H.

11. A watershed in Orlando must reuse 80% of the annual runoff from a 10-acre impervious area. The pond area is included in the impervious area. The maximum reuse storage volume available for the pond is equal to the runoff form a 3-in. rainfall event. At what rate must the runoff be reused?

12. An apartment complex located in Roanoke, Virginia needs to reuse 95% of the runoff from its parking lots. The EIA is equal to the directly connected impervious area and is 4 acres. They want to use 0.26 in. of water per day over the EIA. What must the reuse volume be in order to maintain these conditions?

13. An Orlando apartment complex must reuse, through irrigation, 90% of the annual runoff from a 3.5-acre impervious parking lot. The maximum reuse storage volume available for the pond is equal to the runoff from a 3-in. rainfall event. If the development wants to irrigate at a rate of 1.75 in./week, how much area must be incorporated into the irrigation system?

14. What is the long-term average removal using the data of Problem 2, Section 7.9, Chapter 7?

8.6 COMPUTER-ASSISTED PROBLEMS

Using the same watershed and design rainfall as used for the computer-assisted problems of Chapter 7, solve the following problems.

1. Using off-line retention to infiltrate the runoff from the first inch of rain from a watershed with a runoff coefficient of 0.25 and a detention pond with a stage–discharge relationship of 2 ft^3/s for the first 1 acre-ft of temporary storage, and 8 ft^3/s for 3 acre-ft of a temporary storage, what is the hydrograph attenuation? Use the Santa Barbara hydrograph procedure. The maximum depth of the retention pond is 4 ft with side slopes of 1 vertical to 4 horizontal. *Note*: For the hydrograph attenuation, use three modes in the routing analysis: one each for the watershed, pond, and outlet. Assume a permanent pool of 2 acre-ft at an outlet elevation of 3 ft. Also, what are the off-line and on-line pond volumes?

2. For Problem 1, use only an on-line wet-detention pond with a water quality volume calculated from PIF curves to equal the runoff from 3 in. of rainfall and a runoff coefficient of 0.4. What is the maximum size of the detention pond based on the design rainfall and watershed conditions?

3. Using the data from Problem 1, the designers wish to reduce the size of the detention pond; thus the initial abstraction on the watershed over the impervious area has been increased to 2.0 in. and the time of concentration increased to 60 min. With the same storage–discharge relationship of Problem 1, what is the volume of the detention pond in acre-feet?

4. The water management district wishes to reduce the volume of land required for detention (it is on the floodplain) in Problem 3. The developer's engineer resubmits the design of Problem 3 by adding additional swale area making the time of concentration equal to 75 min and then reduces the percent directly connected impervious area to 10 acres. What is the volume of the wet-detention pond? Use the same storage discharge relationship that was specified in Problem 3.

5. A detention pond is to be designed for a new community of multifamily condominiums in the Miami, Florida area. The watershed informa-

tion is as follows:

40 acres total area

25% directly connected impervious area

No runoff contribution from pervious area for rainfall up to 5.5 in.

80% reuse efficiency criteria

Maximum reuse storage is 4 in. on equivalent impervious area

1.5 in./week of reuse by irrigation

How much land area is needed for irrigation, and what is the volume of the reuse storage?

6. The owners of a golf course are considering a water reuse system to decrease the use of county potable water. They will be using an existing water hazard for reuse storage, the volume of which is 6.5 in. on the EIA. They plan to irrigate at a rate of 0.12 in./day on the EIA. How much water do they need to supplement the reuse volume over 1 year? How much would they need if they reduced the storage volume to 5 in. on the EIA while maintaining the same efficiency?

8.7 REFERENCES

Athayde, D. 1979. *A Statistical Method for the Assessment of Urban Stormwater*, EPA 440/3-79-023, U.S. EPA Nonpoint Source Branch.

Bell, J., and Wanielista, M. P. 1981. "Use of Overland Flow in Stormwater Management on Interstate Highways," in *Transportation Research Record 736*, National Academy of Sciences, Washington, D.C., pp. 13–21.

Camp, Dresser, and McKee. Inc. 1985. *An Assessment of Stormwater Management Programs*, Florida Department of Environmental Regulation, Tallahassee, Fla.

DiToro, D. M., and Small M. J. 1979. "Stormwater Interception and Storage," *Journal of the Environmental Engineering Division*, ASCE, Vol. 105, No. EE1.

Dornbush, J. N. 1981. *Infiltration Land Treatment of Stabilization Pond Effluent.*, EPA 600/S2-81-226, Robert S. Kerr Laboratory, Ada, Okla.

Driscoll, Eugene D. 1983. "Performance of Detention Basins for Control of Urban Runoff Quality," *Proceedings of the International Symposium on Urban Hydrology, Hydraulics and Sediment Control.* University of Kentucky, Lexington, Ky.

Driscoll, Eugene D. 1989. *Analysis of Storm Event Characteristics for Selected Rainfall Gages Throughout the United States*, draft report prepared for U.S. Environmental Protection Agency, Woodward-Clyde Consultants, Oakland, Calif.

Ferrara, R. A., and Hildick-Smith, A. 1982. "A Modeling Approach for Storm Water Quantity and Quality Control via Detention Basins," *Water Resources Bulletin*, Vol. 18, No. 6, pp. 975–981.

Fillos, J. 1976. "Effect of Sediments on the Quality of the Overlying Water," *Proceedings of the International Symposium on Interaction Between Sediments and Fresh Water*, The Netherlands, Sept. 6–10.

Gizzard, T. J., Randall, C. W., Weand, B. L., and Ellis, K. L. 1986. "Effectiveness of Extended Retention Ponds," *ASCE Proceedings of an Engineering Foundation Conference on Urban Runoff Quality*, New England College, Henniker, N.H., June 23-27.

Goforth, G. F., Heaney, J. P., and Huber, W. C. 1983. "Comparison of Basin Performance Modeling Techniques," *Journal of Environmental Engineering, ASCE*, Vol. 109, No. 5, pp. 1082-1098.

Harper, G. 1991. "Reuse Charts for Stormwater Reuse," Master's thesis, University of Central Florida, Orlando, Fla.

Hutchinson, G. E. 1957. *A Treatise on Limnology, Geography, Physics and Chemistry*, Vol. I, Wiley, New York.

Hvitved-Jacobsen, T. H., and Yousef, Y. 1987. "Analysis of Rainfall Series in the Design of Urban Drainage Control Systems," *Water Research*, Vol. 22, No. 4, pp. 491-496.

Loganathan, V. G., Delleur, J. W., and Segarra, R. I. 1985. *Journal of Water Resources Planning and Management, ASCE*, Vol. 111, No. 4, pp. 382-398.

Maass, A., Hufschmidt, M. M., Dorfman, R., Thomas, H. A. Marglin, S. A., and Fair, G. M. 1966. *Design of Water Resource Systems*, Harvard University Press, Cambridge, Mass., p. 592.

Martin, Edward. 1988. "Effectiveness of an Urban Runoff Detention Pond-Wetland Systems," *Journal of Environmental Engineering, ASCE*, Vol. 11, No. 4, pp. 810-827.

Mortimer, C. H. 1971. "Chemical Exchange Between Sediments and Water in the Great Lakes—Speculation on Probably Regulatory Mechanisms," *Limnology and Oceanography*, Vol. 16, pp. 387-404.

Nix, S. J. 1985. "Residence Time in Stormwater Detention Basins," *Journal of Environmental Engineering, ASCE*, Vol. 111, No. 1, pp. 95-100.

Overton, D. E., and Meadows, M. E. 1976. *Stormwater Modeling*, Academic Press, New York.

Padmanabhan, G., and Delleur, J. W. 1978. *Statistical and Stochastic Analyses of Synthetically Generated Urban Storm Drainage Quantity and Quality Data*, TR108, Purdue University, West Lafayette, Ind.

Rawls, W. J., Brakensick, D. L., and Saxton, K. E. 1982. "Estimation of Soil Properties," *Transactions of the American Society of Agricultural Engineers*, Vol. 25, No. 5, pp. 1316-1320.

Riple, W. W., and Lindmark, G. 1978. "Ecosystem Control by Nitrogen on Metals in Sediment," *Vatten*, Vol. 34, No. 2.

Roesner, L. A. 1974. *A Model for Evaluating Runoff Quality in Metropolitan Master Planning*, Urban Water Resources Research Program, TM 23, ASCE, New York.

Schuler, T. B. 1987. *Controlling Urban Runoff: A Practical Manual for Planning and Designing Urban BMPs*, Washington Metropolitan Water Resources Planning Board. Washington, D.C.

Southwest Florida Water Management District. 1987. *Permit Information Manual*, Chapter 40D-4, Florida Administrative Code, Brooksville, Fla.

Stahre, P., and Urbonas, B. 1990. *Stormwater Detention for Drainage, Water Quality, and CSO Management*, Prentice Hall, Englewood Cliffs, N.J.

Theis, T. L., and McCabe, P. J. 1978. "Phosphorus Dynamics in Hypr-eutrophic Lake Sediments," *Water Research*, Vol. 12.

U.S. Environmental Protection Agency. 1986. *Methodology for Analysis of Detention Basins for Control of Urban Runoff Quality*, EPA 440/S-87-001, U.S. EPA Washington, D.C.

Wanielista, M. P. 1990. *Hydrogology and Water Quantity Control*, Wiley, New York.

Wanielista, M. P., and Shannon, E. 1977. *An Evaluation of Best Management Practices for Stormwater*, East Central Florida Regional Planning Council, Winter Park, Fla.

Wanielista, M. P., Yousef, Y. A., and Harper, G. M. 1990. "Precipitation Volumes and Interevent Dry Periods," presented at *15th Biennial International Conference on Water Pollution Research and Control*, Kyoto, Japan, July 29–Aug. 3.

Wanielista, M. P., Yousef, Y. A., and Lineback, T. R. 1991. *Precipitation and Interevent Dry Periods for Selected Areas of Florida*, Florida Department of Environmental Regulation, Tallahassee, Fla.

Wenzel, H. G., and Voorhees, M. L. 1981. *An Evaluation of the Urban Design Storm Concept*, Research Report VILU-WRC-81-0164. Water Resources Center, University of Illinois at Urbana-Champaign.

Yousef, Y. A., and Wanielista, M. P. 1989. *Efficiency Optimization of Wet-Detention Ponds for Urban Stormwater Management*, Florida Department of Environmental Regulation, Tallahassee, Fla.

Yousef, Y. A., Harper, H. H., and Jellerson, D. 1980. *Inactivation of Lake Sediment Release of Phosphorus*, University of Central Florida, Orlando, Fla.

Yousef, Y. A., Hvitved-Jacobsen, T. H., Wanielista, M. P., and Tolbert, R. D. 1986a. "Nutrient Transformation in Retention/Detention Ponds Receiving Highway Runoff," *Journal of the Water Pollution Control Federation*, Vol. 58, No. 8, pp. 838–844.

Yousef, Y. A., Wanielista, M. P., Harper, H. H., and Jacobsen, T. 1986b. *Best Management Practices—Effectiveness of Retention/Detention Ponds for Control of Contaminants in Highway Runoff*, FL-ER-34-86, Florida Department of Transportation, Tallahassee, Fla.

CHAPTER 9

Economic and Fiscal Feasibility

An understanding of economic and fiscal feasibility is required for the selection of stormwater management projects. Economic feasibility is measured by benefits and costs of a project and is achieved if total benefits from a project exceed those that would occur without the project by an amount greater than the cost of the project. Economic feasibility is different than fiscal feasibility. Fiscal feasibility is achieved if the necessary funds can be raised to pay initial and continuing costs. The design and operation of stormwater facilities depends not only on the engineering and science disciplines, but also on economic and fiscal principles. All feasibility criteria are used to aid in determining how much of the system is designed, built, and operated.

9.1 COMPREHENSIVE STORMWATER MANAGEMENT

A comprehensive stormwater management plan includes as a minimum (1) an inventory of watershed features that affect runoff; (2) a formulation of alternative stormwater control plans with objectives; (3) an engineering, economic, and fiscal evaluation procedure to evaluate the feasibility of various alternatives; and (4) a plan for selecting the "best" alternative. The best plan is one from all the feasible alternatives that satisfies the goals and objectives. A flowchart shown in Figure 9.1 identifies other details of a comprehensive plan. If the objective is expressed in monetary units, the best alternative is that one that either minimizes cost or maximizes benefits. National guidelines (U.S. Environmental Protection Agency, 1976) have been established for comprehensive planning which require an identification and formulation of alternatives, constraints to reaching the alternatives, and the measure of the best solution. This measure is frequently a volume of stormwater, peak discharge, cost, benefits, or a combination of cost and benefits. Other combined measures also are possible. The primary measure of a plan is often stated as the maximization of national welfare (Maass et al., 1970). This is an economic measure. There are many monetary and intangible ways to measure national welfare. However, one or a few common measures

314 ECONOMIC AND FISCAL FEASIBILITY

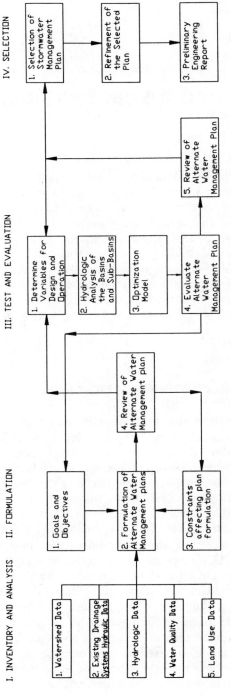

Figure 9.1 Flowchart for a comprehensive stormwater management plan.

are used and most other ways are reduced to the common ones. As an example, consider employment as a common measure. Employment levels measure national welfare and a common measure of employment may be the monetary value of the "paycheck."

Some alternatives may be technologically attractive but are eliminated because of a lack of public acceptance and other factors, such as financial capacity, managerial acceptance, and legal authority. The general public and elected officials should have input on the establishment of all constraints that will make stormwater plan implementation more likely to occur.

9.2 ECONOMIC MEASURES

How much will it cost me? Will it generate jobs? Can we maximize pollution removal? What are the benefits? How much money will it take to operate? The questions are many and relate to economic welfare and fiscal management. The value of stormwater treatment depends on the cost of providing the treatment and the intended benefits.

Poor or unacceptable water quality can reduce the use of water. If the water use is downstream from a stormwater pollution source, and the downstream quality is such that it is not usable, it is necessary to treat the stormwater and a cost results. Water pollution or volume changes by one discharger (stormwater, municipal, and industrial sewage plants) can cause greater economic hardship (higher cost) to another user which has no control of the dischargers. This cause and effect is called a "technological external diseconomy" (Kneese and Bower, 1971). These external costs must be added to all costs to determine the actual cost to the public for stormwater management.

9.2.1 Costs

The cost of a stormwater management system can generally be classified into three basic categories: capital, operating, and risks. Capital cost includes the cost of planning, design, construction, land or easements, surveys, and all others for startup. Operating cost includes those for operation (labor and expenses), replacement, and repairs necessary to operate over the economic life of the project. These costs are abbreviated as ORM costs. Risk is the cost of not providing a level of stormwater protection. It can be measured by damages or restoration costs.

One of the criteria used to help determine the best size is cost-effectiveness or the maximum operational effectiveness for each dollar invested. The procedure for cost-effectiveness analysis is a systematic comparison of alternatives to select the one that minimizes cost while reliably satisfying technical limitations and preferences over the expected time for operation. The total

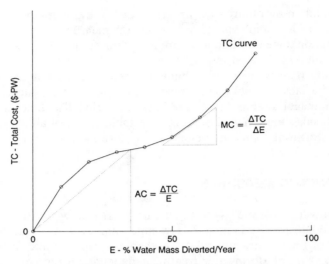

Figure 9.2 Typical cost curve.

cost of a project is related to size and effectiveness. A typical total cost curve for off-line retention ponds for infiltration is shown in Figure 9.2 (Wanielista, 1983). Others have presented similar cost curve information for general water pollution control systems (Tihansky, 1974). Effectiveness for off-line retention ponds can be measured by the percent stormwater runoff diverted to the pond divided by the total stormwater runoff volume.

Total cost is generally calculated as the summation of present and future capital and ORM costs discounted to present worth (PW). Some other examples of cost factors that can be included are (1) resource replacement, (2) social impact, and (3) environmental protection. As the size (effectiveness) of a project increases, total cost usually increases rapidly at first, but the rate of increase or marginal cost (MC) decreases until it reaches a size where the marginal cost increases again very rapidly. Above this size it becomes more expensive to obtain a small increment of effectiveness, and other alternatives become more cost-effective. Average cost (AC) is also shown in Figure 9.2 and is defined as the total cost divided by the level of effectiveness.

Economies of scale result when average cost decreases with the size of project. Where possible it may be cost advantageous to construct a regional facility for stormwater management, rather than a few smaller ones. Conversely, average cost may increase (diseconomies of scale may be present) and regional facilities are not attractive from a cost viewpoint. The total cost of the stormwater management should include reservoirs (ponds) and transport among the smaller sites and the regional site. If the total cost of a larger regional facility plus the transport cost is lower than other alternatives, the economies of scale of a regional facility would indicate a cost-effective solution.

9.2.2 Benefits

More large water resources projects are justified on the basis of the difference between benefits and costs. Benefits are net increases in the value of goods and services that result from the size of a project as compared to conditions without the project. Benefits are both tangible and intangible. Tangible benefits are those that are expressed in monetary units either derived from revenue projections or the cost of alternative means to produce equivalent results. Intangible benefits are those that are not measured in monetary units. An example is the beautification of an area.

Benefits result from stormwater management. Some benefits are measured directly in terms of a quantity of water, such as flood control, domestic water supply, agricultural supplies, navigation, and power generation. Usually, these benefits are converted into monetary units of the primary currency of the country (dollars, pounds, etc.). However, there are some nonmonetary benefits associated with public health and wildlife which are more difficult to quantify. These may become limitations or constraints on achieving a design with maximum benefits or minimum cost. In some areas it became obvious that cleanup limited to point-source discharge alone was not sufficient to meet limitations on water quality concentrations and wildlife habitats. Stormwater discharges also had to be managed.

Benefits that directly result from a project less the cost associated with the project are primary benefits; secondary benefits are those that indirectly result from the project. Frequently, no distinction is made as to primary versus secondary benefits because it depends on what is defined as direct or indirect. Thus all benefits are summed together. An example to illustrate the differences is the construction of a reservoir that directly provides a benefit to agricultural water supply, power generation, and commercial fishing, while indirectly providing for improved transportation, wildlife, and aesthetics. The secondary benefits are frequently much more difficult to quantify in degree and monetary terms.

Some of the more common benefits for large stormwater ponds are:

1. *Flood Control.* A flow of specified magnitude (e.g., 1000 ft^3/s once in 100 years) establishes the flood elevation. If this flood elevation is exceeded and protection is not provided for domestic, commercial, and industrial property, a monetary loss will result. This monetary loss minus the cost of construction to prevent a flood exceedence is one measure of the benefit resulting from flood control. Another means to measure benefits is the insurance premium paid in lieu of the project or in place of a fixed level of

protection. This assumes a certain replacement value if a flood occurs. Flood control benefits are generally nonreimbursable to the project.

2. *Municipal Water Supply.* Consumer groups, such as homeowners and industry, pay for water. The price per quantity used may vary with amount used and on the availability. Economies of scale may or may not be present. The price of water will vary from one region to another. However, it is available for estimating benefits associated with municipal water supply.

3. *Irrigation.* Similar to municipal water supply, the price for irrigation water is established for a region. Market forces which establish type of agricultural activity and yield help determine the price an activity is willing to pay for water.

4. *Power Supply.* The geographic location of a reservoir determines the quantity of water and potential energy (head) to generate electricity. If the size of facility is large enough so that the average cost of power generation will be reduced to a competitive economic position, benefits will result and are measured by the price of electricity.

5. *Fish and Wildlife.* If a stormwater detention pond is constructed to reduce peak flows to a stream, wildlife downstream may be protected and enhanced. Increased fishing, wildlife, and aesthetic values result. Benefits are measured by the revenue from the sale of fishing licenses and the monetary outlay related to equipment and travel to a site. Some benefits are intangible and difficult to measure; however, procedures to assign monetary value have been developed as noted in the following discussion of recreation benefits.

6. *Recreation.* Recreation benefits can account for about 75% of all water quality benefits resulting from reservoir management (Binkley and Haneman, 1978). Add to this the benefits from water quantity (volume and flow rate) control and reservoirs may have a significant benefit. Many of the benefits are intangible, but methods to quantify the monetary value have been done and are based on the concept of a recreation day or visit. The recreation day unit is defined as the presence of one person over a period of 12 h, either continuous or intermittent, at an area providing recreation. The cost procedures for estimating the monetary value per recreation day unit are (1) gross expenditures for recreation, (2) market value of products (i.e., fish), (3) property value increase, (4) income increases for concessionaires of recreation activities in the area, and (5) interviews of people using the recreation facilities to determine how much they are willing to pay. Recreation unit benefits were determined by York et al. (1976), and an example of some of the benefits updated to 1987 dollars is shown in Table 9.1.

7. *Water Quality.* Erosion control and public health improves with improved water quality through reduction of the mass of pollution. The net benefits are in general measured by the cost associated with the avoidance of adverse effects. Soil can be lost from land, which decreases farming yields and thus revenue. Reduction of nutrients being discarded decreases unwanted water plant growth, which improves recreation values. Reduction in

TABLE 9.1 Recreation Unit Benefits Ranges (1987 Dollars)

Activity	Low to High Unit Benefit ($/person-day)
Lake fishing	2.0–8.0
River fishing	2.0–10.0
Boating	4.0–10.5
Swimming	1.5–5.0
Visual/photographic	1.5–4.0
Camping	4.5–8.5
Picnicking	1.5–3.5

salinity to an agricultural use improves the yield and revenue. Decreases in oxygen-demanding material to a water body may increase the dissolved oxygen, which prevents fish kills and structural deterioration. Thus the benefits can be measured by the cost of avoiding fish kills, structural deterioration, siltation by sedimentation, health hazards, and agricultural decline. If nothing is done to prevent deterioration, the cost of restoration is an estimate of the damage.

8. *Multiple Use.* A stormwater management plan has many benefits and is usually a combination of some of the seven benefits above plus others, such as navigation and wastewater assimilation (Loucks et al., 1981). All benefits that can be quantified should be used. Once a benefit curve is determined, the marginal benefits can be developed and compared to marginal cost. In a free-enterprise system, the size of a stormwater management system is determined by the ratio of marginal benefits to marginal cost. When marginal cost exceeds marginal benefits, the project is considered economically infeasible. However, in many cases the marginal benefit curve is an estimate of intangible activities and other methods have to be sought to determine the size of facilities. These other methods take the form of standards and effluent charges. Standards fix the size of a facility and effluent charges are levied upon a discharge for regional facility development or the cost of cleanup. Effluent charges take various forms, some of the most common ones have units of dollars per impervious acre or dollars per pound of material discharged.

9.2.3 Benefit–Cost Comparisons

When the benefits of a stormwater project exceed the cost, economic feasibility has been achieved. This commonly used philosophy and equations for economic feasibility are

$$\text{maximum(benefits} - \text{costs)} \qquad (9.1)$$

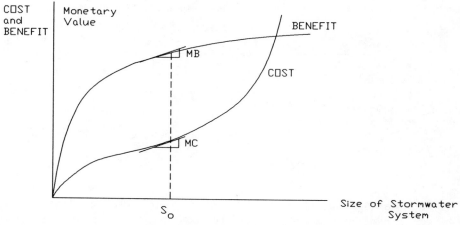

Figure 9.3 Benefit–cost diagrams.

or

$$\text{maximum(benefits/costs)} \qquad (9.2)$$

All the options must be compared and the one with the maximum difference or ratio is the best. The benefit–cost comparison equations can be used to determine the best size of a stormwater system and to specify a system to build and operate from among other systems. The difference between benefits and costs have to be positive (the ratio greater than 1) to justify the expenditures from an economic point of view. The comparison procedure for determining the best one is called *optimization*. A graphical comparison of the procedure is shown in Figure 9.3 for a single stormwater system. The maximum difference between benefits and costs and the maximum ratio of benefits to costs occur where marginal cost (MC) equal marginal benefits (MB). The optimal (best) size occurs at S_0, where $MC = MB$, maximum ($B - C$), and maximum (B/C).

Heaney (1988) reviewed and summarized methods for benefit–cost assessment of urban stormwater management. His analysis suggested that comprehensive water quality improvement should be documented in terms of benefits to a community. These benefits should be compared to the cost of stormwater management. Since there is variability in the effluent quality during dry weather flow in a combined sewered system (General Accounting Office, 1980) and greater variability during dry/wet weather conditions, quantification of the benefits is difficult.

Example Problem 9.1 A stormwater detention system consists of four plans. Plan 1 has the smallest volume that results in the most flooding. The greater the flow attenuation, the less the cost of flooding. Assuming that sufficient funds are available, which of the four detention systems should be built, given that one of them must be built? The annual damages and the costs of each plan are as follows:

Plan	Damages	Costs
1	750,000	250,000
2	650,000	300,000
3	250,000	500,000
4	100,000	800,000
Do nothing	1,100,000	0

SOLUTION: If nothing is done, there are no costs, but if plan 1 is built, the benefits are measured by the difference in damages between the do-nothing option and those that occur with plan 1 or

$$\text{benefits of plan 1} = 1{,}100{,}000 - 750{,}000 = \$350{,}000$$

Also:

$$\text{benefits of plan 2} = 1{,}100{,}000 - 650{,}000 = \$450{,}000$$
$$\text{benefits of plan 3} = 1{,}100{,}000 - 250{,}000 = \$850{,}000$$
$$\text{benefits of plan 4} = 1{,}100{,}000 - 100{,}000 = \$1{,}000{,}000$$

The best plan is the one among the four that maximizes $(B - C)$ or (B/C).

Plan	$B - C$	B/C
1	100,000	1.40
2	150,000	1.50
3	350,000	1.70
4	200,000	1.25

The best plan is 3 because it maximizes $(B - C)$ and (B/C).

9.3 FISCAL RESPONSIBILITY AND FINANCIAL PLANNING

Are sufficient funds available to implement the technical and economic plan? The answer to this question requires a careful estimate of monetary (cash) flow over the project life. To make a choice from many alternatives, a clear definition of cost is necessary. There are capital cost (initial), operating cost, and discounted operating costs. The alternatives should be expressed as far as possible in terms of cash flow (the time value of money).

322 ECONOMIC AND FISCAL FEASIBILITY

All stormwater plans should consider what is financially feasible over the life of the project. Interest rates change with time, and thus the time value of money changes with the choice of interest rate.

9.3.1 Interest Rate and Planning Horizon

Interest is the return obtainable from the productive investment of money. To understand the concept of interest, it is helpful to consider borrowing money. If $9 of interest is payable annually on a borrowed $100, the interest rate for use of the $100 is 0.09 ($9/$100). This is the interest rate expressed per annum. The interest rate is the ratio between the interest chargeable or payable at the end of a time period and the money owed at the beginning of the period. The time period should be specified with the interest rate.

The planning horizon is the period of time the stormwater project is expected to operate. Commonly used time periods are 25, 50, and 100 years. Estimates of stormwater flow rates and volumes may be required for the distant future. A project may last 50 to 100 years, but the political and financial payback period may be less, say 25 to 50 years. This implies that streamflow records and watershed changes must be estimated for long time periods. However, the estimate of interest rates and other financial problems for the future may be as difficult, or additional uncertainities may exist, making financial plans shorter than project plans.

9.3.2 Types of Financial Plans

There are four generally acceptable financial plans for repaying a debt: (1) interest rate only, (2) decreasing principal, (3) uniform payment, and (4) single payment at the end of the borrowed period. The latter two plans compound the interest. Table 9.2 compares these payment plans using a $10,000 loan in 10 years with an interest rate of 6%. The information in Table 9.2 can be expressed in cash flow diagrams as shown in Figure 9.4. Receipts are noted above the line and disbursements (payments) below the line. If benefits and costs are graphed, benefits are plotted above and costs below. The advantages and disadvantages of the various plans depend on the sources of money for repayments, their uncertainty, and the timing for receipts, among other factors.

9.3.3 Plan Equivalency

To determine a financial plan, the concept of equivalence or the timing of payments and income are important. Payment and income at different times may be expressed as an equivalent total investment measured in current monetary value. As an example, consider the four payment plans shown in Table 9.2. They are all equivalent in current monetary units, namely $10,000. However, the future flow of monetary units are different. Obviously, there

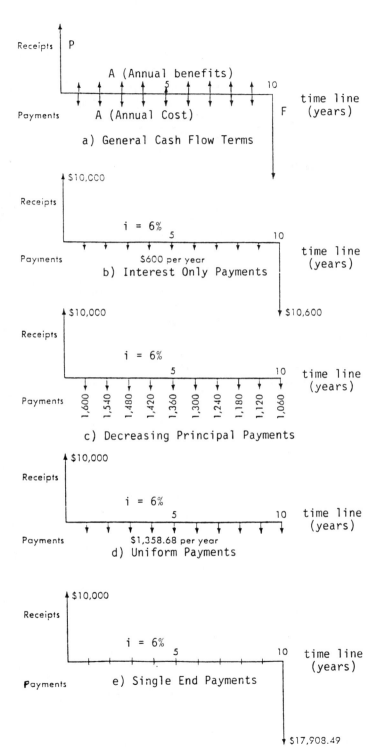

Figure 9.4 Cash flow diagrams.

TABLE 9.2 Comparison of Four Plans of Repayment Based on a $10,000 Initial Cost

End of Year	Interest Due (6% of Money Owed at Start of Year)	Total Money Owed Before Year-End Payment	Year-End Payment	Money Owed After Year-End Payment
		Plan 1 (Interest Only)		
0				$10,000
1	$600	$10,600	$ 600	10,000
2	600	10,600	600	10,000
3	600	10,600	600	10,000
4	600	10,600	600	10,000
5	600	10,600	600	10,000
6	600	10,600	600	10,000
7	600	10,600	600	10,000
8	600	10,600	600	10,000
9	600	10,600	600	10,000
10	600	10,600	10,600	0
		Total payment $16,000		
		Plan 2 (Decreasing Principal)		
0				$10,000
1	$600	$10,600	$ 1,600	9,000
2	540	9,540	1,540	8,000
3	480	8,480	1,480	7,000
4	420	7,420	1,420	6,000
5	360	6,360	1,360	5,000
6	300	5,300	1,300	4,000
7	240	4,240	1,240	3,000
8	180	3,180	1,180	2,000
9	120	2,120	1,120	1,000
10	60	1,060	1,060	0
		Total payment $13,300		
		Plan 3 (Uniform Payment)		
0				$10,000.00
1	$600.00	$10,600.00	$1,358.68	9,251.32
2	554.48	9,795.80	1,358.68	8,437.12
3	506.23	8,943.35	1,358.68	7,584.67
4	455.08	8,039.75	1,358.68	6,681.07
5	400.86	7,081.93	1,358.68	5,723.25
6	343.40	6,066.65	1,358.68	4,707.98
7	282.48	4,990.45	1,358.68	3,631.77
8	217.91	3,849.68	1,358.68	2,491.00
9	149.46	2,640.46	1,358.68	1,281.78
10	76.90	1,358.68	1,358.68	0.00
		Total payment $13,586.80		

9.3 FISCAL RESPONSIBILITY AND FINANCIAL PLANNING

TABLE 9.2 Comparison of Four Plans of Repayment Based on a $10,000 Initial Cost

End of Year	Interest Due (6% of Money Owed at Start of Year)	Total Money Owed Before Year-End Payment	Year-End Payment	Money Owed After Year-End Payment
		Plan 4 (Single End Payment)		
0				$10,000.00
1	$ 600.00	$10,600.00	$ 0.00	10,600.00
2	636.00	11,236.00	0.00	11,236.00
3	674.16	11,910.16	0.00	11,910.16
4	714.61	12,624.77	0.00	12,624.77
5	757.49	13,382.26	0.00	13,382.26
6	802.94	14,185.20	0.00	14,185.20
7	851.11	15,036.31	0.00	15,036.31
8	902.18	15,938.49	0.00	15,938.49
9	956.31	16,894.80	0.00	16,894.80
10	1,013.69	17,908.49	17,908.49	0.00
		Total payment $17,908.49		

may be other cash flow series of payments that would serve to repay the $10,000. If all receipts are expressed in terms of initial monetary units, this present sum is the present worth of any future payments at an assumed interest rate and time period. Cash flow studies are important as a basis for comparison in deciding on alternative stormwater management plans.

The formulas that express the relationship between present worth (PW), annual series payment (A), interest rate (i), and some future worth at the end of n time periods (FW) are as follows:

Future Worth. Given PW, to find FW.

$$FW = PW(1 + i)^n \quad (9.3)$$

Present Worth. Given FW, to find PW.

$$PW = FW\left[\frac{1}{(1+i)^n}\right] \quad (9.4)$$

Sinking Fund. Given FW, to find A.

$$A = FW\left[\frac{i}{(1+i)^n - 1}\right] \quad (9.5)$$

Capital Recovery. Given PW, to find A.

$$A = PW\left[\frac{i(1 \times i)^n}{(1 + i)^n - 1}\right] \quad (9.6)$$

or

$$A = PW\left[\frac{i}{(1 + i)^n - 1} + i\right]$$

Annuity Fund. Given A, to find FW.

$$FW = A\left[\frac{(1 + i)^n - 1}{i}\right] \quad (9.7)$$

Present Worth. Given A, to find PW.

$$PW = A\left[\frac{(1 + i)^n - 1}{i(1 + i)^n}\right] \quad (9.8)$$

or

$$PW = A\left[\frac{1}{\frac{i}{(1 + i)^n - 1} + i}\right]$$

Although a 1-year interest period is assumed in most of the presentation, the formulas apply to any length.

The expression $(1 + i)^n$ is called the single-payment compound amount factor and represents the compounding of money over time periods. The reciprocal $[1/(1 + i)^n]$ is called the present-worth factor and represents a single payment. If PW is invested at interest i, the total amount of money at the end of the first interest period is

$$PW + iPW = PW(1 + i)$$

and at the end of the second investment period is

$$PW(1 + i) + iPW(1 + i) = PW(1 + i)(1 + i) \quad \text{or} \quad PW(1 + i)^2$$

After n years, one obtains

$$PW(1 + i)^n. \tag{9.9}$$

A fund that establishes a series of payments to produce a desired amount at the end of a specified period of time is called a sinking fund and the sinking fund factor is

$$SFF = \frac{i}{(1 + i)^n - 1} \tag{9.10}$$

The capital recovery factor is that which is multiplied by the capital cost to obtain an annual cost over a period of n years at a given interest rate i, or

$$CRF = \frac{i(1 + i)^n}{(1 + i)^n - 1} \tag{9.11}$$

Example Problem 9.2 In 3 years, a water management district must invest $400,000 in a stormwater retention/detention system. What annual amount of money must it save at 10% over the next 3 years to have the $400,000 available for construction?

SOLUTION: This is a sinking fund problem and we must determine A given FW at 10% and 3 years to have the $400,000 available for construction (using equation 9.5).

$$A = FW \left[\frac{i}{(1 + i)^n - 1} \right] = 400,000 \left[\frac{0.10}{(1 + 0.10)^3 - 1} \right] = \$120,845.92$$

9.3.4 Stormwater Financing Alternatives

State, regional, county, and city governments have at least six reasonable alternatives for financing stormwater management planning, construction, and operation:

1. General revenue
2. Municipal service taxing units
3. Independent taxing districts
4. Bonding
5. Impact or connection fees
6. Stormwater utility

General revenue is used when the cost of the projects are not high enough to bring special attention and operating costs are generally low. The projects must be of general benefit to the people of the area. As the cost of systems increases, the use of alternative financing mechanisms must be found.

Municipal service taxing units (MSTUs) have been developed for special well-defined areas. Usually, there is an upper limit to the taxing rate, and thus it may be marginally efficient for solving financial problems. The same may result with independent taxing districts; however, the districts cover a wider area and usually are permitted a higher tax rate.

The bonding power of a government may be used for larger projects and can be very successful in securing necessary funds. The benefits are a stable specific term financial arrangement. As an alternative for developing areas, impact and connection fees can be charged to new developments. This provides an "up-front" source of money for both construction and operation.

Most recently, another funding mechanism, called a stormwater utility, has been developed for local governments to obtain additional stormwater construction and operating funds (Florida Administrative Code, 1986). The utility is usually created by local ordinance, and monies collected have to be specifically designated for stormwater management in well-defined watersheds. All property owners within the watersheds may be assessed a per acreage fee based on common use of the drainage system and quantity of runoff. Provisions for a reduced charge for land use within a specific watershed may be made if stormwater management is built on-site to reduce runoff. The utility is a reasonable option to raising funds using taxing mechanisms.

9.4 UNCERTAINTY AND RISK

Associated with most hydrologic data is uncertainty. However, uncertainty can be quantified in terms of probability distributions and the associated costs. Economic and statistical data are frequently combined to find the optimal or best combination of stormwater management alternatives. There exist a number of procedures for comparison of alternative systems when uncertainty is present. Expected monetary value for an alternative is one of the most frequently used calculations, and is defined as

$$\text{EMV}(x) = \text{Pr}(x)V(x) \qquad (9.12)$$

where $\text{EMV}(x)$ = expected monetary value of alternative (x)
$\text{Pr}(x)$ = probability of alternative (x) occurring, fraction
$V(x)$ = monetary value of alternative (x)

Thus given a probability for a hydrologic event and an associated value if the event occurs, an expected monetary value can be calculated.

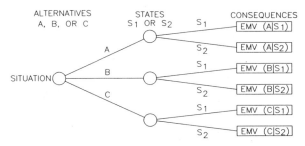

Figure 9.5 Generalized tree diagram for expected monetary value.

Decision trees are helpful in presenting all alternative designs and comparing the final expected monetary values, as shown in Figure 9.5. The expected monetary value of alternative A depends on the uncertainty for the estimates of the parameters affecting the design or the probability of states S_1, S_2, and so on. As an example, consider A as a size of detention pond and the probability of net runoff assigned to two values S_1 and S_2; thus the expected monetary return for alternative A is

$$\text{EMV}^A = \sum_{i=1}^{2} \Pr(S_i)[V_i(A \mid S_i)] \qquad (9.13)$$

where $V_i(A \mid S_i)$ is the monetary value of alternative A if net flow is equal to S_i.

9.4.1 Flood Damage Analysis

Approximately 7% of the land in the United States is classified as floodplain. Floodplains are low areas adjacent to streams, oceans, and lakes that are subject to flooding at least once in 100 years (U.S. Water Resources Council, 1981). In the United States in the early 1980s, about 90% of all losses from natural disasters were caused by floods. The economic impact of floods alone was about $4 billion each year. In addition to economic impacts, health and safety problems are evident. There are in the United States, approximately 200 flood-related deaths per year. Flash floods are the biggest killers: for example, 236 people were killed in Rapid City, South Dakota in 1972 (U.S. Water Resources Council, 1981). There appears to be a reasonable need to control building on floodplains and to plan for the beneficial use of floodplains by doing a complete benefit–cost analysis (Milliman, 1983).

Mathematical procedures for analyzing flood management alternatives depend on an assessment of hydrologic and economic factors. Expected monetary values for the cost of each alternative level of protection is compared to the expected benefits derived from each level of protection. A

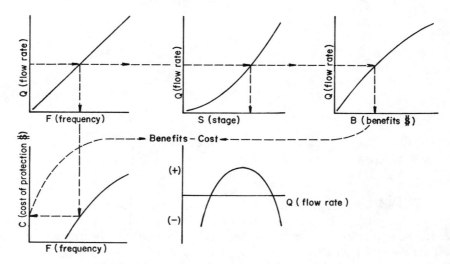

Figure 9.6 Hydrologic and economic relationship.

general procedure is shown in Figure 9.6. A detailed manual for analysis was developed by Davis (1988) and is used by the U.S. Army Corps of Engineers. From basic hydrologic relationships between flow rates and frequency, and depth (stage) versus flow rates, damages can be calculated as a function of flow rate. If these damages are mitigated by building a flood control structure, a benefit of flood control results for the cost of construction. Summation of the expected value of damages for a given frequency distribution is compared to the cost of protection to determine a level of protection.

9.4.2 Expected Values with Benefits and Cost

A criterion for sizing a detention pond to protect against flooding is that the marginal annual costs are equal to the marginal annual benefits of protection. This is equivalent to maximizing the difference between benefits and costs. Floods of different magnitudes exist; thus a probability distribution is used to describe them. The corresponding cost of constructing the pond to protect against each level of flood flow and the expected damages resulting from this flood flow protection can also be determined. The theoretical probability distribution and the empirical probability distribution should be determined using standard procedures (see Chapter 2).

Damages are used to calculate the benefits derived from building a detention pond. A stormwater detention pond cannot be built to protect against all damages. But a level of damage can be assigned to a specific size of detention pond. Monetary values are assigned to damages when a specific size of detention pond is exceeded. Losses may involve damages to buildings,

9.4 UNCERTAINTY AND RISK

TABLE 9.3 Discharge and Benefits

Q Discharge (ft^3/s)	Annual Losses (D) ($1000)	Exceedence Probability (fraction)	Averages		Incremental Expected Benefits[b] ($1000)
			Exceedence (Δ Pr)	Losses[a]	
200	0	0.620			
			0.240	30	7.2
300	60	0.380			
			0.210	130	27.3
600	200	0.170			
			0.120	310	37.2
1500	420	0.050			
			0.026	600	15.6
3000	780	0.024			
			0.017	890	15.3
4900	1000	0.007			
			0.003	1500	4.5
6500	2000	0.004			

[a] $1000[(D_i + D_{i+1})/2]$.
[b] Expected value of benefits risk/interval = $Pr(Q \geq x)[(D_i + D_{i+1})/2]$ = estimate of benefits.

water protection structures, health, other adjacent property, channel structures, and traffic-related activities. The annual cost is the sum of capital cost multiplied by the capital recovery factor, and annual maintenance. Corry et al. (1981) in a Federal Highway Administration publication specifies that these annual costs should be minimized.

Assume that Table 9.3 represents an exceedence probability distribution for a discharge relationship with associated annual losses and expected benefits. Annual losses only occur when discharge exceeds 200 ft^3/s. The benefits resulting from a detention pond design to protect against a discharge of 200 ft^3/s is zero; however, $60,000 per year results when protection is provided at the 300-ft^3/s level of discharge. The average damage in this range is $30,000. The expected benefits due to discharges between 200 and 300 ft^3/s is $7200 (0.240 × 30,000). For each discharge, the benefits are the sum of preceding incremental expected benefits. Table 9.4 illustrates the results of the benefit calculations and compares them to annual cost. The incremental benefits increase up to protection level of 1500 ft^3/s and decrease after 1500 ft^3/s. Incremental annual cost increases at first, then decreases.

Above 4900 ft^3/s the incremental cost exceeds incremental benefits or to protect above the 4900-ft^3/s discharge, requires $11,000 annually, while only $4500 in benefits result. Thus the detention pond size should be calculated for a 4900-ft^3/s peak discharge according to the stated criterion. In Table

TABLE 9.4 Expected Risks/Cost Comparisons

(1)	(2)	(3)	(4)	(5)	(6) = (3) − (4)
Q Discharge (ft³/s)	Incremental Benefits[a] ($1000)	Annual Benefits ($1000)	Pond Annual Costs ($1000)	Incremental Costs ($1000)	Benefits Minus Cost ($1000)
200		0	0		0
	7.2			18	
300		7.2	18		−10.8
	27.3			20	
600		34.5	38		−3.5
	37.2			32	
1500		71.7	70		1.7
	15.6			15	
3000		87.3	85		2.3
	15.3			13	
4900		102.6	98		4.6
	4.5			11	
6500		107.1	109		−1.9

[a] From Table 9.3.

9.4, the benefits minus cost criteria would specify the reservoir size be related to a peak discharge of 4900 ft³/s.

9.4.3 Expected Values with Total Cost

For comparison of expected values with total cost, an example of a stormwater transport system is presented. A transport system can be a closed conduit (pipe, culvert) or an open channel. A stormwater transport (pipe in this example) design is based on a return period that is commonly determined by an expected value analysis that minimizes the total cost. Total cost is the sum of initial cost, maintenance, and damages from undersizing the culvert. The culvert will always be undersized in the sense that there will always be a probability (no matter how small) of a peak discharge exceeding the design discharge. Again there is an uncertainty in sizing the culvert. The empirical distribution for peak discharge with pipe size is shown in Table 9.5. The pipe sizes are those required to pass the peak discharge without causing any ponding behind the structure. As the exceedence probability decreases, the size of pipe increases. Thus damages should decrease while the cost increases. A trade-off between pipe cost (initial and maintenance) and damages results. If the pipe size is too small, excessive damages occurs. However, if the pipe is too large, excessive pipe cost results. There must exist some size of pipe that minimizes total cost.

The total cost associated with each pipe size is calculated and converted to an annual cost. As an example, the initial cost of the 24-in.-diameter pipe is

9.4 UNCERTAINTY AND RISK

TABLE 9.5 Exceedence Probability and Culvert Sizes

Peak Discharge Q_p (ft^3/s)	Exceedence Probability per Year, $\Pr Q \geq Q_p$	Return Period (yr)	Culvert Diameter (in.)
30	0.20	5	24
40	0.10	10	27
55	0.05	20	30
80	0.025	40	36
120	0.01	100	42

$490 for a 20-ft length and head walls. Using a 8% interest rate over 30 years, the annual cost is

$$(A\mid P, 8\%, 30 \text{ yr}) = \text{PW}\left[\frac{i(1+i)^n}{(1+i)^n - 1}\right]$$

$$A = 490\left[\frac{0.08(1.08)^{30}}{(1.08)^{30} - 1}\right] = \$43.53 \text{ per year}$$

The annual maintenance cost will probably not vary with culvert size, and a reasonable value is $15 per year. Next, expected damages must be calculated as the expected monetary benefit, which requires the exceedence probability and damage estimates. Note that the damages are generally difficult to estimate but must be available.

The calculation of damages for a 24-in. culvert is shown in Table 9.6. For the 24-in. culvert, the total cost per year is the sum of initial cost, mainte-

TABLE 9.6 24-Inch-Diameter Culvert Expected Damages

Peak (Q_P) Discharge (ft^3/s)	Exceedence Probability/yr ($\Pr Q \geq Q_P$)	Exceedence Range, $\Delta \Pr$	Average Damages ($)	Incremental Expected Damages ($)
30	0.20			
		0.10	50	5.00
40	0.10			
		0.05	80	4.00
55	0.05			
		0.025	200	5.00
80	0.025			
		0.015	600	9.00
120	0.010			
			Total	$23.00

TABLE 9.7 Annual Cost Comparisons

Culvert Diameter (in.)	Initial Cost ($)	Maintenance ($)	Expected Damages ($)	Total Cost ($)
24	43.53 +	15.00 +	23.00	81.53
27	46.05 +	15.00 +	19.00	80.05
30	49.00 +	15.00 +	14.00	78.00
36	57.50 +	15.00 +	9.00	81.50
42	64.00 +	15.00 +	7.00	86.00

nance, and expected damages, or $43.53 + 15.00 + 23.00$, or $81.53. As pipe size increases, initial cost increases but damages decrease. Table 9.7 illustrates the comparisons. The pipe size that minimizes total cost is 30 in.

Not all locations for culverts or structures need to be evaluated for the uncertainty and risk associated with each site. The analysis is in some cases at least as expensive as the cost of the pipe. Thus, in traditional design, a return-period storm is specified and a single calculation for pipe sizing is done based on the design flow. However, there remains an element of risk and uncertainty. Nevertheless, if risk data are readily available or apparent, a selection of discharges may be analyzed.

Example Problem 9.3 *(Adapted from Ossenbruggen, 1984)* Consider the selection of a pump system to remove ponded water with the uncertainty of pump operation. This example illustrates expected value calculations and pump system reliability. In this example, only parallel pumping systems are considered. The capital costs of the pump systems and economic losses of not removing 20 ft^3/s will be incorporated into the analysis. The alternative plans with reliability estimates are:

TABLE 9.8 Uncertainty of Two Plans—Example Problem 9.3

Plan A: Two 20-ft^3/s Pumps in Parallel Operation		Plan B: Three 10-ft^3/s Pumps in Parallel Operation	
Flow Condition, g (ft^3/s)	Pr($Q = q$)	Flow Condition, g (ft^3/s)	Pr($Q = q$)
0	0.0025[a]	0	0.001
10	0.0	10	0.007
20	0.9975	20	0.992

[a]The probability of one pump failing is 0.05 and the pumps act independent of one another; thus the probability of both failing is $(0.05)(0.05) = 0.0025$.

TABLE 9.9 Annual Damages (Losses)—Example Problem 9.3

Flow, Q (ft^3/s)	Annual Loss (million $)
0	$80
10	$25
20	0

The probabilities cited here are derived from empirical data on the reliability of the pumps.

The capital cost for construction of the pumping station, in millions of dollars, is a function of flow Q in cubic feet per second.

$$C = 0.04 Q^{1.21} \quad \text{for each pump}$$

Failure to remove 20 ft^3/s of water is assumed to cause problems related to flooding downstream of the detention pond. In addition, there is a reuse potential for the ponded water which may be compromised, and the ponded water removed will carry with it some nutrients and other pollutants that could cause damage to downstream users. The annual losses are estimated as follows:

Determine the better pump system. Use an interest rate of 5% and a system design life of 15 years.

SOLUTION: Define the states of nature S_1, S_2, and S_3 in terms of the discrete random variable Q (water removed from the detention pond).

$$Q = (0, 10, 20 \text{ ft}^3/\text{s})$$

or

$S_1 = (\text{pump system delivers no flow}) = (Q = 0)$
$S_2 = (\text{pump system delivers } 10 \text{ ft}^3/\text{s}) = (Q = 10)$
$S_3 = (\text{pump system delivers } 20 \text{ ft}^3/\text{s}) = (Q = 20)$

The probability assignments are

Plan A:
$\Pr(S_1) = \Pr(Q = 0) = 0.0025$
$\Pr(S_2) = \Pr(Q = 10) = 0.0$
$\Pr(S_3) = \Pr(Q = 20) = 0.9975$

Plan B:
$\Pr(S_1) = \Pr(Q = 0) = 0.001$
$\Pr(S_2) = \Pr(Q = 10) = 0.007$
$\Pr(S_3) = \Pr(Q = 20) = 0.992$

336 ECONOMIC AND FISCAL FEASIBILITY

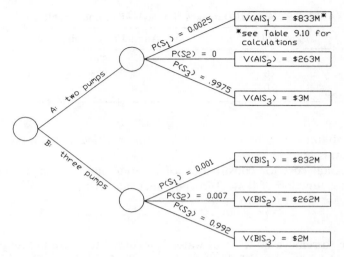

Figure 9.7 Decision tree for Example Problem 9.3.

The capital costs for the systems are

Plan A: $\qquad C = 2(0.04)(20)^{1.21} = \$3.0\text{M}$

Plan B: $\qquad C = 3(0.04)(10)^{1.21} = \$2.0\text{M}$

A minimum present-worth selection approach will be utilized. The annual benefit received from both systems is the same. The benefit is assumed to be greater than the costs. The present worth of annual losses PW for each state of nature or flow is depicted in Figure 9.7. The decision tree for the two proposed systems is shown in Figure 9.7. The present worth of capital plus economic losses is shown in the boxes. They are calculated as follows in

TABLE 9.10

	Present Worth of Capital and Losses	
	State of Nature q	Capital + Losses
Plan A	$S_1 \quad q = 0$	$V(A \mid S_1) = \$3.0\text{M} + \$830\text{M} = \$833\text{M}$
(two pumps)	$S_2 \quad q = 10$	$V(A \mid S_2) = \$3.0\text{M} + \$260\text{M} = \$263\text{M}$
	$S_3 \quad q = 20$	$V(A \mid S_3) = \$3.0\text{M} + \quad 0 = \$\quad 3\text{M}$
Plan B	$S_1 \quad q = 0$	$V(B \mid S_1) = \$2.0\text{M} + \$830\text{M} = \$832\text{M}$
(three pumps)	$S_2 \quad q = 10$	$V(B \mid S_2) = \$2.0\text{M} + \$260\text{M} = \$262\text{M}$
	$S_3 \quad q = 20$	$V(B \mid S_3) = \$2.0\text{M} + \quad 0 = \$\quad 2\text{M}$

Table 9.10. The expected monetary loss value (EMV) for each alternative is:

Plan A   EMV = $\Pr(S_1)V(A|S_1) + \Pr(S_2)V(A|S_2) + \Pr(S_3)V(A|S_3)$
(two pumps)   = (0.0025)(833M) + (0.0)(263M) + (0.9975)(3M)
     = $2.08M + 0 + $2.99M = $5.07M

Plan B   EMV = $\Pr(S_1)V(B|S_1) + \Pr(S_2)V(B|S_2) + \Pr(S_3)V(B|S_3)$
(three pumps)   = (0.001)(833M) + (0.0070)(262M) + (0.992)(2.0M)
     = $0.83 + $1.83M + $1.98M = $4.64M

Plan B is the best.

9.5 SUMMARY

Economic feasibility is obtained if the total benefits from a stormwater facility are greater than those which result without the facility by a measured amount greater than the cost of the facility. If the stormwater facility is to be of value, its selection must be made based on a financial plan and must be optimized with respect to a stated objective and constraints.

- All stormwater facilities should include, to the extent possible, all related benefits and costs.
- Tangible benefits should exceed cost.
- The present value of money is the discounted value of all capital and operating expenditures.
- Since hydrologic events are stochastic, a measure of the value of a project may be done by using expected monetary value.
- As the size of a stormwater system increases, the level of protection increases and the damages decrease. However, the cost of the project increases. The sum of the cost of damages and project cost can be used to determine the size of a project.

9.6 PROBLEMS

1. Explain in your own words for a hypothetical detention pond designed for the control of floods, recreation, and agricultural water supply how to estimate the benefits of the detention pond.

2. For a stormwater detention pond, explain the most probable benefits assuming there is reuse of the stored water and the storage below the invert of the control structure can be lowered before each storm event.

3. A municipality wishes to repay a loan of $1,000,000 for a regional stormwater management plan. The repayment period is 6 years and the interest rate is 10%. Prepare a comparison of four financial plans similar to that of Table 9.2.

4. Prepare a cash flow diagram for Problem 3.

5. Assuming that a water management authority through its taxing receipts has available $500,000 per year for the next 5 years, how much money can it now spend on a 5-year stormwater management system if the interest rate is 8% per year? As another option, the authority can borrow $1.4 million at 8% per year over 5 years for construction in year 1, but incurs a $100,000 per year operating cost. If the authority agrees to pay back the loan on a decreasing principal fiscal plan, what is the total outlay of money? Which plan does the authority use if the selection criteria is to minimize total cost over a 5-year period? Also, which plan is specified if the criterion is to minimize present worth?

6. A water management district borrows a lump sum of $2,000,000 to spend on acquiring land. If a land parcel will become available in 2 years, what is the maximum amount of money available at the end of 2 years if the $2,000,000 is invested at 9.25% per year?

7. If $2,000,000 is available in year 1, what is the constant annual monetary value over a 4-year period if at the end of the 4-year period, a district has been promised $500,000 and if the $2 million is invested in a project with an annual return rate of 7.20%?

8. Revenue from the use of a detention pond over 5 years is estimated at $10,000 per year. At an interest rate of 5.35%, what is the annual maintenance cost that can be used if the initial cost of the detention pond were $25,000? The criterion is that revenue over 5 years must equal cost (initial plus annual).

9. If the marginal cost of a detention pond to remove annually an additional 10% of pollutant mass is $6000 and the same additional 10% can be removed by a street sweeper, what should be the marginal cost of the street sweeper?

10. Determine the optimal size of a 100-ft-long culvert if the criterion is to minimize expected cost. The economic loss and annual cost as a

function of culvert size is:

Culvert Diameter (in.)	Annual Cost ($)	Economic Loss ($/yr) Exceedence Probability					
		0.20	0.10	0.05	0.01	0.001	< 0.001
36	400	0	200	350	600	1200	2000
42	465		0	150	300	800	1500
48	540			0	120	350	800
54	655				0	100	300

Also, if you have access to pipe cost data, calculate your own annual cost. The interest rate is 7.65% with a useful life of 30 years.

11. A city has the option of providing a regional stormwater reservoir or on-site detention for stormwater peak flow attenuation. If the estimated flood damage without any regional or on-site detention is $160,000 per year, which of the alternatives would you use if there is no salvage value to either alternative, an annual interest rate of 10%, and a 60-year design life? Base your decision on two different criteria. The cost and damage data are:

Alternative	Capital Cost	Annual ORM	Annual Damage After Protection
Reservoir	$120,000	$20,000	$50,000
On-site	160,000	10,000	25,000

12. City A will construct a 1-mi storm sewer. The city specifies certain design criteria for engineering feasibility. You must use a velocity of 5 ft/s. Town engineers have evaluated the population and industrial growth of the area and have forecasted the peak flow to increase at a linear rate from 10.8 ft^3/s to 31 ft^3/s in 20 years. Determine the least cost alternative (adapted from Ossenbruggen, 1984):

Alternative A. Build a storm sewer to transport the forecasted 20-year design flow.

Alternative B. Build a storm sewer in two stages. The first-stage plan calls for constructing the storm sewer to transport the forecasted 10-year design flow (assume linear increase) and then a second line to handle additional flow up to the forecasted 20-year design flow.

The construction cost of installing the pipe is a function of pipe diameter and length, $c = \$3$ per inch of pipe diameter per linear foot of pipe. Determine the least cost alternative.

a. Draw cash flow diagrams.

b. Use an annual interest rate of 7.35% and ignore the effects of inflation. Use the present-worth method.

13. An existing storm sewer has become inadequate because of new development. Two alternative plans to transport the water are suggested (adapted from Ossenbruggen, 1984). The new phase must have a life of 50 years.

Alternative A. Replace the existing 30-in. corrugated pipe with a new 60-in. pipe at a construction cost of $35,000 and an annual cost of $1000.

Alternative B. Add a parallel 27-in. pipe next to the existing 30-in. corrugated pipe at a construction cost of $28,000. The existing 30-in. pipe will have to be replaced in 20 years at a cost of $50,000. It will have a salvage value of $10,000 in 50 years.

The maintenance cost will be $800 per year for the first 20 years and $1000 per year for the next 30 years.

a. For each alternative show the cash flow diagrams.

b. Determine the best alternative by use of the net present-worth method. The annual interest rate is 9%.

14. Two dams (A or B) are being considered for construction.

Alternative A. $4M construction cost.

Alternative B. $6M construction cost.

For a serious flood, it is assumed that there is 10% chance that dam A will fail, causing $10M in damages, and a 5% chance that dam B will fail, causing $15M in damages. If a major flood occurs, it is assumed that there is a 15% chance that dam A will fail, causing $15M in damages, and there is an 8% chance that dam B will fail, causing $17M in damages. The probabilities of serious and major floods are estimated to be 5 and 1%, respectively. The design life of the project is 40 years, and the social-opportunity-cost cost interest rate is 8%.

a. Draw a decision tree and label it appropriately.

b. Determine the better alternative to minimize losses.

Hint: Let Pr(FS) = Pr(serious flood occurs in a 1-year period) = 0.05

Pr(FM) = Pr(major flood occurs in a 1-year period) = 0.01

Pr(dam A fails FS) = Pr(FA, FS) = 0.1

Pr(dam B fails FS) = Pr(FB, FS) = 0.05

Pr(dam A fails FM) = Pr(FA, FM) = 0.15

Pr(dam B fails FM) = Pr(FB, FM) = 0.08

15. If the annual losses of Table 9.3 were increased by 10%, how does it affect the choice of designing for 4900 ft³/s? Show all calculations using benefits minus cost criteria.

16. What size detention pond should be specified if the criteria for choice is the difference between expected benefits and cost? The data on benefits and costs as a function of a probability distribution are:

Detention Pond Size (acre-ft)	Exceedence Probability	Benefits ($1000/yr)	Cost ($1000/yr)
9.0	0.30	0	10
10.5	0.16	100	20
12.0	0.05	700	40
13.5	0.02	1500	62
15.0	0.01	2500	87

17. For Problem 16 the benefit function changes as a function of detention pond size and is

Acre-ft	Benefits ($1000/yr)
9	0
10.5	800
12.0	1600
13.5	2000
15.0	2600

Using the expected value function, what is your decision? The cost and probability functions remain the same.

18. a. A stormwater project for water quality improvement consists of three plans, A, B, and C. Plan C is best for reducing pollutant loads of a particulate and dissolved nature, while plan A is the worst, removing only a small fraction of the particulate material. For economic feasibility, which plan is best given the following data?

Plan	Damages	Costs
A	650,000	250,000
B	550,000	300,000
C	150,000	500,000
Do nothing	1,100,000	

b. If it is decided to fund plan C over a 5-year period using a uniform payment plan per year with a 10% yearly interest rate, how much is the yearly payment and the total 5-year cost?

9.7 REFERENCES

Binkley, C. S., and Haneman, W. M. 1978. *The Recreation Benefits of Water Quality Improvements: Analysis of Day Trips in an Urban Setting*, EPA 600/5-78-010, U.S. Environmental Protection Agency, Washington, D.C.

Corry, M. L., et al. 1981. *Design of Encroachments of Flood Plains Using Risk Analysis*, HEC-17, Federal Highway Administration, Washington, D.C.

Davis, S., Ed. 1988. *National Economic Development Procedures Manual: Urban Flood Damage*, IWR Report 88-R-2, U.S. Army Corps of Engineers, Fort Belvoir, Va.

Florida Administrative Code, 1986. *Stormwater Utility*, Section 403.0893, Tallahassee, Fla.

General Accounting Office. 1980. *Costly Wastewater Treatment Plants Fail to Perform as Expected*, Comptroller General Report, U.S. Government Printing Office, Washington, D.C.

Heaney, J. P. 1988. "Cost Effectiveness and Urban Storm Water Quality Criteria," in *Design of Urban Runoff Quality Controls*, L. A. Roesner, B. Urbonas, and M. Sonnen, Eds., American Society of Civil Engineers, New York, pp. 84–99.

Kneese, A. V., and Bower, B. T. 1971. *Managing Water Quality: Economics, Technology, Institutions*, John Hopkins University Press, Baltimore.

Loucks, D. P., Stedinger, J. R., and Haeth, D. A. 1981. *Water Resources Systems Planning and Analysis*, Prentice Hall, Englewood Cliffs, N.J.

Maass, A., Hufschmidt, M. M., Dorfman, R., Thomas, H. A., Marglin, S. A., and Fair, G. M. 1966. *Design of Water Resources Systems*, Harvard University Press, Cambridge.

Milliman, J. W. 1983. "An Agenda for Economic Research on Flood Hazard Mitigation in Changnon, S.A.," *A Plan for Research on Floods and Their Mitigation in the United States*, Illinois Water Survey, Champaign, Ill., pp. 83–104.

Ossenbruggen, P. J. 1984. *Systems Analysis for Civil Engineers*, Wiley, New York, pp. 106–107.

Tihansky, D. P. 1974. "Historical Development of Water Pollution Control Cost Function," *Journal of the Water Pollution Control Federation*, Vol. 46, pp. 813–833.

U.S. Environmental Protection Agency. 1976. *Guidelines for State and Areawide Water Quality Management Program Development*, U.S. EPA, Washington, D.C.

U.S. Water Resources Council. 1981. *Floodplain Management Handbook*, U.S. Government Printing Office, Washington, D.C.

Wanielista, Martin. 1983. *Stormwater Management: Quantity and Quality*, Ann Arbor Science Publishers (Butterworth), Boston.

York, D. W., Cannan, L. W., and Dysart, B. C. 1976. "Modeling Water Related Recreation," *Journal of the Water Resources Management Division, ASCE*, Vol. 102, pp. 409–413.

CHAPTER 10
Optimization

A person may well be curious to know how one stormwater management system is picked from among many possibilities. Certainly, these possibilities must meet the technical, economic, and fiscal criteria of previous chapters. However, a best system remains to be defined as one that has either the lowest capital cost, highest reliability, maximum benefits, lowest expected monetary cost, or other best measure. The determination of the best system from among many can be done if a mathematical statement of the problem with all variables and factors for the interrelationships is developed. Then there exists commonly used mathematical principles and concepts that can be used to help solve for the best system.

10.1 INTRODUCTION

Optimization means to make as perfect as possible. The realization of this definition depends on how well the mathematical relationships among all the important variables defined the situation. The procedures for defining the variables and relationships with subsequent methods for finding the best solution are found in the subject matter associated with the study of operations research.

The term *operations research* was used first during World War II. It is used to define a scientific means of problem solving (Wagner, 1969). However, the term is somewhat misleading because research is not always the goal, and in fact, decision making with the knowledge of many factors affecting decisions must be available. Well... possibly some research must be done to define the problem. However, the general ideas from the field of operations research aid in a problem-solving approach. This approach has basically five major phases (Hillier and Lieberman, 1967):

1. Formulating the problem
2. Constructing a representative mathematical model
3. Deriving a solution from the model
4. Testing the model and solutions
5. Implementing the solution

10.1.1 Formulating the Problem

The first phase is concerned with defining the real-world problem and determining the existence of data required to aid in solving the stormwater problem. Vocal expressions of the problem objective and limitations are communicated in this phase and are most crucial because the problem statement greatly affects the relevancy of the conclusions. A model must be developed from the problem statements. If the model is not an accurate abstraction of the real-world problem, the model should be reformulated. A reexamination of the initial formulation should be performed when new insights are gained during the analysis. All assumptions used to develop a model should be stated and recorded for future use.

The engineer, scientist, and planner should be responsible for the development and precise statement of assumptions, limitations, and objectives. The objective of the stormwater problem should be specific and quantitative, but attainable and consistent with the primary goals of a stormwater management project. There is an objective associated with each problem. Frequently, the objective is the overall goal (i.e., minimize cost, maximize profits), but a combination of many objectives for many problems can be used to obtain the overall goal.

Once the objective has been established, limitations on the attainment of the objective must be considered. These limitations, when expressed in mathematical terms, are known as constraints for the model. If the constraints limit the choice to only one possible course of action, that one action will be done and the problem is finished. If more than one course of action exists and there is doubt about the best choice, the remaining phases of the analysis are continued.

10.1.2 Constructing a Model

A *model* will be defined here as a mathematical abstraction of reality that preserves and uses characteristics of the problem. The mathematical abstraction is expressed in terms of mathematical symbols and expressions. These symbols are classified as variables and constants of the model. If there are n decisions to be made, they are represented by decision variables (say $X_1, X_2, \ldots, X_n$) whose values have to be determined.

The values for the decision variables are evaluated by the objective function, which is a quantifiable measure of the objectives of the study. The values of the decision variables are limited by the constraints of the problem. The constraints are expressed mathematically by means of inequalities and/or equations. An interpretation of the model is to choose the values of the decision variables that optimize the objective function subject to the constraint set. This is *optimization*. Solving the equations is referred to as *maximization* or *minimization* of the objective function.

An advantage of the mathematical model is that it describes a problem more concisely and thus makes the problem more comprehensible, while helping to reveal important cause-and-effect relationships. It also allows one to deal with the problem in its entirety.

It should be remembered that the model is only an abstraction of reality, and simplifying assumptions generally are required to make the model solvable by mathematical techniques. Again, it should be emphasized that all assumptions necessary for a model should be listed and tested before a solution is implemented.

10.1.3 Deriving a Solution

After the model is constructed, a mathematical technique is then used to obtain a solution. The choice of a technique depends on the mathematical formulation of the model, the accuracy of desired solutions, and the resources available for solving the model. If accuracy is not highly desired and little time and money exist for analysis, then approximation or heuristic procedures can be used. Heuristic procedures use intuition and judgment but do not guarantee an optimal solution. However, optimal or near-optimal solutions result.

Sometimes the solution to the model is not the best because the model does not represent real-world conditions. However, the responsiveness of the model to various actions can be examined by the use of sensitivity analysis. Sensitivity analysis refers to varying the parameters of the model and the recording of the results. The analyses determine which parameters of the model results are most sensitive to input data changes, so that more attention can be focused on accurate determination of input data. Also, variability of the results can be recorded for reasonable changes in the input parameters.

Beginning with Section 10.2, methods for finding the best solution are presented. The methods depend on the mathematical formulation of the model.

10.1.4 Testing the Model and Solutions

In this phase of the analysis, methods for evaluating and improving solutions are discussed. One of the best methods of testing is to evaluate the predictability of the model under different conditions. This may be impossible to do physically because of huge costs or time involved, but simulations with the aid of an electronic computer are available. For a simulation, historical data are used to recreate the past and see if the model would perform as expected.

Continued evaluation over time should also be made to determine if the solutions remain valid because of changes in the constants and/or constraints of the system under study. Sensitivity analysis can again be used to satisfy the inquisitive mind. This is also known as a control over system solutions.

10.1.5 Implementing the Solution

Here is where the benefits of optimization are realized. It is important that all decision makers participate, to ensure timely and accurate translation of the analysis. To accomplish implementation, it is generally conceived that there are three important guidelines. First, the model and its solution should be stated in terms that decision makers understand. Second, the responsibility for developing the implementation belongs to everyone. Then the need for future controls should be established.

10.2 LINEAR PROGRAMMING

In Chapter 9 information was presented on the use of economic measures to aid in finding optimal solutions. These optimization procedures use a relatively small number of alternatives and complete enumeration to determine the best alternative. There are some stormwater problems that more realistically can be solved by evaluating a substantial number of alternatives. This choice of the best solution requires more than a few calculations to determine which one of many is the best. One procedure useful for optimization in the hydrology and stormwater management areas is linear programming.

The *linear programming procedure* can be summarized in equation form as follows:

Maximize or minimize
$$F(\bar{x}) = c_1 X_1 + c_2 X_2 + \cdots + c_n X_n \tag{10.1}$$

subject to
$$a_{11} X_1 + a_{12} X_2 + \cdots + a_{1n} X_n \leq, \geq, = b_1$$
$$a_{21} X_1 + a_{22} X_2 + \cdots + a_{2n} X_n \leq, \geq, = b_2$$
$$a_{m1} X_1 + a_{m2} X_2 + \cdots + a_{mn} X_n \leq, \geq, = b_m$$

and
$$X_i \geq 0, \quad X_i \leq X_i^{\max}$$

where X_i = decision variables, $i = 1 \cdots n$
a_{ij}, b_i, c_j = constants

The $F(\bar{x})$ and the constraint equations must be linear equations over the values of the decision variable in question. If only two decision variables are present, a graphical solution technique can be used, but computer programs are available to solve the larger problems. One such program is available with this book.

10.2.1 Objective Function and Constraints

An objective function is a mathematical expression of the specific measures for the objective. The objective function can be written in terms of water volume, cost, benefits, and so on. As an example, it can be a mathematical expression that relates the present worth of a stormwater detention plan for many sites in a watershed to the decision variables on size of detention facility.

The typical mathematical expression includes at least one objective function which is either minimized or maximized. No matter how many decision variables are included in the objective function, the overall measure must be the same. Units of measurement cannot be mixed, or water quality (kilograms) cannot be added to flood control (cubic feet per second). However, the cost of improving water quality can be added to the cost of peak discharge control where the decision variables have units of volume (m^3, acre-ft) and the unit cost of the use is $/volume. An example objective function for stormwater detention may be written as

$$Z = \text{maximize benefits} = c_1 X_1 + c_2 X_2 \qquad (10.2)$$

where Z = objective function, $, present worth
 c_1 = unit return per volume of water provided for quality control, $/acre-ft
 c_2 = unit return per volume of water provided for peak discharge control, $/acre-ft
 X_1 = decision variable 1, volume of water for quality control, acre-ft
 X_2 = decision variable 2, volume of water for peak discharge control, acre-ft

The value of the example objective function (equation 10.2) would be a maximum if the greatest quantity of water can be allocated to that use with the greatest return. However, there are limitations to all allocations. These limitations are expressed in mathematical equations and are called *constraints*. There are two general types of constraints: (1) technical and (2) institutional. The technical constraint cannot be violated in the strictest sense, while the institutional ones are expressions of an implicit objective that may in fact be changed, although the consequences of such change may not be desirable. Examples of technical constraints include those to ensure conservation of mass (inventory equation), volume of storage (capacity), and the flow rate of conduits. Examples of institutional constraints are minimum stream flow for wildlife or water quality, budgetary limitations, and political preferences. Adjustments can be made in these institutional constraints often after an optimal solution is obtained. The initial formulation of constraints can provide valuable information for review of institutional constraints and the consequences of changing these constraints. Examples of constraints to the objective function of equation 10.2 for two decision variables are:

Physical

Pipe size:	$X_1 + X_2 \leq 500$ acre-ft	(10.3)
Size of pond 1:	$X_1 \leq 250$ acre-ft	(10.4)
Size of pond 2:	$X_2 \leq 400$ acre-ft	(10.5)

Institutional

Budget:	$m_1 X_2 \leq \$1,000,000$	(10.6)
Preference:	$2X_1 - X_2 \leq 0$	(10.7)

where m_1 is the cost of control per unit volume (\$/acre-ft).

The preference constraint (Equation 10.7) states that if one unit of pond volume is used for water quality, at least twice as much must be used for peak control. The pipe size constraint must have been developed assuming that both ponds have the same pipeline for the stormwater runoff and it cannot exceed 500 acre-ft. Also note that a specific time frame is implicit in the formulation.

10.2.2 Problem Formulation / Graphical Solution

Once the mathematical model is formulated as an objective function and constraint set with two decision variables, a graphical solution is possible. The graphical procedure is applied by following these steps:

1. Plot on graph paper the constraining equations (Figure 10.1). This will establish the region that satisfies all constraints, called the *feasible region*. An infeasible solution results when one of the constraints is counter to the others or one of the inequalities cannot be achieved without violating another inequality.
2. Determine the optimum solution by first assuming a value for the objective function. The slope of the objective function is determined by the ratio (c_1/c_2). Next, move the objective function value until it touches an extreme point of the feasible region (see Figure 10.1).

The feasible region defined by a plot of the two decision variables forms a convex region (see Figure 10.1). If a straight line is drawn from one point on the edge of the feasible region to another edge, the line will be entirely on or within the region. All linear programming graphical solutions have the properties of convex sets, feasible regions, and extreme points. The location of the optimum solution must lie along the edge of a convex set or at an extreme point.

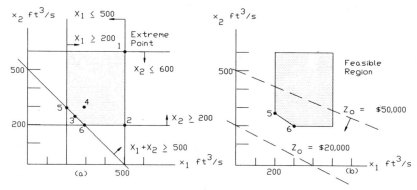

Figure 10.1 Feasible region for Example Problem 10.1.

The graphical procedure is an easy approach to establishing the values of the decision variables that optimize the objective function. However, application is limited to a two-variable problem. Nevertheless, a two-variable plot with an example is useful for illustrating many concepts.

Example Problem 10.1 Two new developments on either side of a drainage canal are considering detention ponds to attenuate peak flows into the canal. A schematic is shown in Figure 10.2. Each development engineer calculates a minimum 200-ft^3/s reduction in peak flow is needed based on pre-equals post-peak conditions. However, the water management district requires a total of 500 ft^3/s reduction based on in-stream protected uses (wildlife). Maximum containment of all runoff can be achieved if 500 and 600 ft^3/s are attenuated at sites 1 and 2, respectively. If the cost of reducing the peak at site 1 is $50 per 1-ft^3/s reduction and the cost of reducing the peak at site 2 is $100 per 1-ft^3/s reduction, what does the water manager suggest in terms of peak attenuation at each site, and how much overall savings are realized?

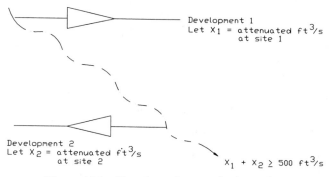

Figure 10.2 Two-detention-pond schematic.

350 OPTIMIZATION

SOLUTION: The objective function is expressed in mathematical terms as a minimum cost, or

Minimize $\quad Z = C_1 X_1 + C_2 X_2 = 50 X_1 + 100 X_2 \quad$ (10.8)

Subject to:

1. Minimum containment from each site

$$X_1 \geq 200 \text{ ft}^3/\text{s} \quad (10.9)$$
$$X_2 \geq 200 \text{ ft}^3/\text{s} \quad (10.10)$$

2. Maximum containment from each site

$$X_1 \leq 500 \text{ ft}^3/\text{s} \quad (10.11)$$
$$X_2 \leq 600 \text{ ft}^3/\text{s} \quad (10.12)$$

3. Total attenuation from both sites

$$X_1 + X_2 \geq 500 \text{ ft}^3/\text{s} \quad (10.13)$$

It may be possible that a minimum cost results when site 1 attenuates the most since its unit cost is the lowest. This can be proven by a graphical solution to the equations above. The constraints are shown in Figure 10.1. The minimum and maximum attenuation based on pre- and postdevelopments are shown as the horizontal and vertical lines in Figure 10.1a. The diagonal line represents the total attenuation required from both sites.

The feasible region is illustrated by the shaded area. All points that lie within or on the edge of the feasible region are called *feasible solutions*. The constraint set has established the solution set or combinations of X_1 and X_2. If $X_1 = 100$ ft^3/s and $X_2 = 400$ ft^3/s, the total reduction of 500 ft^3/s would be achieved, but site 1 must have a minimum reduction of 200 ft^3/s. Thus this solution is infeasible and outside the feasible region.

The next step in the solution requires a minimum-cost combination of peak reductions be found from among the many combinations in the feasible region. As an example, consider the following four combinations being evaluated by the objective function.

Point	X_1 (ft^3/s)	X_2 (ft^3/s)	Objective Function ($)	Comment
1	500	600	50(500) + 100(600) = $85,000	Greatest cost
2	500	200	50(500) + 100(200) = $45,000	Cost reduction
3	250	250	50(250) + 100(250) = $37,500	Lowest of the four
4	300	300	50(300) + 100(300) = $45,000	An "interior" point

If more than 500 ft³/s attenuation is required, cost will be higher. This statement is verified from points 1, 2, and 4. But is extreme point 3 any less expensive than 5 or 6? Consider an objective function with a value of $50,000 and $20,000 as shown in Figure 10.1b. Since the Z_0 = $20,000 line is outside the feasible region, it does not define a solution that meets the constraints. Note, however, that if the objective function line continues to move downward at the same slope, the last extreme point of the feasible region that will be touched is point 6. Thus the minimum-cost solution occurs when X_1 = 300 ft³/s and X_2 = 200 ft³/s. The minimum cost is 300(50) + 200(100) = $35,000. The equality of treatment solution would be for both sites to attenuate the hydrograph by 250 ft³/s for a cost of $37,500. There is a savings of $2500 when the linear programming solution is used.

10.2.3 Regional Stormwater Facilities Operation and Design

Regional stormwater facilities are large holding reservoirs constructed for multiple uses, some of which are in conflict with one another. Conflicts arise due to the need to hold water and release water. There are conflicts as to the priorities for the different uses of water. Water can be released from a reservoir for low-flow augmentation, recreation, power supply, wildlife, navigation, and to decrease storage for flood control. On the other hand, reservoir storage can be used for lake recreation, water supply for many users, flood control, water quality (pollution) management, and wildlife management. Frequently, release rates are set at target levels (i.e., 200,000 acre-ft release during the spring months). Also, storage levels can be set as targets. Targets also are known as firm water levels.

A mathematical model relating release and storage volumes with net inputs would help specify the optimal design and operation plan. Such models have been used to specify quantity of water for each user in contract negotiation and have served as operational guidelines to ensure that contract violations are minimized.

The basic equations for reservoir management are mass balance equations. An example of mass balance is

$$\text{reservoir input} = \text{reservoir output} + \text{storage change} \quad (10.14)$$

beginning storage + surface input − evapotranspiration − releases
= ending storage

$$S_0 + Q(\Delta T) - ET - Y = S_1 \quad (10.15)$$

with consistent set of volume units (i.e., mg, acre-ft, liters, etc.) and where

S_0 = storage level at time zero
$Q(\Delta T)$ = product of flow rate and time is flow volume
ET = evapotranspiration volume
Y = release volume
S_1 = storage level at the end of the time period

352 OPTIMIZATION

If net input is $I = Q(\Delta T) - ET$, equation 10.15 is rewritten for the first two time periods as:

First time period: $\quad S_0 + I_1 - Y_1 = S_1 \quad$ (10.16)
Second time period: $\quad S_1 + I_2 - Y_2 = S_2$

Substituting for S_1 yields

$$S_0 + \sum_{i=1}^{2} I_i - \sum_{i=1}^{2} Y_i = S_2 \quad (10.17)$$

There is always a maximum (S^m) and minimum (S_m) size of reservoir; thus

$$S_1 \leq S^m \quad S_2 \leq S^m \quad (10.18)$$

and

$$S_1 \geq S_m \quad S_2 \geq S_m$$

Given a linear relationship between size of reservoir and net benefits (B_S) and a linear relationship between releases and net benefits (B_Y), or

$$B_S = s_1 S_1 + s_2 S_2 \quad (10.19)$$

and

$$B_Y = s_3 Y_1 + s_4 Y_2$$

where s_1, s_2, s_3, s_4 are the unit values of storage and releases (\$/unit). The mathematical formulation of the linear programming model is:

Maximize benefits $\quad = s_1 S_1 + s_2 S_2 + s_3 Y_1 + s_4 Y_2 \quad$ (10.20)
subject to:
$$S_0 + I_1 - Y_1 = S_1$$

I_1, S_0 are known, thus

(1) $\quad S_1 + Y_1 = S_0 + I_1$
(2) $\quad S_2 + Y_2 - S_1 = I_2; \quad I_2$ is known
(3) $\quad S_1, S_2 \leq S^m \quad$ and $\quad \geq S_m$
(4) $\quad Y_1, Y_2 \geq 0$

10.2.4 Unit Hydrograph Estimation

Linear programming also can be used to estimate the ordinate values of a unit hydrograph. All the ordinate values must be positive; thus if the ordinate values are the decision variables of a linear programming problem, the ordinates will be positive. Another popular method for estimating the ordinate values is the least squares method. However, with the least squares method, the ordinate values cannot be constrained to be positive.

Given a streamflow record and rainfall excess, matrix methods were shown to be useful for estimating the ordinates of the unit hydrograph: namely, designated as U_i

$$U_j = R_j^{-1}(Q_j) \quad \text{for all } j \text{ (symbol is } V_j) \tag{10.21}$$

and the discrete form convolution is

$$Q_j^c = R_i U_{j+i-1} \quad \text{for all } j \text{ and } i \tag{10.22}$$

where Q_j^c = calculated streamflow
R_i = rainfall excess

Given a four-period unit hydrograph and three periods of rainfall excess, equation 10.22 is expanded as

$$Q_1^c = R_1 U_1 \qquad Q_2^c = R_2 U_1 + R_1 U_2$$
$$Q_3^c = R_3 U_1 + R_2 U_2 + R_1 U_3$$
$$Q_4^c = R_4 U_1 + R_3 U_2 + R_2 U_3 + R_1 U_4$$

but $R_4 = 0$; thus

$$Q_4^c = R_3 U_2 + R_2 U_3 + R_1 U_4$$

A further constraint specifies that the area under the unit hydrograph must represent 1 in. of rainfall excess. Thus $1 = (U_1 + U_2 + U_3 + U_4)\Delta t$, and using $\Delta t = 1$ h, the unit conversion is

$$1 \text{ acre-in.} = \frac{(1 \text{ h})[U(\text{ft}^3/\text{s}) \times 3600 \text{ s/h} \times 12 \text{ in./ft}]}{43{,}560 \text{ ft}^2/\text{acre}}$$

or

$$1 \text{ acre-in.} = 1.008 U \text{ ft}^3/\text{s}$$

and for 1 in.:

$$\text{area(acres)} = (U_1 + U_2 + U_3 + U_4)\Delta t \times 1.008 \tag{10.23}$$

The constant 1.008 is frequently dropped from analysis.

The objective is to minimize the sum of the deviations from the calculated and field-recorded streamflow. Each discrete value of the unit hydrograph is estimated using the discrete convolution formula; however, the estimate may be larger or smaller than the field-recorded value. Rewritten, the discrete convolution equations are:

$$\begin{aligned} R_1U_1 + U_5 - U_6 &= Q_1 \\ R_2U_1 + R_1U_2 + U_7 - U_8 &= Q_2 \\ R_3U_1 + R_2U_2 + R_1U_3 + U_9 - U_{10} &= Q_3 \\ R_3U_2 + R_2U_3 + R_1U_4 + U_{11} - U_{12} &= Q_4 \end{aligned} \quad (10.24)$$

and for a unit hydrograph with $t = 1$ h:

$$U_1 + U_2 + U_3 + U_4 = \text{area} \quad (10.25)$$

All decision variables are nonnegative

$$U_1, U_2, U_3, U_4 \geq 0 \quad (10.26)$$

and the objective function is

$$\text{Minimize} \quad U_5 + U_6 + U_7 + U_8 + U_9 + U_{10} + U_{11} + U_{12} \quad (10.27)$$

An additional constraint to preserve the monotonical decrease in the unit hydrograph ordinates beyond the peak could be added.

10.2.5 Optimal Load Reduction Model

When there is more than one source of stormwater loading of a particular chemical or solids to a receiving water body, a stream standard may be violated. These stream standards may be in concentration form or a mass loading. It may be possible to reduce the loadings so that the standard of the stream is met for a series of loadings and flow rate conditions, or a probability distribution may be specified for mass or concentration using various flows and loadings.

Using the linear programming formulation, a mathematical model can be developed that expresses concentration in the receiving stream in terms of stormwater loadings. Let L_i be the loading without treatment (lb) of a pollutant from a particular site j, and X_j the removal fraction at site j. If the flow rate in the river at this site is Q_j, the concentration can be calculated from

$$\text{concentration} = \frac{\text{mass}}{\text{volume}} = \frac{L_i(1 - X_j)}{Q_j(\Delta t)} \quad (10.28)$$

Note that appropriate conversion factors must be used.

For each unit of waste removed at a site, a unit or marginal cost can be estimated. As long as the marginal cost remains constant, the linear programming formulation is appropriate. If the slope of the cost curve changes, a piecewise linear approximation may be necessary and desirable. Let C_j be the unit or marginal cost approximation for site j over the range of percent removals of interest in the problem formulation, which leads to constraints on degree of treatment or

$$X_j \geq LB_j \quad \text{and} \quad X_j \leq UB_j \qquad (10.29)$$

where LB_j = lower bound of treatment at site j
UB_j = upper bound of treatment at site j

Once the chemical load enters the receiving water, the load may decay by either sedimentation, chemical conversion, biological assimilation, or other processes. Let r_j be the composite decay coefficient for a first-order decay curve at site j:

$$L_j^t = L_j(1 - X_j)e^{-r_j t_j} \qquad (10.30)$$

where L_j^t = remaining chemical load at site j, lb
L_j = initial loading without treatment from site j, lb
r_j = decay rate at site j, day^{-1}
t_j = time of flow at site j, days
= D/V, where D is the distance (ft) and V is velocity (ft/day)

Thus as a chemical mass enters an upstream point after treatment at site j, the remaining mass at a downstream location (site $j + 1$) is given by

$$L_{j+1}^t = L_j(1 - X_j)e^{-r_j t_j} \leq B_{j+1}Q_{j+1} \qquad (10.31)$$

where B_{j+1} = site $j + 1$ concentration standard. The concentration standard must also be appropriate at any point of discharge, or

$$L_{j+1}^t + L_{j+1}(1 - X_{j+1}) \leq B_{j+1}Q_{j+1} \qquad (10.32)$$

Example Problem 10.2 A water management district wishes to minimize the quantity of a chemical pollutant in a river system. The system consistent with the notation used in this section is:

Discharge Site	Location Number	Mile Point	Chemical Load[a] (lb/day)	Rate of Decay (day^{-1})	Marginal Cost ($/lb)	Flow (ft^3/s)	River Velocity[b] (ft/s)
City 1 (upstream)	1	45	1000	0.20	0.18	30	2.5
New development	2	29	800	0.15	0.26	38	2.2
New development	3	15	8000	0.25	0.10	50	2.0
City 2 (downstream)	4	0	680		0.25	60	1.6

[a] Before treatment by stormwater management.
[b] Average for the stream reach (e.g., 2.5 ft/s from site 1 to 2).

Each of the discharge sites can be controlled up to 98% removal of the chemical. Write out the mathematical equations necessary to minimize the cost of treatment if the stream standard on the particular pollutant is 1.6 mg/L or 0.0001 lb/ft^3.

SOLUTION: Let X_j be the fraction of chemical load removed at site j.

Minimize $\quad C_1 X_1 L_1 + C_2 X_2 L_2 + C_3 X_3 L_3 + C_4 X_4 L_4$

$$0.18(X_1)1000 + 0.26(X_2)800 + 0.10(X_3)8000 + 0.25(X_4)680$$
$$180 X_1 + 208 X_2 + 800 X_3 + 170 X_4 \qquad (10.33)$$

subject to:

Site 1: $\quad L_1(1 - X_1) \leq (0.0001 \text{ lb/ft}^3)(30 \text{ ft}^3/\text{s})(86{,}400 \text{ s/day})$

$$1000 - 1000 X_1 \leq 259 \text{ lb/day}$$
$$X_1 \geq 0.741 \qquad (10.34)$$

Site 2:*

$$L_1(1 - X_1)e^{-0.20 t_{12}} + L_2(1 - X_2) \leq 0.0001(38)(86{,}400)$$

where

$$t_{12} = \frac{(45 - 29) \times 5280}{2.5 \times 86{,}400} = 0.39 \text{ days}$$

$$(1 - X_1)1000 e^{-0.20(0.39)} + 800(1 - X_2) \leq 328$$
$$925 - 925 X_1 + 800 - 800 X_2 \leq 328$$
$$925 X_1 + 800 X_2 \geq 1397$$
$$1.1 X_1 + 1.0 X_2 \geq 1.75 \qquad (10.35)$$

*Subscript "12" refers to time from discharge site 1 to 2.

Site 3:
$$(1 - X_1)1000e^{-0.20(0.78)} + (1 - X_2)800e^{-0.15(0.39)} + 8000(1 - X_3)$$
$$\leq 0.0001(50)86{,}400$$

where

$$t_{23} = \frac{(29 - 15) \times 5280}{2.2 \times 86{,}400} = 0.39 \text{ day}$$

$$856 - 856X_1 + 755 - 755X_2 + 8000 - 8000X_3 \leq 432$$
$$856X_1 + 755X_2 + 8000X_3 \geq 9179$$
$$0.11X_1 + 0.09X_2 + 1.0X_3 \geq 1.15 \tag{10.36}$$

Site 4:
$$(1 - X_1)1000e^{-0.20(1.24)} + (1 - X_2)800e^{-0.15(0.85)}$$
$$+ (1 - X_3)8000e^{-0.20(1.24)} + (1 - X_r)680 \geq 0.0001(60)86{,}400$$

where

$$t_{34} = \frac{(15 - 0) \times 5280}{2.0 \times 86{,}400} = 0.46 \text{ day}$$

$$780 - 780X_1 + 704 - 704X_2 + 7140 - 7140X_3 + 680 - 680X_4 \geq 518$$
$$780X_1 + 704X_2 + 7140X_3 + 680X_4 \geq 8798$$
$$1.15X_1 + 1.04X_2 + 10.5X_3 + X_4 \geq 12.94 \tag{10.37}$$
$$X_1; X_2; X_3; X_4 \leq 0.98 \tag{10.38}$$

10.3 MULTIOBJECTIVE PROGRAMMING

With increasing frequency, a decision maker is faced with two or more objectives in determining the best for the organization. Also, when more than one decision maker is involved, there are most likely more than one objective. Similarly, with large-scale systems, which cross political boundaries, there are several objectives, some of which are conflicting and noncommensurable. For example, a reservoir that serves many political areas may need to minimize benefits to certain regions while maximizing environmental quality. On a smaller scale, a detention pond built for a community may need to achieve a maximum flow rate attenuation at a minimum cost (Water Resources Bulletin, 1992).

Traditionally, the multiobjective problem considers as a variable one of the objectives (i.e., economics) with the other objective (i.e., hydrograph attenuation) as a constraint. In this way the constraint level can be changed and a new minimum cost solution obtained and compared to previous solutions. This is one method for solving the multiobjective problem. However, one is always left with the question: Is a change in a constraint proportional to the stated objective function change? Are the solutions commensurate or in proper proportion? This question is particularly important when one of the objectives is difficult to quantify in monetary units. Because of noncommensurate and nonmonetary problems, society has developed laws and rules. Some examples are effluent charges and treatment standards (runoff from first inch of rainfall). These charges and standards are important and necessary and can be further supported where situations permit proportional emphasis on each objective; optimization should be done.

The use of a vector of objective functions allows proportionality or weighting of each objective function. Some of the noncommensurable objectives are set equal to the cost and benefit functions, or the many decision makers can be asked to place a proportional value of one objective in terms of another; that is, each mg/L of environmental quality is worth $1000. Thus a surrogate measure (relative importance) is substituted for the decision variables. Values of the decision variables can be determined consistent with the relative importance placed on each objective function. Also note that relative importance and mathematical models utilize logical argument to eliminate a large number of possible decisions (interior points) and then proceed to the best combinations of values for the decision variables.

The general formulation is:

Minimize or maximize $f_1(\bar{x}), f_2(\bar{x}), \ldots, f_n(\bar{x})$ (10.39)

subject to: $g_k(\bar{x}) \leq, \geq, = 0 \quad k = 1, 2, \ldots, m$

where $\bar{x}$ is an n-dimensional vector of decision variables. Using surrogate weights (w_i), we have

$$Z = w_1 f_1(\bar{x}) + w_2 f_2(\bar{x}) + \cdots + w_n f_n(\bar{x}) \qquad (10.40)$$

The general procedure is to identify the feasible region using the constraint set and then for each extreme point evaluate the objective function values. Then by weighting the objective function, the extreme points of the function space convex set can be evaluated: thus the term *surrogate worth trade-off method*.

Example Problem 10.3 Consider a problem of allocating stormwater for irrigation and recharge from a reuse pond with two objective functions and two constraints. The first objective is to maximize the volume of water for irrigation, while the second is to maximize releases from the pond. There is a total of 20 acre-ft of water available in the pond and a maximum of 10 can be used for irrigation or released. The first objective function is 1.5 times as important as the second.

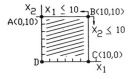

Figure 10.3 Decision space.

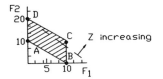

Figure 10.4 Function space.

SOLUTION: Let

X_1 = volume of stormwater for irrigation
X_2 = volume of stormwater for release
f_1 = volume of stormwater for irrigation
f_2 = volume of stormwater released

Maximize $[f_1 = X_1, f_2 = 20 - X_1 - X_2]$
subject to: $X_1 \leq 10$
$X_2 \leq 10$

First, develop the decision space (Figure 10.3) and then calculate the value of each objective function and plot in the function space (Figure 10.4). Objective function 1 is 1.5 times as important as the second:

Maximize $Z = 1.5 F_1 + F_2$

Plot the Z function: $Z = 15$, $F_1 = 0$, $F_2 = 15$; $F_2 = 0$, $F_1 = 10$ (see Figure 10.4), Z is increasing until extreme point C, which maximizes Z or $X_1 = 10$, $X_2 = 0$.

10.4 OTHER OPTIMIZATION TECHNIQUES

When the objective function and constraints do not follow a strict linear mathematical relationship, other methods must be used to search for the optimal combination of decision variables provided that the equations are not assumed linear. These methods can be categorized into nonlinear programming. Nonlinear methods can be classified as classical or search procedures. The classical techniques include differential calculus, Lagrange multipliers, and geometric programming. Search procedures frequently use a gradient technique that makes use of trend characteristics of the objective function (Beard, 1966; Davis, 1975).

10.4.1 Classical Optimization

Most of the classical optimization procedures find the optimal by solving a system of equations or finding the "root" of an equation. The usefulness depends on the computational effort. If a single function of X is specified, the minimum or maximum can be determined by setting the first derivative of the function equal to zero. If the second derivative is positive, a minimum or convex function exists; if negative, a maximum or concave function is defined; and if zero, a linear function is defined.

Example Problem 10.4 Given a fixed area for a trapezoidal canal with cross-sectional area equal to A and side slope defined by the angle with the horizon as shown in Figure 10.5, find the minimum wetted perimeter defined as

$$p = b + 2d \csc \Phi$$
$$A = db + d^2 \cot \Phi$$

SOLUTION: We have two equations but three unknowns (d, b, Φ); thus reduce the number of unknowns and eliminate b.

$$p = \frac{A}{d} - d \cot \Phi + 2d \csc \Phi \qquad (10.41)$$

Take partial derivatives with respect to the two unknowns and set equal to zero:

$$\frac{\partial p}{\partial d} = \frac{-A}{d^2} - \cot \Phi + 2 \csc \Phi = 0 \qquad (10.42)$$

$$\frac{\partial p}{\partial \Phi} = d \csc^2 \Phi - 2d \csc \Phi \cot \Phi = 0 \qquad (10.43)$$

Divide equation 10.43 by d and solve for $\Phi = \pi/3 = 60°$.

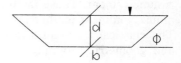

Figure 10.5 Channel geometry.

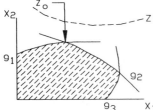

Figure 10.6 Nonlinear decision space.

10.4.2 Nonlinear Optimization

The general optimization model is a more general form of equation 10.1 and can be rewritten as:

Optimize: $\quad Z = f(x_1, x_2, \ldots, x_n)$ (10.44)

subject to: $\quad g_i(x_1, x_2, \ldots, x_n) \geq \leq = b_1 \quad \forall j$

$\quad x_j \geq 0 \quad \forall j$

A general two-dimensional nonlinear optimization graphical representation for maximization is shown in Figure 10.6. Thus, if the feasible region can be established from the constraint set, the optimal combination of values for the decision variables can be obtained. For the nonlinear decision space shown in Figure 10.6 the best value Z^0 is identified. Conditions of optimality also must be satisfied.

Example Problem 10.5 The flow rate (q in gpm) developed by a pump is a function of the horsepower, (x_1), which is a reflection of pump design and the head it must pump against (x_2 in feet). An empirical relationship is $q = (x_1)(x_2)^{1.6}$. The cost of production is given by $C \leq x_1^{2.5} + x_2^{2.0}$. What is the maximum flow rate using a graphical approach if the flow cannot exceed 300 gallons per minute (gpm)?

SOLUTION:

a. *Formulation.* The mathematical model is:

Maximize $\{q = x_1 x_2^{1.6}\} \quad x_1^{2.5} + x_2^{2.0} \leq 300$

$x_1 \geq 0, \quad x_2 \geq 0$ (10.45)

b. *Graphical Solution.* The maximum output will occur when the strict inequality $x_1^{2.5} + x_2^{2.0} \leq 300$ is imposed. Since the objective function is a monotonically increasing function, the solution will occur at a tangency point. The feasible region is drawn (Figure 10.7). The optimum output q^* is $405. This is shown as the tangency point on the graph. The optimum level of production is

$\quad x_1^* = 5$ units $\quad x_2^* = 15.6$ units

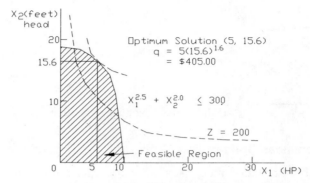

Figure 10.7 Nonlinear graphical solution.

The graphical solution is only as accurate as the grid on the paper and the thickness of the line. In many cases the solution is more accurate if intersection of the objective function and the constraint can be solved for algebraically. To plot the feasible region and the objective function on Figure 10.7, assume a value of one of the variables and solve for the value of the other.

Feasible Region			Objective Function		
Value	x_1	x_2	Value	x_1	x_2
300	0	17.3	200	1.0	27.4
	9.8	0		21.8	4.0
	8.3	10.0		5.0	10.0
	5.0	15.6		2.5	15.5

10.4.3 Piecewise Linear Approximations

When the objective function is nonlinear and the constraint set is linear, it may be possible to piecewise linearize the nonlinear terms of the objective function. Nonlinear constraints may also be linearized. If the objective function given as $f(x_1, x_2, \ldots, x_n)$ is convex or for any pair of points (x_1 and x_2) and all weights λ, $0 \leq \lambda \leq 1$, the following results:

$$f[x_2 + (1 - \lambda)x_1] \leq \lambda f(x_2) + (1 - \lambda)f(x_1) \quad (10.46)$$

Equation 10.46 states that if the segment joining two points lies entirely above or on the graph of $f(x_1, x_2)$, the function is convex. Also, the function is concave if the inequality sign is reversed, or the line segment lies below the graph of $f(x_1, x_2)$. Figure 10.8 illustrates the type of functions.

If the objective function is convex and the function is divided into linear piecewise segments, the function can be minimized by linear programming

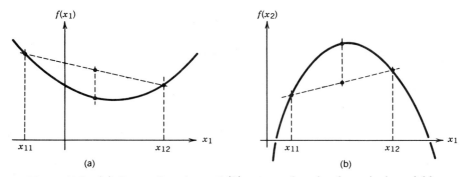

Figure 10.8 (*a*) Convex function and (*b*) concave function for a single variable.

techniques. The function cannot be maximized using linear programming. Consider the convex cost curve shown in Figure 10.9. To be physically possible, the units of X_1 must be used before units of X_2, and X_2 before X_3. As long as the slope of the linear approximations are such that $s_1 < s_2 < s_3$, X_1 will always be used before X_2, and X_2 before X_3, because the linear programming algorithm works to allocate to the least cost first. Next, the allocations must be constrained so that only X_1 units are allocated as associated with the unit cost (s_1). One programming formulation for the minimization of the convex piecewise curve is:

Minimize $$f(x) = s_1 x_1 + s_2 x_2 + s_3 x_3 \tag{10.47}$$
subject to:
$$a_1 + x_1 + x_2 + x_3 = x$$
$$x_1 \leq a_2 - a_1$$
$$x_2 \leq a_3 - a_2$$
$$x_3 \leq a_4 - a_3$$
$$x_j \geq 0 \quad \forall j;$$

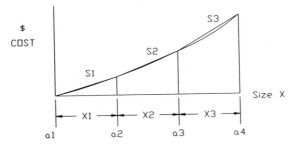

Figure 10.9 Piecewise linear approximations to convex curve.

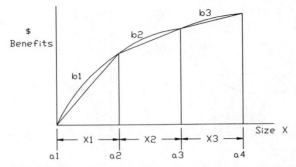

Figure 10.10 Piecewise linear approximations to concave curve.

If the objective function is concave and its function is divided into linear piecewise segments, the function can be maximized by programming techniques, but not minimized. Consider the concave benefits curve of Figure 10.10. The units of x_1 have to be allocated before x_2, and x_2 before x_3. If maximizing, the linear programming algorithm will always allocate to the largest return consistent with the constraint set and with the function described by Figure 10.10, $b_1 > b_2 > b_3$; thus x_1 will be allocated before x_2 and x_2 before x_3. One linear programming formulation for the maximization of the concave piecewise curve is:

Maximize $\qquad f(x) = b_1 x_1 + b_2 x_2 + b_3 x_3 \qquad (10.48)$

subject to:
$$a_1 + x_1 + x_2 + x_3 = x$$
$$x_1 \leq a_2 - a_1$$
$$x_2 \leq a_3 - a_2$$
$$x_3 \leq a_4 - a_3$$
$$x_j \geq 0 \qquad \forall j$$

10.4.4 Dynamic Programming

When a multiple variable problem can be decomposed into a few variables for which optimal values can be solved for in one "stage" and then the properties of the solution transferred to another stage, a recursive relationship exists and the problem can be formulated into a general optimization procedure called dynamic programming. A common problem that fits the dynamic programming definition is the one for operating a regional stormwater facility (Section 10.2.3). Releases and storage volumes for the regional facility must be determined for various time intervals (i.e. monthly, seasonal, etc.). Thus using monthly time periods, there are 24 variables (12 releases and 12 storage volumes). However, the release in any month is related to the storage in the previous month and this relationship is recurring. If the release and storage is feasible in one month, the decisions in the next month will carry forward the feasibility. That which is carried forward is defined as the *state* of the system.

Another recursive-type problem is the capacity expansion of a detention pond serving an expanding land development. The basic questions are: how large a detention facility and at what time do we build the capacity? The capacity of the pond is transferred from one possible building period to another possible building period. The state of the problem is the capacity of the facility and the stages are the building periods.

Still another recursive-type problem involved the construction of storm sewers. Storm sewer pipe and excavation costs have been minimized using dynamic programming (Walters, 1978; Fetter, 1988). Other procedures can also be used (Miles and Heaney, 1988). There exists a trade-off between the depth of excavation and pipe sizing. The smaller the pipe, the greater the slope; thus the deeper the excavation, and vice versa. The pipe diameter for each section of pipe is constrained by acceptable velocities (frequently, 2 to 10 ft/s), and excavation cover provides another constraint. The section of pipe is considered as the stage where the flow for each section is the state. The decision variables are both excavation depth and pipe diameter. The computer program OPSEW provided with this book can be used to solve the storm sewer minimum cost construction problem.

The notion of recursion implied in the examples above can be expressed in a principle of optimality (Bellman, 1957) that provides the structure for dynamic programming, or "an optimal policy has the property that whatever the initial state and decisions are, the remaining decisions must constitute an optimal and feasible policy with regard to the state resulting from the first decision."

The specific optimization procedure that is used to determine the values of the decision variables is irrelevant. The mathematical form of the objective function, mathematical form of constraints, and the number of decision variables at each stage determine the optimization routine to use. Enumeration and differential calculus procedures are commonly used.

The basic solution procedure examines the first stage and allocates to that the optimum quantity, then examines the second stage and allocates based on the first stage allocation. Thus the equations are solved recursively and all the feasible solutions in stage 1 are carried on to stage 2. Mathematically, for a maximization problem,

$$f_1(x) = \max_{x_1} g_1(x_1) \tag{10.49}$$

$$f_2(x) = \max_{x_2} [g_2(x_2) + f_1(x - x_1)] \tag{10.50}$$

and for *n* stages:

$$f_n(x) = \max_{x_n} [g_n(x_n) + f_{n-1}(x - x_{n-1})] \tag{10.51}$$

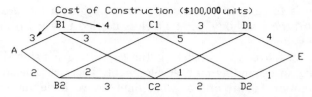

Figure 10.11 Minimum-cost pathways.

Many problems can be formulated for solution by general dynamic programming procedures. The following examples illustrate the formulation and solution procedures.

Example Problem 10.6 A storm sewer system must be built for a city and its direction must follow specific rights-of-way. Each right-of-way has a cost for construction. The various pathways are shown in Figure 10.11. Find the least-cost pathway.

SOLUTION: Once at a stage, say C, the remaining decisions must constitute an optimal policy. The system is recursive in that solutions are carried from one stage to another. Evaluation of the pathways depends on cost.

Stage 1: $A \to B$

$$f_1(x) = \min_{x_1}\{3, 2\} = 2$$

Thus

$$A \longrightarrow B_1 \ (3) \text{ evaluation values}$$
$$\searrow B_2 \ (2) \text{ optimal}$$

Stage 2: $B \to C$

$$f_2(x) = \min_{x_2}\{(4,3) + 3 \text{ and } (2,3) + 2\}$$
$$= \min_{x_2}\{(7,6) \text{ and } (4,5)\} = 4$$

Thus

$$A \quad B_1 \quad C_1 \ (4)$$
$$\searrow \quad \nearrow$$
$$B_2 \quad C_2 \ (5)$$

10.4 OTHER OPTIMIZATION TECHNIQUES

Stage 3: $C \rightarrow D$

$$f_3(x) = \min_{x_3}\{(3,5) + 4 \text{ and } (1,2) + 5\} = 6$$

Thus

$$
\begin{array}{cccc}
A & B_1 & C_1 & D_1\,(6) \\
 & \searrow & & \nearrow \\
 & B_2 & \longrightarrow C_2 & D_2\,(7)
\end{array}
$$

Note: Since the history of the preceding stage has been maintained, the optimal pathway can change from one stage to the next.

Stage 4: $D \rightarrow E$

$$f_4(x) = \min_{x_4}\{6 + 4 \text{ and } 7 + 1\} = 8$$

Thus

$$
\begin{array}{ccccc}
A & B_1 & C_1 & D_1 & E \\
 & \searrow & & & \nearrow \\
 & B_2 & \longrightarrow C_2 & \longrightarrow D_2 &
\end{array}
$$

The problem could also be done by complete enumeration in a tree diagram (Figure 10.12). The tree diagram further illustrates the recursive nature of the problem. For a small number of stages and decision variables, complete enumeration may be easier for calculating the alternatives. Dynamic programming required 14 calculations and comparisons for a solution, while complete enumeration requires 27 calculations and comparisons.

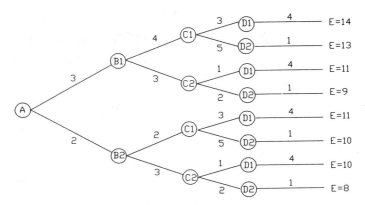

Figure 10.12

Example Problem 10.7 Consider the possibility of expanding the storage of a detention pond. The present storage is $5(10^5)$ ft^3. Eighteen years from the present year, the storage capacity must be expanded to $20(10^5)$ ft^3. The financial plan specifies that every 6 years it is possible to add capacity to the detention pond. The needed storage as a function of time is:

Construction Time (yr)	Needed Storage (10^5 ft^3)
0	5
6	10
12	15
18	20

The construction cost in present-worth terms ($1000) is a function of the construction storage, or letting X = constructed storage (10^5 ft^3).

$$\text{cost} = 50X - 2(X)^2$$

The maintenance cost in present worth ($1000) is a function of the storage provided for a 6-year period, and define S as the storage in 10^5 ft^3 units. The annual maintenance cost discounted to present worth is

$$\text{cost} = 9.0S$$

What storage should be provided, and when should it be built to minimize cost?

SOLUTION: First a complete enumeration solution according to the tree diagram (Figure 10.13). By complete enumeration:

Plan A: 5 → 10 $ = 9(10) + 50(5) − 2(25)
          10 → 15   +9(15) + 50(5) − 2(25)
          15 → 20   +9(20) + 50(5) − 2(25) = $1005

Plan B: 5 → 10 $ = 9(10) + 50(5) − 2(25)
          10 → 20   +9(20) + 50(10) − 2(100)
          20 → 20   +9(20) = $950

Plan C: 5 → 15 $ = 9(15) + 50(10) − 2(100)
          15 → 15   +9(15)
          15 → 20   +9(20) + 50(10) − 2(25) = $950

Plan D: 5 → 15 $ = 9(15) + 50(10) − 2(100)
          15 → 20   +9(20) + 50(5) − 2(25)
          20 → 20   +9(20) = $950

10.4 OTHER OPTIMIZATION TECHNIQUES

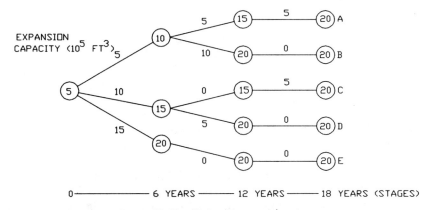

Figure 10.13 Capacity expansion tree.

Plan E: $5 \to 20$ \$ = 9(20) + 50(15) − 2(225)

$20 \to 20$ +9(20)

$20 \to 20$ +9(20) = \$840   *best*

Now structure and solve by dynamic programming as shown in the schematic of Figure 10.14, with S = storage, X = constructioned storage (10^5 ft^3).

Year 0: S_0 X_0 S_6 $g(S_6, X_0)$

         5   5   10   9(10) + 50(5) − 2(25) = 290

             10   15   9(15) + 50(10) − 2(100) = 435

             15   20   9(20) + 50(15) − 2(225) = 480

Year 6: S_6 X_6 S_{12} $g(s_{12}, X_6) + f(S_6, *)(* = $ optimal$)$

         10   5   15   335 + 290 = 625

             10   20   480 + 290 = 770

         15   5   20   380 + 435 = 815

             0   15   135 + 435 = 570

         20   0   20   180 + 480 = 660

Year 12: S_{12} X_{12} S_{18} $g(S_{18}, X_{12}) + f(S_{12}, *)$

          15   5   20   380 + 570 = 950

          20   0   20   180 + 660 = 840   *best*

As in Example Problem 10.6, the use of dynamic programming reduces the number of calculations (in this example from 15 to 10). There is a recursive

370 OPTIMIZATION

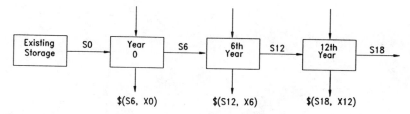

Figure 10.14 Dynamic programming schematic for capacity expansion.

equation and the evaluator can take any functional form. Some may even wish to use dynamic programming for their primary method for optimization analysis. However, unlike linear programming, a new computer program or significant modification to an existing one is frequently necessary.

Example Problem 10.8 Find the least cost for the construction of a gravity storm sewer where the cost of construction is the sum of the pipe cost and excavation cost. A computer program is available to aid you in selecting the slope of the pipes, pipe sizes, invert elevations, and excavation depths. The program is resident on the stormwater management diskette and is called **OPSEW** for "OPtimum SEWer design." Find the least cost design for the storm sewer profile shown in Figure 10.15 with the runoff coefficients for the impervious area and pervious areas equal to 0.98 and 0.30, respectively. The manhole loss coefficient is 0.5 and Manning's friction coefficient is 0.013. The invert elevation at the discharge must be at 60 ft, the pipe velocity must be between 2 and 10 ft/s with a minimum cover of 3 ft, and the design return period is 25 years for zone 7 in Florida.

SOLUTION: The peak discharge must be determined to size the pipes. The watershed areas are small and the time of concentration is most likely less than about 20 to 30 min, or the rainfall intensity is about constant from time zero up to the time of concentration. Thus the rational formula can be used

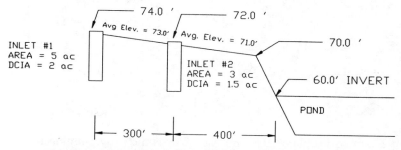

Figure 10.15 Storm sewer profile for Example Problem 10.8.

to reasonably estimate peak discharge. To use the rational formula requires estimates of rainfall intensity, watershed area, and runoff coefficient. As the runoff water proceeds to the next inlet, additional time is required and the time of concentration increases. The computer program OPSEW will do the peak flow calculations using the rational formula. The output from the computer program is given in Appendix E. Note that the optimal solution was to use a 24-in. pipe between inlet 1 and inlet 2, and a 27-in. pipe between inlet 2 and the pond. Also, if a 21-in. pipe can be used in place of the 24-in. pipe, (the 21-in. pipe is generally not commercially available) there results a savings of $3900 ($54,200 − $50,300).

10.5 COST-EFFECTIVENESS

In the evaluation of stormwater management alternatives for water quality control, the combination that produces the required concentration reduction and mass removal at least cost is considered to be cost-effective. Effectiveness is measured by either percent concentration change or mass removal per time period for specified measures of pollutants. Examples for the urban environment are considered in this chapter. The quantification of costs are in terms of construction (capital), operation replacement and maintenance (ORM), or present value (PV), given an interest rate and time period. Economic and fiscal feasibility must be attainable. Costs are site specific, depending on labor rates, construction materials costs, availability of money, land costs where appropriate, and other. In preliminary screening, average cost data are usually appropriate. Presented in this chapter are cost-effective analysis factors for some stormwater management methods. Also presented is an optimization routine as a method to determine the least cost combination of management practices to achieve a desired mass removal.

10.5.1 Cost-Effectiveness Defined

Cost-effectiveness is defined for a stormwater practice as a ratio of net favorable consequences to the cost of the practice. The net favorable consequences should be measured by a consistent set of units, say the monetary units of the government. The cost of the practice over a period of years should reflect all costs, both initial and operational. The practice must be defined as a best management practice to be eligible for cost-effective analysis.

A best management practice (BMP) was defined by federal code (Federal Regulations, 1976) as "a means of practice or combination of practices that is determined by a state (or designated area-wide planning agency) after problem assessment, examination of alternative practices, and appropriate public participation to be the most effective practicable (including technological, economic, and institutional considerations) means of preventing or reducing the amount of pollution generated by nonpoint sources to a level compatible with water quality goals." Thus a BMP meets the feasibility criteria except that an optimization analysis has not been completed to determine which alternative from among many is the most cost-effective. The methods associated with optimization, economic, and fiscal analysis are used to compare BMPs so that the "best" can be specified.

10.5.2 Broom Sweeping of Mall Areas

The efficiencies of cleaning parking areas depend on the use of the parking area and frequency or number of passes of the broom sweepers. The usual frequency of cleaning mall areas is one pass per day, during the early morning hours. At some mall areas, the available material for pickup are very low in terms of nitrogen, phosphorus, and BOD_5 (Wanielista, 1977). On a national average, the removal efficiencies should be higher because the availability of these materials is higher (Table 10.1). Also, shown in Table 10.1 are the capital and ORM cost data per impervious acre. The analysis that follows illustrates the calculation of the cost-effective ratio. Cost-effectiveness is the ratio of the average yearly percent removal to the present value. For solids at 80% removal and a present value of $3200 per acre of area, the cost-effective factor is 0.025 percent per present-value cost per acre. If pounds of solids removed per year (800 lb/acre) were used in place of percentage, the cost-effective ratio would be (800/3200) or 0.25 lb solids removed per acre per present-value cost. This ratio is compared to other practices and the most cost-effective would be the one with the highest ratio. Note that all practices must at least meet the water quality goal and that the

TABLE 10.1 Broom Sweeping Efficiencies, Low Versus Average Availability

	Efficiencies (%)[c]	
Parameters	Low Values (Mall Area)[a]	National Average Data[b]
Solids	80–85	80–90
BOD_5	0–2	20–40
N	0–5	0–12
P	0–3	0–4
Parking Area Cost Data (Once/Day Frequency)		
Capital cost (1985 dollars)	$955/acre	
ORM (1985 dollars)	$14.28/acre-month	
Present value (20 yr, $6\frac{3}{8}$%)	$3200/acre	

[a] From Shannon (1977).
[b] From American Public Works Association (1969).
[c] Site specific data may indicate higher or lower values.

cost and efficiency data are site specific. Thus generalizations cannot be made for all "parking" areas from this example.

10.5.3 Combined Sewer Flushing

Some solids will accumulate in a combined sewer until removed by a storm flow. When the point-source treatment is bypassed and storage/treatment of the bypassed fluid is not available, these deposited pollutants plus the stormwater pollutants are discharged directly into a receiving water body. It was estimated that about 6 to 10% of daily dry weather flow solids are deposited in the sewer lines. Sewer lines with adequate slope or conduits for flushing should produce no or little deposition. Heaney and Nix (1977) have shown that sewer flushing efficiency decreases rapidly when more than 20% of the sewers are flushed. Costs are highly variable, ranging from about $2.50 per foot for pipes less than 12 in. in diameter to about $7.00 per foot for pipes 3 ft in diameter (1985 dollars).

10.5.4 Cost-Effectiveness of Diversion Systems

The first flush of stormwaters from small watersheds usually carries more pollutants than do later flows. This assumes that the sewer system and watershed will flush with early runoff. However, in larger systems, runoff at longer times of concentration will add more pollutants to runoff waters from areas closer to the diversion point. Thus first-flush effects can continue throughout most of the runoff event. In combined systems, overflow controls can be set to treat the first flush of each storm and then reduce mass of discharge.

A diversion to an infiltration pond is shown in Figure 10.16. This type of diversion system was evaluated to aid in establishing design criteria and costs as a function of mass removed for treatment.

If the cost of the infiltration basin is calculated for each volume of storage and plotted against the resulting average yearly efficiency, a cost-efficiency curve results (Figure 10.17). Essentially, one can estimate the cost of treatment (construction) using Figure 10.17 if an efficiency or volume of treatment is given. It should be noted that most cost-effective curves do not increase rapidly up to about 70% efficiency. Then, for additional treatment, the unit cost (marginal cost) increases rapidly. When treating 1 in. of runoff from 4.6 acres, the data of Figure 10.17 indicate 99% plus removal efficiencies. The unit cost data used to construct this curve are considered low or related to rural or large contract construction work. In addition, the diversion structure was not included in the cost calculation.

The soil infiltration rates affect the size of a percolation pond. For a pond with excellent percolation (approximately 10 in./h), drainage will probably occur before the next storm event. However, a poor soil with a percolation

374 OPTIMIZATION

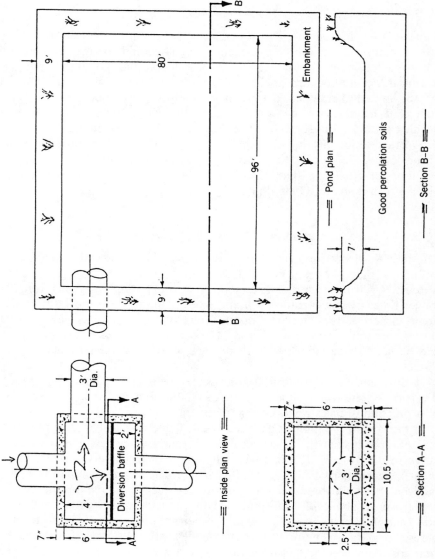

Figure 10.16 Diversion structure/percolation pond.

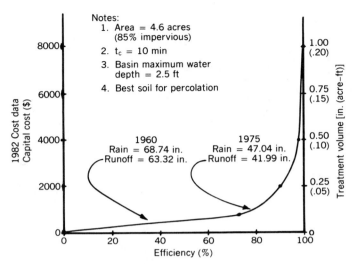

Figure 10.17 Cost/treatment/efficiency graph showing 1960 versus 1975 rainfall.

rate of 0.2 in./h takes longer to drain, and thus the size of the pond increases.

A comparison is shown in Figure 10.18. For 0.5 in. of runoff, the pond size increased so significantly that it became impracticable to build the pond, because area would not be available or the depth of the basin was below the water table. From this comparison it appears that soil infiltration is an important variable for design.

Another form of a diversion practice is the exfiltration trench. Stormwater is stored in a horizontal perforated or slotted pipe underground and allowed to recharge (exfiltrate from standing water) the surficial acquifer (Figure 10.19). Clogging or reduced exfiltration rates were found to occur in the filter wrap and a limiting exfiltration rate of 1–2 in./h through the fabric should be used for design (Wanielista et al., 1991).

Groundwater Effects. If percolation basins are used for treatment after diversion, direct discharge of pollutants to surface waters is minimized. However, stormwater pollutants may not be removed by soil filtration or other mechanisms. This can result in a deterioration of groundwater or indirect discharge of pollutants into the surface waters by means of groundwater transmission. In sandy soils with < 5% silt, discharges from an underdrain system were measured to determine soil removal potential. In addition, the groundwater in the vicinity of a percolation pond was measured to determine soil removal potential. A comparison of selected water quality measures are shown in Table 10.2. The drinking water aquifer values were from a well within 6 mi of the percolation pond. Iron was one element that,

376 OPTIMIZATION

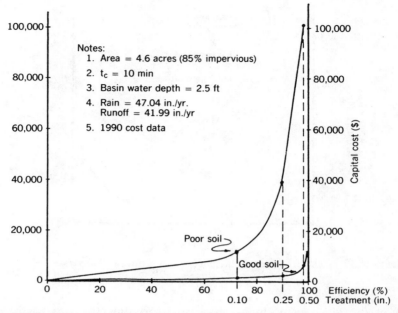

Figure 10.18 Efficiency curves for percolation ponds using two soil conditions. (From Wanielista, 1977.)

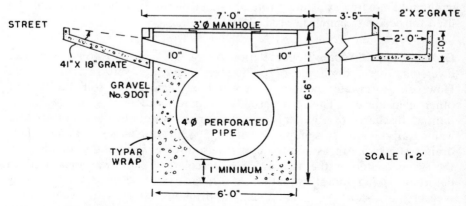

Figure 10.19 Typical section of underground exfiltration trench.

TABLE 10.2 Groundwater Quality Underlying a Percolation Pond Compared to Other Groundwater Data (mg / L Unless Indicated Otherwise)

Parameter	Percolation Pond Average[a]	Percolation Pond Maximum	Drinking Water Standard Maximum Contaminant Level	Upper Florida Aquifer[b]
Chromium	0.005	0.006	0.005	—
Copper	0.069	0.120	1.00	0.0025
Mercury	0.0005	0.0005	0.002	—
Nitrate-N	0.14	0.48	10.0	0.25
Phosphorus-O	0.102	0.204	—	0.032[c]
Zinc	0.21	0.33	5.0	< 0.2
BOD_5	6.5	20.0	—	—
Cadmium	< 0.003	< 0.004	0.01	< 0.003
Cyanide	< 0.004	< 0.004	—	—
Iron	0.7	0.8	0.3	0.048
Lindance, g/L	< 0.01	< 0.01	4.0	—
2, 4-D	< 0.10	< 0.10	100.0	—

Source: Shannon (1977).
[a] Three-sample average.
[b] One sample, 200 ft deep.
[c] Deep well (1080 ft below surface).

with high probability, could violate drinking water standards. Removals of pollutants can be enhanced by using a loamy sand material but filtration rates are decreased.

Example Problem 10.9 A watershed has an area of 12.5 acres with a soil classification of type A. The directly connected impervious area is 2.33 acres.

(a) Determine the weighted or composite curve number for the area based on a type A soil for pervious areas, which has a curve number of 50.

(b) What is the area and depth of diversion/retention ponds constructed at Atlanta and Roanoke if 80% average annual removal of runoff water is required? Assume no first flush, storage from the first 1/2 in. of rainfall, and a limiting exfiltration rate of 4 in./hour.

SOLUTION: (a)

$$\text{Watershed area} = 12.50 \text{ acres} \quad 100.00\%$$
$$\text{Impervious area} = 2.33 \text{ acres} \quad 18.64\%$$
$$\text{Pervious area} = 10.17 \text{ acres} \quad 81.36\%$$

For an impervious area, a curve number of 100 is used. For pervious areas, soil type A, curve number of 50 is used. Thus we have the following computation:

Area	%	CN	Composite CN
Impervious	18.64	100	18.64
Pervious	81.36	50	40.68
			59.32
			USE 60

(b) The runoff volume is only from the directly connected impervious area because the pervious area does not contribute runoff from the first one-half inch of rainfall. The composite curve number cannot be used to estimate runoff because of the directly connected impervious area. The pond volume is:

$$\text{Volume} = (0.5 \text{ in.})(2.33 \text{ acres})/(12 \text{ in./ft})$$
$$= 0.097 \text{ acre-feet } (4225 \text{ ft}^3)$$

The exfiltration rates for both locations are obtained from Figures 7-5 and 7-6 and are 0.08 and 0.06 ft^3/s for one equivalent impervious acre (EIA) respectively. The pond area and depth calculations for Atlanta are:

$$\text{Area} = (0.08 \text{ cfs/EIA})(2.33 \text{ EIA})(3600 \text{ s/h})(12 \text{ in./ft})/(4 \text{ in./h})$$
$$= 2013 \text{ ft}^2$$
$$\text{Depth} = \text{Volume/area (assuming vertical sides)}$$
$$= 4225/2013$$
$$= 2.1 \text{ feet}$$

Using similar calculations for Roanoke:

$$\text{Area} = 1510 \text{ ft}^2$$
$$\text{Depth} = 2.8 \text{ feet}$$

Note that the depth is for the exfiltration volume only and does not include freeboard or additional depth for other stormwater management reasons, such as flood control.

10.5.5 Alternative Comparisons

In this section seven different stormwater management practices are compared. For a site-specific location, the capital, ORM, and efficiencies in terms of reduced direct surface-water discharge pollutant mass per year were determined from field-collected data (Wanielista, 1977). The diversion/infiltration system was designed for 1 in. of diversion volume. The percolation

10.5 COST-EFFECTIVENESS

TABLE 10.3 Comparative Data per Impervious Acre, (Land Cost Not Included)

Management Practice	Impervious Area (% of Total)	Overall (%) Efficiency[a]	ORM ($/acre-month)	Average Costs ($/acre/% removal) Capital	(1980 dollars) PV (7%, 20 yr)
Diversion/percolation[b]	70	99	16.00	25.00	45.5
Percolation pond[c]	42	99+	35.00	36.30	80.8
Swales with infiltration[d]	23	92	30.00	28.40	62.9
Residential swales[d]	20	80	20.00	26.08	60.3
Sedimentation[e]	50	50	29.00	19.20	59.0
Fabric bag[f]	30	25	26.00	1.00	81.8
Advanced sweeping[g]	70	68	26.00	30.40	80.5

[a] Yearly average of BOD_5, N, P, and SS not discharged to surface waters.
[b] Designed 1-in. runoff diversion.
[c] Designed for 4-in. runoff diversion.
[d] 80% of the runoff infiltrates.
[e] Designed for 0.65 in. of runoff water.
[f] Fabric bag replacement every 2 years.
[g] Assumed 60% nitrogen in particulate form.

TABLE 10.4 Unit Cost Data for Construction of Management Practices

Activity	Unit Cost Range ($ 1985)	Per
Clearing and grubbing	150–200	acre
Selective clearing and grubbing	1400–2000	acre
Sand excavation by dozer	1.50–3.00	yd^3
Sand excavation and offsite disposal	2.00–5.00	yd^3
Drag line excavation	5.00–7.00	yd^3
Demucking and backfilling	4.00–8.00	yd^3
Limestone removal	6.00–9.00	yd^3
Swale preparation with dozer	2.00–4.00	yd^3
Seeding and mulching	0.15–0.45	yd^2
Sodding	1.50–2.00	yd^2
Irrigation (permanent sprinkler with pump)	1000–2000	acre
Tile field piping, 4 in.	1.00–4.00	ft
Tile field piping, 5 in.	2.00–5.00	ft
Fencing (44-in. woven + two-strand barbed wire)	1.00–1.50	ft
Fencing (5-in. cyclone + two-strand barbed wire)	2.00–3.00	ft
Final grading (motor)	100–200	acre
Final grading (hand)	500–800	acre
Riprap placement	100–150	ton
Asphaltic concrete surface, 1 in.	1.00–1.50	gal
Concrete curb and walk, 4 ft	5.00–7.00	yd^2
Concrete (in place) for walls and open channels	200–260	yd^3
Concrete spillway, 4 in.	6.00–10.00	yd^2
Asphalt spillway, 3 in.	5.00–8.00	yd^2
Concrete with reinforcings, in place	200–400	yd^3
Sidewalk grates, 2 ft × 3 ft	200–250	each
Sidewalk grates, 4 ft × 6 ft	300–350	each
End wall, up to 24-in. pipe	700–1200	each
End wall, 24 to 48-in. pipe	1000–1500	each
Corrugated metal pipe, 36 in.	36.00–49.00	ft
Reinforced concrete pipe, 4 in.	6.00–9.00	ft
Reinforced concrete pipe, 15 in.	15.00–20.00	ft
Reinforced concrete pipe, 24 in.	30.00–40.00	ft
Overflow pipe spillway	400–500	each

pond was designed for 4 in. of runoff over a 24-h period. No overflow is expected with this volume of discharge. The swale systems were designed to remove by infiltration approximately 80% of the runoff. A fabric bag is used to screen solids in a catch basin. Approximately 80 ft^2 of materials were required per bag. Bag replacement was assumed every 2 years. The sedimentation basin (detention) was designed to provide a volume of 0.4 gal/ft^2 of impervious acre, or 0.65 in. of runoff with 360 ft^2 of pond area per impervious acre. Generally, a design using 0.65 in. of runoff will not produce high efficiencies. The advanced sweeping was done on parking (mall) areas. Table 10.3 compares the results of this site-specific study. The table illustrates the type of data and the general range for comparative purposes. If land cost

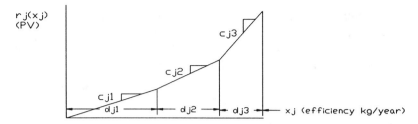

Figure 10.20 Convex cost curves.

data were added, the sweeping operation on a present-value comparison may be more competitive. The individual investigator, however, must determine site-specific data for cost-effective analyses. Unit cost data for construction costs are shown in Table 10.4. The lower costs were used to construct Table 10.3. Other investigators reported similar cost data (Walesh, 1989).

10.6 OPTIMIZATION OF STORMWATER MANAGEMENT PRACTICES

The objective of the optimization is to find the combination of stormwater management practices that removes a specified amount of pollutant mass at a minimum cost. Cost-efficiency curves for mass removal can be estimated and are frequently nonlinear, convex functions, as illustrated in Figure 10.20. Axis X_j is that of efficiency, or kilograms removed per year, and axis $r_j(x_j)$ is present-value cost or other suitable measure.

10.6.1 Linear Programming Formulation

Consider the following problem:

Minimize $\quad r(x) = \sum_{j=1}^{n} r_j(x_j)$ are convex

subject to: $\quad \sum_{j=1}^{n} a_{ij} x_j \geq b_i \quad i = 1, \ldots, m \quad\quad (10.52)$

$$x_j \geq 0 \quad j = 1, \ldots, n$$

First reduce those $r_j(x_j)$ that are nonlinear to their piecewise linear approximations by fitting linear segments to the original curve as indicated in Figure 10.20. This approximation can be carried out to any degree of accuracy by dividing the curve into a suitable number of intervals for linear representation. Now assume that each $r_j(x_j)$ is divided into p_j intervals marked by $x_j = 0, d_{j1}, \ldots, d_{jp}$, and that the slopes of the linear segments occupying these intervals are correspondingly $c_{j1}, \ldots, c_{jp}$. Since $r_j(x_j)$ is convex by assumption, it follows $c_{j1} < c_{j2} < c_{jp}$, which makes it possible to substitute:

$$x_j = \sum_{t=1}^{n} x_{jt} \quad \text{where } 0 \le x_{jt} \le d_{jt} - d_{jt-1} \quad (10.53)$$

into equation 10.52 and obtain the following equivalent problem:

Minimize
$$r(x) = \sum_{j=1}^{n} \sum_{t=1}^{p} c_{jt} x_{jt}$$

subject to:
$$\sum_{j=1}^{n} a_{ij} \left(\sum_{t=1}^{p} x_{jt} \right) \ge b_i \quad (10.54)$$

$$x_{jt} \ge 0 \quad i = 1, \ldots, m; \quad j = 1, \ldots, n; \quad t = 1, \ldots, p_j$$

This now gives a bounded-variable problem that can be solved by the simplex method (linear programming).

10.6.2 Two Subwatershed Linear Programming Form

A sketch of two subwatersheds is shown in Figure 10.21. Street patterns are included, representing street sweeping operations, while the catchbasin sites represent catchbasin technology. Diversion, percolation, and storage treatment may also be included in the subwatersheds. The nodal diagram representative of the two subwatersheds is illustrated in Figure 10.22. The formulation of the linear programming problem for the two subwatersheds is as follows:

Variables

M_i = lb of pollutant i/year (for all pollutant i)

R_i = lb of pollutant i removed/year (for all pollutant i)

F_i = mass of pollutant i/year remaining ($F_i = M_i - R_i$)

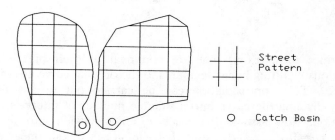

Figure 10.21 Sketch of dual subwatersheds.

10.6 OPTIMIZATION OF STORMWATER MANAGEMENT PRACTICES 383

For pollutant "i"

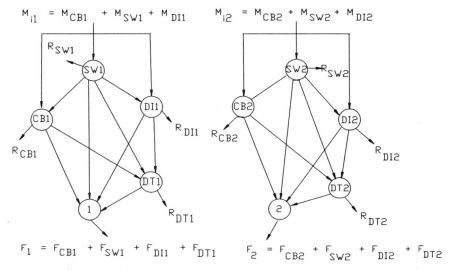

Figure 10.22 Nodal diagram of dual subwatersheds.

Constraints

1. Continuity of mass must be maintained. Thus one can write nodal equations for each decision point of Figure 10.22 with the following convention using node CB1 in subwatershed one as a partial example.

(Left-hand side of equation)

$$\xrightarrow[\text{inputs }(+)]{F_{\text{SWCB1}}} \boxed{\text{CB}} \xrightarrow[\text{outputs }(-)]{R_{\text{CB1}}}$$

The subscripts are:

CB1 = catchbasin, watershed 1;
DT1 = detention pond, watershed 1
SW1 = street sweeping, watershed 1;
DT2 = detention pond, watershed 2
DI1 = diversion/retention, watershed 1;
SW2 = street sweeping, watershed 2
CB2 = catchbasin, watershed 2;
DI2 = diversion/retention, watershed 2

Node	Equation
CB1	$M_{CB1} + F_{SWCB1} - R_{CB1} - F_{CB1} - F_{CBST1} = 0$
SW1	$M_{SW1} - F_{SWCB1} - F_{SWDI1} - F_{SW1} - F_{SWDT1} - R_{SW1} = 0$
DI1	$M_{DI1} + F_{SWDI1} - F_{DI1} - F_{DIDT1} - R_{DI1} = 0$
DT1	$F_{SWDT1} + F_{CBDT1} + F_{DIDT1} - F_{DT1} - R_{DT1} = 0$
CB2	$M_{CB2} + F_{SWCB2} - R_{CB2} - F_{CB2} - F_{CBST2} = 0$
SW2	$M_{SW2} - F_{SWCB2} - F_{SWDI2} - F_{SW2} - F_{SWDT2} - R_{SW2} = 0$
DI2	$M_{DI2} + F_{SWDI2} - F_{DI2} - F_{DIDT2} - R_{DI2} = 0$
DT2	$F_{SWDT2} + F_{CBDT2} + F_{DIDT2} - F_{DT2} - R_{DT2} = 0$

For pollutant "i"

$$M_{i1} = M_{CB1} + M_{SW1} + M_{DI1} \tag{10.55}$$

$$M_{i2} = M_{CB2} + M_{SW2} + M_{DI2} \tag{10.56}$$

$$F_1 = F_{CB1} + F_{SW1} + F_{DI1} + F_{DT1} \tag{10.57}$$

$$F_2 = F_{CB2} + F_{SW2} + F_{DI2} + F_{DT2} \tag{10.58}$$

2. Removals by each process cannot exceed the total available for each watershed:

Watershed 1: $\quad R_{CB1} + R_{SW1} + R_{DI1} + R_{DT1} \leq M_1 \quad (10.59)$

Watershed 2: $\quad R_{CB2} + R_{SW2} + R_{DI2} + R_{DT2} \leq M_2 \quad (10.60)$

3. Minimum removal is required before discharge to the lake:

$$R_{CB1} + R_{CB2} + R_{SW1} + R_{SW2} + R_{DI1} + R_{DI2} + R_{DT1} + R_{DT2} \geq R^{\min} \tag{10.61}$$

4. Maximum removal attainable by each process (in general terms):

$$R_{ij} \leq R_{ij}^{\max} \qquad \forall i, j \tag{10.62}$$

Generalized Form (Two Subwatersheds — All Treatment)

$$\sum_{j=1}^{n} R_{ij1} + \sum_{j=1}^{n} R_{ij2} \geq R_i^{\min} \qquad \forall i \quad \text{where } j = \text{management practice} \tag{10.63}$$

Generalized Form (Multisubwatersheds — All Treatment).

For k subwatersheds the generalized form of the linear programming problem is

$$\sum_{j=1}^{n} \sum_{k=1}^{1} R_{ijk} \geq R_i^{\min} \qquad \forall i \tag{10.64}$$

where i = pollutant type
  j = management practice
  k = subwatershed
  l = total subwatersheds
  n = total possible management practices in a subwatershed

and

$$R_{ij} \leq R_{ij}^{\max} \quad \forall i, j \tag{10.65}$$

The generalized form states that the quantity of pollutants removed from all subwatersheds (R_{ijk}) cannot exceed the difference between available runoff mass and allowable mass discharged into the receiving water body ($M_{ik} - F_{ik}$).

10.7 SUMMARY

The choice of the best solution depends on how well the problem is defined. The use of operations research concepts and ideas helps to formulate, construct, derive, test, and implement a solution to a stormwater management problem.

- Optimization means to make as perfect as possible.
- Linear programming is an optimization procedure that uses a linear objective function and constraints. It is useful for resource allocation and hydrograph shape determination problems.
- Multiobjectives can also be assessed by assigning a weight to each of objective functions.
- Nonlinear functions can be solved for optimal values using classical calculus techniques. Also, piecewise linear approximations can be made for convex or concave functions and then minimized or maximized, respectively, by linear programming procedures.
- Computer programs exist to solve both linear and nonlinear programming problems. Computer programs for the general linear programming problem and one for storm sewer design are provided with this book.
- Dynamic programming is useful for finding optimal solutions for problems with recursive relationships.
- Cost-effectiveness can be quantified as a ratio of net favorable consequences to the cost of each practice. The practice must meet water quality goals and other feasibility measures, and can be called BMPs.
- The most cost-effective system is one that achieves a desirable water quality condition at minimum cost. Off-line diversion for infiltration is generally most cost-effective. However, infiltration capacity must be available.

- Exfiltration pipes are off-line systems placed underground. They are used when land cost is a major factor for site selection and infiltration capacity is available.
- Wet-detention ponds are most cost-effective when infiltration capacity is not available.
- Computer programs can be used to aid in determining the least-cost combination of treatment to meet a mass discharge limitation.

10.8 PROBLEMS

1. For the detention pond of Problem 2 of Section 9.6, formulate in written terms suitable objectives and constraints.

2. Write in mathematical terms an optimization model for Problem 1. Define all variables.

3. An industry requires at least 4.0 MGD more water than it is currently using. The metal concentration cannot exceed 10 μg/L. A detention pond can provide up to 2 MGD with 20 μg/L of metals, while a potable supply can provide up to 10 MGD at a metal concentration of 5 μg/L. The cost of potable water is \$1000/MGD, and from the detention pond it is \$500/MGD. The goal is to minimize cost. You can blend the waters. Find the best solution using graphical linear programming.

4. An off-line retention facility for stormwater management is being built for a 306-acre area. The rainfall excess from the area is expected at 9 in./yr. The stored water can either recharge the groundwater or be used to water a golf course. There is a revenue gain of \$60 per 10,000 ft^3 if water is used on the golf course, and no revenue from the use of water for recharge. However, there is some recharge gain from watering the golf course, or 0.01 in. per 10,000 ft^3 of golf course watering. The recharge facility acts to recharge 0.05 in. per 10,000 ft^3 of recharge. The water management district (good guys) wish to maximize recharge; the developer wishes to maximize revenue. The district specifies that at least 1,000,000 ft^3 must be used from the recharge pond but no more than 5,000,000 ft^3 from the recharge pond. Also, from the recharge pond and the gold course watering at least 8 in. of recharge must be obtained. In an open meeting the developer of the pond said: "I believe that \$10,000 is as important as 1 inch of recharge."

   a. At most, how many cubic feet of water is available for off-line storage based on a 9 in./yr rainfall excess?

   b. What do you suggest that will maximize both recharge and revenue?

   c. If after hearing the answer to part (b), the developer says, "I meant to say that 1 inch of recharge is equivalent to only \$1000." What is the solution? (*Note:* Less importance is being given to recharge.)

5. Stormwater is collected in a large detention structure for reuse after a suitable settling time and chemical coagulation. For the rainy season, the total quantity of water detained is 8000 acre-ft. There are two potential users of the stormwater, a sod-farming operation and a saltwater intrusion control barrier. The water management district wishes to maximize recoverable fresh water and is pushing for the injection of the detained stormwaters. The owner of the detention pond has the sod farm and expects to make money reusing the detained water for the sod farming. The sod farming operation can use up to 6000 acre-ft of water and the intrusion barrier can use up to 4000 acre-ft. Political pressure states that for every 4000 acre-ft allocated to sod farming, no less than 1000 acre-ft must be allocated to injection (the county needs the potable water supply).

   a. What allocation do you make if the revenue function is

   $$\text{Maximize} \quad f_1(\bar{x}) = 5X_1 - 2X_2 \text{ dollars}$$

   where X_1 is the sod farm volume and X_2 is the saltwater control volume.

   b. What allocation do you make if the revenue and potable water objective are:

   $$\text{Maximize } f_1(\bar{x}) = 5X_1 - 2X_2 \text{ dollars}$$
   $$f_2(\bar{x}) = -X_1 + 4X_2 \text{ acre-ft recovered}$$

   and potable water is 1.5 times more important than revenue.

6. Specify the radius and height of a stormwater filtration cylindrical tank given a design flow rate from a detention pond as 0.4 MGD. The flow rate is kept relatively constant. The filtration tank volume (V) is a function of the flow rate (Q) and tank detention time (t) and is given by

   $$t = \frac{V}{Q}$$

   The cost to construct the walls and flooring of the tank is $3.50/ft^2 of wall surface and $6.50/ft^2 of floor surface, respectively. Formulate a model to minimize the cost of the cylindrical tank if the radius of the tank is r in feet and the height of the tank is h in feet. The detention time must be at least 20 min.

7. If the cost of water is $5/acre-ft in the linear programming formulation of equations 10.2 through 10.7, find the optimal allocation by graphical techniques given the benefits for quality and discharge con-

trol are $6/acre-ft and $3/acre-ft, respectively. Also, what is the allocation if the benefits changed to $4/acre-ft and $8/acre-ft, respectively? Which constraints are redundant?

8. Assume that a municipality wishes to store excess runoff waters for distribution to three other holding areas, such as lakes within the corporate boundaries. The lakes are along a proposed main flood control canal. The first lake is 10 mi from the excess storage, lake 2 is 25 mi, and lake 3 is 35 mi. The municipality wishes to know how large the flood control canal should be to maximize the difference between the benefits of flood control and the cost of the system. You may transport up to 1800 acre-ft. The cost of the canal is a function of flow volume transfer and is shown below.

Volume (Acre-ft)	Cost (10^3 $/mi)
600	26.0
1200	38.5
1800	50.0

There are additional site-specific benefits associated with additional water storage at the three lakes. These benefits plus the flood control benefits results in the following benefits.

Volume	Benefits (10^3 $)		
600	400	440	840
1200	700	580	1230
1800	1000	720	1510

9. Four (3-month period) operating curves must be established for a reservoir; thus the average storage volume and releases for agricultural use, municipal water supply, and excess releases must be determined for the four seasons of a year. The benefits from storage and the two releases are:

Activity	Benefits ($100/acre-ft) for Each Season			
	Winter	Spring	Summer	Fall
Water supply	1.2	1.5	2.0	1.5
Agriculture	0.2	2.0	3.0	1.0
Storage	−0.4	1.0	2.5	1.0
Excess releases	−0.5	−1.0	−1.5	−3.0

Write out a specific mathematical description of the objective function (maximize benefits) and the constraint set for a 1-year operation. You have available inflow per season and the net seepage and evaporation per season. The beginning reservoir volume is 100,000 acre-ft. The maximum storage is 150,000 acre-ft. The minimum storage volume is 50,000 acre-ft. Identify your decision variables first.

10. The target level for water in a retention pond is 5 acre-ft. The target level for water removed from the pond for irrigation is 2 acre-ft during the first season and 8 acre-ft during the next season. There are only two seasons per year. If the allocations are below or above the target acre-ft, there is an economic loss expressed as

$$\text{loss} = (5 - S_t)^2 + (2 - R_1)^2 + (8 - R_2)^2$$

where S_t = ending period storage (acre-ft)
R_1 = releases for irrigation (acre-ft in period 1)
R_2 = releases for irrigation (acre-ft in period 2)

The maximum size of retention pond is 7 acre-ft with a minimum size of 4 acre-ft. If the retention pond starting volume is 5 acre-ft and the average seasonal inputs are 4 acre-ft the first season and 6 acre-ft the second season, how should the pond be operated to minimize loss, assuming that you can operate in incremental values of 1 acre-ft? Only do two seasonal iterations and start the next year with 5 acre-ft. Do this problem using dynamic programming.

11. You must determine the optimal release policy over four seasons per year from a reservoir. It is decided to judge the policy by minimizing the sum of the square of the differences between actual and a target release level or

$$\text{Minimize } f(S, R) = (50 - S_t)^2 + (60 - R_t)^2 \quad t = 1, 2, 3$$

where S_t = storage level, season t, 10^5 acre-ft
R_t = release level, season t, 10^5 acre-ft

and for $t = 4$;

$$\text{Minimize } f(S, R) = (70 - S_t)^2 + (30 - R_t)^2 \quad t = 4$$

The maximum capacity for the reservoir is $100(10^5)$ acre-ft, and the minimum capacity is $30(10^5)$ acre-ft. The history of net input to the reservoir is to be used and the average seasonal values are

$$I_t = (60, 90, 20, 40)$$

The reservoir can be operated per season in units of $10(10^5)$ acre-ft. The starting reservoir volume is $50(10^5)$ acre-ft. If the average seasonal values are to repeat every year, what is the optimal stationary operating policy for the given input?

12. The evaluation formula for storage and reuse of rainfall excess in a wet-detention pond is given by a function of deviation from a storage and reuse policy, or

$$V(S_t, r_t) = (20 - S_t)^2 + (25 - r_t)^2$$

where S_t = storage in acre-feet per time period t
r_t = reuse in acre-feet per time period t

The maximum capacity of the wet-detention pond is 30 acre-ft, the target storage is 20 acre-ft, and the reuse target is 25 acre-ft. The runoff volume into the pond per time period is 10, 50, and 20 acre-ft for three time periods. What is the optimal storage and reuse policy to minimize deviations from the targets?

13. Rainfall excess is available from a 100-acre multifamily development with 48 acres of directly connected impervious surfaces. The pervious area has a curve number of 80 with an initial abstraction equal to 20% of the rainfall. You are required to size a reuse water volume in a wet-detention pond as the runoff from 4 in. of rainfall. The reuse water volume is used for a golf course and a sod farm with the value of water obtained from another source being \$250/acre-ft and \$100/acre-ft, respectively. For every acre-foot of water used on the golf course, at least 2 acre-ft must be used on the sod farm. What is the best "maximum value" solution? The cost of supply water to either use is the same. At least 10 acre-ft of water must be given to the sod farm.

14. What are the capital, ORM, PV, and cost data for a broom sweeping operation in a city with a population density of 30 people/acre? List all your assumptions.

15. For a residential land use of 100 contributing acres, what are the yearly removal mass, removal efficiency, and present-value costs for broom sweeping, diversion/percolation, percolation at 4 in. of runoff, swales, advanced sweeping, and sedimentation? Assume A-type soils, 40% impervious area, a land cost of \$5000/acre, diversion volume of 0.5 in., and once-per-day frequency of sweeping. List all assumptions.

16. Using the constraint set that was developed for Example Problem 10.2, can a feasible solution (constraints all hold) be obtained with a percent treatment level of 0.90 for each site? Show by numerical calculations with each constraint and then by running the computer program.

17. For a total watershed (24.9 acre), develop the capital and present-value cost curves for a diversion/percolation basin as a function of pounds suspended solids removed. Assume that the watershed is 100% impervious and the pond is in type D soils. Also, develop the marginal cost curves.

18. Formulate a mathematical model (objective function and constraints) to minimize the cost of stormwater management for a lake that has two discharges (watersheds), considering two types of stormwater management, 80% removal of yearly mass, linear cost functions, and average removal effectiveness for the stormwater management practices. Be specific, define all variables, and write in mathematical terms.

10.9 COMPUTER-ASSISTED PROBLEMS

1. Using your computer programs, do Problems 3 and 5(a) of Section 10.8.

2. A detention pond can store water for on-site irrigation and release water for low-flow augmentation and another release for downstream irrigation. A total of no more than 16 acre-ft can be released for downstream users, and on-site irrigation can use only 10 acre-ft. Legal requirements state that for any 3 acre-ft released from the reservoir, at least 1 acre-foot must be used for low-flow augmentation. The benefits from on-site irrigation is $200/acre-ft, and for downstream irrigation it is $150/acre-ft. The low-flow augmentation benefit is $50/acre-ft. If 20 acre-ft are available for the three beneficial uses, what allocation do you recommend for maximizing benefits?

3. It is desirable to estimate the size of a detention pond for one watershed out of two which feed an agricultural area. The two inputs (net surface waters) are shown in Figure 10.23. Net surface water is the sum of stream flow minus evaporation and seepage per period of time. There are two seasons (or the time periods we feel confident in estimating average flow volumes). At the start of the wet season, the detention pond is dry. All units are 10^3 acre-ft. The benefits to the agricultural area and the present worth of the detention pond are given by a piecewise linear equation of net benefits and costs ($1000 units) (Figure 10.24). Consider the following mathematical expression to be maximized.

Net benefits = gross benefits − cost of reservoir

$$Z = 36X1 + 28X2 + 10X3 - 6Y1 - 7.5Y2$$

392 OPTIMIZATION

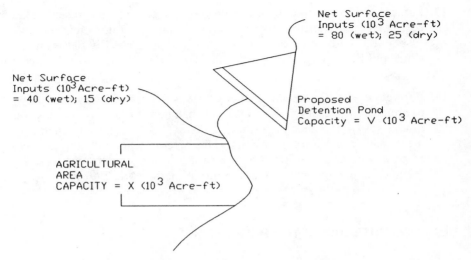

Figure 10.23 Schematic for Problem 3.

where

$$X1 + X2 + X3 = X$$
$$V1 + V2 = V$$
$$X1 < 20$$
$$X2 < 45$$
$$X3 < 15$$
$$V1 < 30$$
$$V2 < 120$$

What is the agricultural allocation and the detention pond size if the total yearly amount of water (X) given to the agricultural area requires 80% of the total yearly water in the dry season and 20% in the wet

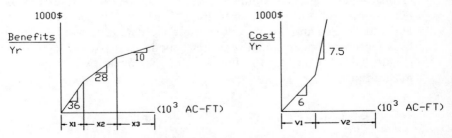

Figure 10.24 Cost curves for Problem 3.

season? *Note:* You may wish to keep the problem resident on computer and save it for the next problem.

4. Solve Problem 3 assuming that irrigation needs can be increased to $165(10^3)$ acre-ft of water or $X3 \leq 100(10^3)$ acre-ft. Actually, this was indicated from the range of constraining values of Problem 3 along with the opportunity cost. Also, is $165(10^3)$ acre-ft more than can be used from the system?

5. For each month of a year, determine the storage level and releases from a reservoir if the objective were to maximize benefits. The benefits associated with storage and releases are given below with the net inflow.

Month	Net Benefit Coefficients ($1000/1000 acre-ft)		Net Inflow (1000 acre-ft)
	Storage	Release	
Jan.	0	2	10
Feb.	0	3	40
Mar.	0	−1	80
Apr.	0	−4	110
May	0	2	60
June	5.3	7	40
July	4.2	8	20
Aug.	3.0	6	20
Sept.	0	5	80
Oct.	0	1	60
Nov.	2.5	4	40
Dec.	0	2	20

The maximum and minimum reservoir size is $200(10^3)$ and $100(10^3)$ acre-ft with an initial size of $100(10^3)$ acre-ft.

6. For Problem 5, if there were no benefits associated with storage, what is the optimal release policy to maximize benefits associated with the releases?

7. Given a runoff hydrograph measured at three 1-h intervals as 11, 30, and 21 ft^3/s from an 80-acre area, find the unit hydrograph ordinates if the rainfall excess for the first and second hour were 0.5 and 0.3 in., respectively. Use a linear programming algorithm.

8. Using the SM program disk, initiate the pipe system optimization program (Problem 5). Review the main menu and check the cost data file. When in the 24-in.-diameter cost data file, change the cost per linear foot ($/LF) of pipe and excavation at the 8- to 10-ft depth from $74/LF to $76/LF.

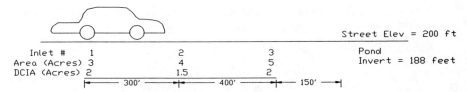

Figure 10.25 Schematic of storm sewer.

9. The sizing of storm sewer pipe for a small subdivision is required. The finish-grade elevation of the street is constant at 200 ft. The schematic of the system is shown in Figure 10.25. What is the minimum-cost design keeping a minimum velocity of 2 ft/s and a maximum of 10 ft/s. Manning friction factor = 0.013, head loss in manholes = 0.50. The C factors for impervious and pervious area are 0.9 and 0.2, respectively. You must maintain a 3-ft cover on the pipes.

10. Using your OPSEW computer program, solve for the minimum-cost sewer line construction considering the following data:

    Ground elevation is constant at 60.0 ft.

    There are three inlets, all in a line leading to a pond with invert elevation equal to 50.0 ft.

    The three watersheds servicing the inlets have the watershed characteristics shown in Figure 10.26. The impervious C factor is 0.98 and the pervious C factor is 0.4.

    The Manning friction coefficient is 0.013 and the manhole loss coefficient is 0.5.

    Use the 25-year, zone 7 rainfall curve.

11. If the objective function for Problem 18 of Section 10.8 with two linear segments each were $50RD1,1 + 80RD1,2 + 60RD2,1 + 70RD2,2 +$

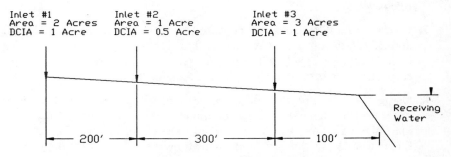

Figure 10.26 Schematic for Problem 10.

65RS1 + 64RS2, and the constraints were as follows, find the optimal solution.

(1) $\quad$ RD1,1 + RD1,2 + RD2,1 + RD2,2 + RS1 + RS2 $\geq$ 240

(2) $\quad\quad\quad\quad\quad\quad\quad\quad\quad\quad$ RD1,1 + RD1,2 + RS1 $\leq$ 100

(3) $\quad\quad\quad\quad\quad\quad\quad\quad\quad\quad\quad\quad\quad\quad\quad\quad$ RD1,1 $\leq$ 80

$\quad\quad\quad\quad\quad\quad\quad\quad\quad\quad\quad\quad\quad\quad\quad\quad\quad\quad$ RD1,2 $\leq$ 15

$\quad\quad\quad\quad\quad\quad\quad\quad\quad\quad\quad\quad\quad\quad\quad\quad\quad\quad\quad$ RS1 $\leq$ 30

(4) $\quad\quad\quad\quad\quad\quad\quad\quad\quad\quad\quad$ RD2,1 + RD2,2 + RS2 $\leq$ 200

(5) $\quad\quad\quad\quad\quad\quad\quad\quad\quad\quad\quad\quad\quad\quad\quad\quad\quad$ RD2,2 $\leq$ 160

$\quad\quad\quad\quad\quad\quad\quad\quad\quad\quad\quad\quad\quad\quad\quad\quad\quad\quad$ RD2,2 $\leq$ 30

$\quad\quad\quad\quad\quad\quad\quad\quad\quad\quad\quad\quad\quad\quad\quad\quad\quad\quad\quad$ RS2 $\leq$ 50

12. Increase the right-hand side of constraint (1) in Problem 11 from 240 to 260 and comment on the change without running the program again.

13. For Example Problem 10.2, why is the solution infeasible if the maximum treatment fraction is 0.9? Execute a computer program to solve the original problem, then discuss what options are available to meet the stream standard.

14. For Problem 13, change the discharge at site 3 to 800 lb, which reflects additional on-site treatment. Also, the treatment fraction cannot be exceeded.

15. For Problem 14, constrain each location to a minimum treatment level of 80%. Compare the best solution to that obtained by Problem 14.

16. Change the discharge load at site 3 to 2000 lb and find the minimum-cost solution with a maximum treatment fraction of 0.9.

17. Reduce the load at site 3 to 800 lb and increase the river flow (low flow augmentation) by 50 ft^3/s at each site. The velocity for each section changes to 3.0, 2.6, 2.2, and 2.0 ft/s.

10.10 REFERENCES

American Public Works Association. 1969. *The Causes and Remedies of Water Pollution From Surface Drainage of Urban Areas*, Report No. WA66-23, Chicago.

Beard, L. R. 1966. *Optimization Techniques for Hydrologic Engineering*, HEC Paper 2, Hydrologic Engineering Center, Davis, Calif.

Bellman, R. 1957. *Dynamic Programming*, Princeton University Press, Princeton, N.J.

Davis, D. W. 1975. "Optimal Sizing of Urban Flood-Control Systems," *Journal of the Hydraulics Division, ASCE*, Vol. 101, No. HY8, pp. 1077–1092.

Federal Regulations. 1976. Title 40, 130.2, *Protection of the Environment*, U.S. Government Printing Office, Washington, D.C.

Fetter, D. 1988. "Minimizing the Construction Cost of Sewer Systems," Master's thesis, University of Central Florida, Orlando, Fla.

Heaney, J. P., and Nix, S. J. 1977. *Stormwater Management Model: Level I—Comparative Evaluation of Storage-Treatment and Other Management Practices*, EPA 600/2-77-083, U.S. Environmental Protection Agency, Washington, D.C.

Hillier, F. S., and Lieberman, G. J. 1967. *Introduction to Operations Research*, Holden-Day, San Francisco.

The International Conference on Urban Storm Drainage, Pentech Press, Devon, England, p. 63.

Miles, S. Wayne, and Heaney, James P. 1988. "Better Than Optimal Methods for Designing Drainage Systems," *Journal of Water Resources Planning and Management*, Vol. 114, No. 5, pp. 477–499.

Shannon, E. E. 1977. *A Preliminary Assessment of the Drainage Well Situation in the Orlando Area*, BC&E/CH2MHill, Gainesville, Fla.

Wagner, H. M. 1969. *Principles of Operations Research*, Prentice Hall, Englewood Cliffs, N.J.

Walesh, S. G. 1989. *Urban Surface Water Management* Wiley, New York, pp. 476–477.

Walters, G. A. 1978. "Designing New Sewer Networks to a Minimum Cost by the Use of Dynamic Programming," *Proceedings of the International Conference on Urban Storm Drainage*, Pentech Press, Devon, England, pp. 636–647.

Wanielista, M. P. 1977. "Off-Line Retention Pond Design," in *Proceedings of the Stormwater Retention/Detention Basins Seminar*, Y. A. Yousef, ed., University of Central Florida, Orlando, Fla.

Wanielista, M. P. 1990. *Hydrology and Water Quality Control*, Wiley, New York.

Wanielista, M. P., and Shannon, E. E. 1977. *Stormwater Management Practices Evaluation*, East Central Florida Regional Planning Council, Orlando Metropolitan 208 Study.

Wanielista, M. P., Yousef, Y. A., and Bell, J. 1978. *Shallow-Water Roadside Ditches for Stormwater Purification*, Report EESEI-78-11, University of Central Florida, Orlando, Fla.

Wanielista, M. P., Gauthier, M. J., and Evans, D. L. 1991. *Design and Performance of Exfiltration Systems*, Florida Department of Transportation, Tallahassee, Fla.

Water Resources Bulletin. 1992. (19 articles), American Water Resources Association. Jan./Feb. Vol. 28, No. 1, 232 pages.

Yingling, J. 1987. "Economic Impact Statement," *Southwest Florida Water Management District Basis of Review*, Brooksville, Fla.

CHAPTER 11

Rural Area Stormwater Management

Agricultural activities account for the largest percentage of nonpoint-source pollution in the United States (General Accounting Office, 1990). Soil erosion and runoff of pesticides are the major problems and focus on agricultural controls is becoming more important. Soil losses from rural environments have been considered to be of major agricultural significance. Regardless of pollution control needs, some abatement of rural nonpoint-source effects has been common because of the need to minimize soil losses, and when soil loss is controlled, some chemicals also are controlled. The universal soil loss equation (USLE) (Wischmeir and Smith, 1972) and water yield model (Williams and LaSeru, 1976) are presented in this chapter with applications to illustrate stormwater management for the rural environment.

11.1 EROSION AND SEDIMENTATION

Erosion is the release of particles from the land by wind, water, ice, or other natural cause, while sedimentation is the transport and deposition of the eroded particles. Some cropland conditions can result in a high-sediment-yield potential because of long, sloping land farmed without terraces or runoff diversion, no cover between harvest and new crop growth, intense farming close to surface waters, formation of gullies, and unstabilized roadways. Rural areas can have high potential erosion and sediment problems and are generally associated with unstabilized roadway surfaces, croplands, silviculture, unstabilized streambanks or ditches, surface mining, and unplanted areas.

Erosion and sediment controls for the rural environments are typically on-site methods, but can include off-site ponding with or without irrigation systems to return ponded water to agriculture and silviculture activities. Procedures for pond sizing that were used in the urban environment are again used in the rural environment. Some typical erosion and runoff control procedures are as follows:

1. *Ponding with or without Irrigation:* helps control surface runoff and can also provide a source of irrigation water
2. *No-Till Plant in Prior-Crop Residues:* helps reduce erosion by placing seeds in the soil without tillage and maintains previous plant residues
3. *Sod-Based Rotation:* sod most frequently planted in 2- to 4-year rotations
4. *Winter-Crop Cover:* shredded plants left on the fields until the next planting
5. *Tail Water Recovery:* large reuse ponds collecting runoff and seepage (also called runoff farming and irrigation ponds)
6. *Contour Plowing:* plowing and crop rows follow field contours
7. *Strip-Cropping:* alternates row crops with close-grown crops along contours
8. *Terraces:* reduces the slope of the land to minimize erosion and runoff
9. *Pesticide–Herbicide Substitutes:* using mechanical and biological substitutes
10. *Reservoir Tillage:* small pits with small dikes between crop rows to reduce runoff and promote infiltration (Garvin et al., 1986; Kincaid et al., 1990)
11. *Computer-Base Irrigation Scheduling:* using sensing and forecast methods (Camp et al., 1990)

The U.S. Environmental Protection Agency (1973) summarizes many of the predictive methods for erosion and sediment control. Of particular note is the agricultural chemical transport model (Free et al., 1975), the agricultural runoff management model (Donigian and Crawford, 1976), and the nonpoint simulation model (Donigian and Crawford, 1976). An economic assessment of runoff farming using irrigation and ponds is available (Frasier and Scrimgeour, 1990).

Contours, terraces, and minimum tillage were evaluated using the agricultural runoff management (ARM) model. The results of a 10-year simulation showed that pollutant loadings decreased significantly when contouring or terracing were used. The ARM model is primarily an overland flow version of the hydrocomp simulation program (HSP) (Crawford, 1970; Novotny et al., 1979). The ARM model simulates sediment, pesticide, and nutrient loadings. It uses the adsorption model and first-order degradation models of Chapter 6.

Although it appears that there are many possible sediment prediction methods that might be adapted in nonpoint studies, many of the methods have limited applicability. Many simulation methods are proprietary and others are very complex and not readily amenable to quick studies and interpretation. Other statistical and simulation studies are limited in scope, and although very good for certain specific cases, are not readily adaptable to a variety of sediment loss situations (Flaxman, 1972; Onstad, 1973).

Many pollution analysis programs such as STORM (U.S. Army Corp of Engineers, 1975) use the universal soil equation empirical method (Wischmeir and Smith, 1972). Other reports and studies also indicate its general acceptability. The model is easy to use and interpret from simple hand calculations to more sophisticated computer programs as well as its readily understandable parameters. In the next section we deal with this empirical model.

11.2 UNIVERSAL SOIL LOSS EQUATION

The universal soil loss equation (USLE), an empirical method, was developed in the latter 1950s at the Runoff and Soil Loss Data Center of the Agricultural Research Service (ARS) at Purdue University. The equation was a modification of earlier empirical equations (Musgrave, 1947; Zingg, 1949; Smith, 1941) which were found to be too localized for general use. Because of the more general applicability of the equation developed at Purdue, it was called "universal." Although originally developed for soil conservation work in cropland areas, the equation has since been adapted and interpreted for other erosion loss problems. The original definitive publication on this equation was printed in 1965 by Wischmeier and Smith.

The USLE is the product of six factors:

$$A = (R)(K)(L)(S)(C)(P) \tag{11.1}$$

where A = calculated soil loss, tons/acre/time period
R = rainfall factor/time period
K = soil-erodibility factor
L = slope length
S = slope gradient factor
C = cropping management factor
P = erosion-control practice factor

Equation 11.1 can be modified to include a sediment delivery factor that adjusts the estimated sediment loading based on deposition within the area. The sediment deposition is frequently estimated from direct field measurements.

11.2.1 Rainfall Factor (*R*)

Early research on soil loss indicated that when all factors other than rainfall were held constant, the soil loss was directly proportional to the product of the kinetic energy of the storm times its maximum 30-min intensity. This factor has been called the erosion index (EI) factor. Data have been accumulated throughout the United States for suggested EI values to use, and several maps and charts give suggested data (Stewart et al., 1975). For an annual calculation, the rainfall factor (R) equals the number of erosion index units in a normal year's rain. Obviously, there will be variations from year to year as well as variations within the pattern of rainfall during a given year. Adjustment for these variations can be made when selecting or modifying the rainfall factor within the equation. Examples of such modifications will be given.

The isoerodent map as originally presented by Wischmeier and Smith is illustrated in Figure 11.1 (Stewart et al., 1975). Depending on the variation of regional rainfall, the R value used will accumulate in a different fashion in different parts of a region. The suggested erosion-index distribution curves for six areas in Florida have been presented by Griffin (1975). Figure 11.2 shows the suggested erosion-index distribution curve for the north central region in Florida. The curves plot the percentage rainfall factor accumulation as a function of time. For example, in the north central region 30% of the R factor has accumulated by June 1 in an average year, while the remaining 70% will accumulate over the remainder of the year.

11.2.2 Soil Erodibility Factor (K)

The soil erodibility factor (K) is an experimentally determined quantitative factor. Many variables influence the erodibility of a soil, including its particle-size distribution, organic content, structure, profile, and so on. Measured values on soils studied at erosion research stations have indicated a K range of 0.03 to 0.69. Additional studies on various soil types have resulted in tables of suggested K values (U.S. Environmental Protection Agency, 1973). An indication of the general magnitude of the soil-erodibility factor, K, for different soil textures and organic matter content is shown in Table 11.1. The nomograph method is shown in Figure 11.3. Judgment must be used when soils seem to fit across several areas. When there are a variety of soils, individual areal calculations must be made or an average K value must be selected. Specific area values of K are found in SCS publications and special publications (Soil Conservation Society, 1977).

11.2.3 Slope Length (L) and Slope Gradient (S) Factor

Obviously, the slope gradient and its length on a given plot or watershed will be a major factor in erosion potential, with steep slopes being more susceptible to soil loss. These two factors are usually combined in what is referred to as the soil loss ratio, or topographic factor, LS. A recently suggested and

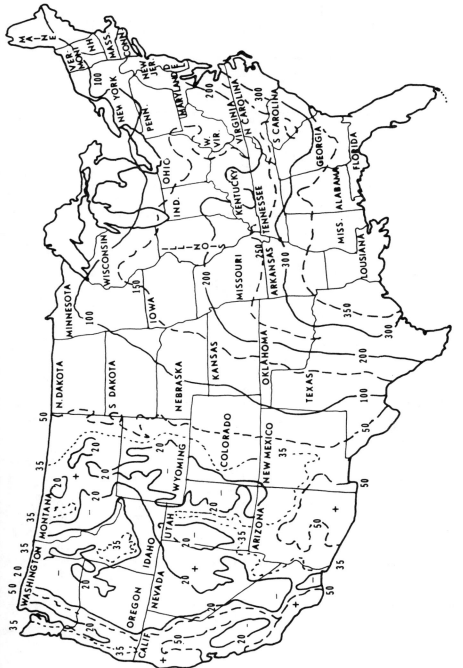

Figure 11.1 Average annual values of the rainfall erosivity factor, R.

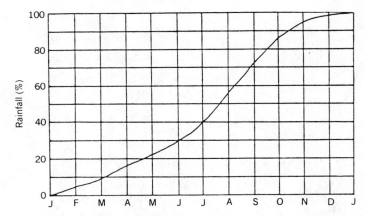

Figure 11.2 Rainfall distribution curve, north central Florida.

TABLE 11.1 Indications of the General Magnitude of the Soil-Erodibility Factor, K

	Organic Matter Content		
Texture Class	< 0.5% K	2% K	4% K
Sand	0.05	0.03	0.02
Fine sand	0.16	0.14	0.10
Very fine sand	0.42	0.36	0.28
Loamy sand	0.12	0.10	0.08
Loamy fine sand	0.24	0.20	0.16
Loamy very fine sand	0.44	0.38	0.30
Sandy loam	0.27	0.24	0.19
Fine sandy loam	0.35	0.30	0.24
Very fine sandy loam	0.47	0.41	0.33
Loam	0.38	0.34	0.29
Silt loam	0.48	0.42	0.33
Silt	0.60	0.52	0.42
Sandy clay loam	0.27	0.25	0.21
Clay loam	0.28	0.25	0.21
Silty clay loam	0.37	0.32	0.26
Sandy clay	0.15	0.13	0.12
Clay	0.25	0.23	0.19

Source: U.S. Environmental Protection Agency (1973).

11.2 UNIVERSAL SOIL LOSS EQUATION **403**

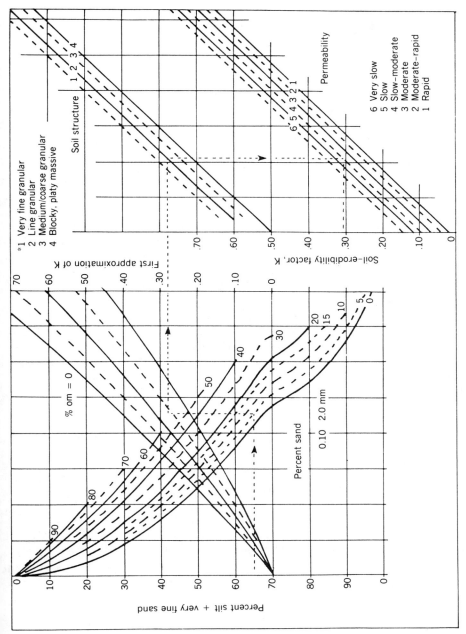

Figure 11.3 Nomograph for determining soil erodibility factor, K, for U.S. mainland soils.

used formula for the topographic factor is (U.S. Department of Agriculture, 1975).

$$\text{LS} = \left(\frac{\lambda}{72.6}\right)^m \left(\frac{430x^2 + 30x + 0.43}{6.57415}\right) \tag{11.2}$$

where λ = field slope length, ft
m = 0.5 if slope equals 5% or greater
  = 0.4 if slope equals 4%
  = 0.3 if slope equals 3% or less
x = sine of the slope angle

The topographic factor can thus be calculated directly; presented in tabular form, as in Table 11.2; or presented in graphic form, as in Figure 11.4.

The slope length requires some additional discussion. It is defined as the distance from the point of origin of overland flow either to (1) the point where the slope decreases to the extent that deposition might occur, or (2) the point where runoff enters a well-defined channel. Whichever point is limiting becomes the basis for selecting the slope length. In many cases, of course, there is a variety of slope gradients and lengths. Under such conditions, the area must be broken up for separate calculations or some average value can be used.

Foster and Wischmeier (1974) proposed an approach for determining the topographic factor when dealing with irregular slopes, including the effects of concavity and convexity. The resulting modifications can cause variations of from 30 to 40% in the predicted soil loss. For his gross assessment method, True (1974) uses a simplified topographic factor equation for all slopes and slope lengths:

$$\text{LS} = \left(\frac{L}{75}\right)^{0.6} \left(\frac{S}{9}\right)^{1.4} \tag{11.3}$$

where L is the slope length in feet and S is the slope in percent. The use of equation 11.3 can also cause variations of from 30 to 40% or more in the soil loss prediction when compared to equation 11.1 and used for shallow slopes.

Depending on the sophistication of the desired calculations, it is clear that there are several approaches in using the topograhpic factor. The analyst must use judgment to determine how much time and effort must be expended on this particular factor, especially in light of the effects of the other factors in the equation.

11.2.4 Cropping Management Factor (*C*)

The cropping management factor, C, also called the cover factor, is the ratio of the soil loss using certain cover and cropping conditions compared to the

11.2 UNIVERSAL SOIL LOSS EQUATION

TABLE 11.2 Topographic Factor, LS, Slope-Effect Table

Percent Slope	Slope (Length) in Feet									
	60	80	100	110	120	130	140	150	160	180	200
0.20	0.07	0.07	0.08	0.08	0.08	0.08	0.09	0.09	0.09	0.09	0.10
0.30	0.07	0.08	0.08	0.09	0.09	0.09	0.09	0.09	0.10	0.10	0.10
0.40	0.07	0.08	0.09	0.09	0.09	0.10	0.10	0.10	0.10	0.11	0.11
0.50	0.08	0.09	0.09	0.10	0.10	0.10	0.10	0.11	0.11	0.11	0.12
1.00	0.11	0.12	0.12	0.13	0.13	0.14	0.14	0.14	0.14	0.15	0.15
2.00	0.17	0.18	0.20	0.20	0.21	0.21	0.22	0.22	0.23	0.24	0.24
3.00	0.24	0.26	0.28	0.29	0.30	0.31	0.31	0.32	0.33	0.34	0.35
4.00	0.32	0.36	0.40	0.41	0.43	0.44	0.45	0.47	0.48	0.50	0.52
5.00	0.42	0.47	0.53	0.56	0.58	0.61	0.63	0.65	0.67	0.71	0.75
6.00	0.52	0.60	0.67	0.70	0.73	0.76	0.79	0.82	0.85	0.90	0.95
8.00	0.76	0.88	0.99	1.04	1.08	1.13	1.17	1.21	1.25	1.33	1.40
10.00	1.06	1.22	1.36	1.43	1.50	1.56	1.62	1.67	1.73	1.83	1.93
12.00	1.39	1.61	1.80	1.89	1.97	2.05	2.13	2.21	2.28	2.42	2.55
14.00	1.77	2.05	2.29	2.40	2.51	2.61	2.71	2.81	2.90	3.07	3.24
16.00	2.19	2.53	2.83	2.97	3.11	3.23	3.35	3.47	3.59	3.80	4.01
18.00	2.66	3.07	3.43	3.60	3.76	3.91	4.06	4.20	4.34	4.60	4.85
20.00	3.16	3.64	4.07	4.27	4.46	4.65	4.82	4.99	5.16	5.47	5.76

Percent Slope	Slope (Length) in Feet									
	300	400	500	600	700	800	900	1000	1100	1200	1300
0.20	0.11	0.12	0.13	0.14	0.14	0.15	0.15	0.16	0.16	0.17	0.17
0.30	0.12	0.13	0.14	0.15	0.15	0.16	0.16	0.17	0.18	0.18	0.18
0.40	0.12	0.14	0.15	0.15	0.16	0.17	0.18	0.18	0.19	0.19	0.20
0.50	0.13	0.14	0.16	0.16	0.17	0.18	0.19	0.19	0.20	0.20	0.21
1.00	0.17	0.19	0.20	0.22	0.23	0.24	0.25	0.25	0.26	0.27	0.27
2.00	0.27	0.30	0.32	0.34	0.36	0.37	0.38	0.40	0.41	0.42	0.43
3.00	0.39	0.43	0.46	0.49	0.51	0.53	0.55	0.57	0.59	0.60	0.62
4.00	0.62	0.69	0.76	0.81	0.87	0.91	0.96	1.00	1.04	1.08	1.11
5.00	0.92	1.07	1.19	1.31	1.41	1.51	1.60	1.69	1.77	1.85	1.93
6.00	1.16	1.34	1.50	1.64	1.78	1.90	2.01	2.12	2.23	2.33	2.42
8.00	1.71	1.98	2.21	2.42	2.62	2.80	2.97	3.13	3.29	3.43	3.57
10.00	2.37	2.73	3.06	3.35	3.62	3.87	4.10	4.33	4.54	4.74	4.93
12.00	3.12	3.60	4.03	4.42	4.77	5.10	5.41	5.70	5.98	6.25	6.50
14.00	3.97	4.58	5.13	5.62	6.07	6.49	6.88	7.25	7.61	7.95	8.27
16.00	4.91	5.67	6.34	6.95	7.51	8.02	8.51	8.97	9.41	9.83	10.23
18.00	5.94	6.86	7.68	8.41	9.08	9.71	10.31	10.86	11.39	11.89	12.38
20.00	7.06	8.15	9.12	9.99	10.79	11.53	12.23	12.90	13.53	14.13	14.70

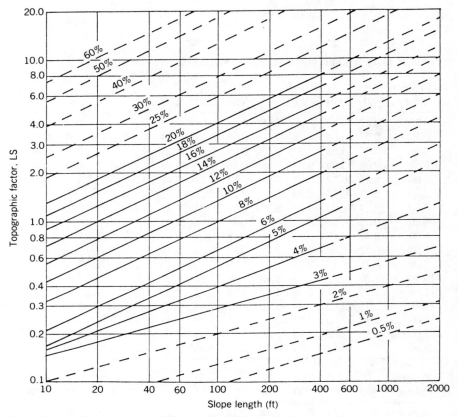

Figure 11.4 Slope-effect chart (topographic factor, LS).
*The dashed lines represent estimates for slope dimensions beyond the range of lengths and steepnesses for which data are available. The curves were derived by the formula

$$\mathrm{LS} = \left(\frac{\lambda}{72.6}\right)^m \left(\frac{430x^2 + 30x + 0.43}{6.57415}\right)$$

where λ = field slope length in feet and m = 0.5 if s = 5% or greater, 0.4 if s = 4%, and 0.3 if s = 3% or less; and $x = \sin \theta$. σ is the angle of slope in degrees.

corresponding loss assuming continuous fallow. Obviously, if cropping practice varies annually, the C factor will also. Wischmeier and Smith propose a very complicated technique for determining the C factor for a variety of construction site conditions. It can be seen that this factor can vary through several orders of magnitude, from 0.01 for sodding and permanent seeding to 1.0 for fallow ground, as shown in Table 11.3.

The 1973 Council on Environmental Quality (CEQ) report to the President recommended typical cropping management factors for the following

TABLE 11.3 Cover Index Factor, C for Construction Sites

		Factor C
None (fallow ground)		1.0
Temporary seedings (90% stand)—after 60 days		
Ryegrass (perennial type)		0.05
Ryegrass (annulus)		0.1
Small grain		0.05
Millet or sudan grass		0.05
Permanent seedings (90% stand)		
First 60 days		0.40
60–365 days		0.05
After 365 days		0.01
Sod (laid immediately)		0.01
Mulch (rate of application, tons/acre)		
Hay	0.50	0.25
	1.00	0.13
	1.50	0.07
	2.00	0.02
Small grain straw	2.00	0.02
Wood chips	6.00	0.06
Wood cellulose	1.75	0.1
Fiberglass	0.50	0.05
Asphalt emulsion (1250 gal/acre)		0.02

general land-use categories:

Land Use	C Factor
Cropland	0.08
Pastureland	0.01
Forestland	0.005
Urbanland	0.01

Recent SCS suggestions for the cover factor for pasture, rangeland, and woodland are documented in the literature (U.S. Department of Agriculture, 1975). Obviously, because of the range of C factors encountered, the analyst must use considerable judgment in assigning this variable.

11.2.5 Erosion-Control Practice Factor (P)

The erosion-control practice factor, P, enables the analyst to account for various mechanical erosion control practices such as contouring, terracing, and strip-cropping. Recommended P factors, as a function of land slope, are given in Table 11.4.

TABLE 11.4 Erosion-Control Practice Factors, P

Land Slope (%)	P Factor				
	Strip-Cropping or		Contour Strip-Cropping[a]		
	Ridge Planting	Terracing	R–W	R–R–M–M	R–S
1.1–2	0.30	0.12	0.52	0.30	0.60
2.1–7	0.25	0.10	0.44	0.25	0.50
7.1–12	0.30	0.12	0.52	0.30	0.60
12.1–18	0.40	0.16	0.70	0.40	0.80
18.1–24	0.45	0.18	0.90	0.45	0.90

[a]R, row crop; W, fall-seeded grain; S, spring-seeded grain; M, meadow.

It is possible for the P factor to be greater than 1.0 if certain practices actually aid erosion. In conjunction with the 1973 Council of Environmental Quality report, True (1974) recommends the following P factors for the broad land-use designations described previously under the C factor:

Land Use	P Factor
Cropland	0.50
Pastureland	1.0
Forestland	1.0
Urbanland	1.0

As with the cover factor, C, considerable judgment must be used in applying the erosion control practice factor, P.

Example Problem 11.1   As an example of the use of the USLE, an Alabama country club watershed has the following parameters:

Area	240 acres
Soil type	sand, rapid permeability ($K = 0.17$)
Average slope length	600 ft
Average slope gradient	1.0% (from Table 9.3, LS = 0.22)
Predominant land use	cropland (groves); ($C = 0.08$, $P = 0.5$)

Then the USLE would give the following hand calculations:

$$A = (R)(K)(L)(S)(C)(P)$$
$$= (350)(0.17)(0.22)(0.08)(0.5)$$
$$= 0.524 \text{ ton/acre, annual basis}$$
$$= 126 \text{ tons, annual basis}$$

11.2 UNIVERSAL SOIL LOSS EQUATION

TABLE 11.5 Approximate Weights of Soils (lb / ft^3) and Factors for Converting Soil Losses (Air-Dry) from Tons to Cubic Yards

	Volume Weight (lb/ft^3)	Tons to Cubic Yards
Sands and loamy sands	110	0.67
Sandy loams	105	0.71
Fine sandy loam	100	0.74
Loam	90	0.82
Silty loam	85	0.87
Silty clay loam	80	0.93
Clay loam	75	0.99
Silty, sandy clay, and clay	70	1.06
Aerated sediment	80^a	0.93
Saturated sediment	60^a	1.24

^{a}Approximate saturated weight.

To convert this into cubic yards, the SCS uses the factors shown in Table 11.5. Thus, for a sand:

$$\text{volume} = 126.0 \text{ tons} \times 0.67 = 84 \text{ yd}^3$$

Example Problem 11.2 An improvement on this single-area hand calculation would be to break up the watershed into two areas. The following might be a reasonable breakdown:

Pasturelands (Fairways, etc.)	Croplands (Groves, etc.)
30 acres	210 acres
$K = 0.17$	$K = 0.17$
$C = 0.01$	$C = 0.08$
$P = 1.0$	$P = 0.5$

Then the total annual yield from this watershed (assuming same slope data) would be

$$\text{total annual soil loss (in tons)} = (350)(0.17)(0.22)(0.01)(1.0) \times 30 \text{ acres}$$
$$+ (350)(0.17)(0.22)(0.08)(0.5) \times 210 \text{ acres}$$
$$= 4.6 + 110.0 = 114.6 \text{ tons (annual basis)}$$

This should be compared with 126 tons from Example Problem 11.1.

To make further modifications and include other variables would simply require an extension of the foregoing hand calculations with careful bookkeeping to ensure that all areas are properly categorized and calculated.

11.2.6 Conclusions on USLE

One can be as sophisticated as one desires with the USLE, even with hand calculations. If a watershed has a great variety of slope gradients and lengths, soil types, land uses, and so on, the entire watershed could be divided into small areas. Each area could be calculated individually and then the sum of the results used for the entire watershed.

It is obvious from the examples above that using the USLE requires considerable judgment, and the numbers resulting from any calculations can cover a broad spectrum. For considering the various factors involved in soil loss, however, the USLE can provide broad numerical guidelines, which can assist planners and analysts in formulating erosion control and cover practices to cut down on soil loss.

The USLE does not include any factors to account for rill or gully erosion. Areas with considerable gully erosion must use other predictive methods to complement or supplement the results of USLE calculations (U.S. Environmental Protection Agency, 1973).

11.3 WATER YIELD MODEL

The water yield model uses the SCS cover complex method (curve number, CN) with modifications for water storage (Williams and LaSeru, 1976). Usually, two water yield models will be used: the calibration model and record extending model. The calibration model utilizes daily rainfall, monthly flow data, and lake evaporation (LE) and predicts the long-term CN and soil moisture depletion coefficient (B). The record extended model uses daily rainfall, starting CN, long-term CN, B, and lake evaporation to predict daily runoff from the watershed. The program was modified by Smoot (1977) to include water quality.

In reality, the CN varies continuously with soil moisture. Runoff prediction accuracy can be improved by using a soil moisture accounting procedure to estimate the curve number for each storm. The soil moisture index, SM, is related to the potential abstraction (S) by

$$\text{SM} = V - S' \qquad (11.4)$$

where V is the maximum soil storage equal to 20 in. or 50.8 cm. The soil moisture maximum value was picked because it provides ample storage to allow a wide range of curve number ($33\frac{1}{3} - 100$) and yet is small enough to allow daily rainfall to influence SM properly. Substituting for S', SM becomes

$$\text{SM} = V - S' = V - \frac{b}{\text{CN}} + c \qquad (11.5)$$

where b = 1000 in. (English) or 2540 cm (metric)
c = 10 in. or 25.4 cm

and

$$SM = 30 - 1000/CM \text{ (English)} \quad \text{or} \quad 76.2 - 2540/CN \text{ (metric)}$$

Soil moisture is usually depleted continuously between storms by evapotranspiration, deep seepage, and other net changes. Depletion is greater when soil moisture and evaporation are high, and most rapid immediately after a storm (high SM). Thus a depletion relationship is assumed as a second-order equation or

$$\frac{d(SM)}{dt} = -B \times LE \times SM^2 \tag{11.6}$$

where t = time
B = depletion coefficient
LE = lake evaporation

This assumption should be verified if calibration results are not acceptable. Integrating equation 11.6 yields

$$SM_t = \frac{SM_{t-1}}{1.0 + B \times SM_{t-1} \sum_{t=1}^{T} LE_t} \tag{11.7}$$

where SM = soil moisture indexes at the beginning of the first storm
SM_t = soil moisture index at time t
LE_t = average monthly lake evaporation for day, t
T = number of days between beginning of storms.

During a storm, the amount that infiltrates, $P - R$, must be added to the soil moisture. Thus the soil moisture index depletion equation becomes

$$SM_t = \frac{SM_{t-1} + P - R}{1.0 + B \times (SM_{t-1} + P) \sum_{t=1}^{T} LE_t} \tag{11.8}$$

The soil moisture coefficient is found by an iteration technique. To estimate B initially, the average daily depletion is calculated by subtracting average annual runoff from average rainfall and dividing by the number of days in a year.

$$DP + \frac{AVP - AVQ}{365} \qquad (11.9)$$

where DP = average daily depletion
AVP = average annual rainfall
AVQ = average annual runoff

If SM is computed as the average curve number, soil moisture condition, and rainfall and runoff are zero for a time period, the following equations result:

$$DP = SM_a - SM_t \qquad (11.10)$$

and

$$DP = SM_a - \frac{SM_a}{1.0 + B \times SM_a \times LE_t} \qquad (11.11)$$

or substituting and rearranging,

$$B = \frac{-DP}{LE_t \times SM_a(DP - SM_a)} \qquad (11.12)$$

The model above is self-calibrating such that if the initial SM_a is too low, predicted runoff is low and SM builds up rapidly; if initial SM_a is too high, predicted runoff is high and SM decreases.

Using this modified water yield model, Smoot (1977) simulated the quality and quantity responses on an approximate 60,000-acre rural watershed, Spruce Creek. The main drainage channel was 10 mi long with about 15% of the land in cropland and 12% in grasslands for cattle and horses. The remaining land was predominantly undeveloped with a small percentage classified as residential. An example of the simulated flow rates is shown in Figure 11.5. Not all simulations can be expected to produce these accurate results. Loading rates were computed for the Spruce Creek watershed and are compared to other watersheds within a radius of 50 mi. These comparisons are shown in Table 11.6.

Example Problem 11.3 Consider a watershed of 0.75 km^2 (0.29 mi^2) in Mississippi, which has an average volume of rainfall per year of 90 cm and an average volume of runoff per year of 20 cm. Using the water yield model,

11.3 WATER YIELD MODEL   413

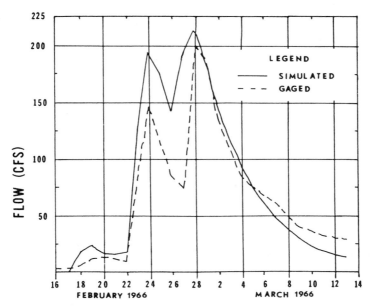

Figure 11.5 Comparison of selected hydrographs, simulated versus gaged. (From Smoot, 1977.)

predict runoff volume if the average curve number is 80, the monthly lake evaporation is 12 cm, and the following rainfall data are available:

Date	Precipitation (mm)
June 1, 1977	0.0
June 5, 1977	24.0
June 24, 1977	18.0

TABLE 11.6 Comparison of Mass Loading Rates (lb/acre/yr) of Selected Watersheds

Watershed	Type	TOC	SS	TN	TP
Spruce Creek	Agricultural	120.71[a]	7.91	4.51	0.18
Shingle Creek	Agricultural	32.85	9.27	1.05	2.30[a]
Big Econlockhatchee	Agricultural	61.71	15.45	1.16	0.19
Lake Eola	Urban	119.67	211	5.07	2.32

Sources: Smoot (1977) and Wanielista (1976).
[a] High background (nonrunoff conditions).

SOLUTION: The solution requires calculating the starting SM and B values from the given data:

$$DP = \frac{AVP - AVQ}{365} = \frac{90 - 20}{365} = 0.25 \text{ cm/day}$$

$$S' = \frac{1000}{80} - 10 = 2.5 \text{ in. } (6.35 \text{ cm})$$

$$SM_a = 20 - 2.5 = 17.5 \text{ in. } (44.45 \text{ cm})$$

$$B = -\frac{DP}{LE_t \times SM_a(DP - SM_a)}$$

$$= -\frac{0.25}{12 \times 44.45(0.25 - 44.45)} = 1.06 \, (10^{-5})/\text{cm}$$

Using a tabular form the solution is:

Date	Precipitation (cm)	SM (cm)	CN = 2540/(76.2 − SM)	S' = 50.8 − SM	Q(cm)
June 1, 1977	0.0	44.45	80	6.35	0.00
June 5, 1977	2.4	44.41[a]	79.9	6.39	0.17[b]
June 24, 1977	1.8	46.48[c]	86.5	4.42	0.15[d]

[a] $SM_5 = \dfrac{44.45}{1.0 + 1.06(10^{-5})(44.45)(12)(5/30)} = 44.41 \text{ cm.}$

[b] $Q_5 = \dfrac{(P - 0.2S')^2}{P + 0.8S'} = \dfrac{[2.4 - 0.2(6.39)]^2}{2.4 + 0.8(6.39)} = 0.17 \text{ cm.}$

[c] $SM_{24} = \dfrac{44.41 + 2.4 - 0.17}{1 + 1.06(10^{-5})(44.41 + 2.4)(12)(20/30)} = 46.48 \text{ cm.}$

[d] $Q_{24} = \dfrac{[1.8 - 0.2(4.42)]^2}{1.8 + 0.8(4.42)} = 0.15 \text{ cm.}$

11.4 SOLIDS DISPOSAL

Under the National Pollutant Discharge Elimination System (NPDES), some rural nonpoint-source activities were converted to point sources because of their obvious pollution potential. However, many activities are still excluded from the permit system. These exclusions apply to small, nonconcentrated land uses. Confined feeding lots, which usually require a permit, are those that hold animals for 30 or more days per year. Also, the permit is issued if the number of animals exceed a specific limit. These activities with exceedance numbers are shown in Table 11.7.

Management of highly concentrated waste is discussed in this chapter. Some characteristics of feedlot runoff are shown in Table 11.8. The range of

TABLE 11.7 NPDES Permit Limits

Activity	Units	Required Permit if Equal to or Greater Than
Slaughter steers and heifers	lb	1,000
Dairy cattle	lb	700
Swine over 55 lb	lb	2,500
Sheep	lb	10,000
Turkeys (open lots only)	lb	55,000
Laying hens or broilers		
Constant flow watering	lb	100,000
Liquid manure handling system	lb	30,000
Ducks	lb	5,000
Irrigation return flow	acres	3,000

TABLE 11.8 Characteristics of Feedlot Runoff[a]

	Runoff (mg/L)			
	Austin Co., Texas	Bushland, Texas	Kansas	Nebraska
Total solids	9000	—	8450	—
	2080–42,500	5000–50,000	214–19,250	24,000–17,400
Volatile solids	4500	—	3890	—
	800–14,000	—	36–9500	1200–7300
Nitrogen (N)	50	—	675	—
	4–125	600–2400	165–1580	39–455
Phosphorus (P)	85	—	79	—
	5–305	100–500	9–242	14–47
Potassium (K)	340	—	—	—
	20–740	900–2100	—	—
Sodium	230	—	—	—
	65–700	400–1000	—	—
Chloride	410	—	—	—
	30–890	1250–2200	—	—
Chemical oxygen demand (COD)	4000	—	7600	—
	500–14,000	—	800–16,000	13,000–8250

Source: U.S. Department of Agriculture (1975).

[a] Upper figure is average; lower figures represent the range; dashes indicate data not available.

concentrations varies considerably and are a function of rainfall, runoff, antecedent dry conditions, mass of animal/feedlot, and others. These numbers should be used for guideline purposes.

11.4.1 Land Sprinkling

The solids content of manure is particularly important because of disposal by standard land spreading, and specifically, by sprinkling equipment. Before sprinkling equipment is used, the percentage solids should be determined in the raw manure and that percent solids that can be disposed of through standard sprinkling equipment. Usually, water must be added to dry manure to decrease the solids content before sprinkling. Figure 11.6 can be used to aid in estimating the makeup water.

Manure has high nutrient values and is considered for land spreading provided that the nutrients do not contribute to water pollution or resource decay in general. The approximate fertilizer value of manure is shown in Table 11.9.

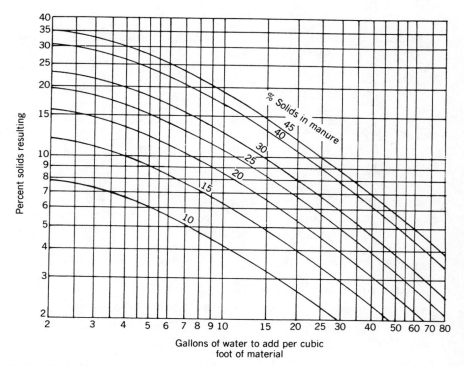

Figure 11.6 Gallons of water required per cubic foot of material for dilution to pumping consistency.

TABLE 11.9 Approximate Fertilizer Value of Raw Manure

Animals	Nutrients in Manure (lb/ton)		
	Nitrogen	Phosphorus	Potassium
Dairy cattle	11	2	10
Beef cattle	14	4	9
Swine	10	3	8
Horses	14	2	12
Sheep	28	4	20
Poultry	31	8	7

Source: U.S. Department of Agriculture (1975).

Some of these fertilizer values can be lost due to biological or weathering processes. Therefore, the applied fertilizer value is most likely less than the raw manure.

Example Problem 11.4 It is required to dispose of cattle manure containing 25% solids. How much makeup water is required if there are 20 ft^3 of manure and the resulting sprinkler water must be 4% solids or less?

SOLUTION: From Figure 11.6, at 4% residual and 25% manure solids, 39 gal of water must be added per cubic foot of manure; therefore, for 20 ft^3 of manure, 780 gal of water is needed. This assumes that the makeup water has little or no solids.

11.4.2 Storage Reservoirs

The location of feedlots depends on many factors: prevailing winds, drainage slope, convenience to water, and space for water storage. Slopes of 4 to 6% are common in the midwestern United States, while 1 to 2% are more common in the southern United States. Locations near sinkholes or surface-water bodies are discouraged unless diversion of runoff and groundwater pollution can be avoided. The type of collection conduits for the wastewaters depends on the amount and type of material to be collected. The total runoff of an area is the sum of precipitation, wash water, and wastewaters from the animals. Approximate daily production of manure is shown in Table 11.10.

A decision must be made on the time for storage of the runoff materials and the final disposal of the waste materials. A holding pond for 1-month storage of runoff and wash water may be specified. The wastewaters in the pond may then be used for land spreading to grow corn and/or other crops for silage.

If percolation is not permitted, a computer program can be used with a treatment rate (in./week land spreading) to estimate storage volume consid-

TABLE 11.10 Approximate Daily Production of Manure per 450 kg Weight

Animal	Solids and Liquids[a] (ft^3/day)	Water (%)
Dairy cattle	1.4	80–90
Beef cattle	1.0	80–90
Horses	0.9	65
Swine	1.0	75
Sheep	0.7	70
Poultry	1.0	55–75

Source: U.S. Department of Agriculture (1975).
[a] A generally accepted conversion value is that 1 ton of manure = 32 to 34 ft^3.

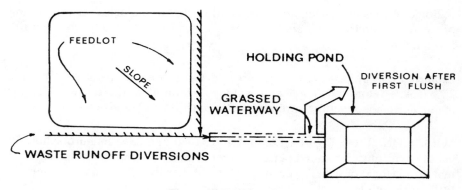

Figure 11.7 Diversions.

ering the antecedent dry conditions and land spreading rates. A typical diversion system for feedlot wastewaters is shown in figure 11.7.

Example Problem 11.5 Assume that a storage reservoir is required for feedlot wastewater. The feedlot operates for 10 straight months and then is cleaned and allowed to stand idle for 2 months. During the two idle months, the storage reservoir also is drained and cleaned. If rainfall during the 10-month feeding period is 80 cm with runoff of 74 cm, what size storage reservoir should be built if the feedlot is 1 acre and there are an average of 400 head of beef cattle of average weight 400 kg in the 10-month (304-days) period? The wash water is estimated at 2 cm/month.

SOLUTION: The solids and liquids produced per day are calculated as

$$400 \text{ head at } 400 \text{ kg/head} = 160{,}000 \text{ kg live wt}$$

$$\text{ft}^3 \text{ solids + liquid} = 1.0 \text{ ft}^3/\text{day} \times 160{,}000/450 = 356 \text{ ft}^3/\text{day}$$

Manure storage for 10 months is

$$\text{manure storage} = 356 \times 204 = 108{,}224 \text{ ft}^3 \text{ or } 2.48 \text{ acre-ft of storage}$$

Runoff storage for 10 months is

$$\text{runoff storage} = 74 \text{ cm/days} \times 1 \text{ in.}/2.54 \times 1 \text{ ft}/12 \text{ in.} \times \text{acres}$$
$$= 2.43 \text{ acre-ft}$$

Wash water storage for 10 months is

$$\text{washwater Storage} = 10 \text{ months} \times 2 \text{ cm/month} \times 1/2.54 \times 1/12 \times \text{acres}$$
$$= 0.66 \text{ acre-ft}$$

Therefore,

$$\text{total storage} = 2.48 + 2.43 + 0.66 = 5.57 \text{ acre-ft}$$

It should be noted that infiltration is not allowed and evaporation rates are equal to rainfall on the storage basin. If evaporation rates are not equal to precipitation, adjustments must be made in the required storage.

11.5 SUMMARY

Mathematical models of rural hydrology and water quality follow the same basic concepts of hydrology and hydraulics for urban models. However, the emphasis appears to be on soil loss rather than other water quality parameters. It should be emphasized that urban models can be used to simulate rural conditions as long as the basic hydrology remains the same. Calibration and verification is necessary to improve the credibility where it can be done.

- The most widely used soil loss model is the universal soil loss equation. It is empirical and can be used for comparative studies. Extensive data are available for its use.
- The water yield model accounts for soil moisture, which changes the infiltration volume potential.
- Manure disposal and treatment is one of the biggest water quality problems in the rural environment. The loading rate data of Chapter 5 illustrates the magnitude of the problem.
- Recycling of rural stormwater may be beneficial because of the high nutrient value of some of the waste.
- Feedlots or other animal concentration areas should be considered for a permit to discharge, thus removing the variable discharge characteristics and converting the problem to a point-source treatment system.

11.6 PROBLEMS

1. Using the USLE, estimate the tons of soil lost from a 4-acre pasture land with a slope of 4% in northern Missouri over an average year of rainfall. The soil is a silty loam with 2% organic content. The length of land in the direction of flow is 600 ft.

2. Assume that the cattle manure of Example Problem 11.5 is 20% solids before rainfall. If irrigation equipment can land-spread 3% solid materials, do you need to add water to the storage reservoir assuming 50% solids settling from solution? If so, how much water? Show calculations.

3. For Example Problem 11.3, the yearly rainfall/runoff values were not recorded correctly. Change the average yearly rainfall to 110 cm and the average yearly runoff to 50 cm, then calculate the discharge on a land area with average CN equal to 90 given the June precipitation data of Example Problem 11.3.

11.7 REFERENCES

Camp, C. R., Sadler, E. J., and Harvey, T. P. 1990. "Computer-Based Irrigation Scheduling Method for Humid Areas," in *Proceedings of the 3rd National Irrigation Symposium, ASCE*, Phoenix, Ariz., pp. 606–611.

Crawford, N. H. 1970. "What Is Simulation?" *Hydrocomp Simulation Network Newsletter*, Palo Alto, Calif. (Aug.).

Donigian, A. S., and Crawford, N. H. 1975. *Modeling Pesticides and Nutrients on Agricultural Lands*, EPA 600/2-76/043, U.S. Environmental Protection Agency, Washington, D.C.

Donigian, A. S., and Crawford, N. H. 1976. *Nonpoint Pollution from the Land Surface*, EPA 600/3-76/083, U.S. Environmental Protection Agency, Washington, D.C.

Flaxman, E. M. 1972. "The Use of Suspended Sediment Load Measurements and Equations for Evaluation of Sediment Yield in the West (revised)," *Proceedings of the Sediment Yield Workshop*, Oxford, Miss.

Fleming, G. 1972. "Sediment Erosion Transport–Deposition Simulation: State of the Art," *Proceedings of the Sediment Yield Workshop*, Oxford, Miss.

Foster, G. R., and Wischmeier, W. H. 1974. "Evaluating Irregular Slopes for Soil Loss Prediction," *Transactions of the American Society of Agricultural Engineers*, Vol. 17, No. 2, pp. 305–309.

Frasier, G. W., and Scrimgeour, J. 1990. "Supplemental Irrigation by Runoff Farming: An Economic Assessment, *Proceedings of the Watershed Management Symposium, ASCE, Irrigation and Drainage Division*, pp. 10–20.

Free, M. H., Onstad, C. A., and Holtan, N. H. 1975. *ACTMO—An Agricultural Chemical Transport Model*, ARS-H-3, USDA Agricultural Research Service Washington, D.C.

Garvin, P. C., Busch, J. R., and Kincaid, D. C. 1986. *Reservoir Tillage for Reducing Runoff*, ASAE Paper 86-2093, American Society of Agricultural Engineers, St. Joseph, Mich.

General Accounting Office. 1990. *Water Pollution*, GAO/RCED-91-10, GAO, Washington, D.C., p. 8.

Griffin, J. D. 1975. "Predicting Soil Loss on Construction and Other Similar Nonvegetative Areas," *Proceedings of the Storm-Water Management Workshop*, Gainesville, Fla., Feb.

Kincaid, D. C., McCann, I., Busch, J. R., and Hasheminia, M. 1990. "Low Pressure Center Pivot Irrigation and Reservoir Tillage," *Proceedings: Third National Irrigation Symposium*, American Society of Agricultural Engineers, St. Joseph, Mich., pp. 54–60.

Musgrave, G. W. 1947. "The Quantitative Evaluation of Factors in Water Erosion—A First Approximation," *Journal of Soil, and Water Conservation*, Vol. 2, pp. 133–138.

Negev, N. 1967. *A Sediment Model on a Digital Computer*, T.R.R. Transportation Research Record, 76, Washington, D.C. Mar.

Novotny, V., Chin, M., and Tran, H. V. 1979. *LANDRUN—An Overland Flow Mathematical Model*, International Joint Commission on the Great Lakes, Windsor, Ontario.

Onstad, C. A. 1973. "ACIMO: What Does It Mean?" *The Morris Tribune—Focus on Farming*, Mar. 22.

Smith, D. D. 1941. "Interpretation of Soil Conservation Data for Field Use," *Agricultural Engineering*, Vol. 22, pp. 173–175.

Smoot, J. L. 1977. *Spruce Creek Watershed Nonpoint Source Loading Model*, Research Report, Florida Technological University, Orlando, Fla.

Snyder, W. M. 1972. "Developing a Parametric Hydrologic Model Useful for Sediment Yield," *Proceedings of the Sediment Yield Workshop*, USDA Sedimentation Laboratory, Oxford, Miss.

Soil Conservation Society. 1977. "Soil Erosion: Prediction and Control," *Proceedings of a National Conference on Soil Erosion*, May 25–26, 1976, Purdue University, Indiana; a publication of the Soil Conservation Service, Ankeny, Iowa.

Stewart, B. A., et al. 1975. *Control of Water Pollution from Cropland*, Vol. I, *A Manual for Guideline Development*, Agricultural Research Services and Environmental Protection Agency, Washington, D.C.

True, H. A. 1974. *Erosion, Sedimentation and Rural Runoff: A Gross Assessment Process*, Environmental Protection Agency, Surveillance and Analysis Division, Athens, Ga.

U.S. Army Corp of Engineers. 1975. *Urban Stormwater Runoff STORM*," Hydrologic Engineering Center, Davis, Calif.

U.S. Department of Agriculture. 1975. *Agricultural Waste Management Field Manual*, USDA Soil Conservation Service, Washington, D.C. (with updated materials).

U.S. Environmental Protection Agency. 1973. *Methods for Identifying and Evaluating the Nature and Extent of Nonpoint Sources of Pollutants*, EPA 430/9-73-014, U.S. Environmental Protection Agency, Washington, D.C.

Wanielista, M. P. 1976. *Nonpoint Source Effects*, Florida Technological University, Orlando, Fla.

Williams, J. R., and LaSeru, W. V. 1976. "Water Yield Model Using SCS Curve Numbers," *Journal of the Hydraulics Division, ASCE*, Vol. 102, No. HY9, pp. 1241–1253.

Wischmeir, W. H., and Smith, D. D. 1972. *Predicting Rainfall-Erosion Losses from Cropland East of the Rocky Mountains*, Agriculture Handbook 282, USDA Agricultural Research Service, Washington, D.C.

Woolhiser, D. A. 1971. "Deterministic Approach to Watershed Modeling," *Nordic Hydrology II*, Munksgaard, Copenhagen, pp. 146–166.

Zingg, A. W. 1949. "Degree and Length of Land Slope As It Affects Soil Loss in Runoff," *Agricultural Engineering*, Vol. 21, pp. 59–64.

APPENDIX A
Notation

A	annual value of money	BO	boundary output
A	watershed area	BOD	biochemical oxygen demand
A	area		
A	soil hydrologic group	b	limiting value
A	soil loss (USLE)	b	constant
A_o	area of orifice	b	regression coefficient
A_R	reuse (irrigation) area	b	stage–storage coefficient
A_s	water surface area	b	bottom width of channel
$\overline{A}_s$	average surface area	C	constant shape factor
A_x	cross-sectional area	C	soil hydrologic group
AC	average cost	C	cost
ADP	antecedent dry period	C	concentration (water quality)
ADT	average daily traffic		
ALOSS	potential rain loss rate	C	runoff coefficient
AMC	antecedent moisture condition	C	cropping management factor in USLE
AVP	average annual rainfall (water yield model)	$\overline{C}$	composite concentration
		$\overline{C}$	event mean concentration
AVQ	average annual runoff (water yield model)	C'	constituent concentration
a	treatment rate over watershed (infiltration pond)	C_0	storage coefficient
		C_1	storage coefficient
		C_1	constituent input to a system
a	regression constant		
a	stage–discharge coefficient	C_2	storage coefficient
		C_2	constituent output from a system
B	soil hydrologic group		
B	benefits	C_{60}	60° triangular weir coefficient
B	depletion coefficient (water yield model)	C_{90}	90° triangular weir coefficient
B	subsurface flow		
BI	boundary input	C_d	discharge coefficient

423

NOTATION

C_e	equilibrium concentration	E	evaporation
C_I	runoff coefficient—impervious surface	E	treatment effectiveness
		$E°$	relative flow fraction
C_o	initial organic concentration	E_L	lake evaporation
		E_p	pan evaporation
C_p	runoff coefficient—pervious surface	E_s	depth of evaporation from snow
C_s	saturation concentration of dissolved oxygen	EFF	effectiveness
		EI	rainfall factor
CA	contributing area	EMC	event mean concentration
CN	rainfall excess curve number	EMV	expected monetary value
		ERAIN	watershed coefficient
CN_c	composite curve number	ET	evapotranspiration
CR	cation ratio	EVAP	long-term evaporation
CRF	capital recovery factor	e	vapor pressure
CUML	cumulative rain loss	e	symbol for exponential of e
C_V	coefficient of variation		
c	wave celerity	e_0	saturation vapor pressure
D	duration of rainfall	e_a	vapor pressure in air
D	soil hydrologic group	e_s	vapor pressure of snowpack
D	dissolved oxygen deficit		
D	depth of pond or water	F	accumulated mass infiltration
D	annual losses		
D_t	water depth at time t	F	lb pollutant/lb dust and dirt
D_w	dustfall on watershed		
DA	drainage area	F	cumulative distribution factor
DCIA	directly connected impervious area		
		F	flood event
DD	dust and dirt accumulation	F_D	fraction not settled during runoff
DI	diversion volume	F_Q	fraction not settled during nonrunoff
DLTK	initial rain loss		
DO	dissolved oxygen	Fr	Froude number
DP	depletion (water yield model)	$F(t)$	infiltration rate as a function of time
d	depth of flow	$F(x)$	cumulative distribution function
d	depth of gutter flow		

APPENDIX A

Symbol	Description	Symbol	Description
$F(x, y)$	cumulative joint distribution function	$i(t - \tau)$	rainfall intensity at time $(t - \tau)$
FW	future worth	K	recession constant (infiltration curve)
f	infiltration rate		
f	relative humidity	K	frequency factors
f	function symbol	K	peak attenuation factor
f	pipe friction factor	K	hydraulic conductivity
f'	first derivative of function	K	Muskingum storage coefficient
f_0	initial infiltration rate		
f_c	ultimate infiltration rate	K	soil erodibility factor (in USLE)
$f(x)$	probability density function	K	swale constant (Chapter 7)
G	population skew coefficient	K	permeability
G_s	specific gravity	K	storage coefficient
$G(x)$	exceedance probability	K'	composite storage coefficient
g	acceleration constant of gravity	K_3	BOD sediment coefficient
g	skew coefficient		
g	grams	K_a	reaeration coefficient
H	crest depth of swale	K_i	recession limb coefficient
H	head	K_L	BOD decay coefficient
ΔH	energy change	K_N	nitrogenous decay coefficient
h	head		
h	height of swale	K_r	routing coefficient for Santa Barbara method
h_0	depth of settling pond		
I	available depression storage	K_s	settling rate for materials
		K_T	thermal conductivity
I	detention pond inflow volume	k	consumptive use coefficient
I_A	initial rainfall abstraction	k	proportionality constant
$I(t) = Q_I$	instantaneous flow rate	k	convolution recession constant
IR	instantaneous runoff hydrograph		
		k	Pearson type III deviate
IUH	instantaneous unit hydrograph	k_s	seasonal consumptive use coefficient
i	rainfall intensity	L	stream length
i	interest rate	L	length of swale

426 NOTATION

L	lag time
L	length of overland flow
L	loading (pollutant mass)
L	channel length
L_f	latent heat of fusion
L_p	depth of percolating water
L_t	length of slotted pipe
L_u	depth of unsaturated soil column
LE	lake evaporation
LC	lethal concentration
LK	lake area
M	number of months in a rainfall event
M	mass
MB	marginal benefits
MC	marginal cost
MSL	mean sea level
m	kinematic parameter
m	mass
m	stage–storage coefficient
m	regression coefficient
m	plot position
max	maximum
min	minimum
N	outflow–storage relationship
N	mass of algae
N	Manning's roughness (overland flow)
N	number of rainfall events
N_D	number of days without runoff
N_R	Reynolds number
N_u	number of stream segments of order u
n	number of storm hydrograph ordinates
n	number of years of observations
n	number of events/time period
n	regression parameter
n	stage–discharge coefficient
n	Manning roughness coefficient
n	shape parameter
n	empirical constant
n_p	porosity
O	outflow volume
O	outflow rate
O_i	ith outflow rate
O_p	peak outflow rate
OR	overflow rate
P	precipitation volume
P	erosion control practice factor in USLE
P	photosynthetic oxygen production rate
P	loading
$\bar{P}$	mean of event rain depths
P_0	initial amount of pollutant
P_a	antecedent precipitation index
P_{0w}	annual wet weather loading
P_{po}	pollutant remaining
P_t	amount of pollutant remaining on ground at time t
PD	population density

PIF	precipitation interevent frequency curve	R	removal load (Chapter 10)
PMF	probable maximum flood	R	runoff depth
Pr	probability	$\bar{R}$	weighted hydraulic radius of the main sewer flowing full
PW	present worth		
p	permeability		
p	monthly daytime hours	R_I	runoff rate from impervious surface
p_c	pan coefficient		
Q	streamflow	R_L	efficiency for all events
Q	discharge	R_M	efficiency during runoff
$\bar{Q}$	average discharge	RA	ratio of impervious area to total area
Q_{10}	10-year peak flow rate		
Q_{233}	mean annual flood rate	REV	reuse rate efficiency volume curve
Q_{50}	50-year flow rate		
Q_{100}	100-year flow rate	RU	reuse volume
Q_I	instantaneous runoff rate	r	rainfall excess rate
Q_i	ith flow in sequence	r	correlation coefficient
Q_i^0	observed hydrograph flow rate	r	reaction rate
		r_{avg}	average street runoff rate
Q_p	peak discharge	S	storage
Q_t	discharge at time t	S	storage potential of soil
Q_w	overland flow discharge rate	S	energy gradient
		S	overland or swale slope
		S'	potential maximum watershed retention
$Q(t)$	surface runoff rate at time t		
		S_0	average streambed slope
q	discharge per unit area	S_0	initial sediment concentration
q	runoff rate		
q	flow rate	S_B	benthic oxygen demand
q	overland flow discharge rate	S_c	groundwater aquifer storage coefficient
q	specific adsorption capacity	S_n	standard deviation of annual maximum rain depths
q_0	initial discharge		
R	risk	S_t	lake storage at time t
R	rainfall excess	S_u	sediment upstream concentration
R	rainfall factor (USLE)		
R	hydraulic radius	S_w	sediment waste concentration
R	algal respiration		

$S_x = S_T$	mixed sediment concentration	t_d	detention time
		t_d	pond detention time
S_{yx}	standard error of estimated runoff	t_e	time of equilibrium of runoff rate
SFF	sinking fund factor	t_o	time between outflow events
SL_d	shoreline length at depth d		
		t_p	time to peak
SM	soil moisture index	t_r	recession limb time
SS	suspended solids	t_t	travel time
s	linear slope parameter	t_w	hydraulic residence time (lakes)
s	standard deviation		
s^2	variance	U	advective velocity
T	transpiration	USLE	universal soil loss equation
T	temperature		
T	transmissivity	V	volume of pond
T	width of gutter flow	V	volume of water
T	travel time overland	V	maximum soil storage
T_0	water surface temperature	V	velocity
		$\bar{V}$	average velocity
T_A	return period, annual series	V_B	wet-pond permanent volume
T_a	air temperature	V_D	volume of pond at depth D
T_{max}	daily maximum temperature		
		V_E	pond infiltration volume
T_p	partial series return period	V_P	volume of pond
		V_R	reuse pond volume
T_r	return period	V_R	runoff volume from mean storm
T_r	recurrence interval		
TC	total cost	V_s	volume of swale
TSI	trophic state index	$V(A)$	monetary value of (A)
t	time	v_0	design settling velocity
Δt	time interval	v_s	settling velocity
t_1	lag time	W	BOD discharge
t_b	time base of the hydrograph	W	width of swale
		W	width of overland flow
t_c	time of concentration	W_i	weighted area
t_c	time to minimum dissolved oxygen	w	scour load

w_s	width of swale	β	inverse of average runoff duration
X	random variable		
X	weight of organics adsorbed	Γ	gamma function
		γ	specific weight
x	falling limb factor	Δ	interevent time
x	weighting factor	δ	standard deviation parameter
$\bar{x}$	first moment about the origin (mean)		
		δ	time between midpoint of storm events
Y	release volume		
y	depth of flow	ε	flow capture efficiency
Z	side slope of a channel	λ	average number of rainfall events
Z	standard normal deviate		
$\bar{Z}$	mean depth (lake)	λ	exponential distribution constant
Z_o	optimal solution		
z	elevation	λ	field slope length, ft
z	standard deviation	μ	mean value
z_0	roughness parameter	μ	dynamic viscosity
		μ'_r	rth moment about the origin
α	loading coefficient		
α	order of reaction	ν	kinematic viscosity
α	inverse of average runoff volume	ν_s	seepage velocity
		$\rho = \rho_w$	density of water
α_p	portion of advective energy	σ	standard deviation
		σ^2	variance
β	loading coefficient	τ	time parameter
β	loading factor (Chapter 5)	Φ	infiltration index
		Ω	pond empty rate

APPENDIX B

Metric Units with English Equivalents

Length

Metric Units

millimeter	(mm)	10 mm = cm
centimeter	(cm)	100 cm = m
meter	(m)	1000 m = km
kilometer	(km)	

English equivalents

meters × 39.37 = inches × 0.0254 = meters
meters × 3.28 = feet × 0.3049 = meters
kilometers × 0.62 = miles × 1.6129 = kilometers
millimeters × 0.039 = inches × 25.4 = millimeters
centimeters × 0.394 = inches × 2.54 = centimeters

Example: Convert 3 m to feet. 3 m × 3.28 ft/m = 9.84 ft

Area

Metric Units

square millimeter	(mm^2)	10^2 mm^2 = cm^2
square centimeter	(cm^2)	10^4 cm^2 = m^2
square meter	(m^2)	10^6 m^2 = km^2
hectare	(ha)	10^2 ha = km^2

English Equivalents

$mm^2 \times 0.00155 = in^2 \times 645.16 = mm^2$
$cm^2 \times 0.155 = in.^2 \times 6.45 = cm^2$
$m^2 \times 10.764 = ft^2 \times 0.093 = m^2$
$km^2 \times 0.384 = mi^2 \times 2.605 = km^2$
$km^2 \times 247.10 = ac \times 0.004 = km^2$
$ha \times 2.471 = ac \times 0.405 = ha$
$ha \times 0.00386 = mi^2 \times 259 = ha$

Mass

Metric Units

milligram	(mg)	1000 mg = g
gram	(g)	1000 g = kg
kilogram	(kg)	1000 kg = 1 tonne (t)

English Equivalents

milligram × 0.01543 = grains × 64.809 = mg
gram × 0.0022 = pounds × 453.6 = gram
gram × 15.43 = grains × 0.065 = gram
kilogram × 2.205 = pounds × 0.454 = kg
kilogram × 0.0011 = ton × 907.20 = kg
tonne (t) × 1.1023 = ton × 0.907 = tonne

(using 2000 pounds ton (short ton))

Volume

Metric Units

cubic centimeter	(cm^3)	10^6 cm^3 = m^3
cubic meter	(m^3)	10^3 L = m^3
liter	(L)	10^3 cm^3 = L

English Equivalents

$cm^3 \times 0.061 = in.^3 \times 16.393 = cm^3$
$m^3 \times 35.314 = ft^3 \times 0.028 = m^3$
$L \times 1.057 = qt. \times 0.946 = L$

English Equivalents (Continued)

L	× 0.264	= gal × 3.788 = L
L	× 0.81(10^{-6})	= ac-ft × 1.235(10^6) = L
m^3	× 0.41(10^{-3})	= SFD × 2.45(10^3) = m^3
	SFD = second foot day	

Time

Metric Units = English Units

second (s)	86,400 s = 1 day
day (day)	365 day = yr or a
year (yr or a)	(366 days every 4 years)
hour (h)	

Force

Metric Unit with English Equivalent

newton (N) ×0.22481 = lb (weight)
             ×7.24 = lb (force)

Commonly Used Conversion Factors

Linear Velocity

$$m/s \times 3.280 = fps \times 0.305 = m/s$$
$$km/s \times 2.230 = mph \times 0.448 = km/s$$
$$km/h \times 0.621 = mph$$
$$km/h \times 0.540 = knots$$

Flow or Discharge

$m^3/s \times 15.850(10^3)$ = gpm × 0.063(10^{-3}) = m^3/s
$m^3/s \times 2.12(10^3)$ = cfm × 0.472(10^{-3}) = m^3/s
$m^3/s \times 35.314$ = cfs × 0.0283 = m^3/s
L/s × 15.850 = gpm × 0.063 = L/s
$m^3/s \times 22.82$ = MGD × 0.0438 = m^3/s
L/day × 0.264 = GPD × 3.788 = L/day

Loading

kg/km × 3.576 = lb/mi × 0.280 = kg/km
kg/ha × 0.892 = lb/ac × 1.21 = kg/ha
kg/ha × 0.286 = ton/mi^2 × 3.50 = kg/ha
kg/ha × 0.446(10^{-3}) = ton/ac × 2.24(10^3) = kg/ha
$kg/m^3 \times 0.065$ = lb/ft^3 × 15.38 = kg/m^3
$m^3/m^2 \times 3.28$ = ft^3/ft^2 × 0.305 = m^3/m^2

Density

$kg/cm^3 \times 0.0624 = lb/ft^3 \times 16.026 = kg/cm^3$

$lb/gal \times 1.2 \times 10^5 = mg/L \times 8.33 \times 10^{-6} = lb/gal$

$kg/m^3 \times 0.065 = lb/ft^3 \times 15.38 = kg/m^3$

Commonly Used Conversions

Area

$43{,}560 \ ft^2 = 1 \ ac$

$4{,}840 \ yd^2 = 1 \ ac$

$144 \ in.^2 = 1 \ ft^2$

$640 \ ac = 1 \ mi^2$

Volume

$7.48 \ gal = 1 \ ft^3$

$1728 \ in.^3 = 1 \ ft^3$

$1 \ MGD = 694.4 \ gpm$

$8.34 \ lb = 1 \ gal$ (of water)

$62.43 \ lb = 1 \ ft^3$ (of water)

Mass

$2000 \ lb = 1 \ ton$

$454 \ g = 1 \ lb$

$7000 \ gr = 1 \ lb$

$2240 \ lb = 1 \ long \ ton$

Other Conversions

Pressure

$2.307 \ ft \ H_2O = 1 \ lb/in.^2$

$2.036 \ in. \ Hg = 1 \ lb/in.^2$

$14.70 \ psia = 1 \ atm$

$29.92 \ in \ Hg = 1 \ atm$

$33.93 \ ft \ H_2O = 1 \ atm$

$76.0 \ cm \ Hg = 1 \ atm$

$0.205 \ kg/m^2 = 1 \ lb/ft^2$

Miscellaneous

$in. \cdot mi^2 \times 26.9 = SFD \quad \times 2.45(10^3) = m^3$

$in. \cdot mi^2 \times 53.3 = ac\text{-}ft \quad \times 1.235(10^3) = m^3$

$SFD/mi^2 \times 0.0372 = in. \quad \times 25.4 = mm$

$CFS \times 0.992 = ac\text{-}in./hour \times 101.6 = m^3/hr$

$lb/gal \times 120(10^3) = mg/L$

Miscellaneous (Continued)

$g = 9.806 \text{ m/s}^2 = 32.174 \text{ ft/s}^2$

standard conditions 4°C, 706 mm Hg

ρ water $= 1.94 \text{ slugs/ft}^3 = 1000 \text{ kg/m}^3$ (4°C)

γ water $= 62.43 \text{ lb/ft}^3 = 9806 \text{ N/m}^3$ (4°C)

for permeability:

$1 \text{ ft/day} = 0.305 \text{ m/day} = 7.48 \text{ gal/day-ft}^2$

for transmissivity:

$1 \text{ ft}^2/\text{day} = 0.0929 \text{ m}^2/\text{day} = 7.48 \text{ gal/day-ft}$

Also

$\log_e 10 = 2.30259$

$e = 2.71828$

Physical Properties of Water
Source: Hydraulic Models ASCE Manual 25, New York, 1942.

TEMPERATURE (°F)	SPECIFIC WEIGHT γ (lb/ft^3)	DENSITY ρ (slugs/ft^3)	VISCOSITY $\mu \times 10^5$ lb(s/ft^2)	KINEMATIC VISCOSITY $\nu \times 10^5$ (ft^2/s)
32	62.42	1.940	3.746	1.931
40	62.42	1.940	3.229	1.664
50	62.41	1.940	2.735	1.410
60	62.36	1.938	2.359	1.217
70	62.29	1.936	2.050	1.059
80	62.22	1.934	1.799	0.930
90	62.13	1.931	1.595	0.826
100	62.00	1.927	1.424	0.739
110	61.87	1.923	1.284	0.667

SI Unit Prefixes

PREFIX	SYMBOL	MULTIPLES	PREFIX	SYMBOL	MULTIPLES
tera	T	10^{12}	deci	d	10^{-1}
giga	G	10^9	centi	c	10^{-2}
mega	M	10^6	milli	m	10^{-3}
kilo	k	10^3	micro	μ	10^{-6}
hecto	h	10^2	nano	n	10^{-9}
deka	da	10	pico	p	10^{-12}

APPENDIX C

Nondimensional Rainfall and Frequency–Intensity–Duration Curves

TABLE C.1   SCS Type II—Rainfall Distribution (24-hours)

NONDIMENSIONAL		NONDIMENSIONAL		NONDIMENSIONAL		NONDIMENSIONAL	
TIME	RAINFALL	TIME	RAINFALL	TIME	RAINFALL	TIME	RAINFALL
.000	.000	.521	.735	.094	.026	.615	.849
.010	.002	.531	.758	.104	.029	.625	.856
.021	.005	.542	.776	.115	.032	.635	.863
.031	.008	.552	.791	.125	.035	.646	.869
.042	.011	.563	.804	.135	.038	.656	.875
.052	.014	.573	.815	.146	.041	.667	.881
.063	.017	.583	.825	.156	.044	.677	.887
.073	.020	.594	.834	.167	.048	.688	.893
.083	.023	.604	.842	.177	.052	.698	.898
.188	.056	.708	.903	.354	.133	.875	.965
.198	.060	.719	.908	.365	.140	.885	.968
.208	.064	.729	.913	.375	.147	.896	.971
.219	.068	.740	.918	.385	.155	.906	.974
.229	.072	.750	.922	.396	.163	.917	.977
.240	.076	.760	.926	.406	.172	.927	.980
.250	.080	.771	.930	.417	.181	.938	.983
.260	.085	.781	.934	.427	.191	.948	.986
.271	.090	.792	.938	.438	.203	.958	.989

TABLE C.1 *(Continued)*

NONDIMENSIONAL TIME	NONDIMENSIONAL RAINFALL	NONDIMENSIONAL TIME	NONDIMENSIONAL RAINFALL	NONDIMENSIONAL TIME	NONDIMENSIONAL RAINFALL	NONDIMENSIONAL TIME	NONDIMENSIONAL RAINFALL
.281	.095	.802	.942	.448	.218	.969	.992
.292	.100	.813	.946	.458	.236	.979	.995
.302	.105	.823	.950	.469	.257	.990	.998
.313	.110	.833	.953	.479	.283	1.000	1.000
.323	.115	.844	.956	.490	.387		
.333	.120	.854	.959	.500	.663		
.344	.126	.865	.962	.510	.707		

TABLE C.2 SCS Type III—Rainfall Distribution (24-hours)

NONDIMENSIONAL TIME	NONDIMENSIONAL RAINFALL	NONDIMENSIONAL TIME	NONDIMENSIONAL RAINFALL	NONDIMENSIONAL TIME	NONDIMENSIONAL RAINFALL	NONDIMENSIONAL TIME	NONDIMENSIONAL RAINFALL
.000	.000	.521	.702	.135	.034	.656	.878
.010	.002	.531	.729	.146	.037	.667	.886
.021	.005	.542	.751	.156	.040	.677	.893
.031	.007	.552	.769	.167	.043	.688	.900
.042	.010	.563	.785	.177	.047	.698	.907
.052	.012	.573	.799	.188	.050	.708	.911
.063	.015	.583	.811	.198	.053	.719	.916
.073	.017	.594	.823	.208	.057	.729	.920
.083	.020	.604	.834	.219	.060	.740	.925
.094	.023	.615	.844	.229	.064	.750	.929
.104	.026	.625	.853	.240	.068	.760	.933
.115	.028	.635	.862	.250	.072	.771	.936
.125	.031	.646	.870	.260	.076	.781	.940
.271	.080	.792	.944	.396	.167	.917	.981
.281	.085	.802	.947	.406	.178	.927	.983
.292	.089	.813	.951	.417	.189	.938	.986
.302	.094	.823	.954	.427	.202	.948	.988
.313	.100	.833	.957	.438	.216	.958	.991

TABLE C.2 *(Continued)*

NONDIMENSIONAL		NONDIMENSIONAL		NONDIMENSIONAL		NONDIMENSIONAL	
TIME	RAINFALL	TIME	RAINFALL	TIME	RAINFALL	TIME	RAINFALL
.323	.107	.844	.960	.448	.232	.969	.993
.333	.115	.854	.963	.458	.250	.979	.996
.344	.122	.865	.966	.469	.271	.990	.998
.354	.130	.875	.969	.479	.298	1.000	1.000
.365	.139	.885	.972	.490	.339		
.375	.148	.896	.975	.500	.500		
.385	.157	.906	.978	.510	.662		

TABLE C.3 Corps of Engineers Design Storms

Hr	(vol. ≈ 7.90 in.)		(vol. ≈ 9.00 in.)		(vol. ≈ 11.00 in.)	
	P_{inc}	ΣP	P_{inc}	ΣP	P_{inc}	ΣP
0.0	—	—	—	—	—	—
0.5	0.06	0.06	0.06	0.06	0.08	0.08
1.0	0.06	0.12	0.06	0.12	0.08	0.16
1.5	0.06	0.18	0.06	0.18	0.08	0.24
2.0	0.06	0.24	0.06	0.24	0.08	0.32
2.5	0.06	0.30	0.06	0.30	0.08	0.40
3.0	0.06	0.36	0.07	0.37	0.09	0.49
3.5	0.06	0.42	0.07	0.44	0.09	0.58
4.0	0.06	0.48	0.07	0.51	0.09	0.67
4.5	0.08	0.56	0.09	0.60	0.11	0.78
5.0	0.08	0.64	0.09	0.69	0.11	0.89
5.5	0.09	0.73	0.11	0.80	0.13	1.02
6.0	0.09	0.82	0.11	0.91	0.13	1.15
6.5	0.09	0.91	0.11	1.02	0.13	1.28
7.0	0.09	1.00	0.11	1.13	0.13	1.41
7.5	0.13	1.13	0.15	1.28	0.19	1.60
8.0	0.13	1.26	0.15	1.43	0.19	1.79
8.5	0.13	1.39	0.15	1.58	0.19	1.98

TABLE C.3 *(Continued)*

Hr	(vol. ≈ 7.90 in.)		(vol. ≈ 9.00 in.)		(vol. ≈ 11.00 in.)	
	P_{inc}	ΣP	P_{inc}	ΣP	P_{inc}	ΣP
9.0	0.13	1.52	0.15	1.73	0.19	2.17
9.5	0.14	1.66	0.16	1.89	0.20	2.37
10.0	0.14	1.80	0.16	2.05	0.20	2.57
10.5	0.14	1.94	0.16	2.21	0.20	2.77
11.0	0.14	2.08	0.16	2.37	0.20	2.97
11.5	0.16	2.24	0.18	2.55	0.22	3.19
12.0	0.16	2.40	0.18	2.73	0.22	3.41
12.5	0.19	2.59	0.22	2.95	0.26	3.67
13.0	0.21	2.80	0.23	3.18	0.29	3.96
13.5	0.29	3.09	0.33	3.51	0.41	4.37
14.0	0.30	3.39	0.34	3.85	0.42	4.79
14.5	0.46	3.85	0.52	4.37	0.64	5.43
15.0	0.46	4.31	0.52	4.89	0.64	6.07
15.5	0.73	5.04	0.83	5.72	1.01	7.08
16.0	0.81	5.85	0.92	6.64	1.12	8.20
16.5	0.34	6.19	0.39	7.03	0.47	8.67
17.0	0.33	6.52	0.38	7.41	0.46	9.13
17.5	0.18	6.70	0.21	7.62	0.25	9.38
18.0	0.17	6.87	0.20	7.82	0.24	9.62
18.5	0.11	6.98	0.13	7.95	0.15	9.77
19.0	0.11	7.09	0.13	8.08	0.15	9.92
19.5	0.09	7.18	0.11	8.19	0.13	10.05
20.0	0.09	7.27	0.11	8.30	0.13	10.18
20.5	0.09	7.36	0.10	8.40	0.12	10.30
21.0	0.09	7.45	0.10	8.50	0.12	10.42
21.5	0.08	7.53	0.09	8.59	0.11	10.53
22.0	0.08	7.61	0.09	8.68	0.11	10.64
22.5	0.09	7.70	0.08	8.76	0.10	10.74
23.0	0.07	7.77	0.08	8.84	0.10	10.84
23.5	0.07	7.84	0.08	8.92	0.10	10.94
24.0	0.06	7.90	0.08	9.00	0.06	11.00

APPENDIX C   **439**

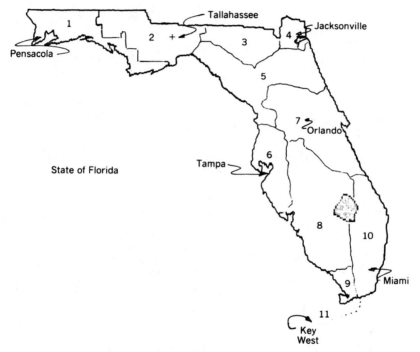

Figure C.1 Zones for precipitation intensity–duration–frequency (IDF) curves developed by the department.

440 NON-DIMENSIONAL RAINFALL AND FREQUENCY – INTENSITY – DURATION CURVES

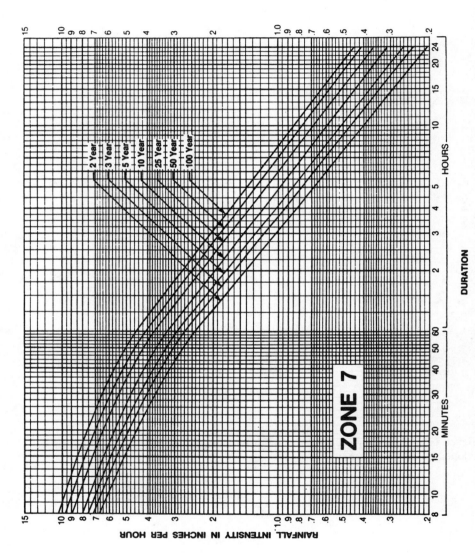

Figure C.2 Rainfall intensity–duration–frequency curves for zone 7.

APPENDIX C **441**

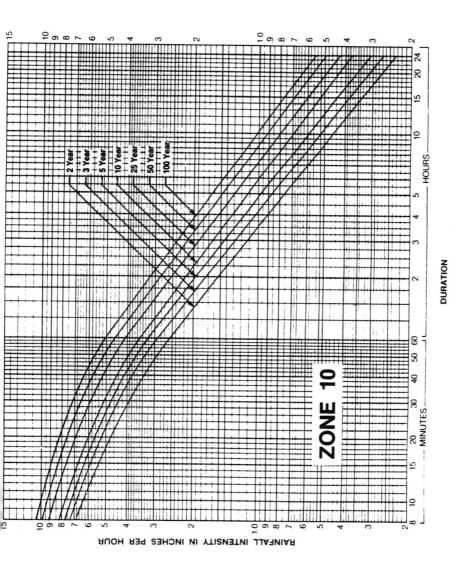

Figure C.3 Rainfall intensity–duration–frequency curves for zone 10.

442 NON-DIMENSIONAL RAINFALL AND FREQUENCY – INTENSITY – DURATION CURVES

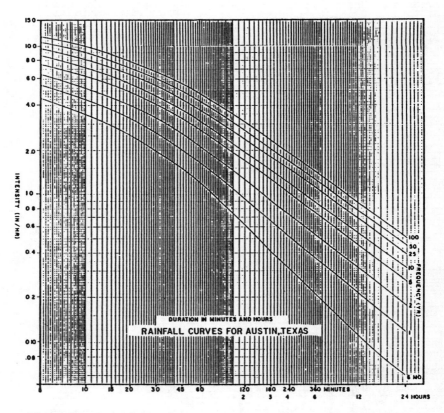

Figure C.4 Rainfall intensity–duration–frequency curves for Austin, Texas.

APPENDIX D
Statistical Tables

TABLE D.1 Normal Distribution Function Table

$$F(z) = \frac{1}{\sqrt{2\pi}} \int_{-\infty}^{z} e^{-(1/2)t^2} dt$$

z	.0	.0100	.0200	.0300	.0400	.0500	.0600	.0700	.0800	.0900
.0	.5000	.5040	.5080	.5120	.5160	.5199	.5239	.5279	.5319	.5359
.10	.5398	.5438	.5478	.5517	.5557	.5596	.5636	.5675	.5714	.5733
.20	.5793	.5832	.5871	.5910	.5948	.5987	.6026	.6064	.6103	.6141
.30	.6179	.6217	.6255	.6293	.6331	.6368	.6406	.6443	.6480	.6517
.40	.6554	.6591	.6628	.6664	.6700	.6736	.6772	.6808	.6844	.6879
.50	.6915	.6950	.6985	.7019	.7054	.7088	.7123	.7157	.7190	.7224
.60	.7257	.7291	.7324	.7356	.7389	.7422	.7454	.7486	.7517	.7549
.70	.7580	.7611	.7642	.7673	.7703	.7734	.7764	.7793	.7823	.7852
.80	.7881	.7910	.7939	.7967	.7995	.8023	.8051	.8078	.8106	.8133
.90	.8159	.8186	.8212	.8238	.8264	.8289	.8315	.8340	.8365	.8389
1.00	.8413	.8437	.8461	.8485	.8508	.8531	.8554	.8577	.8599	.8621
1.10	.8643	.8665	.8686	.8708	.8729	.8749	.8770	.8790	.8810	.8830
1.20	.8849	.8869	.8888	.8906	.8925	.8943	.8962	.8980	.8997	.9015
1.30	.9032	.9049	.9066	.9082	.9099	.9115	.9131	.9147	.9162	.9177
1.40	.9192	.9207	.9222	.9236	.9251	.9265	.9278	.9292	.9306	.9319

	.00	.01	.02	.03	.04	.05	.06	.07	.08	.09
1.50	.9332	.9345	.9357	.9370	.9382	.9394	.9406	.9418	.9429	.9441
1.60	.9452	.9463	.9474	.9484	.9495	.9505	.9515	.9525	.9535	.9545
1.70	.9554	.9564	.9573	.9582	.9591	.9599	.9608	.9616	.9625	.9633
1.80	.9641	.9648	.9656	.9664	.9671	.9678	.9686	.9693	.9699	.9706
1.90	.9713	.9719	.9726	.9732	.9738	.9744	.9750	.9756	.9761	.9767
2.00	.9772	.9778	.9783	.9788	.9793	.9798	.9803	.9808	.9812	.9817
2.10	.9821	.9826	.9830	.9834	.9838	.9842	.9846	.9850	.9854	.9857
2.20	.9861	.9864	.9868	.9871	.9874	.9878	.9881	.9884	.9887	.9890
2.30	.9893	.9895	.9898	.9901	.9904	.9906	.9909	.9911	.9913	.9916
2.40	.9918	.9920	.9922	.9924	.9926	.9928	.9930	.9932	.9934	.9936
2.50	.9938	.9940	.9941	.9943	.9944	.9946	.9949	.9948	.9951	.9952
2.60	.9953	.9955	.9956	.9957	.9958	.9960	.9961	.9962	.9963	.9964
2.70	.9965	.9966	.9967	.9968	.9969	.9970	.9971	.9972	.9973	.9974
2.80	.9974	.9975	.9976	.9977	.9977	.9978	.9979	.9979	.9980	.9981
2.90	.9981	.9982	.9982	.9983	.9983	.9984	.9985	.9985	.9985	.9986
3.00	.9986	.9987	.9987	.9988	.9988	.9988	.9989	.9989	.9990	.9990
3.10	.9990	.9991	.9991	.9991	.9991	.9992	.9992	.9992	.9990	.9993
3.20	.9993	.9993	.9993	.9994	.9994	.9994	.9994	.9995	.9993	.9993
3.30	.9995	.9995	.9995	.9996	.9996	.9996	.9996	.9996	.9995	.9995
3.40	.9997	.9997	.9997	.9997	.9997	.9997	.9997	.9997	.9997	.0097

TABLE D.2 Binomial Distribution Function

$$B(x; N, p) = \sum_{k=0}^{x} \binom{N}{x} p^k (1-p)^{N-k}$$

N	x	.05	.10	.15	.20	.25	.30	.35	.40	.45	.50
2-											
	0	.9025	.8100	.7225	.6400	.5625	.4900	.4225	.3600	.3025	.2500
	1	.9975	.9900	.9775	.9600	.9375	.9100	.8775	.8400	.7975	.7500
3-											
	0	.8574	.7290	.6141	.5120	.4219	.3430	.2746	.2160	.1664	.1250
	1	.9928	.9720	.9393	.8960	.8437	.7840	.7183	.6480	.5748	.5000
	2	.9999	.9990	.9966	.9920	.9844	.9730	.9571	.9360	.9089	.8750
4-											
	0	.8145	.6561	.5220	.4096	.3164	.2401	.1785	.1296	.0915	.0625
	1	.9860	.9477	.8905	.8192	.7383	.6517	.5630	.4752	.3910	.3125
	2	.9995	.9963	.9880	.9728	.9492	.9163	.8735	.8208	.7585	.6875
	3	1.0000	.9999	.9995	.9984	.9961	.9919	.9850	.9744	.9590	.9375
5-											
	0	.7738	.5905	.4437	.3277	.2373	.1681	.1160	.0778	.0503	.0313
	1	.9774	.9185	.8352	.7373	.6328	.5282	.4284	.3370	.2562	.1875
	2	.9988	.9914	.9734	.9421	.8965	.8369	.7648	.6826	.5931	.5000
	3	1.0000	.9995	.9978	.9933	.9844	.9692	.9460	.9130	.8688	.8125
	4	1.0000	1.0000	.9999	.9997	.9990	.9976	.9947	.9898	.9815	.9688

APPENDIX D

n	k	.05	.10	.15	.20	.25	.30	.35	.40	.45	.50
6	0	.7351	.5314	.3771	.2621	.1780	.1176	.0754	.0467	.0277	.0156
	1	.9672	.8857	.7765	.6554	.5339	.4202	.3191	.2333	.1636	.1094
	2	.9978	.9842	.9527	.9011	.8306	.7443	.6471	.5443	.4415	.3438
	3	.9999	.9987	.9941	.9830	.9624	.9295	.8826	.8208	.7447	.6563
	4	1.0000	.9999	.9996	.9984	.9954	.9891	.9777	.9590	.9308	.8906
	5	1.0000	1.0000	1.0000	.9999	.9998	.9993	.9982	.9959	.9917	.9844
7	0	.6983	.4783	.3206	.2097	.1335	.0824	.0490	.0280	.0152	.0078
	1	.9556	.8503	.7166	.5767	.4449	.3294	.2338	.1586	.1024	.0625
	2	.9962	.9743	.9262	.8520	.7564	.6471	.5323	.4199	.3164	.2266
	3	.9998	.9973	.9879	.9667	.9294	.8740	.8002	.7102	.6083	.5000
	4	1.0000	.9998	.9988	.9953	.9871	.9712	.9444	.9037	.8471	.7734
	5	1.0000	1.0000	.9999	.9996	.9987	.9962	.9910	.9812	.9643	.9375
	6	1.0000	1.0000	1.0000	1.0000	.9999	.9998	.9994	.9984	.9963	.9922
8	0	.6634	.4305	.2725	.1678	.1001	.0576	.0319	.0168	.0084	.0039
	1	.9428	.8131	.6572	.5033	.3671	.2553	.1691	.1064	.0632	.0352
	2	.9942	.9619	.8948	.7969	.6785	.5518	.4278	.3154	.2201	.1445
	3	.9996	.9950	.9786	.9437	.8862	.8059	.7064	.5941	.4770	.3633
	4	1.0000	.9996	.9971	.9896	.9727	.9420	.8939	.8263	.7396	.6367
	5	1.0000	1.0000	.9998	.9988	.9958	.9887	.9747	.9502	.9115	.8555
	6	1.0000	1.0000	1.0000	.9999	.9996	.9987	.9964	.9915	.9819	.9648
	7	1.0000	1.0000	1.0000	1.0000	1.0000	.9999	.9998	.9993	.9983	.9961

TABLE D.2 (Continued)

N	x	.05	.10	.15	.20	.25	.30	.35	.40	.45	.50
9-	0	.6302	.3874	.2316	.1342	.0751	.0404	.0207	.0101	.0046	.0020
	1	.9288	.7748	.5995	.4362	.3003	.1960	.1211	.0705	.0385	.0195
	2	.9916	.9470	.8591	.7382	.6007	.4628	.3373	.2318	.1495	.0898
	3	.9994	.9917	.9661	.9144	.8343	.7297	.6089	.4826	.3614	.2539
	4	1.0000	.9991	.9944	.9804	.9511	.9012	.8283	.7334	.6214	.5000
	5	1.0000	.9999	.9994	.9969	.9900	.9747	.9464	.9006	.8342	.7461
	6	1.0000	1.0000	1.0000	.9997	.9987	.9957	.9888	.9750	.9502	.9102
	7	1.0000	1.0000	1.0000	1.0000	.9999	.9996	.9986	.9962	.9909	.9805
	8	1.0000	1.0000	1.0000	1.0000	1.0000	1.0000	.9999	.9997	.9992	.9980
10-	0	.5987	.3487	.1969	.1074	.0563	.0282	.0135	.0060	.0025	.0010
	1	.9139	.7361	.5443	.3758	.2440	.1493	.0860	.0464	.0233	.0107
	2	.9885	.9298	.8202	.6778	.5256	.3828	.2616	.1673	.0996	.0547
	3	.9990	.9872	.9500	.8791	.7759	.6496	.5138	.3823	.2660	.1719
	4	.9999	.9984	.9901	.9672	.9219	.8497	.7515	.6331	.5044	.3770
	5	1.0000	.9999	.9986	.9936	.9803	.9527	.9051	.8338	.7384	.6230
	6	1.0000	1.0000	.9999	.9991	.9965	.9894	.9740	.9452	.8980	.8281
	7	1.0000	1.0000	1.0000	.9999	.9996	.9984	.9952	.9877	.9726	.9453
	8	1.0000	1.0000	1.0000	1.0000	1.0000	.9999	.9995	.9983	.9955	.9893
	9	1.0000	1.0000	1.0000	1.0000	1.0000	1.0000	1.0000	.9999	.9997	.9990

n	k	.05	.10	.15	.20	.25	.30	.35	.40	.45	.50
11	0	.5688	.3138	.1673	.0859	.0422	.0198	.0088	.0036	.0014	.0005
	1	.8981	.6974	.4922	.3221	.1971	.1130	.0606	.0302	.0139	.0059
	2	.9848	.9104	.7788	.6174	.4552	.3127	.2001	.1189	.0652	.0327
	3	.9984	.9815	.9306	.8389	.7133	.5696	.4256	.2963	.1911	.1133
	4	.9999	.9972	.9841	.9496	.8854	.7897	.6683	.5328	.3971	.2744
	5	1.0000	.9997	.9973	.9883	.9657	.9218	.8513	.7535	.6331	.5000
	6	1.0000	1.0000	.9997	.9980	.9924	.9784	.9499	.9006	.8262	.7256
	7	1.0000	1.0000	1.0000	.9998	.9988	.9957	.9878	.9707	.9390	.8867
	8	1.0000	1.0000	1.0000	1.0000	.9999	.9994	.9980	.9941	.9852	.9673
	9	1.0000	1.0000	1.0000	1.0000	1.0000	1.0000	.9998	.9993	.9978	.9941
	10	1.0000	1.0000	1.0000	1.0000	1.0000	1.0000	1.0000	1.0000	.9998	.9995
12	0	.5404	.2824	.1422	.0687	.0317	.0138	.0057	.0022	.0008	.0002
	1	.8816	.6590	.4435	.2749	.1584	.0850	.0424	.0196	.0083	.0032
	2	.9804	.8891	.7358	.5583	.3907	.2528	.1513	.0834	.0421	.0193
	3	.9978	.9744	.9078	.7946	.6488	.4925	.3467	.2253	.1345	.0730
	4	.9998	.9957	.9761	.9274	.8424	.7237	.5833	.4382	.3044	.1938
	5	1.0000	.9995	.9954	.9806	.9456	.8822	.7873	.6652	.5269	.3872
	6	1.0000	.9999	.9993	.9961	.9857	.9614	.9154	.8418	.7393	.6128
	7	1.0000	1.0000	.9999	.9994	.9972	.9905	.9745	.9427	.8883	.8062
	8	1.0000	1.0000	1.0000	.9999	.9996	.9983	.9944	.9847	.9644	.9270
	9	1.0000	1.0000	1.0000	1.0000	1.0000	.9998	.9992	.9972	.9921	.9807
	10	1.0000	1.0000	1.0000	1.0000	1.0000	1.0000	.9999	.9997	.9989	.9968
	11	1.0000	1.0000	1.0000	1.0000	1.0000	1.0000	1.0000	1.0000	.9999	.9998

TABLE D.2 *(Continued)*

N	x	.05	.10	.15	.20	.25	.30	.35	.40	.45	.50
13-											
	0	.5133	.2542	.1209	.0550	.0238	.0097	.0037	.0013	.0004	.0001
	1	.8646	.6213	.3983	.2336	.1267	.0637	.0296	.0126	.0049	.0017
	2	.9755	.8661	.6920	.5017	.3326	.2025	.1132	.0579	.0269	.0112
	3	.9969	.9658	.8820	.7473	.5843	.4206	.2783	.1686	.0929	.0461
	4	.9997	.9935	.9658	.9009	.7940	.6543	.5005	.3530	.2279	.1334
	5	1.0000	.9991	.9925	.9700	.9198	.8346	.7159	.5744	.4268	.2905
	6	1.0000	.9999	.9987	.9930	.9757	.9376	.8705	.7712	.6437	.5000
	7	1.0000	1.0000	.9998	.9988	.9944	.9818	.9538	.9023	.8212	.7095
	8	1.0000	1.0000	1.0000	.9998	.9990	.9960	.9874	.9679	.9302	.8666
	9	1.0000	1.0000	1.0000	1.0000	.9999	.9993	.9975	.9922	.9797	.9539
	10	1.0000	1.0000	1.0000	1.0000	1.0000	.9999	.9997	.9987	.9959	.9888
	11	1.0000	1.0000	1.0000	1.0000	1.0000	1.0000	1.0000	.9999	.9995	.9983
	12	1.0000	1.0000	1.0000	1.0000	1.0000	1.0000	1.0000	1.0000	1.0000	.9999
14-											
	0	.4877	.2288	.1028	.0440	.0178	.0068	.0024	.0008	.0002	.0001
	1	.8470	.5846	.3567	.1979	.1010	.0475	.0205	.0081	.0029	.0009
	2	.9699	.8416	.6479	.4481	.2811	.1608	.0839	.0398	.0170	.0065
	3	.9958	.9559	.8535	.6982	.5213	.3552	.2205	.1243	.0632	.0287
	4	.9996	.9908	.9533	.8702	.7415	.5842	.4227	.2793	.1672	.0898
	5	1.0000	.9985	.9885	.9561	.8883	.7805	.6405	.4859	.3373	.2120
	6	1.0000	.9998	.9978	.9884	.9617	.9067	.8164	.6925	.5461	.3953
	7	1.0000	1.0000	.9997	.9976	.9897	.9685	.9247	.8499	.7414	.6047

APPENDIX D

n	k	.05	.10	.15	.20	.25	.30	.35	.40	.45	.50
14	8	1.0000	1.0000	1.0000	.9996	.9978	.9917	.9757	.9417	.8811	.7880
	9	1.0000	1.0000	1.0000	1.0000	.9997	.9983	.9940	.9825	.9574	.9102
	10	1.0000	1.0000	1.0000	1.0000	1.0000	.9998	.9989	.9961	.9886	.9713
	11	1.0000	1.0000	1.0000	1.0000	1.0000	1.0000	.9999	.9994	.9978	.9935
	12	1.0000	1.0000	1.0000	1.0000	1.0000	1.0000	1.0000	.9999	.9997	.9991
	13	1.0000	1.0000	1.0000	1.0000	1.0000	1.0000	1.0000	1.0000	1.0000	.9999
15	0	.4633	.2059	.0874	.0352	.0134	.0047	.0016	.0005	.0001	.0000
	1	.8290	.5490	.3186	.1671	.0802	.0353	.0142	.0052	.0017	.0005
	2	.9638	.8159	.6042	.3980	.2361	.1268	.0617	.0271	.0107	.0037
	3	.9945	.9444	.8227	.6482	.4613	.2969	.1727	.0905	.0424	.0176
	4	.9994	.9873	.9383	.8358	.6865	.5155	.3519	.2173	.1204	.0592
	5	.9999	.9978	.9832	.9389	.8516	.7216	.5643	.4032	.2608	.1509
	6	1.0000	.9997	.9964	.9819	.9434	.8689	.7548	.6098	.4522	.3036
	7	1.0000	1.0000	.9994	.9958	.9827	.9500	.8868	.7869	.6535	.5000
	8	1.0000	1.0000	.9999	.9992	.9958	.9848	.9578	.9050	.8182	.6964
	9	1.0000	1.0000	1.0000	.9999	.9992	.9963	.9876	.9662	.9231	.8491
	10	1.0000	1.0000	1.0000	1.0000	.9999	.9993	.9972	.9907	.9745	.9408
	11	1.0000	1.0000	1.0000	1.0000	1.0000	.9999	.9995	.9981	.9937	.9824
	12	1.0000	1.0000	1.0000	1.0000	1.0000	1.0000	.9999	.9997	.9989	.9963
	13	1.0000	1.0000	1.0000	1.0000	1.0000	1.0000	1.0000	1.0000	.9999	.9995
	14	1.0000	1.0000	1.0000	1.0000	1.0000	1.0000	1.0000	1.0000	1.0000	1.0000
16	0	.4401	.1853	.0743	.0281	.0100	.0033	.0010	.0003	.0001	.0000

STATISTICAL TABLES

TABLE D.2 (Continued)

N	x	.05	.10	.15	.20	.25	.30	.35	.40	.45	.50
	1	.8108	.5147	.2839	.1407	.0635	.0261	.0098	.0033	.0010	.0003
	2	.9571	.7893	.5614	.3518	.1971	.0994	.0451	.0183	.0066	.0021
	3	.9930	.9316	.7899	.5981	.4050	.2459	.1339	.0651	.0281	.0106
	4	.9991	.9830	.9209	.7982	.6302	.4499	.2892	.1666	.0853	.0384
	5	.9999	.9967	.9765	.9183	.8103	.6598	.4900	.3288	.1976	.1051
	6	1.0000	.9995	.9944	.9733	.9204	.8247	.6881	.5272	.3660	.2272
	7	1.0000	.9999	.9989	.9930	.9729	.9256	.8406	.7161	.5629	.4018
	8	1.0000	1.0000	.9998	.9985	.9925	.9743	.9329	.8577	.7441	.5982
	9	1.0000	1.0000	1.0000	.9998	.9984	.9929	.9771	.9417	.8759	.7728
	10	1.0000	1.0000	1.0000	1.0000	.9997	.9984	.9938	.9809	.9514	.8949
	11	1.0000	1.0000	1.0000	1.0000	1.0000	.9997	.9987	.9951	.9851	.9616
	12	1.0000	1.0000	1.0000	1.0000	1.0000	1.0000	.9998	.9991	.9965	.9894
	13	1.0000	1.0000	1.0000	1.0000	1.0000	1.0000	1.0000	.9999	.9994	.9979
	14	1.0000	1.0000	1.0000	1.0000	1.0000	1.0000	1.0000	1.0000	.9999	.9997
	15	1.0000	1.0000	1.0000	1.0000	1.0000	1.0000	1.0000	1.0000	1.0000	1.0000
17	0	.4181	.1668	.0631	.0225	.0075	.0023	.0007	.0002	.0000	.0000
	1	.7922	.4818	.2525	.1182	.0501	.0193	.0067	.0021	.0006	.0001
	2	.9497	.7618	.5198	.3096	.1637	.0774	.0327	.0123	.0041	.0012
	3	.9912	.9174	.7556	.5489	.3530	.2019	.1028	.0464	.0184	.0064
	4	.9988	.9779	.9013	.7582	.5739	.3887	.2348	.1260	.0596	.0245
	5	.9999	.9953	.9681	.8943	.7653	.5968	.4197	.2639	.1471	.0717

APPENDIX D **453**

6	1.0000	.9992	.9917	.9623	.8929	.7752	.6188	.4478	.2902	.1662
7	1.0000	.9999	.9983	.9891	.9598	.8954	.7872	.6405	.4743	.3145
8	1.0000	1.0000	.9997	.9974	.9876	.9597	.9006	.8011	.6626	.5000
9	1.0000	1.0000	1.0000	.9995	.9969	.9873	.9617	.9081	.8166	.6855
10	1.0000	1.0000	1.0000	.9999	.9994	.9968	.9880	.9652	.9174	.8338
11	1.0000	1.0000	1.0000	1.0000	.9999	.9993	.9970	.9894	.9699	.9283
12	1.0000	1.0000	1.0000	1.0000	1.0000	.9999	.9994	.9975	.9914	.9755
13	1.0000	1.0000	1.0000	1.0000	1.0000	1.0000	.9999	.9995	.9981	.9936
14	1.0000	1.0000	1.0000	1.0000	1.0000	1.0000	1.0000	.9999	.9997	.9988
15	1.0000	1.0000	1.0000	1.0000	1.0000	1.0000	1.0000	1.0000	1.0000	.9999
16	1.0000	1.0000	1.0000	1.0000	1.0000	1.0000	1.0000	1.0000	1.0000	1.0000
0	.3972	.1501	.0536	.0180	.0056	.0016	.0004	.0001	.0000	.0000
1	.7735	.4503	.2241	.0991	.0395	.0142	.0046	.0013	.0003	.0001
2	.9419	.7338	.4797	.2713	.1353	.0600	.0236	.0082	.0025	.0007
3	.9891	.9018	.7202	.5010	.3057	.1646	.0783	.0328	.0120	.0038
4	.9985	.9718	.8794	.7164	.5187	.3327	.1886	.0942	.0411	.0154
5	.9998	.9936	.9581	.8671	.7174	.5344	.3550	.2088	.1077	.0481
6	1.0000	.9988	.9882	.9487	.8610	.7217	.5491	.3743	.2258	.1189
7	1.0000	.9998	.9973	.9837	.9431	.8593	.7283	.5634	.3915	.2403
8	1.0000	1.0000	.9995	.9957	.9807	.9404	.8609	.7368	.5778	.4073
9	1.0000	1.0000	.9999	.9991	.9946	.9790	.9403	.8653	.7473	.5927
10	1.0000	1.0000	1.0000	.9998	.9988	.9939	.9788	.9424	.8720	.7597
11	1.0000	1.0000	1.0000	1.0000	.9998	.9986	.9938	.9797	.9463	.8811
12	1.0000	1.0000	1.0000	1.0000	1.0000	.9997	.9986	.9942	.9817	.9519

TABLE D.2 (Continued)

N	x	.05	.10	.15	.20	.25	.30	.35	.40	.45	.50
	13	1.0000	1.0000	1.0000	1.0000	1.0000	1.0000	.9997	.9987	.9951	.9846
	14	1.0000	1.0000	1.0000	1.0000	1.0000	1.0000	1.0000	.9998	.9990	.9962
	15	1.0000	1.0000	1.0000	1.0000	1.0000	1.0000	1.0000	1.0000	.9999	.9993
	16	1.0000	1.0000	1.0000	1.0000	1.0000	1.0000	1.0000	1.0000	1.0000	.9999
	17	1.0000	1.0000	1.0000	1.0000	1.0000	1.0000	1.0000	1.0000	1.0000	1.0000
19	0	.3774	.1351	.0456	.0144	.0042	.0011	.0003	.0001	.0000	.0000
	1	.7547	.4203	.1985	.0829	.0310	.0104	.0031	.0008	.0002	.0000
	2	.9335	.7054	.4413	.2369	.1113	.0462	.0170	.0055	.0015	.0004
	3	.9868	.8850	.6841	.4551	.2631	.1332	.0591	.0230	.0077	.0022
	4	.9980	.9648	.8556	.6733	.4654	.2822	.1500	.0696	.0280	.0096
	5	.9998	.9914	.9463	.8369	.6678	.4739	.2968	.1629	.0777	.0318
	6	1.0000	.9983	.9837	.9324	.8251	.6655	.4812	.3081	.1727	.0835
	7	1.0000	.9997	.9959	.9767	.9225	.8180	.6656	.4878	.3169	.1796
	8	1.0000	1.0000	.9992	.9933	.9713	.9161	.8145	.6675	.4940	.3238
	9	1.0000	1.0000	.9999	.9984	.9911	.9674	.9125	.8139	.6710	.5000
	10	1.0000	1.0000	1.0000	.9997	.9977	.9895	.9653	.9115	.8159	.6762
	11	1.0000	1.0000	1.0000	1.0000	.9995	.9972	.9886	.9648	.9129	.8204
	12	1.0000	1.0000	1.0000	1.0000	.9999	.9994	.9969	.9884	.9658	.9165
	13	1.0000	1.0000	1.0000	1.0000	1.0000	.9999	.9993	.9969	.9891	.9682
	14	1.0000	1.0000	1.0000	1.0000	1.0000	1.0000	.9999	.9994	.9972	.9904

APPENDIX D

	.05	.10	.15	.20	.25	.30	.35	.40	.45	.50
15	1.0000	1.0000	1.0000	1.0000	1.0000	1.0000	1.0000	.9999	.9995	.9978
16	1.0000	1.0000	1.0000	1.0000	1.0000	1.0000	1.0000	1.0000	.9999	.9996
17	1.0000	1.0000	1.0000	1.0000	1.0000	1.0000	1.0000	1.0000	1.0000	1.0000
18	1.0000	1.0000	1.0000	1.0000	1.0000	1.0000	1.0000	1.0000	1.0000	1.0000

$n = 20$

x	.05	.10	.15	.20	.25	.30	.35	.40	.45	.50
0	.3585	.1216	.0388	.0115	.0032	.0008	.0002	.0000	.0000	.0000
1	.7358	.3917	.1756	.0692	.0243	.0076	.0021	.0005	.0001	.0000
2	.9245	.6769	.4049	.2061	.0913	.0355	.0121	.0036	.0009	.0002
3	.9841	.8670	.6477	.4114	.2252	.1071	.0444	.0160	.0049	.0013
4	.9974	.9568	.8298	.6296	.4148	.2375	.1182	.0510	.0189	.0059
5	.9997	.9887	.9327	.8042	.6172	.4164	.2454	.1256	.0553	.0207
6	1.0000	.9976	.9781	.9133	.7858	.6080	.4166	.2500	.1299	.0577
7	1.0000	.9996	.9941	.9679	.8982	.7723	.6010	.4159	.2520	.1316
8	1.0000	.9999	.9987	.9900	.9591	.8867	.7624	.5956	.4143	.2517
9	1.0000	1.0000	.9998	.9974	.9861	.9520	.8782	.7553	.5914	.4119
10	1.0000	1.0000	1.0000	.9994	.9961	.9829	.9468	.8725	.7507	.5881
11	1.0000	1.0000	1.0000	.9999	.9991	.9949	.9804	.9435	.8692	.7483
12	1.0000	1.0000	1.0000	1.0000	.9998	.9987	.9940	.9790	.9420	.8684
13	1.0000	1.0000	1.0000	1.0000	1.0000	.9997	.9985	.9935	.9786	.9423
14	1.0000	1.0000	1.0000	1.0000	1.0000	1.0000	.9997	.9984	.9936	.9793
15	1.0000	1.0000	1.0000	1.0000	1.0000	1.0000	1.0000	.9997	.9985	.9941
16	1.0000	1.0000	1.0000	1.0000	1.0000	1.0000	1.0000	1.0000	.9997	.9987
17	1.0000	1.0000	1.0000	1.0000	1.0000	1.0000	1.0000	1.0000	1.0000	.9998
18	1.0000	1.0000	1.0000	1.0000	1.0000	1.0000	1.0000	1.0000	1.0000	1.0000
19	1.0000	1.0000	1.0000	1.0000	1.0000	1.0000	1.0000	1.0000	1.0000	1.0000

TABLE D.3 *K*-Standard Deviates—Log Pearson Type III Positive Skew Values (+G)

G \ P	.999	.990	.975	.950	.900	.500	.100	.050	.025	.010	.001
.0	−3.090	−2.326	−1.960	−1.645	−1.282	.000	1.281	1.645	1.960	2.326	3.090
.1	−2.948	−2.253	−1.912	−1.616	−1.270	−0.166	1.292	1.673	2.007	2.400	3.233
.2	−2.808	−2.178	−1.864	−1.586	−1.258	−0.330	1.301	1.700	2.053	2.472	3.377
.3	−2.670	−2.104	−1.814	−1.555	−1.245	−0.500	1.309	1.726	2.098	2.544	3.521
.4	−2.533	−2.030	−1.764	−1.523	−1.231	−0.665	1.317	1.750	2.142	2.615	3.666
.5	−2.400	−1.954	−1.714	−1.491	−1.216	−0.830	1.323	1.774	2.185	2.686	3.810
.6	−2.688	−1.880	−1.662	−1.458	−1.200	−0.099	1.330	1.797	2.227	2.755	3.955
.7	−2.140	−1.806	−1.611	−1.423	−1.835	−0.116	1.333	1.818	2.268	2.823	4.100
.8	−2.017	−1.733	−1.560	−1.388	−1.656	−0.132	1.336	1.840	2.308	2.891	4.244
.9	−1.899	−1.660	−1.507	−1.353	−1.147	−0.148	1.339	1.858	2.346	2.957	4.388
1.0	−1.786	−1.588	−1.455	−1.317	−1.280	−0.164	1.340	1.877	2.383	3.022	4.531
1.5	−1.313	−1.256	−1.200	−1.130	−1.018	−0.240	1.333	1.951	2.552	3.330	5.233
2.0	−0.999	.989	−0.975	.949	−0.895	−0.307	1.303	1.995	2.688	3.605	5.907
2.5	−0.800	−0.799	−0.797	−0.790	−0.770	−0.360	1.250	2.012	2.793	3.845	6.548
3.0	−0.667	−0.666	−0.666	−0.665	−0.660	−0.396	1.180	2.003	2.867	4.051	7.152
4.0	−0.500	−0.500	−0.500	−0.499	−0.499	−0.413	1.000	1.920	2.933	4.367	8.253
5.0	−0.400	−0.400	−0.400	−0.400	−0.400	−0.379	.795	1.773	2.909	4.457	9.219

APPENDIX D **457**

6.0	−0.333	−0.333	−0.333	−0.333	−0.330	.589	1.585	2.817	4.687	10.068	
7.0	−0.285	−0.285	−0.285	−0.285	−0.285	.400	1.377	2.676	4.726	10.813	
8.0	−0.250	−0.250	−0.250	−0.250	−0.249	.239	1.163	2.500	4.705	11.468	
.0	−3.090	−2.326	−1.960	−1.645	−1.281	.000	1.281	1.645	1.960	2.326	3.090
−0.1	−3.233	−2.400	−2.007	−1.673	−1.291	.166	1.270	1.616	1.912	2.252	2.948
−0.2	−3.377	−2.472	−2.053	−1.700	−1.301	.033	1.258	1.586	1.863	2.178	2.808
−0.3	−3.352	−2.544	−2.098	−1.726	−1.309	.050	1.245	1.555	1.814	2.104	2.670
−0.4	−3.666	−2.615	−2.142	−1.750	−1.317	.066	1.231	1.523	1.764	2.029	2.532
−0.5	−3.810	−2.685	−2.185	−1.774	−1.323	.083	1.216	1.491	1.713	1.954	2.400
−0.6	−3.956	−2.755	−2.220	−1.800	−1.330	.099	1.200	1.457	1.662	1.880	2.268
−0.7	−4.100	−2.823	−2.268	−1.818	−1.332	.116	1.183	1.423	1.611	1.806	2.350
−0.8	−4.244	−2.891	−2.307	−1.839	−1.336	.132	1.166	1.388	1.560	1.732	2.184
−0.9	−4.388	−2.957	−2.346	−1.858	−1.338	.148	1.147	1.353	1.507	1.660	2.030
−1.0	−4.531	−3.022	−2.383	−1.876	−1.340	.164	1.127	1.317	1.455	1.588	1.884
−1.5	−5.233	−3.330	−2.552	−1.950	−1.333	.240	1.018	1.130	1.200	1.256	1.312
−2.0	−5.907	−3.605	−2.689	−1.995	−1.303	.306	.894	.949	.975	.990	.999
−2.5	−6.548	−3.845	−2.793	−2.012	−1.250	.360	.770	.790	.797	.799	.800
−3.0	−7.152	−4.051	−2.867	−2.003	−1.180	.395	.660	.665	.666	.666	.666
−4.0	−8.252	−4.367	−2.933	−1.920	−1.000	.413	.499	.500	.500	.500	.500
−5.0	−9.219	−4.573	−2.909	−1.773	−0.795	.380	.400	.400	.400	.400	.400
−6.0	−10.068	−4.686	−2.817	−1.585	−0.590	.329	.333	.333	.333	.333	.333
−7.0	−10.813	−4.726	−2.676	−1.377	−0.400	.285	.285	.285	.285	.285	.285

TABLE D.4 Parameters δ for Standard Error of Normal Distribution

EXCEEDENCE PROBABILITY IN PERCENT						
50.0	20.0	10.0	4.0	2.0	1.0	.2
CORRESPONDING RETURN PERIOD IN YEARS						
2	5	10	25	50	100	500
1.0000	1.1637	1.3496	1.5916	1.7634	1.9253	2.2624

TABLE D.5 Parameter δ for Standard Error of Log-Normal Distribution

	EXCEEDENCE PROBABILITY IN PERCENT						
	50.0	20.0	10.0	4.0	2.0	1.0	0.2
COEFFICIENT OF VARIATION	CORRESPONDING RETURN PERIOD IN YEARS						
	2	5	10	25	50	100	500
.05	.9983	1.2162	1.4323	1.7105	1.9087	2.0968	2.4939
.10	.9932	1.2698	1.5222	1.8453	2.0766	2.2979	2.7714
.15	.9848	1.3241	1.6187	1.9956	2.2676	2.5298	3.0993
.20	.9733	1.3784	1.7211	2.1613	2.4819	2.7940	3.4820
.25	.9589	1.4323	1.8289	2.3423	2.7202	3.0917	3.9241
.30	.9420	1.4855	1.9417	2.5383	2.9829	3.4246	4.4305
.35	.9229	1.5378	2.0591	2.7496	3.2708	3.7942	5.0065
.40	.9021	1.5890	2.1811	2.9762	3.5845	4.2023	5.6574
.45	.8801	1.6389	2.3074	3.2184	3.9251	4.6508	6.3890
.50	.8575	1.6876	2.4382	3.4766	4.2935	5.1418	7.2076
.55	.8351	1.7351	2.5735	3.7514	4.6910	5.6774	8.1196
.60	.8138	1.7814	2.7134	4.0435	5.1190	6.2604	9.1322
.65	.7945	1.8266	2.8583	4.3535	5.5790	6.8934	10.2529
.70	.7784	1.8709	3.0085	4.6826	6.0729	7.5794	11.4897
.75	.7669	1.9143	3.1644	5.0316	6.6024	8.3217	12.8513
.80	.7615	1.9570	3.3264	5.4018	7.1698	9.1238	14.3468
.85	.7635	1.9991	3.4949	5.7945	7.7773	9.9894	15.9861
.90	.7746	2.0408	3.6705	6.2109	8.4272	10.9225	17.7796
.95	.7959	2.0821	3.8536	6.6524	9.1221	11.9272	19.7381
1.00	.8284	2.1232	4.0449	7.1206	9.8646	13.0081	21.8734

TABLE D.6 Parameter δ for Standard Error of Gumbel Extreme-Value Distribution

SAMPLE SIZE n	EXCEEDENCE PROBABILITY IN PERCENT						
	50.0	20.0	10.0	4.0	2.0	1.0	0.2
	CORRESPONDING RETURN PERIOD IN YEARS						
	2	5	10	25	50	100	500
10	.9305	1.8540	2.6200	3.6275	4.3870	5.1460	6.9103
15	.9270	1.7695	2.4756	3.4083	4.1127	4.8173	6.4565
20	.9250	1.7249	2.3990	3.2919	3.9670	4.6427	6.2154
25	.9237	1.6968	2.3507	3.2183	3.8748	4.5322	6.0626
30	.9229	1.6772	2.3169	3.1667	3.8103	4.4547	5.9556
35	.9223	1.6627	2.2919	3.1286	3.7624	4.3973	5.8763
40	.9218	1.6514	2.2725	3.0990	3.7253	4.3528	5.8147
45	.9214	1.6424	2.2569	3.0752	3.6955	4.3171	5.7653
50	.9211	1.6350	2.2441	3.0555	3.6707	4.2874	5.7242
55	.9208	1.6288	2.2333	3.0390	3.6502	4.2626	5.6900
60	.9206	1.6235	2.2241	3.0249	3.6325	4.2414	5.6607
65	.9204	1.6190	2.2163	3.0130	3.6175	4.2234	5.6357
70	.9202	1.6149	2.2092	3.0022	3.6039	4.2071	5.6132
75	.9200	1.6114	2.2032	2.9929	3.5923	4.1932	5.5939
80	.9199	1.6083	2.1977	2.9846	3.5818	4.1806	5.5765
85	.9198	1.6055	2.1929	2.9771	3.5725	4.1694	5.5610
90	.9197	1.6030	2.1885	2.9704	3.5640	4.1592	5.5468
95	.9196	1.6007	2.1845	2.9643	3.5563	4.1500	5.5341
100	.9195	1.5986	2.1808	2.9586	3.5492	4.1414	5.5222

TABLE D.7 Parameter δ for Standard Error of Log-Pearson Type III Distribution

	EXCEEDENCE PROBABILITY IN PERCENT						
	50.0	20.0	10.0	4.0	2.0	1.0	0.2
COEFFICIENT	CORRESPONDING RETURN PERIOD IN YEARS						
OF SKEW	2	5	10	25	50	100	500
.0	1.0801	1.1698	1.3748	1.8013	2.1992	2.6369	3.7212
.1	1.0808	1.2006	1.4368	1.9092	2.3429	2.8174	3.9902
.2	1.0830	1.2310	1.4990	2.0229	2.4990	3.0181	4.3001
.3	1.0866	1.2610	1.5611	2.1414	2.6661	3.2373	4.6486
.4	1.0918	1.2906	1.6228	2.2639	2.8428	3.4732	5.0336
.5	1.0987	1.3200	1.6840	2.3898	3.0283	3.7247	5.4534
.6	1.1073	1.3493	1.7442	2.5182	3.2215	3.9905	5.9066
.7	1.1179	1.3786	1.8033	2.6486	3.4215	4.2695	6.3920
.8	1.1304	1.4083	1.8611	2.7802	3.6274	4.5607	6.9085
.9	1.1449	1.4386	1.9172	2.9123	3.8383	4.8631	7.4550
1.0	1.1614	1.4701	1.9717	3.0442	4.0532	5.1756	8.0303
1.1	1.1799	1.5032	2.0243	3.1751	4.2711	5.4969	8.6335
1.2	1.2003	1.5385	2.0751	3.3043	4.4909	5.8259	9.2631
1.3	1.2223	1.5767	2.1242	3.4311	4.7115	6.1613	9.9177
1.4	1.2457	1.6186	2.1718	3.5546	4.9319	6.5017	10.5959
1.5	1.2701	1.6649	2.2182	3.6741	5.1507	6.8456	11.2957
1.6	1.2951	1.7164	2.2640	3.7891	5.3669	7.1915	12.0155
1.7	1.3202	1.7741	2.3097	3.8989	5.5792	7.5378	12.7231
1.8	1.3450	1.8385	2.3562	4.0029	5.7865	7.8829	13.5064
1.9	1.3687	1.9104	2.4046	4.1008	5.9875	8.2252	14.2731
2.0	1.3907	1.9904	2.4560	4.1922	6.1812	8.5629	15.0508

APPENDIX D 461

TABLE D.8 Confidence Limit Deviate Values for Normal and Log-Normal Distributions

CONFIDENCE LEVEL	SYSTEMATIC RECORD LENGTH n	EXCEEDANCE PROBABILITY								
		.002	.010	.020	.040	.100	.200	.500	.800	.990
.05	10	4.862	3.981	3.549	3.075	2.355	1.702	.580	−.317	−1.563
	15	4.304	3.520	3.136	2.713	2.068	1.482	.455	−.406	−1.677
	20	4.033	3.295	2.934	2.534	1.926	1.370	.387	−.460	−1.749
	25	3.868	3.158	2.809	2.425	1.838	1.301	.342	−.497	−1.801
	30	3.755	3.064	2.724	2.350	1.777	1.252	.310	−.525	−1.840
	40	3.608	2.941	2.613	2.251	1.697	1.188	.266	−.592	−1.896
	50	3.515	2.862	2.542	2.188	1.646	1.146	.237	−.592	−1.936
	60	3.448	2.807	2.492	2.143	1.609	1.116	.216	−.612	−1.966
	70	3.399	2.765	2.454	2.110	1.581	1.093	.199	−.629	−1.990
	80	3.360	2.733	2.425	2.083	1.559	1.076	.186	−.642	−2.010
	90	3.328	2.706	2.400	2.062	1.542	1.061	.175	−.652	−2.026
	100	3.301	2.684	2.380	2.044	1.527	1.049	.166	−.662	−2.040

continued

TABLE D.8 (Continued)

CONFIDENCE LEVEL	SYSTEMATIC RECORD LENGTH n	EXCEEDENCE PROBABILITY							
		.002	.010	.020	.040	.100	.200	.500	.800	.990
.95	10	1.989	1.563	1.348	1.104	.712	.317	−.580	−1.702	−3.981
	15	2.121	1.677	1.454	1.203	.802	.406	−.455	−1.482	−3.520
	20	2.204	1.749	1.522	1.266	.858	.460	−.387	−1.370	−3.295
	25	2.264	1.801	1.569	1.309	.898	.497	−.342	−1.301	−3.158
	30	2.310	1.840	1.605	1.342	.928	.525	−.310	−1.252	−3.064
	40	2.375	1.896	1.657	1.391	.970	.565	−.266	−1.188	−2.941
	50	2.421	1.936	1.694	1.424	1.000	.592	−.237	−1.146	−2.862
	60	2.456	1.966	1.722	1.450	1.022	.612	−.216	−1.116	−2.807
	70	2.484	1.990	1.745	1.470	1.040	.629	−.199	−1.093	−2.765
	80	2.507	2.010	1.762	1.487	1.054	.642	−.186	−1.076	−2.733
	90	2.526	2.026	1.778	1.500	1.066	.652	−.175	−1.061	−2.706
	100	2.542	2.040	1.791	1.512	1.077	.662	−.166	−1.049	−2.684

TABLE D.9 Cumulative Poisson Probabilities

$$F(x;\lambda) = \sum_{x=0}^{x} \frac{e^{-\lambda}\lambda^x}{x!}$$

x \ λ	0.5	1.0	2.0	3.0	4.0	5.0	6.0	7.0	8.0	9.0	10.0	15.0	20.0
0	.607	.368	.135	.050	.018	.007	.002	.001	.000	.000	.000	.000	.000
1	.910	.736	.406	.199	.092	.040	.017	.007	.003	.001	.000	.000	.000
2	.986	.920	.677	.423	.238	.125	.062	.030	.014	.006	.003	.000	.000
3	.998	.981	.857	.647	.433	.265	.151	.082	.042	.021	.010	.000	.000
4	1.000	.996	.947	.815	.629	.440	.285	.173	.100	.055	.029	.001	.000
5		.999	.983	.916	.785	.616	.446	.301	.191	.116	.067	.003	.000
6		1.000	.995	.966	.889	.762	.606	.456	.313	.207	.130	.008	.000
7			.999	.988	.949	.867	.744	.599	.453	.324	.220	.018	.001
8			1.000	.996	.979	.932	.847	.729	.593	.456	.333	.037	.002
9				.999	.992	.968	.916	.830	.717	.587	.458	.070	.005
10				1.000	.997	.986	.957	.901	.816	.706	.583	.118	.011
11					.999	.995	.980	.947	.888	.803	.697	.185	.021
12					1.000	.998	.991	.973	.936	.876	.792	.268	.039
13						.999	.996	.987	.966	.926	.864	.363	.066
14						1.000	.999	.994	.983	.959	.917	.466	.105
15							.999	.998	.992	.978	.951	.568	.157
16							1.000	.999	.996	.989	.973	.664	.221
17								1.000	.998	.995	.986	.749	.297
18									1.000	.999	.993	.819	.381
19										1.000	.997	.875	.470
20											.998	.917	.559
21											.999	.947	.644
22											1.000	.967	.721
23												.981	.787
24												.989	.843
25												.994	.888
30												1.000	.987

Note: Entries of zero and one are rounded to 4 place significant digits.

APPENDIX E

Computer Programs, Descriptions and Example Outputs

E.1 CONTENT AND ORGANIZATION

In this appendix, details are presented on the content, organization, computer hardware, and use of computer programs resident with this book. The programs have been designed to be used by people with a limited understanding of personal computers; however, they must have some knowledge of hydrology and stormwater management. Students of hydrology and stormwater management together with people who practice in the planning, science, and engineering areas will find this appendix and the programs very beneficial. The benefits result because the appendix will not only aid in understanding theory and practice but will also reduce the time for learning and solving practical problems.

There is one diskette with the book. It includes three subdirectories: SMADA, RETEN, and S. SMADA is a hydrograph generation and detention pond design and analysis program. RETEN is a combination of programs that aid in the design and analysis of retention systems and pollution loading estimates. These programs are similar to others used to solve hydrology and stormwater management problems and are written to be user friendly and for learning experiences. User friendly with these programs implies "prompts" or questions that logically are used to enter data or make decisions regarding the use of data. The learning experience can be gained by repetitively solving problems. Repetitive runs are made relatively fast by the computer programs because most of the input and output data can be saved and comparisons made without extensive reentry of data. As with all computation aids, the programs are specific to certain types of practical problems and theory. Thus each program is described and example input data and solutions developed.

At the end of most chapters, there are computer-assisted problem sections. These problems can be solved with a minimum of other reference material, but it is recommended that the reader consult this and other texts and reference material on hydrology, hydraulics, and stormwater management. These references will aid in understanding the data input and interpretation of results.

E.2 COMPUTER CONFIGURATION

The computer diskette with programs are designed to operate on IBM Personal Computers or 100% equivalent compatibles with $3\frac{1}{2}$-in. diskettes. They are low-density formatted. The minimum characteristics for the hardware are:

1. IBM PCs or 100% compatibles
2. Two floppy diskette drives or one floppy drive with a hard disk
3. Color monitor with graphics display or enhanced graphics adapter
4. 512K bytes of RAM (memory)
5. Disk Operating System (DOS) version 3.2 or higher

If you plan on using a diskette for data storage, format the diskette now for future use.

E.3 CARING FOR AND COPYING THE DISKETTE

All diskettes must be stored in an area not subjected to extreme heat, magnetic forces, fluid spills, or rough treatment. Keeping a diskette in a hot automobile has been known to erase or alter the program codes. A hard-covered case for caring and storing the diskettes is valuable protection. A copy of a program diskette should be made immediately after obtaining the diskette. You may have been caring properly for your diskettes, but accidents and pilferage do occur.

Floppy System. Please make a backup copy of the program disk and format another disk that will be used to store the programs. You will need to have graphics resident when using the S program directly and wish to obtain graphs. A commonly used procedure to install the the programs on a floppy high density diskette is:

1. Place DOS diskette in drive A.
2. Place unformatted diskette in drive B (high density).
3. A > FORMAT B:
4. Place the original program disk in drive A.
6. A > INSTALL A: B:

You are finished. Three directories have been created under the main directory called SWATER. The three directories are named SMADA, RETEN, and S.

Hard Disk System. Usually, the hard disk is preferable for storage because the programs will run faster from the hard disk than from the floppy. The install program will create a main program and three subdirectories. A commonly used procedure to install the programs on a hard drive is:

1. Place program diskette in drive A.
2. A > INSTALL A: C:

You are finished. If you have not made a backup copy of the program, please do it now.

Additional copies of the diskettes can be obtained by contacting Dr. M. Wanielista, University of Central Florida, College of Engineering, Orlando, Florida 32816, or by using the order form with two of the programs.

E.4 INITIATING THE PROGRAMS

Your computer must be operational and "booted up" in a DOS format. After the system is up and running, an A > (known as A prompt) will appear. The prompt will appear for either the A, B, C, or other drive. Place the program diskette in drive A, or from your hard disk, and change to the appropriate subdirectory. Then type in one of the following commands:

Program: SMADA   RETEN   S
Command: A:> SMADA   A:> RETEN   A:> S

E.5 SMADA (STORMWATER MANAGEMENT AND DESIGN AID)

Once the command SMADA is keyed and entered, the screen shown in Figure E.1 appears and the user simply enters a return or with a mouse "clicks" on the highlighted area: namely, ⟨OK⟩ for this screen. Immediately, the computer will operate to move the program into memory and display another menu or questions. Some of the programs are large; thus it may take more than a few seconds to load the program. Note that you may exit a screen by pressing the ESC (escape) key.

The SMADA and RETEN programs are written in BASIC and compiled using Microsoft Professional Version 7. Pull-down menus are available with various options on keyboard use or mouse use. Each of these two programs has a help menu that can be customized by a user group. The existing help

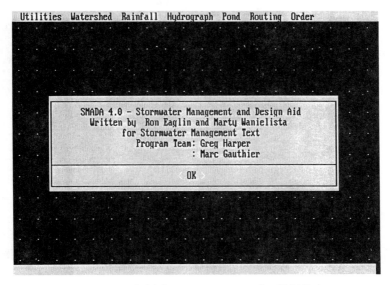

Figure E.1 Initial computer screen for SMADA.

menu will aid in keyboard movement and mouse clicking. In addition, program explanations also are coded in the help menu. The S program is written in the BASIC language and is interactive but without pull-down menus.

The SMADA program is used for both hydrograph generation and stormwater management. SMADA is the acronym for "stormwater management and design aid." The SMADA program menu appears as shown in Figure E.2.

Rain and watershed data files can be stored after entry, and they may be used again and again. A directory of these files may be necessary before the program is initiated; thus the user can use the mouse or keyboard to selectively retrieve files. Frequently, the user may wish to store the program results for printing at the end of a run; thus the user will be asked to specify a drive other than the one presently being used. If the program disk is in drive A, one may wish to specify B:, or C: [note: the colon (:) is necessary for drive specification].

The hydrograph and stormwater management program (SMADA) is one that can be used in a multitude of ways. It is an interactive stormwater management package originally written in the BASIC language. It has an internal directory for file storage. Rainfall and watershed files can be created and used over and over again. Another diskette or a hard drive should be used for file storage. The output is in color and written to the screen; however, one may exercise an option to generate a "hard" copy. The

Figure E.2 Pull-down menu.

program is a useful aid for:

1. Hydrograph generation via rational, SCS, discrete convolution, or Santa Barbara methods
2. Determining detention and off-line retention volumes
3. Sizing pipes and open channels
4. Designing and evaluating detention pond outlet structures
5. Evaluation of watershed conditions as they affect volume and rate of runoff

SMADA can evaluate single or multiple land-use watersheds. For the same rainfall event, each specified subwatershed is evaluated for pollutant and hydrograph effects. The options of pre- versus postdevelopment, pollutant/retention, and peak flow/detention type analyses are available for use on each subwatershed.

E.5.1 Starting SMADA

When the program begins execution, you will see a beginning prompt screen; hit "enter" to continue. The second screen will ask you to enter your access code. This code can be found on the label of your installation disk. It can also be found on the front inside cover of this documentation. Type in this four-digit code and the program will continue execution. If the program is not a properly installed version of SMADA, the program will not execute. The program should be installed using the procedure outlined in Section E.3, page 465.

E.5.2 Using Pull-down Menus

After entering into the SMADA execution field you will see at the top of the screen a series of pull-down menu prompts. These are labeled.

Utilities
Watershed
Rainfall
Hydrograph
Routing
Order

If you have a mouse, the mouse can be used to click any of the pull-down menus. If you do not have a mouse, the Alt key on the keyboard will turn on the pull-down menu field. With the field turned on you can use the cursor keys and the Enter keys to make a selection in any of the pull-down menu fields.

⟨*Utilities Menu*⟩

Using On-Line Help. The Help function contains brief descriptions on selected topics: watersheds, hydrographs, nodal analysis, rainfall, and keyboard movement.

Change Current Directory. This function allows the user to save and retrieve watershed, rainfall, and hydrograph files from locations other than the disk default directory (in which SMADA was started from). If you desire to retrieve files from a disk or another directory, you should use this option. If the current directory is changed for the purpose of retrieving data files, any data files saved during the execution of SMADA will be saved to this directory.

Send Form Feed to Printer. This option allows you to clear the printer buffers to remove any unwanted text from other programs from appearing on your printouts.

Exit. This allows the user to exit SMADA.

⟨Watershed Menu⟩

Creating a Watershed. You can create a watershed by choosing the Create option under Watershed. The watershed infiltration can be estimated by one of two methods: SCS-CN (Soil Conservation Service Curve Number) or the Horton infiltration estimation method. These methods are presented in Chapter 3. Also, descriptions of these methods should also be available in most hydrology textbooks.

Using the Tab key (or mouse) to toggle between the buttons, you should select the method of infiltration estimation that you will use on the watershed. When this selection is made the edit screen will appear allowing you to enter pertinent data about your watershed. When entering these data certain notes should be taken into consideration. Additional abstraction represents that volume of water which will be diverted prior to the watershed outlet. You may have a different additional abstraction on your pervious and impervious watershed regions.

The Edit/View option further explains the input data.

Edit/View Watershed. This routine will allow you to edit any watershed that is currently in memory. If you have not created or retrieved a watershed, you will not be able to use this routine. The following parameters are required:

Watershed Area (acres): total area that will contribute runoff for the design storm

Time of Concentration (minutes): maximum flow time for the design storm

Impervious Area (acres): total impervious area

Percent Directly Connected Impervious Area: for the impervious area, the percentage that is directly connected

Additional Abstraction (inches): that volume retained on site for both the impervious and pervious areas

Maximum Infiltration Capacity (inches): if the soil capacity is limited because of a high water table, if no limitation, enter 999

SCS Pervious Curve Number: for the pervious area only, because the rainfall excess is calculated separately for the pervious and impervious areas

SCS Initial Abstraction Factor: I_A in the rainfall excess formula that has a default value of 0.2

Plus infiltration parameters of the Horton equation if this choice is made.

Retrieve Watershed. This routine will allow you to retrieve any watershed that you have previously saved into the working directory.

Save a Watershed. This routine will save a watershed in the SMADA directory for later use. The name of the watershed should be a legal filename with fewer than eight characters. Do not put an extension on your file, as SMADA will place its own extention on your file, which it will use later to recognize the type of files you have saved.

Delete a Watershed. This routine will allow you to delete a watershed file from the disk when you are done using it.

Print Current Watershed. This routine will allow you to print the current watershed on your printer. Please be sure to have your printed turned on if you use this routine.

⟨ Rainfall Menu ⟩

Creating a Rainfall File. Creating a rainfall file can be done by one of two methods. You can either specify Create Rainfall File or Use Dimensionless Curves from the Rainfall pull-down menu. Either of the two choices will ask you to specify a time increment for use in the rainfall analysis. Recommended values range from 5 to 30 min, depending on the accuracy you require in your analysis. You will also be prompted for the total duration of the rainfall. If you choose "use dimensionless curves," you will be prompted for the total rainfall, which occurs during the rainfall event and the type of dimensionless curve to use. These curves were developed by the Soil Conservation Service, and one was modified for Florida conditions. You will, in either case, be allowed to edit the rainfall data. The total number of rainfall data must not exceed 96 points.

⟨ Hydrograph Menu ⟩

Creating a Hydrograph. After creating (or retrieving) both a watershed and a rainfall file, you are now ready to generate your hydrograph. There are three possible methods that SMADA allows you to use to generate the hydrograph: Santa Barbara method, discrete unit convolution, and rational method. The discrete unit convolution will further prompt you for an attenuation factor. Typical attenuation factors for various circumstances are included in this procedure. When the hydrograph has been generated the hydrograph view function will appear on the screen. To see the entire hydrograph you may scroll through this screen using the cursor buttons or you can use the mouse to click the scroll field. When you are finished viewing the hydrograph you can exit by clicking the = in the upper right-hand corner or by hitting "escape."

Plot Hydrograph. This routine will create a graphics plot of your hydrograph on the screen. If you wish to print this graphics plot on a printer or use a screen capture of this plot, this is possible. You must load graphics.com, which will come with your DOS, in your machine to make printouts of these plots. To do this, type "graphics" from the DOS prompt before executing the program. For ease of use it is sometimes desirable to copy graphics.com to your SMADA directory and create a batch file for running SMADA which executes the graphics.com before executing SMADA for each run. To do this you should copy graphics.com from your DOS directory to SMADA directory. Then using Edlin or any other text editor, create a batch file that includes the following.

GRAPHICS

SMADA

Whenever the batch file is executed from the DOS command line, graphics will be loaded automatically and SMADA will be executed. If using a commercial screen capture routine, follow the direction included with the documentation of the routine.

⟨Routing Menu⟩

Define Node and Type. Before performing a nodal analysis, all of the nodes to be used in the analysis must be defined using this procedure. There are four-types of possible nodes: watershed, pond, junction, and final outlet.

1. *Watershed.* If Watershed is chosen, the current watershed will be defined as the node number that you choose during the procedure.
2. *Pond.* The stage–storage–discharge method requires the user to input a stage storage–discharge relationship. These data are entered in 1-ft increments from the initial stage. It allows the user the capability of entering up to 10 different relationships, at equal depth intervals. The data entered at the values of stage are the storage (in acre-feet) and the discharge (in ft^3/s). SMADA will use these data to route the pond input and produce a value of pond output based on the relationships. To produce this routing the pond should be linked to either an outlet node or a junction node.

   If the users specify a weir condition at the outlet of the pond, they will be asked to choose the type of weir. SMADA allows for two types of weirs: 90° V-notch weirs and sharp-crested rectangular weirs. If a sharp-crested rectangular weir is chosen, the user will also be prompted for a base length in feet. For either weir certain pond conditions must be specified. These conditions are permanent and reuse pool, initial

volume, and pond surface area. The permanent and reuse pool is that volume of water below the invert of the discharge weir structure.
3. *Junction.* A junction node may be used to link multiple watersheds if pipe or channels leading from the watersheds connected downstream before continuing to a pond or outlet. It may also be used to link ponds with outlet structures (pipes, channels) to other nodes.
4. *Final Outlet.* This is a dummy node, which should be specified as the most downstream node in the routing procedure.

Link Defined Node. This is the procedure in which the nodes are actually routed to each other. The linking should always start with the most upstream nodes and proceed downstream. The procedure will ask the user to specify a source node and a destination node. Certain other information may be requested from the user based on the type of node specified as the source node.

In the case of a watershed node or a junction node, the outlet structure must be specified. SMADA allows for two possible outlet structures: circular pipes and trapezoidal channels. In addition to these outlet structures, the user can also specify that no outlet structure exists. The program will design the pipe or channel as it performs the routing. Users may input custom conditions in the case of an outlet pipe. The information that must be specified to perform the various routing functions are longitudinal slope, Manning's n, and structure length. In the case of a pipe, a diameter must be specified; in the case of a trapezoidal channel, a side slope must be specified.

When routing pond nodes the inlet will refer to the water flowing into the pond and the outlet to the water flowing out of the pond. When routing hydrographs "inlet" will refer to the flow from the watershed and "outlet" will refer to the flow at the far end of the channel or pipe. If "no transport structure" is specified, the inlet and outlet flows will be equal.

Nodes defined using nodal analysis should be linked in consecutive order, with the most upstream nodes being defined first. All nodes in the nodal analysis should be defined before any of the nodes are linked.

Change Active Node. In nodal analyses containing multiple nodes, only one node may be viewed on-screen at a given time. This procedure allows users to choose the node they wish to view.

Plot Defined Node. This procedure allows for multiple nodes to be plotted at one time. When prompted for a node to plot, select one of the nodes, and then the user can specify a second (or more) node to plot on the same plot.

Display Defined Nodes. This procedure allows the user to view which nodes have been defined and which of the defined nodes have been linked.

474 COMPUTER PROGRAMS, DESCRIPTIONS AND EXAMPLE OUTPUTS

Example Problem E.1: Nodal Analysis A powerful capability of SMADA is its capability to perform complex nodal analyses. These analyses can become quite complex, and it is recommended that users outline their nodal network on paper before proceeding with the analysis using SMADA. This procedure will be demonstrated with the use of an example (provided by the St. Johns River Water Management District).

Given: Rainfall distribution: SCS type II
Rainfall duration: 24 hours
Total rainfall: 8.5 inches

1. Predevelopment:

   Drainage area = 35.2 acres
   CN = 70, TC = 96 min
   Peak rate factor = 484

2. Postdevelopment:

   Area subbasin 1 = 16 acres
   CN = 89, TC = 60 min
   Peak rate factor = 484

   Area of subbasin 2 = 19.2 acres
   CN = 92, TC = 48 min
   Peak rate factor = 484

Stage–storage–discharge data for structure 1:

Stage (ft)	Storage (acre-ft)	Discharge (ft^3/s)
61.0	0.0	0.0
62.0	0.4	0.0
63.0	0.8	0.0
64.0	1.3	6.0
65.0	1.9	15.1
66.0	2.6	24.2
67.0	3.4	32.0
68.0	4.3	37.2

Initial stage = 61.0 ft.

Stage–storage–discharge data for structure 2;

Stage (ft)	Storage (acre-ft)	Discharge (ft^3/s)
50.0	0.0	0.0
51.0	0.6	0.0
52.0	1.4	0.0
53.0	2.3	9.3
54.0	3.4	24.5
55.0	4.6	41.5
56.0	6.0	58.6
57.0	7.7	74.5

Initial stage = 50.0 ft.

Subwatershed 1 flows directly into pond 1.

Pond 1 connected to pond 2 by 25-ft-long circular concrete pipe on a 0.5% grade.

Subwatershed 2 flows directly into pond 2.

Determine:

1. Predevelopment discharge hydrograph including peak flow rate.
2. Postdevelopment:
   a. Inflow hydrograph into pond 1
   b. Outflow hydrograph from pond 1
   c. Inflow hydrograph into pond 2
   d. Outflow hydrograph from pond 2
   e. Peak stage and storage in ponds 1 and 2

SOLUTION: We will first generate the precondition hydrograph:

1. Choose Create Watershed from Watershed pull-down menu (if not using mouse, use Alt key to access pull-down menus. Choose SCS-CN method of watershed generation and enter data as shown in Figure E.3.

Notice that there is no number entered for the impervious area. The method we are using to solve this problem is the composite curve number method. The 70 which is used for SCS curve number is a composite found by using a weighted average of the impervious and pervious curve numbers over their areas.

476 COMPUTER PROGRAMS, DESCRIPTIONS AND EXAMPLE OUTPUTS

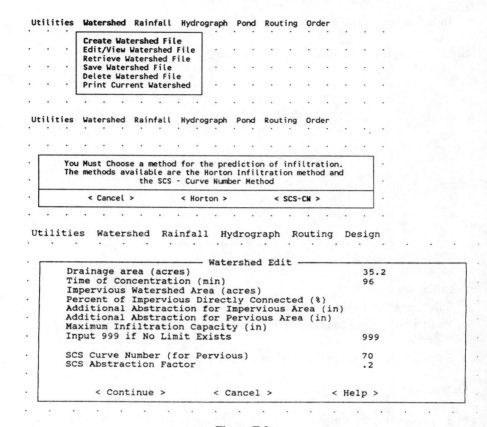

Figure E.3

2. Choose Use Dimensionless Curves from the Rainfall pull-down menu (Figure E.4). Select 30-minute time step and 24-hour duration where prompted, then select type II storm from the storm menu. After selection has been made, the rainfall file will appear on the screen, allowing the user to edit. Hit Esc to leave the edit routine.

3. To generate the hydrograph choose Generate Hydrograph from the Hydrograph pull-down menu. Select the discrete unit convolution method from the generation techniques and select an attenuation factor of 484 as shown in Figure E.5.

4. This hydrograph can be printed to a printer by using the print hydrograph routine. The watershed information can be printed using the print watershed routine. A printout of this watershed is shown in Figure E.6. The print options allow the user to print the respected information to either the

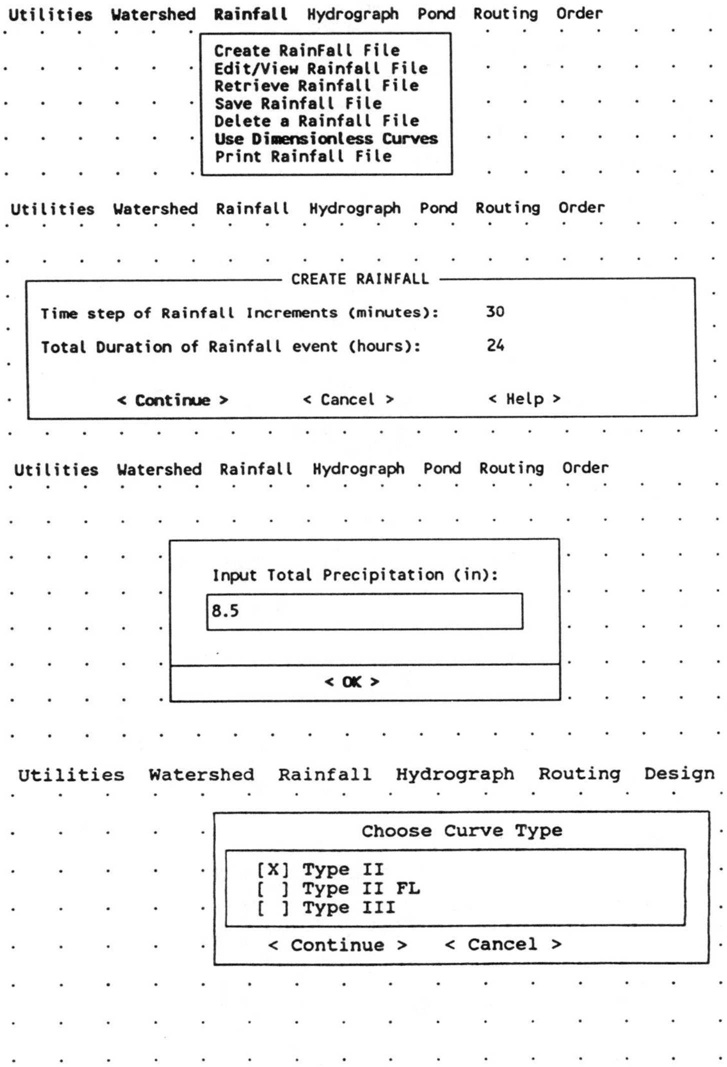

APPENDIX E   **477**

```
Utilities  Watershed  Rainfall  Hydrograph  Pond  Routing  Order
                     ┌─────────────────────────┐
                     │ Create Rainfall File    │
                     │ Edit/View Rainfall File │
                     │ Retrieve Rainfall File  │
                     │ Save Rainfall File      │
                     │ Delete a Rainfall File  │
                     │ Use Dimensionless Curves│
                     │ Print Rainfall File     │
                     └─────────────────────────┘

Utilities  Watershed  Rainfall  Hydrograph  Pond  Routing  Order

          ┌──────────────── CREATE RAINFALL ────────────────┐
          │ Time step of Rainfall Increments (minutes):  30 │
          │ Total Duration of Rainfall event (hours):    24 │
          │                                                 │
          │    < Continue >      < Cancel >     < Help >    │
          └─────────────────────────────────────────────────┘

Utilities  Watershed  Rainfall  Hydrograph  Pond  Routing  Order


              ┌─────────────────────────────────────┐
              │ Input Total Precipitation (in):     │
              │ ┌─────────────────────────────────┐ │
              │ │ 8.5                             │ │
              │ └─────────────────────────────────┘ │
              │                                     │
              │               < OK >                │
              └─────────────────────────────────────┘


Utilities  Watershed  Rainfall  Hydrograph  Routing  Design

              ┌─────────────────────────────────────┐
              │         Choose Curve Type           │
              │   [X] Type II                       │
              │   [ ] Type II FL                    │
              │   [ ] Type III                      │
              │      < Continue >   < Cancel >      │
              └─────────────────────────────────────┘
```

Figure E.4   Rainfall screen

478 COMPUTER PROGRAMS, DESCRIPTIONS AND EXAMPLE OUTPUTS

Utilities Watershed Rainfall Hydrograph Pond Routing Order
─────────────────── Rainfall Edit ───────────────────

Time min	Rain inch	Time min	Rain inch	Time min	Rain inch	Time min	Rain inch	Time min	Rain inch	Time min	Rain inch
30	.04	600	.08	1170	.07						
60	.04	630	.08	1200	.07						
90	.05	660	.38	1230	.06						
120	.06	690	1.31	1260	.05						
150	.06	720	2.63	1290	.05						
180	.06	750	.78	1320	.04						
210	.06	780	.16	1350	.03						
240	.06	810	.16	1380	.03						
270	.07	840	.15	1410	.02						
300	.07	870	.14	1440	.01						
330	.07	900	.14								
360	.07	930	.13								
390	.07	960	.12								
420	.07	990	.11								
450	.07	1020	.11								
480	.07	1050	.1								
510	.08	1080	.09								
540	.08	1110	.09								
570	.08	1140	.08								

Total Rain   8.5 in   < Complete >

Figure E.4 *(Continued)*

Utilities Watershed Rainfall **Hydrograph** Pond Routing Order

```
Generate Hydrograph
View Hydrograph
Retrieve Hydrograph
Save Hydrograph
Plot Hydrograph
Print Hydrograph
```

Utilities Watershed Rainfall Hydrograph Pond Routing Order

```
Choose Generation Method:

[ ] Santa Barbara
[X] Discrete Unit Convolution
[ ] Rational
[ ] SCS 484 Curvilinear Unit
[ ] SCS 256 Curvilinear Unit

   < Continue >    < Cancel >
```

Figure E.5 Hydrograph screen.

APPENDIX E   479

```
Utilities  Watershed  Rainfall  Hydrograph  Routing  Design
┌──────────────────────── Attenuation ────────────────────────┐
│  You must choose a value for the Peak Attenuation Factor.   │
│  This factor is used to simulate a triangular shape for     │
│  hydrograph. As this factor decreases the recession limb    │
│  of the hydrograph lengthens, this physically simulates     │
│  flatter terrains where the water flows slower.             │
│       RL = Recession Limb Time     TP = Time to Peak        │
│                                                             │
│  645: Rational Hydrograph , Steep Terrain                   │
│  Note that Rain Duration must = Time of Concentration       │
│  580: RL = 1.25 * TP                                        │
│  484: RL = 1.67 * TP, Typical SCS Curve                     │
│  400: RL = 2.25 * TP                                        │
│  300: RL = 3.33 * TP, Minimum Slope ,Urban                  │
│  200: RL = 5.50 * TP                                        │
│  100: RL = 12.0 * TP, Rural, Flat Terrain                   │
│                                                             │
│               484               < Cancel >                  │
└─────────────────────────────────────────────────────────────┘
```

```
 Utilities  Watershed  Rainfall  Hydrograph  Pond  Routing  Order
═══════════════════════════ HYDROGRAPH ═══════════════════════════
 Time     Rainfall   Infiltration     Hydrographs in cfs
 Hours    (in)       (in)             Instantaneous    Outflow

 12.5     .78        .16              44.14            34.06
 13       .16        .03               9.27            46.38
 13.5     .16        .03               9.33            46.03
 14       .15        .03               8.81            39.04
 14.5     .14        .02               8.27            30.77
 15       .14        .02               8.31            22.05
 15.5     .13        .02               7.75            13.3
 16       .12        .02               7.19             6.74
 16.5     .11        .02               6.62             7.28
 17       .11        .02               6.64             7.6
 17.5     .1         .01               6.06             7.18
 18       .09        .01               5.47             6.78
 18.5     .09        .01               5.48             6.39
 19       .08        .01               4.88             5.98
 19.5     .07        .01               4.28             5.59

 Total  --- 8.5                       625502.1 cf     625484.3 cf
 Rational Coefficient = .576                Peak Flow = 46.38 cfs
 Current Hydrograph :Unnamed Discrete Unit Convolution
```

Figure E.5 *(Continued)*

screen, the printer (LPT1), or an ASCII file. If you choose to print to a file, you will be prompted for a path and a file name.

Now we generate the postdevelopment conditions:

1. To generate the postdevelopment conditions we will use the routing routines. Each of the postdevelopment hydrographs can be generated in the same manner as described above. For simplicity if following these examples, generate both of the postdevelopment hydrographs similarly to the way the predevelopment hydrograph was generated and save them under the names sub1 and sub2 using the hydrograph save routine. These hydrographs will be used by the routing routine to generate postcondition hydrographs for the pond inflows and outflows. A schematic of the routing is shown in Figure E.7.

COMPUTER PROGRAMS, DESCRIPTIONS AND EXAMPLE OUTPUTS

```
------------------------------------------------------SMADA   PAGE 1
PRE-CONDITION WATERSHED     Hydrograph Type :Discrete Unit Convolution
------------------------------------------------------------------------
```

Time (hr)	Rain (in)	Cumulative (in)	Infiltration (in)	Instantaneous (cfs)	Outflow (cfs)
0.50	0.04	0.04	0.04	0.00	0.00
1.00	0.04	0.08	0.04	0.00	0.00
1.50	0.05	0.13	0.05	0.00	0.00
2.00	0.06	0.19	0.06	0.00	0.00
2.50	0.06	0.25	0.06	0.00	0.00
3.00	0.06	0.31	0.06	0.00	0.00
3.50	0.06	0.37	0.06	0.00	0.00
4.00	0.06	0.43	0.06	0.00	0.00
4.50	0.07	0.50	0.07	0.00	0.00
5.00	0.07	0.57	0.07	0.00	0.00
5.50	0.07	0.64	0.07	0.00	0.00
6.00	0.07	0.71	0.07	0.00	0.00
6.50	0.07	0.78	0.07	0.00	0.00
7.00	0.07	0.85	0.07	0.00	0.00
7.50	0.07	0.92	0.07	0.06	0.00
8.00	0.07	0.99	0.07	0.22	0.03
8.50	0.08	1.07	0.07	0.43	0.08
9.00	0.08	1.15	0.07	0.61	0.17
9.50	0.08	1.23	0.07	0.79	0.30
10.00	0.08	1.31	0.07	0.95	0.46
10.50	0.08	1.39	0.06	1.11	0.63
11.00	0.38	1.77	0.28	7.19	1.25
11.50	1.31	3.08	0.71	42.50	4.94
12.00	2.63	5.71	0.81	129.00	18.20
12.50	0.78	6.49	0.16	44.14	34.06
13.00	0.16	6.65	0.03	9.27	46.38
13.50	0.16	6.81	0.03	9.33	46.03
14.00	0.15	6.96	0.03	8.81	39.04
14.50	0.14	7.10	0.02	8.27	30.77
15.00	0.14	7.24	0.02	8.31	22.05
15.50	0.13	7.37	0.02	7.75	13.30
16.00	0.12	7.49	0.02	7.19	6.74
16.50	0.11	7.60	0.02	6.62	7.28
17.00	0.11	7.71	0.02	6.64	7.60
17.50	0.10	7.81	0.01	6.06	7.18
18.00	0.09	7.90	0.01	5.47	6.78
18.50	0.09	7.99	0.01	5.48	6.39
19.00	0.08	8.07	0.01	4.88	5.98
19.50	0.07	8.14	0.01	4.28	5.59
20.00	0.07	8.21	0.01	4.29	5.21
20.50	0.06	8.27	0.01	3.68	4.82
21.00	0.05	8.32	0.01	3.07	4.41
21.50	0.05	8.37	0.01	3.08	4.02
22.00	0.04	8.41	0.01	2.47	3.62
22.50	0.03	8.44	0.00	1.85	3.20
23.00	0.03	8.47	0.00	1.85	2.81
23.50	0.02	8.49	0.00	1.24	2.40
24.00	0.01	8.50	0.00	0.62	1.98
24.50	0.00	8.50	0.00	0.00	1.53

Figure E.6

```
                                                 -------SMADA   PAGE 2
PRE-CONDITION WATERSHED      Hydrograph Type :Discrete Unit Convolution
-------------------------------------------------------------------
Time       Rain      Cumulative  Infiltration Instantaneous  Outflow
(hr)       (in)        (in)         (in)         (cfs)        (cfs)
-------------------------------------------------------------------
25.00      0.00        8.50         0.00         0.00          1.08
25.50      0.00        8.50         0.00         0.00          0.66
26.00      0.00        8.50         0.00         0.00          0.36
26.50      0.00        8.50         0.00         0.00          0.16
-------------------------------------------------------------------
Total       8.50 in                            625502 cf    625484 cf
Rational Coefficient   = 0.576    Peak Flow =   46.38 cfs
Attenuation Factor = 484
Watershed File: SPRE.SHD      Rainfall File: EXE_1.RNF
-------------------------------------------------------------------
```

Figure E.6 *(Continued)*

2. Once the postcondition hydrographs are generated and saved, we must also generate and save the pond information. This information should be generated using the pond routines (Figure E.8). For simplicity the pond generation for pond 1 will be shown in detail, including the save procedure. The user should follow the same procedure for the second pond but should use a different name for the pond.

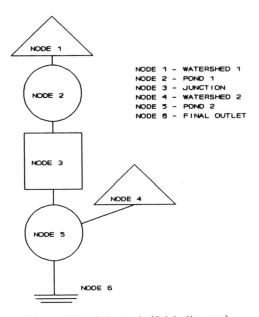

Figure E.7 Schematic (Stick diagram).

482 COMPUTER PROGRAMS, DESCRIPTIONS AND EXAMPLE OUTPUTS

```
Utilities  Watershed  Rainfall  Hydrograph  Pond  Routing  Order
                                            ┌─────────────────────────────┐
                                            │ Create Pond relationship    │
                                            │ Edit Stage Storage Discharge│
                                            │ Retrieve Pond Relationship  │
                                            │ Save Pond Relationship      │
                                            └─────────────────────────────┘

Utilities  Watershed  Rainfall  Hydrograph  Pond  Routing  Order
           ┌──────────────────────────────────────┐
           │       Define Discharge Method        │
           │  ┌────────────────────────────────┐  │
           │  │ [X] Storage Discharge Relation │  │
           │  │ [ ] Outlet Weir Relationship   │  │
           │  └────────────────────────────────┘  │
           │     < Continue >    < Cancel >       │
           └──────────────────────────────────────┘

Utilities  Watershed  Rainfall  Hydrograph  Pond  Routing  Order

           ┌──────────────────────────────────────┐
           │   Minimum stage for relationships    │
           │  ┌────────────────────────────────┐  │
           │  │ 61                             │  │
           │  └────────────────────────────────┘  │
           │              < OK >                  │
           └──────────────────────────────────────┘

Utilities  Watershed  Rainfall  Hydrograph  Pond  Routing  Order

           ┌──────────────────────────────────────┐
           │         How Many Relationships       │
           │  ┌────────────────────────────────┐  │
           │  │ 8                              │  │
           │  └────────────────────────────────┘  │
           │              < OK >                  │
           └──────────────────────────────────────┘
```

Figure E.8

APPENDIX E **483**

Utilities Watershed Rainfall Hydrograph Pond Routing Order

```
Input Increment of Relationships (ft)
[ 1 ]

         < OK >
```

Utilities Watershed Rainfall Hydrograph Routing Design

Stage(ft)	Storage (acre-feet)	Discharge (cfs)
61	0	0
62	.4	0
63	.8	0
64	1.3	6
65	1.9	15.1
66	2.6	24.2
67	3.4	32
68	4.3	37.2

< Continue >   < Cancel >   < Help >

Utilities Watershed Rainfall Hydrograph Pond Routing Order

```
Input initial pond stage (ft)
[ 61 ]

         < OK >
```

Utilities Watershed Rainfall Hydrograph **Pond** Routing Order

```
Create Pond relationship
Edit Stage Storage Discharge
Retrieve Pond Relationship
Save Pond Relationship
```

Utilities Watershed Rainfall Hydrograph Pond Routing Order

```
Enter a File Name (no Extension)
[ C:\SWATER\SMADA\POND1 ]

         < OK >
```

Figure E.8 *(Continued)*

484 COMPUTER PROGRAMS, DESCRIPTIONS AND EXAMPLE OUTPUTS

Utilities Watershed Rainfall Hydrograph Pond **Routing** Order

```
                                        ┌─────────────────────────┐
                                        │ Define Node and Type    │
                                        │ Link Defined Nodes      │
                                        │ Output Batch Data File  │
                                        │ Display Defined Nodes   │
                                        │ Plot Defined Nodes      │
                                        │ Print Nodal Analysis    │
                                        │ View Active Node        │
                                        │ Unlink a Node           │
                                        └─────────────────────────┘
```

Utilities Watershed Rainfall Hydrograph Pond Routing Order

```
        ┌──────────────────────────────────────┐
        │         Choose Node Type             │
        │  ┌────────────────────────────────┐  │
        │  │ [X] Watershed Node             │  │
        │  │ [ ] Pond Node                  │  │
        │  │ [ ] Junction Node              │  │
        │  │ [ ] Final Outlet               │  │
        │  │ [ ] Batch Data File            │  │
        │  └────────────────────────────────┘  │
        │     < Continue >    < Cancel >       │
        └──────────────────────────────────────┘
```

Utilities Watershed Rainfall Hydrograph Pond Routing Order

```
        ┌──────────────────────────────────────┐
        │   Choose Number to assign to node    │
        │  ┌────────────────────────────────┐  │
        │  │ [X] Node  1                    │  │
        │  │ [ ] Node  2                    │  │
        │  │ [ ] Node  3                    │  │
        │  │ [ ] Node  4                    │  │
        │  │ [ ] Node  5                    │  │
        │  │ [ ] Node  6                    │  │
        │  │ [ ] Node  7                    │  │
        │  │ [ ] Node  8                    │  │
        │  │ [ ] Node  9                    │  │
        │  │ [ ] Node 10                    │  │
        │  └────────────────────────────────┘  │
        │     < Continue >    < Cancel >       │
        └──────────────────────────────────────┘
```

Figure E.9

APPENDIX E   **485**

```
Utilities  Watershed  Rainfall  Hydrograph  Pond  Routing  Order


              ┌─────────────────────────────────────┐
              │      Enter a Name for Node #  1     │
              │    ┌───────────────────────────┐    │
              │    │ Watershed Node 1          │    │
              │    └───────────────────────────┘    │
              │                                     │
              ├─────────────────────────────────────┤
              │              < OK >                 │
              └─────────────────────────────────────┘


Utilities  Watershed  Rainfall  Hydrograph  Pond  Routing  Order


              ┌─────────────────────────┐
              │  SUB1.HYD            X  │
              │  SUB2.HYD            ▌  │
              │                         │
              │                         │
              │                         │
              │                         │
              │                         │
              │                      X  │
              │   < OK >   < Cancel >   │
              └─────────────────────────┘


Utilities  Watershed  Rainfall  Hydrograph  Pond  Routing  Order

          ┌──────────────────────────────────────────┐
          │     Choose Number to assign to node      │
          │   ┌────────────────────────────────────┐ │
          │   │  [ ] Node  1   Watershed Node 1    │ │
          │   │  [ ] Node  2   Pond Node 1         │ │
          │   │  [ ] Node  3   Junction Node       │ │
          │   │  [ ] Node  4   Watershed Node 2    │ │
          │   │  [ ] Node  5   Pond Node 2         │ │
          │   │  [X] Node  6                       │ │
          │   │  [ ] Node  7                       │ │
          │   │  [ ] Node  8                       │ │
          │   │  [ ] Node  9                       │ │
          │   │  [ ] Node 10                       │ │
          │   │                                    │ │
          │   │   < Continue >   < Cancel >        │ │
          │   └────────────────────────────────────┘ │
          └──────────────────────────────────────────┘

Current Hydrograph :SUB2.HYD  Current Active Node: Pond Node 2
```

Figure E.9 *(Continued)*

486 COMPUTER PROGRAMS, DESCRIPTIONS AND EXAMPLE OUTPUTS

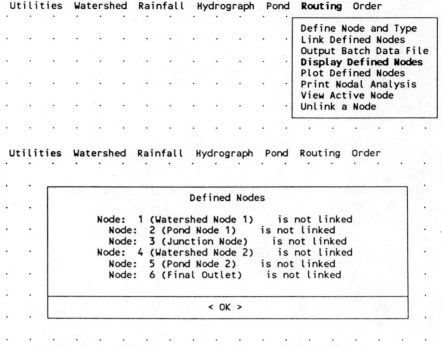

Figure E.9 *(Continued)*

3. The next step in the routing procedure is to define all the nodes. The nodes will be defined as follows:

Node 1	Watershed node:	SUB1.SHD
Node 2	Pond node:	POND1.PND
Node 3	Junction node	
Node 4	Watershed node:	SUB2.SHD
Node 5	Pond node:	POND2.PND
Node 6	Final outlet	

Some of steps used in the defining of the nodes are shown in Figure E.9.

4. The nodes must now be linked using the link nodes routine in the routing pull-down menu (Figure E.10). Link node 1 to node 2 using no outlet structure. Link node 2 to node 3. Link node 3 to node 5 using a circular outlet (pipe), a slope of 0.005, a Manning's of 0.013 (typical for concrete pipes), and a length of 25 ft. Let SMADA select a standard pipe diameter for

APPENDIX E 487

Utilities Watershed Rainfall Hydrograph Pond Routing Order

```
┌─────────────────────────────────────┐
│       Enter Source Node Number      │
│  [X] Node  1   Watershed Node 1     │
│  [ ] Node  2   Pond Node 1          │
│  [ ] Node  3   Junction Node        │
│  [ ] Node  4   Watershed Node 2     │
│  [ ] Node  5   Pond Node 2          │
│  [ ] Node  6   Final Outlet         │
│                                     │
│     < Continue >    < Cancel >      │
└─────────────────────────────────────┘
```

Current Hydrograph :SUB2.HYD Current Active Node: Final Outlet

Utilities Watershed Rainfall Hydrograph Pond Routing Order

```
┌─────────────────────────────────────┐
│      Enter Reception Node Number    │
│  [X] Node  2   Pond Node 1          │
│  [ ] Node  3   Junction Node        │
│  [ ] Node  4   Watershed Node 2     │
│  [ ] Node  5   Pond Node 2          │
│  [ ] Node  6   Final Outlet         │
│                                     │
│     < Continue >    < Cancel >      │
└─────────────────────────────────────┘
```

Current Hydrograph :SUB2.HYD Current Active Node: Final Outlet

Utilities Watershed Rainfall Hydrograph Pond Routing Order

```
┌─────────────────────────────────────┐
│       Choose Transport Geometry     │
│  [ ] Circular                       │
│  [ ] Trapezoidal                    │
│  [X] No Transport Structure         │
│                                     │
│     < Continue >    < Cancel >      │
└─────────────────────────────────────┘
```

Utilities Watershed Rainfall Hydrograph Pond Routing Order

```
┌─────────────────────────────────────┐
│    Required Diameter = 30.51151 in  │
│  ┌───────────────────────────────┐  │
│  │ [X] Use Standard Diameter     │  │
│  │ [ ] Input Custom Diameter     │  │
│  │                               │  │
│  │   < Continue >   < Cancel >   │  │
│  └───────────────────────────────┘  │
└─────────────────────────────────────┘
```

Current Hydrograph :SUB2.HYD

Figure E.10

Utilities   Watershed   Rainfall   Hydrograph   Pond   Routing   Order

```
                    Defined Nodes

Node:  1 (Watershed Node 1)    Flows to Node: 2 (Pond Node 1)
Node:  2 (Pond Node 1)     Flows to Node: 3 (Junction Node)
Node:  3 (Junction Node)   Flows to Node: 5 (Pond Node 2)
Node:  4 (Watershed Node 2)    Flows to Node: 5 (Pond Node 2)
Node:  5 (Pond Node 2)     Flows to Node: 6 (Final Outlet)
       Node:  6 (Final Outlet)    is not linked

                        < OK >
```

Current Hydrograph :SUB2.HYD   Current Active Node: Pond Node 2

Utilities   Watershed   Rainfall   Hydrograph   Pond   Routing   Order

================================ HYDROGRAPH ================================

Time Hours	Rain (in)	Node Inflow (cfs)	Node Outflow (cfs)
12.5	.78	71.93	40.7
13	.16	64.48	51.49
13.5	.16	51.12	51.55
14	.15	36.88	44.44
14.5	.14	26.64	34.38
15	.14	20.08	25.69
15.5	.13	15.44	19.62
16	.12	12.32	15.45
16.5	.11	10.32	12.64
17	.11	9.04	10.72
17.5	.1	8.17	9.36
18	.09	7.46	8.55
18.5	.09	6.88	7.86
19	.08	6.39	7.25
19.5	.07	5.89	6.69

Node : 5   Pond Node
Maximum Volume = 5.48 ac-ft   Maximum Stage = 55.62924 ft

Current Hydrograph :SUB2.HYD

Figure E.10   *(Continued)*

Pond Node 2 Flowing To Final Outlet (Node 6)   SMADA PAGE 1
Pond Node:Output by Storage Discharge Relationship
Maximum Volume =   5.40 acre-feet   Maximum Stage =   55.57 ft

Stage - Storage - Discharge Relationship

Stage (ft)	Storage (Acre-ft)	Discharge (cfs)
50.0	0.00	0.00
51.0	0.60	0.00
52.0	1.40	0.00
53.0	2.30	9.30
54.0	3.40	24.50
55.0	4.60	41.50
56.0	6.00	58.60
57.0	7.70	74.50

Node :   5 - Pond Node 2   SMADA PAGE   2

Time(hr)	Inflow (cfs)	Outflow(cfs)	Stage (ft)
0.50	0.00	0.00	50.00
1.00	0.00	0.00	50.00
1.50	0.00	0.00	50.00
2.00	0.00	0.00	50.00
2.50	0.00	0.00	50.00
3.00	0.02	0.00	50.00
3.50	0.08	0.00	50.01
4.00	0.20	0.00	50.02
4.50	0.35	0.00	50.04
5.00	0.52	0.00	50.08
5.50	0.68	0.00	50.12
6.00	0.83	0.00	50.18
6.50	0.95	0.00	50.24
7.00	1.06	0.00	50.32
7.50	1.15	0.00	50.39
8.00	1.23	0.00	50.48
8.50	1.35	0.00	50.57
9.00	1.49	0.00	50.67
9.50	1.61	0.00	50.78
10.00	1.71	0.00	50.90
10.50	1.78	0.00	51.02
11.00	3.22	0.00	51.17
11.50	14.74	0.00	51.88
12.00	46.73	16.64	53.48
12.50	71.93	40.70	54.95
13.00	64.48	50.62	55.53
13.50	51.12	51.25	55.57
14.00	36.88	45.89	55.26
14.50	26.64	37.99	54.79
15.00	20.08	30.10	54.33
15.50	15.44	23.66	53.94
16.00	12.32	18.71	53.62
16.50	10.32	15.05	53.38
17.00	9.04	12.42	53.21
17.50	8.17	10.56	53.08
18.00	7.46	9.23	52.99
18.50	6.88	8.42	52.90
19.00	6.39	7.71	52.83
19.50	5.89	7.08	52.76
20.00	5.43	6.51	52.70
20.50	5.02	5.99	52.64
21.00	4.56	5.50	52.59
21.50	4.13	5.03	52.54
22.00	3.73	4.58	52.49
22.50	3.28	4.13	52.44
23.00	2.86	3.69	52.40
23.50	2.46	3.27	52.35
24.00	2.01	2.84	52.31
24.50	1.52	2.39	52.26

Figure E.11   Final Outlet

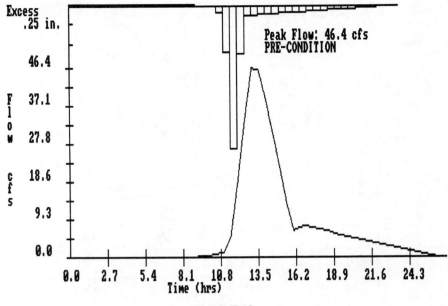

Figure E.12

routing. Link node 4 to node 5 using no outlet structure. Link node 5 to node 6.

5. You can print the results of your nodal analysis. The results of this nodal analysis for node 5 are presented in Figure E.11 along with sample graphs (Figure E.12) for each of the pond nodes.

E.6 RETEN or Retention Design and Evaluation with Pollutant Loadings

For this program, the pull-down menu options are

- Utilities
- Pollutant
- Retention
- Reuse
- Swale
- Exfiltration
- Order

Each can be executed by key entry or by mouse click.

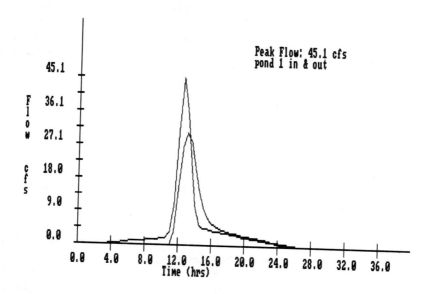
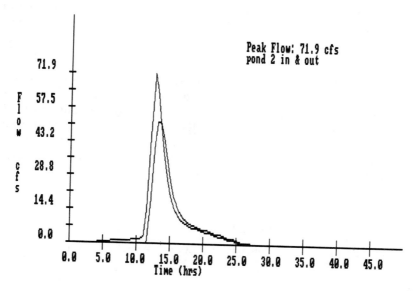

Figure E.12 *(Continued)*

The program was developed and is used to design and evaluate retention systems. The menu item *retention* refers to off-line surface pond design, while the menu item *exfiltration* refers to off-line underground trenches. The menu item *reuse* is used to specify the reuse pond holding volume and reuse rate for a given location and percent runoff not discharged. All three systems are called retention because they prevent direct discharge to a surface water body and hold the runoff water for long periods of time. The menu item *swale* is used to calculate the length of swale for infiltration and ponding for infiltration and is considered as a retention system.

E.6.1. Starting RETEN

To execute RETEN directly from the DOS command line, type RETEN from the SWATER/RETEN directory. The following pages illustrate and explain the capabilities of RETEN.

⟨*Pollutant Menu*⟩. In the pollutant analysis routine, the user can calculate an annual (or other time period) pollutant mass loading based on an event mean concentration. A land use is selected from among six with associated pollutant concentrations of suspended solids (SS), BOD_5, total nitrogen (TN), total phosphorus (TP), lead, copper, and zinc. The concentrations used in the program are shown in Table E.1. The user can also change the concentration data.

Edit Loading Data. This routine allows users to specify their own concentration data, and allows users to save these data to disk. If this routine is not used, program-defined default values will be used for all pollutant loadings. The pollutant loadings calculated by RETEN are biochemical oxygen demand (BOD), suspended solids (SS), total nitrogen (TN), total phosphorus (TP), copper, lead, and zinc. The concentrations used as default data are consistent with those presented in Chapter 5.

TABLE E.1 Concentration Data Used in Retention (mg/L)

	Residential	Commercial	Pasture	Woodland	Citrus	Highway
BOD	7.00	8.70	5.00	4.00	6.40	11.60
SS	84.00	85.00	137.20	68.00	11.20	132.00
TN	1.32	1.46	2.48	2.40	9.20	1.84
TP	0.36	0.28	0.20	0.08	0.38	0.36
Lead	0.08	0.09	0.04	0.04	0.02	0.07
Copper	0.02	0.03	0.01	0.01	0.01	0.01
Zinc	0.08	0.14	0.03	0.02	0.02	0.06

TABLE E.2 Average Yearly Loading Rates and Land Use (kg/ha/yr) Based on Total Area

Land Use	BOD$_5$	Suspended Solids	Total Nitrogen	Total P Phosphorus	Lead	Copper	Zinc
Urban	50	460	8.5	2.0	0.50	0.20	0.40
$\frac{1}{2}$-Acre residential	35	420	6.6	1.8	0.30	0.10	0.25
$\frac{1}{4}$-Acre residential	40	450	7.5	1.8	0.40	0.10	0.40
Commercial	87	840	14.5	2.7	0.85	0.24	1.35
Pasture	11.5	343	6.2	0.50	0.10	0.02	0.08
Cultivated	18	450	26.0	1.05	—	—	—
Citrus	15	25	4.0	0.2	—	—	—
Woodland	5	85	3.0	0.10	0.05	0.01	0.03
Wetlands	14	29	4.9	0.40	—	—	—
Gold course	10	150	4.5	0.78	—	—	—
Highway	87	990	13.8	0.7	0.50	0.08	0.47

Sources: East Central Florida Regional Planning Council (1984), Colston (1974), Kluesener and Lee (1974); U.S. Environmental Protection Agency (1973), Sherwood and Mattraw (1975), Wanielista et al. (1981), Priede-Sedgwick (1983), Driscoll et al. (1990), and Woodward-Clyde (1990).

Calculate Annual Load. This routine calculates the pollutant load for a land use of a given area using the default or user-specified concentrations, runoff coefficients, and rainfall. The results are expressed in annual time units, but the user can specify any rainfall volume for any period of time and the loadings will be for that time period.

Change Land Use. The six land uses in the program are urban residential, urban commercial, pasture, woodland, citrus, and highway. This option allows the user to specify a different land use for calculation of pollutant loadings. The other possible land uses are $\frac{1}{4}$-acre urban residential, cultivated land, wetlands, $\frac{1}{2}$-acre residential, and golf course. Table E.2 lists loading data for 50 in. of rainfall for some land uses. Mixed land-use loadings can also be calculated.

Retrieve Loading Data. This routine allows the user to retrieve a loading data file that was previously saved from the Edit Loading Data routine. Once data are changed, the old data will be lost unless a copy of the old program was saved.

⟨*Retention Menu*⟩

Off-Line Retention Surface Ponds. An off-line infiltration pond (or other system to remove stormwater) is defined by a pond volume to hold diverted stormwater (usually the first flush) and prevent it from direct discharge by infiltration, reuse, or evapotranspiration. The off-line retention volume for water quality control is generally relatively small (≤ 1 in. over watershed) compared to on-line detention ponds ($\geq$ few inches over watershed). However, significant hydrograph attenuation for storms producing over 1 in. of runoff usually cannot be obtained using off-line retention.

To size the off-line retention pond, one can specify either (1) stochastic equations or (2) rational formula. The stochastic equations incorporate the time variable nature of rainfall and the size of watershed to calculate pond volume, while the rational formula used the largest runoff coefficient for a particular land use, or

$$\text{volume} = \frac{C(1)(A)}{12} \tag{E.1}$$

where volume = pond volume, acre-ft
C = runoff coefficient (largest for land use)
1 = 1 in. of rainfall
12 = 12 in./ft
A = area, acres

When the stochastic equations are used to calculate volume of pond, the diversion volume for every storm is specified with a corresponding percent removal (based on areas with 30 to 60 in. of rainfall per year). The depth of pond and slide slopes must be specified, and an accurate calculation is done using trucated pyramids to estimated the volume.

Stochastic Equations. This option is used to design off-line retention ponds for watershed areas less than 100 acres and located in central Florida (Wanielista, 1990). The user must specify a diversion depth to allow for pollutant removal. These diversion depths range from 0.25 to 1.25 in. The user will then be prompted for the land area, side slope of retention pond, and a composite curve number. The routine will output the design volumes and areas for the retention ponds with pond depths of 1 to 5 ft and limiting infiltration rates of 0.2 and 1.0 in./h.

Rational Equation. This routine allows for the design of off-line retention ponds using the rational equations. The user will be prompted for the land area, the pond side slopes, the rational coefficient, and the design precipitation. The program outputs design surface area and volume. Pond volume recovery is assumed after each precipitation event; thus depth of pond has no effect on pond volume. The design precipitation is obtained from PIF curves.

⟨*Reuse*⟩

Design System. The design of a reuse system requires knowledge of the geographic region, total area of the watershed, runoff coefficient, reuse area (if irrigation is to be used), desired percentage of runoff to be reused (efficiency), the reuse volume or temporary storage, and the reuse rate. There are 17 geographical regions for which reuse equations are available. New regions can be developed using the information on long-term pond simulations developed in Chapter 8 as REV curves.

The effective impervious area is calculated by multiplying the total area by the runoff coefficient. The runoff coefficient is defined as the volume of runoff from the watershed divided by the rainfall volume. If the reuse volume is set at 3 in., runoff from the impervious and pervious areas should be calculated based on a rainfall of 3 in. The reuse return percentage is the percentage of reused water that returns to storage. An example is an irrigation system that irrigates some impervious areas that are hydraulically connected to storage. The reuse return is usually set to 0.

The REV curve data requires knowledge of two or three parameters: efficiency: reuse volume, and reuse rate. If the efficiency is specified, the numbers are 50, 60, 70, 80, 90, and 95%. The reuse volume can be any number between 0.25 and 7.0 in. over the effective impervious area. The reuse rate can vary between 0.06 and 0.3 in./day. Mark the parameter (one of the three) you wish to calculate with an ×. All calculated parameters will appear in red on the screen.

Other screen information of interest is an annual budget of inputs and outputs to a pond. A help menu is available with this design routine. When you are finished you may either hit "Escape" to exit the routine or click the = sign in the upper right-hand corner with the mouse.

⟨Swale⟩

Using Hydrograph File. In this routine swale design is performed for the specified hydrograph file. The user will be prompted for the longitudinal slope of the swale, the infiltration rate of the soil in the swale, and the side slope of the swale. RETEN will then calculate the required length of the swale to infiltrate all of the water. If this length of swale is not available, the user has the option of specifying the available length and the program will calculate the storage required for the remaining volume.

Using User Input. This routine is similar to the routine described above; however, the user must specify the average flow from the watershed and the duration of the rainfall event in addition to the input described above.

⟨*Exfiltration*⟩

Design System. The user must specify one of the Florida regions: Jacksonville, Miami, Orlando, Parrish, Tallahassee, or Tampa; one removal efficiency: 80, 85, 90, or 95%; and site-specific data on watershed area (acres), runoff coefficient, limiting exfiltration rate (in./h), depth to seasonal high water (ft), required soil cover (ft), and rock porosity. There are 20 different trench cross sections resident in the program. All assume that runoff water stored in the trench can reach an elevation over the top of the pipe equal to one-third the rock thickness.

Edit. Any new data may be entered or an existing data set may be changed with this routine. However, once the data are edited, the user must return to the design option to obtain answers.

E.7 S OR STORMWATER FREQUENCY DISTRIBUTIONS AND OPTIMIZATION

The menu for this program is obtained by key entry of the letter S, which results in the menu shown in Figure E.13.

E.7.1 Probability Frequency Distributions

Computer programs for statistical analysis were written to aid in the evaluation of both volume and intensity data. A printer is not required but "hard copies" can be made. Each frequency distribution program has an internal menu to allow the user to list, change, and save data as needed with capability of viewing any directory. The maximum number of points to be read is 200 and the data must be greater than zero when using the log-based distributions. Of course, rainfall volumes will always be positive, but zero values cannot be used with distributions assuming a logarithm transformation.

The programs are based on the work of Kite (1985). This program uses the log-Pearson type III distribution to estimate the three parameters necessary

```
              STORMWATER MANAGEMENT MENU

        PROGRAM MENU:  ' Esc ' key to exit system

        1  PEARSON DISTRIBUTION: parameter by moments and likelihood estimates.

        2  ERROR ESTIMATION : residuals, standard error, and randomness test.

        3  LINEAR PROGRAMMING: minimizes or maximizes an objective function.

        4  Editor: create or change data files for probability distributions.

        5  OPSEW : pipe system optimization for minimum construction costs.

        6  LINEAR : bi-variate linear regression.

        MAKE SELECTION (1 - 6)
```

Figure E.13 Main menu—S programs.

to fit the empirical frequency distribution. The method of moments (direct and indirect) and the maximum likelihood procedure are used to estimate the parameters. Also calculated are the mean, standard deviation, variance, skewness, and the chi-square distribution statistic at the 95% confidence level. The parameters for the method of moments (indirect) and the maximum likelihood procedure are used to estimate the storm's volume and standard error for six return periods.

E.7.2 Leasts

Leasts is a program that involves six different distributions: (1) truncated normal, (2) two-parameter log-normal, (3) three-parameter log-normal, (4) Gumbel, (5) Pearson, and (6) log-Pearson. The distributions all give the user the following information: parameter estimation, random test statistics, standard error, equation values, residual values, and graph of the probability versus actual and equation values. In addition, sorted recorded events are developed.

E.7.3 Optimization Using Linear Programming

In a mathematical sense, optimization is a procedure for either minimizing or maximizing an objective statement subject to limitations on the attainment of the objective. The mathematical form of the objective function and the limitations (constraints) determine the mathematical procedure used for optimization.

One of the more classical optimization procedures is linear programming. The objective function and constraints must have a linear form. There then exists a set of mathematical and logical procedures (algorithm) for determining the best solution. The best solution is the one that minimizes or maximizes the objective function. This procedure can be applied to fitting a hydrograph shape to existing data so as to minimize the sum of difference subject to flow rates always being positive. Another application is one that chooses the best combination of treatment procedures to minimize the over all cost subject to the physical system removal constraints. Both of these applications and others can be solved using menu item 3 in the stormwater management programs. When accessing the linear programming package, the user is asked to specify either a minimization or a maximization, followed by identification of the number of decision variables and values for the parameters of the model. Thus the complete model development should be available.

Example Problem E.2: Linear Programming Consider two options for the storage of stormwater, either in large underground pipes (let $X1$ = pipe capacity in acre-feet) or in a surface pond (let $X2$ = pond capacity in acre-feet). The cost curve of each option as a function of limited capacity is assumed linear. Because of space limitations, only 3.0 acre-ft of storage can be provided by the pipe, and 5.0 acre-ft in the pond. A total of at least 6.0 acre-ft is required. However, for each acre-foot storage in the pipe, the surface pond must have at least 0.5 acre-ft.

SOLUTION: The linear programming problem is set up as:

Minimize $\quad\quad\quad\quad 80X1 + 70X2 \quad\quad\quad\quad$ (E.2)

where the cost (80, 70) are in units of thousand dollars per acre-feet

subject to: $\quad\quad\quad\quad X1 \leq 3 \quad\quad\quad\quad$ (E.3)

$$X2 \leq 5 \quad\quad\quad\quad \text{(E.4)}$$

$$X1 + X2 \geq 6 \quad\quad\quad\quad \text{(E.5)}$$

$$-X1 + 2X2 \geq 0 \quad\quad\quad\quad \text{(E.6)}$$

$$X1;\ X2 \geq 0 \quad\quad\quad\quad \text{(E.7)}$$

OBJECTIVE FUNCTION:

Z= 80 (V 1) + 70 (V 2)

ENTER NO. OF VARIABLE TO BE CORRECTED ('0' TO CONTINUE)?

CONSTRAINT 1 :

1 (V 1) + 0 (V 2)<= 3

ENTER NO. OF VARIABLE TO BE CORRECTED

('0' TO CONTINUE, '-1' TO CORRECT RHS)?

CONSTRAINT 2 :

0 (V 1) + 1 (V 2)<= 5

ENTER NO. OF VARIABLE TO BE CORRECTED

('0' TO CONTINUE, '-1' TO CORRECT RHS)?

CONSTRAINT 3 :

1 (V 1) + 1 (V 2)=> 6

ENTER NO. OF VARIABLE TO BE CORRECTED

('0' TO CONTINUE, '-1' TO CORRECT RHS)?

Figure E.14 Example linear programming problem.

CONSTRAINT 4 :

-1 (V 1) + 2 (V 2)=> 0

ENTER NO. OF VARIABLE TO BE CORRECTED

('0' TO CONTINUE, '-1' TO CORRECT RHS)?

VARIABLE NUMBER	VARIABLE TYPE	VALUE	OPPORTUNITY COST
1	DECISION	1	0
2	DECISION	5	0
3	SLACK CONSTRAINT 1	2	0
4	SLACK CONSTRAINT 2	0	10
5	SURPLUS CONSTRAINT 3	0	80
6	SURPLUS CONSTRAINT 4	9	0

HIT 'ENTER' TO CONTINUE?

TOTAL $ = 430 THOUSAND

HIT 'ENTER' TO CONTINUE?

SELECT ONE OF THE FOLLOWING OPTIONS:

1. ENTER NEW DATA

2. EDIT EXISTING DATA

3. PRINT RESULTS

4. EXIT PROGRAM

ENTER YOUR OPTION #(1,2,3, OR 4)?

Figure E.14 *(Continued)*

There are two decision variables (a maximum of 30 is permitted). There are two "less than or equal to" constraints and two "greater than or equal to" constraints. The program automatically sets the "greater than or equal to zero" constraints. Once the data are entered, the user has a chance for corrections. The correction routine and optimal solution with ending routine are shown in Figure E.14.

E.7.4 Optimum Gravity Flow Sewer Design

In stormwater sewer design, pipe size and excavation depths are specified to transport runoff. There exists a cost trade-off between the size of pipe and depth of excavation. Given a flow rate and flow full design conditions, a small pipe has a greater slope relative to a large pipe (see Table E.3). From Table

TABLE E.3 Gravity Sewer Line Sizing for $Q = 12.5 \text{ ft}^{3/s}$ and 500-ft Pipe

Pipe Size (in.)	Velocity (ft/s)	Slope[a] (ft/ft)	Upstream Invert (ft)	Downstream Invert[b] (ft)
24	4	0.0015	100.00	99.25
18	7	0.0138	100.00	93.10

[a] With Manning's $n = 0.013$.
[b] Downstream invert = upstream invert − slope (distance).

TABLE E.4 FID Curve File Creation Example on Drive B from Drive C

C: > EDLIN B:6	File name "6"
New File	Response
*I	Key "I" for Input
1:* 8,6.0	Key in Time (min) and Intensity (in/hr)
37:* Time, Intensity	(37 entries)
38:* ^C	end
*E	Key "E" for exit

E.3, the downstream invert is about 6 ft deeper using the smaller pipe (18 in.) relative to the larger pipe (24 in.). The smaller pipe cost less but requires more excavation, thus a higher excavation cost. The objective is to minimize the cost of excavation and pipe.

The sizing of pipes requires data on peak flow rates into each pipe. Using the computer program, the peak rates are estimated with the rational formula and thus require data for watershed area, impervious area, rainfall return period, and the rational coefficient. The program will prompt the user for the information and will retain the watershed data for sensitivity analysis or for further use.

A rainfall intensity data file must be developed to determine the intensity associated with a watershed time of concentration. Resident in your S program is a 25-year rainfall intensity file for zone 7 in Florida. With the current diskette, it is the only file that can be accessed by keying the number 7 after appropriate prompts. A new intensity file (FID curve) may be developed using the DOS-EDLIN command similar to that shown in Table E.4.

Example Problem E.3: Sewer Optimization There are two inlets for stormwater flow in series and leading to a pond. The pond invert is at elevation 60. The first inlet receives flow for a 5-acre area, 2 acres impervious, and the second inlet receives flow from a 3-acre, 1.5-acre impervious watershed. The runoff coefficients for the impervious and pervious areas are 0.98 and 0.30,

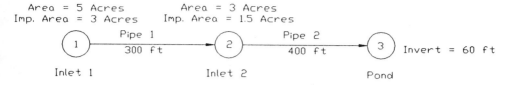

Figure E.15 Schematic for Example Problem E.3.

respectively. Using zone 7 intensity (25-year return period) curve, velocity of flow limitations between 3 and 10 ft/s, and 3-ft minimum cover with a surface elevation of 75 ft, what is the minimum sewer system construction cost?

SOLUTION: First develop a stick diagram to summarize known data (see Figure E.15). Number the nodes from the upstream end starting with the number 1 and increase by 1 as the flow proceeds downstream. Assume that the available commercial size pipes are 24, 27, 30, 36, 42, 48, 54, 60, and 72 in. The computer printout is shown in Table E.5. Note that the optimal solution is a 27- and 24-in. pipe system at a cost of $54,200. Also note the summary information on pipe velocity, slope, and invert elevations.

Next assume that a 21-in. diameter is made available (generally, this is not a commercial size). Will the cost be lower, and can a 21-in.-diameter pipe be used without violating cover and velocity constraints?

If the cover is less than 3 ft or the velocity of 10 ft/s is exceeded using a smaller diameter, the pipe cannot be used. The solution is shown in

TABLE E.5

```
GLOBAL OPTIMUM COST = 54200 DOLLARS
  OPTIMAL CONFIGURATION:
PIPE 2  27 in.
PIPE 1  24 in.
PRESS RETURN TO CONTINUE."."?
```

PIPE #2	PIPE #1
Diameter (in) 27	Diameter (in) 24
Length (ft) 400	Length (ft) 300
Velocity (fps) 7.952567	Velocity (fps) 6.531724
Slope 1.042574E-02	Slope 8.229727E-03
Headwater Pipe Invert (ft) 64.1703	Headwater Pipe Invert (ft) 67.38023
Tailwater Pipe Invert (ft) 60	Tailwater Pipe Invert (ft) 64.91132
Hydraulic Grade Line Elev (ft) at Headwater 66.91132	Hydraulic Grade Line Elev (ft) at Headwater 69.71
Cost of this Pipe Link 32000	Cost of this Pipe Link 22200
	PRESS RETURN TO CONTINUE...?
PRESS RETURN TO CONTINUE...?	

TABLE E.6 Pipe Size Change for Example Problem E.3 (taken from computer program)

```
GLOBAL OPTIMUM COST = 50300 DOLLARS
  OPTIMAL CONFIGURATION:
PIPE 2  27 in.
PIPE 1  21 in.
              PIPE #2                              PIPE #1
Diameter (in) 27                    Diameter (in) 21
Length (ft) 400                     Length (ft) 300
Velocity (fps) 7.952567             Velocity (fps) 8.531232
Slope 1.042574E-02                  Slope 1.677701E-02
Headwater Pipe Invert(ft) 64.1703   Headwater Pipe Invert(ft) 70.19442
Tailwater Pipe Invert (ft) 60       Tailwater Pipe Invert (ft) 65.16132
Hydraulic Grade Line Elev (ft)      Hydraulic Grade Line Elev (ft)
  at Headwater 66.91132               at Headwater 72.5095
Cost of the Pipe Link 32000         Cost of this Pipe Link 18300
                                    PRESS RETURN TO CONTINUE...

PRESS RETURN TO CONTINUE...?
```

Table E.6 with a cost reduction from $54,200 to $50,300. The velocity and cover constraints were not violated.

E.7.5 Bivariate Linear Regression

A hydrograph shape and equation form can be estimated using a least-squares bivariate regression computer program. The criterion of best fit is to minimize the sum of the square of the differences between the measured and predicted values of the flow rates. In general, a polynomial of second or third degree is a good fit to hydrograph data. Other mathematical forms may also be tried. The user should also consider the use of nonlinear regression forms.

E.8 REFERENCES

Colston, N. V. 1974. "Characterization of Urban Land Runoff," Reprint 2135, *ASCE National Meeting*, Los Angeles, Jan.

Driscoll, Eugene D., Shelley, P. E., and Strecker, E. E. 1990. *Pollutant Loadings and Impacts from Highway Stormwater Runoff*, Vol. I; *Design Procedure*, FHWA-RD-88-006, Federal Highway Administration, Washington, D.C.

East Central Florida Regional Planning Council. 1984. *Lake Tohopekalega Drainage Area Agricultural Runoff Management Plan*, Winter Park, Fla.

Kite, G. W. 1985. *Frequency and Risk Analyses in Hydrology*, Water Resources Publication, Littleton, Colo.

Kluesener, J. W., and Lee, G. F. 1974. "Nutrient Loading from a Separate Storm Sewer in Madison, Wisconsin," *Journal of the Water Pollution Control Federation*, Vol. 46, No. 5, p. 932.

Livingston, E., McCarron, E., Cox, J., and Sanzone, P. 1988. *The Florida Development Manual: A Guide to Sound Land and Water Management*, Florida Department of Environmental Regulation, Tallahassee, Fla.

Priede-Sedgwick, Inc. 1983. "Runoff Characterization: Water Quality and Flow," prepared for the Nationwide Urban Runoff Program Phase II.

Sherwood, C. B., and Mattraw, H. C. 1975. "Quantity and Quality of Runoff from a Residential Area near Pompano Beach, Florida," in *Proceedings: Stormwater Management Workshop*, University of Central Florida, Orlando, Fla., pp. 147–157.

U.S. Environmental Protection Agency. 1973. *Methods for Identifying the Nature and Extent of Nonpoint Sources of Pollutants*, EPA 430/9-73-014, U.S. EPA, Washington, D.C.

Wanielista, M. P. 1990. *Hydrology and Water Quantity Control*, Wiley, New York.

Wanielista, M. P., Yousef, Y. A., and Taylor, J. S. 1981, *Stormwater Management to Improve Lake Water Quality*, EPA 600/12-82-084, U.S. Environmental Protection Agency, Washington, D.C.

Woodward-Clyde Consultants. 1990. *Pollutant Loadings and Impacts from Highway Stormwater Runoff*, Vols. I, II, III, and IV, Federal Highway Administration, McLean, Va.

APPENDIX F
Selected Hydrologic Soil Classifications and Curve Numbers

TABLE F.1 Curve Numbers for Urban Land Uses.[a]

Cover Description	Average Percent Impervious Area[b]	Curve Numbers for Hydrologic Soil Group			
Cover Type and Hydrologic Condition		A	B	C	D
Fully Developed Urban Areas (Vegetation Established)					
Open space (lawns, parks, golf courses, cemeteries, etc.)[c]					
Poor condition (grass cover < 50%)		68	79	86	89
Fair condition (grass cover 50 to 75%)		49	69	79	84
Good condition (grass cover > 75%)		39	61	74	80
Impervious areas					
Paved parking lots, roof, driveways, etc. (excluding right-of-way)[d]		98	98	98	98
Streets and roads					
Paved: curbs and storm sewers (excluding right-of-way)		98	98	98	98
Paved: open ditches (including right-of-way)		83	89	92	93
Gravel (including right-of-way)		76	85	89	91
Dirt (including right-of-way)		72	82	87	89
Western desert urban areas					
Natural desert landscaping (pervious areas only)		63	77	85	88
Artificial desert landscaping (impervious weed barrier, desert shrub with 1- to 2-inch sand or gravel mulch and basin borders)		96	96	96	96

TABLE F.1 (Continued)

Cover Description		Curve Numbers for Hydrologic Soil Group			
Cover Type and Hydrologic Condition	Average Percent Impervious Area[b]	A	B	C	D
Urban districts[e]					
Commercial and business	85	89	92	94	95
Industrial	72	81	88	91	93
Residential districts by average lot size[e]					
$\frac{1}{8}$ acre or less (town houses)	65	77	85	90	92
$\frac{1}{4}$ acre	38	61	75	83	87
$\frac{1}{3}$ acre	30	57	72	81	86
$\frac{1}{2}$ acre	25	54	70	80	85
1 acre	20	51	68	79	84
2 acres	12	46	65	77	82
Developing Urban Areas					
Newly graded areas (pervious areas only, no vegetation)		77	86	91	94

Source: U.S. Department of Agriculture (1986). *Urban Hydrology for Small Watersheds*, Technical Release 55, USDA, Washington, D.C.

[a] Average runoff condition, AMC II, and $I_a = 0.2S^c$.
[b] The average percent impervious area shown was used to develop the composite CNs. Other assumptions are as follows: impervious areas are directly connected to the drainage system, impervious areas have a CN of 98, and pervious areas are considered equivalent to open space in good hydrologic condition.
[c] CNs shown are equivalent to those of pasture. Composite CNs may be computed for other combinations of open-space cover type.
[d] In some warmer climates a curve number of 95 may be used.
[e] Composite curve numbers.

TABLE F.2 Runoff Curve Numbers for Hydrologic Soil-Cover Complexes (Antecedent Moisture Condition II)

Land Use	Cover Treatment or Practice	Hydrologic Condition	Hydrologic Soil Group			
			A	B	C	D
Fallow	Straight row		77	86	91	94
Row crops	Straight row	Poor	72	81	88	91
	Straight row	Good	67	78	85	89
	Contoured	Poor	70	79	84	88
	Contoured	Good	65	75	82	86
	Contoured and terraced	Poor	66	74	80	82
	Contoured and terraced	Good	62	71	78	81
Small grain	Straight row	Poor	65	76	84	88
	Straight row	Good	63	75	83	87
	Contoured	Poor	63	74	82	85
	Contoured	Good	61	73	81	84
	Contoured and terraced	Poor	61	72	79	82
		Good	59	70	78	81
Close-seeded legumes[a] or rotation meadow	Straight row	Poor	66	77	85	89
	Straight row	Good	58	72	81	85
	Contoured	Poor	64	75	83	85
	Contoured	Good	55	69	78	83
	Contoured and terraced	Poor	63	73	80	83
	Contoured and terraced	Good	51	67	76	80
Pasture or range		Poor	68	79	86	89
		Fair	49	69	79	84
		Good	39	61	74	80
	Contoured	Poor	47	67	81	88
	Contoured	Fair	25	59	75	83
	Contoured	Good	6	35	70	79
Meadow		Good	30	58	71	78
Woods		Poor	45	66	77	83
		Fair	36	60	73	79
		Good	25	55	70	77
Farmsteads			59	74	82	86
Roads[b] (Dirt)			72	82	87	89
Hard surface			74	84	90	92

Source: V. Mockus, 1969 *National Engineering Handbook*, Section 4, Chapter 9, "Hydrologic Soil Cover Complexes," USDA, Washington, D.C.

[a] Closed-drilled or broadcast.
[b] Including right-of-way.

SELECTED HYDROLOGIC SOIL CLASSIFICATIONS AND CURVE NUMBERS

TABLE F.3 CN Adjustments

CN	Corresponding CNs	
Condition 2	Condition 1[a]	Condition 3[b]
100	100	100
95	87	98
90	78	96
85	70	94
80	63	91
75	57	88
70	51	85
65	45	82
60	40	78
55	35	74
50	31	70
45	26	65
40	22	60
35	18	55
30	15	50

Source: U.S. Department of Agriculture (1986). *Urban Hydrology for Small Watersheds*, Technical Release 55, USDA, Washington, D.C.

[a] Dry Condition.
[b] Wet Condition.

TABLE F.4 Soil Names and Hydrologic Classifications (Examples)

Aaberg	C	Aberdeen	D	Abscota	B
Aastad	B	Abes	D	Absher	D
Abac	D	Abilene	C	Absted	D
Abajo	C	Abington	B	Acacio	C
Abbott	D	Abiqua	C	Academy	C
Abbottstown	C	Abo	B/C	Acadia	D
Abcal	D	Abok	D	Acana	D
Abegg	B	Abra	C	Acasco	D
Abela	B	Abraham	B	Aceitunas	B
Abell	B	Absarkee	C	Acel	D

TABLE F.4 (*Continued*)

Acker	B	Aeneas	B	Akela	C
Ackmen	B	Aetna	B	Aladdin	B
Acme	C	Afton	D	Alae	A
Aco	B	Agar	B	Alaeloa	B
Acolita	B	Agassiz	D	Alaga	A
Acoma	C	Agate	D	Alakai	D
Acove	C	Agawam	B	Alama	B
Acree	C	Agency	C	Alamance	B
Acrelane	C	Ager	D	Alamo	D
Acton	B	Agner	B	Alamosa	C
Acuff	B	Agnew	B/C	Alapaha	D
Acworth	B	Agnus	B	Alapai	A
Acy	C	Agua	B	Alban	B
Ada	B	Agua Dulce	C	Albano	D
Adair	D	Agua Fria	C	Albany	C
Adams	A	Aguadilla	B	Albaton	D
Adamson	B	Agualt	B	Albee	C
Adamstown		Agueda	B	Albemarle	B
Adamsville	C	Aguilita	B	Albertville	C
Adaton	D	Aguirre	D	Albia	C
Adaven	D	Agustin	B	Albion	B
Addielou	C	Ahatone	D	Albrights	C
Addison	D	Ahl	C	Alcalde	C
Addy	C	Ahlstrom	C	Alcester	B
Ade	A	Ahmeek	B	Alcoa	B
Adel	A	Aholt	D	Alcona	B
Adelaide	D	Ahtanum	C	Alcova	B
Adelanto	B	Ahwahnee	C	Alda	C
Adelino	B	Aibonito	C	Aldax	D
Adelphia	C	Aiken	B/C	Alden	D
Adena	C	Aikman	D	Alder	B
Adger	D	Ailey	B	Alderdale	C
Adilis	A	Ainakea	B	Alderwood	C
Adirondack		Airmont	C	Aldino	C
Adiv	B	Airotsa	B	Aldwell	C
Adjuntas	C	Airport	D	Aleknagik	B
Adkins	B	Aits	B	Alemeda	C
Adler	C	Ajo	C	Alex	B
Adolph	D	Akaka	A	Alexandria	C
Adrian	A/D	Akaska	B	Alexis	B

TABLE F.4 (Continued)

Alford	B	Alpon	B	Ammon	B
Algansee	B	Alpowa	B	Amole	C
Algerita	B	Alps	C	Amor	B
Algiers	C/D	Alsea	B	Amos	C
Algoma	B/D	Alspaugh	C	Amsden	B
Alhambra	B	Alstad	B	Amsterdam	B
Alice	A	Alstown	B	Amtoft	D
Alicel	B	Altamont	D	Amy	D
Alicia	B	Altavista	C	Anacapa	B
Alida	B	Altdorf	D	Anahuac	D
Alikchi	B	Altmar	B	Anamite	D
Aline	A	Alto	C	Anapra	B
Alko	D	Altoga	C	Anasazi	B
Allagash	B	Alton	B	Anatone	D
Allard	B	Altus	B	Anaverde	D
Allegheny	B	Altvan	B	Anawalt	D
Allemands	D	Alum	B	Ancho	B
Allen	B	Alusa	D	Anchor Bay	D
Allendale	C	Alvin	B	Anchor Point	D
Allens Park	B	Alvira	C	Anchorage	A
Allensville	C	Alviso	D	Anclote	D
Allentine	D	Alvor	C	Anco	C
Allenwood	B	Amador	D	Anderly	C
Allessio	B	Amagon	D	Anders	C
Alley	C	Amalu	D	Anderson	B
Alliance	B	Amana	B	Andes	C
Alligator	D	Amargosa	D	Andorinia	C
Allis	D	Amarillo	B	Andover	D
Allison	C	Amasa	B	Andreen	B
Allouez	C	Amberson		Andreeson	C
Alloway		Amboy	C	Andres	B
Almac	B	Ambraw	C	Andrews	C
Almena	C	Amedee	A	Aned	D
Almont	D	Amelia	B	Aneth	A
Almy	B	Amenia	B	Angelica	D
Aloha	C	Americus	A	Angelina	B/D
Alonso	B	Ames	C	Angelo	C
Alovar	C	Amesha	B	Angie	C
Alpena	B	Amherst	C	Angle	A
Alpha	C	Amity	C	Anglen	B

TABLE F.4 (Continued)

Angola	C	Appling	B	Arlando	B
Angostura	B	Apron	B	Arle	B
Anhalt	D	Apt	C	Arling	D
Aniak	D	Aptakisic	B	Arlington	C
Anita	D	Araby		Arloval	C
Ankeny	A	Arada	C	Armagh	D
Anlauf	C	Aransas	D	Armijo	D
Annabella	B	Arapien	C	Armington	D
Annandale	C	Arave	D	Armjohee	D
Anniston	B	Araveton	B	Armour	B
Anoka	A	Arbela	C	Armster	C
Anones	C	Arbone	B	Armstrong	D
Ansari	D	Arbor	B	Arnegard	B
Ansel	B	Arbuckle	B	Arnhart	C
Anselmo	A	Arcata	B	Arnheim	C
Anson	B	Arch	B	Arno	D
Ant Flat	C	Archabal	B	Arnold	B
Antelope Springs	C	Archer	C	Arnot	C/D
Antero	C	Archin	C	Arny	A
Antho	B	Arco	B	Arosa	C
Anthony	B	Arcola	C	Arp	C
Antigo	B	Ard	C	Arrington	B
Antilon	B	Arden	B	Arritola	D
Antioch	D	Ardenvoir	B	Arrolime	C
Antler	C	Ardilla	C	Arron	D
Antoine	C	Ardostook		Arrow	B
Antrobus	B	Aredale	B	Arrowsmith	B
Anty	B	Arena	C	Arroyo Seco	B
Anvik	B	Arenales		Arta	C
Anway	B	Arendsa	A	Artois	C
Anza	B	Arendtsville	B	Arvada	D
Anziano	C	Arenzville	B	Arvana	C
Apache	D	Argonaut	D	Arveson	D
Apakuie	A	Arguello	B	Arvilla	B
Apishapa	C	Argyle	B	Arzell	C
Apison	B	Arho	B	Asa	B
Apopka	A	Ariel	C	Asbury	B
Appian	C	Arizo	A	Aschoff	B
Applegate	C	Arkabutla	C	Ash Springs	C
Appleton	C	Arkport	B	Ashby	C

TABLE F.4 (Continued)

Ashcroft	B	Atsion	C	Azarman	C
Ashdale	B	Atterberry	B	Azeltine	B
Ashe	B	Attewan	A	Azfield	B
Ashkum	C	Attica	B	Aztalan	B
Ashlar	B	Attleboro		Aztec	B
Ashley	A	Atwater	B	Azule	C
Ashton	B	Atwell	C/D	Azwell	B
Ashue	B	Atwood	B	Babb	A
Ashuelot	C	Au Gres	B	Babbington	B
Ashwood	C	Aubbeenaubbe	B	Babcock	C
Askew	C	Auberry	C/D	Babylon	A
Aso	C	Auburn	D	Baca	C
Asotin	C	Auburndale	B	Bach	D
Aspen	B	Audian	C	Bachus	C
Aspermont	B	Augsburg	B	Backbone	A
Assalon	B	Augusta	C	Baculan	A
Assinniboine	B	Auld	D	Badenaugh	B
Assumption	B	Aura	B	Badger	C
Astatula	A	Aurora	C	Badgerton	B
Astor	A/D	Austin	C	Bado	D
Astoria	B	Austwell	D	Badus	C
Atascadero	C	Auxvasse	D	Bagard	C
Atascosa	D	Auzqui	B	Bagdad	B
Atco	B	Ava	C	Baggott	D
Atencio	B	Avalanche	B	Bagley	B
Atepic	D	Avalon	B	Bahem	B
Athelwold	B	Avery	B	Baile	D
Athena	B	Avon	C	Bainville	C
Athens	B	Avonburg	D	Baird Hollow	C
Atherly	B	Avondale	E	Bajura	D
Atherton	B/D	Awbrey	D	Bakeoven	D
Athmar	C	Axtell	D	Baker	C
Athol	B	Ayar	D	Baker Pass	B
Atkinson	B	Aycock	B	Balaam	A
Atlas	D	Ayr	B	Balch	D
Atlee	C	Ayres	D	Balcom	B
Atmore	B/D	Ayrshire	C	Bald	C
Atoka	C	Aysees	B	Balder	C
Aton	B	Ayun	B	Baldock	B/C
Atrypa	C	Azaar	C	Baldwin	D

TABLE F.4 *(Continued)*

Baldy	B	Barishman	C	Bassfield	B
Bale	C	Barker	C	Bassler	D
Ballard	B	Barkerville	C	Bastian	D
Baller	D	Barkley	B	Bastrop	B
Ballinger	C	Barlane	D	Bata	A
Balm	B/C	Barling	C	Batavia	B
Balman	B/C	Barlow	B	Bates	B
Balon	B	Barnard	D	Bath	C
Baltic	D	Barnes	B	Batterson	D
Baltimore	B	Barneston	B	Battle Creek	C
Balto	D	Barney	A	Batza	D
Bamber	B	Barnhardt	B	Baudette	B
Bamforth	B	Barnstead		Bauer	C
Bancas	B	Barnum	B	Baugh	B/C
Bancroft	B	Baron	C	Baxter	B
Bandera	B	Barrada	D	Baxtervillle	B
Bango	C	Barrett	D	Bayamon	B
Bangston	A	Barrington	B	Bayard	A
Bangur	B	Barron	B	Baybord	D
Bankard	A	Barronett	C	Bayerton	C
Banks	A	Barrows	D	Baylor	D
Banner	C	Barry	D	Bayshore	B/C
Bannerville	C/D	Barstow	B	Bayside	C
Bannock	B	Barth	C	Bayucos	D
Banquete	D	Bartine	C	Baywood	A
Barabou	B	Bartle	D	Bazette	C
Baraga	C	Bartley	C	Bazile	B
Barbary	D	Barton	B	Bead	C
Barboor	B	Bartonflat	B	Beadle	C
Barbourville	B	Bascom	B	Beales	A
Barclay	C	Basehor	D	Bear Basin	B
Barco	B	Bashaw	D	Bear Creek	C
Barcus	B	Basher	B	Bear Lake	C
Bard	D	Basile	D	Bear Prairie	C
Barden	C	Basin	C	Beardall	C
Bardley	C	Basinger	C	Bearden	D
Barela	C	Basket	C	Beardstown	A
Barfield	D	Bass	A	Bearmough	B
Barfuss	B	Bassel	B	Bearpaw	B
Barge	C	Bassett	B	Bearskin	D

TABLE F.4 *(Continued)*

Beasley	C	Belfast	B	Benz	D
Beasun	C	Belfield	B	Beotia	B
Beaton	C	Belfore	B	Beowawe	D
Beatty	C	Belgrade	B	Bercail	C
Beaucoup	B	Belinda	D	Berda	B
Beauford	D	Belknap	C	Berea	C
Beaumont	D	Bellamy	C	Bereniceton	B
Beauregard	C	Bellavista	D	Berent	A
Beausite	B	Belle	B	Bergland	D
Beauvais	B	Bellefontaine		Bergstrom	B
Beaverton	A	Bellicum	B	Berino	B
Beck	C	Bellingham	C	Berkeley	
Becker	B	Bellpine	C	Berks	C
Becket	C	Belmont	B	Berkshire	B
Beckley	B	Belmore	B	Berlin	C
Beckton	D	Belt	D	Bermaldo	B
Beckwith	C	Belted	D	Bermesa	C
Beckwourth	B	Belton	C	Bermudian	B
Becreek	B	Beltrami	B	Bernal	D
Bedford	C	Beltsville	C	Bernard	D
Bedington	B	Beluga	D	Bernardino	C
Bedner	C	Belvoir	C	Bernardston	C
Beebe	A	Ben Hur	B	Bernhill	B
Beecher	C	Ben Lumond	B	Bernice	A
Beechy		Benclare	C	Berning	C
Beehive	B	Benevola	C	Berrendos	D
Beek	C	Benewah	C	Berryland	D
Beenom	D	Benfirld	C	Bertelson	B
Beezar	B	Benge	B	Berthoud	B
Begay	B	Benin	D	Bertie	C
Begoshian	C	Benito	D	Bertolotti	B
Behanin	B	Benjamin	D	Bertrand	B
Behemotosh	B	Benman	A	Berville	D
Behring	D	Benndale	B	Beryl	B
Beirman	D	Bennett	C	Bessemer	B
Bejucos	B	Bennington	D	Bethany	C
Belcher	D	Benoit	D	Bethel	D
Belden	D	Benson	C/D	Betteravia	C
Belding	B	Benteen	B	Betts	B
Belen	C	Bentonville	C	Beulah	B

APPENDIX F   515

TABLE F.4 (*Continued*)

Bevent	B	Birchwood	C	Blackwater	D
Beverly	B	Birdow	B	Blackwell	B/D
Bew	D	Birds	C	Bladen	D
Bewleyville	B	Birdsall	D	Blago	D
Bewlin	D	Birdsboro	B	Blaine	B
Bexar	C	Birdsley	D	Blair	C
Bezzant	B	Birkbeck	B	Blairton	C
Bibb	B/D	Bisbee	A	Blake	C
Bibon	A	Biscay	C	Blakeland	A
Bickelton	B	Bishop	B/C	Blakeney	C
Bickleton	C	Bisping	B	Blakeport	B
Bickmore	C	Bissell	B	Blalock	D
Bicondoa	C	Bisti	C	Blamer	C
Biddeford	D	Bit	D	Blanca	B
Biddleman	C	Bitter Spring	C	Blanchard	A
Bidman	C	Bitteron	A	Blanchester	B/D
Bidwell	B	Bitterroot	C	Bland	C
Bieber	D	Bitton	B	Blandford	C
Bienville	A	Bixby	B	Blanding	B
Big Blue	D	Bjork	C	Blaney	B
Big Horn	C	Blachly	C	Blanket	C
Big Timber	D	Black Butte	C	Blanton	A
Bigel	A	Black Canyon	D	Blanyon	C
Bigelow	C	Black Mountain	D	Blasdell	A
Bigetty	C	Black Ridge	D	Blasingame	C
Biggs	A	Blackburn	B	Blazon	D
Biggsville	B	Blackcap	A	Blencoe	C
Bignell	B	Blackett	B	Blend	D
Bigwin	D	Blackfoot	B/C	Blendon	B
Bijou	A	Blackhall	D	Blethen	B
Billett	A	Blackhawk	D	Blevins	B
Billings	C	Blackleaf	B	Blevinton	B/D
Bindle	B	Blackleed	A	Blichton	D
Binford	B	Blacklock	D	Bliss	D
Bingham	B	Blackman	C	Blockton	C
Binnsville	D	Blackoak	C	Blodgett	A
Bins	B	Blackpipe	C	Blomford	B
Binton	C	Blackrock	B	Bloom	C
Bippus	B	Blackston	B	Bloomfield	A
Birch	A	Blacktail	B	Blooming	B

TABLE F.4 (Continued)

Bloor	D	Bombay	B	Bosco	B
Blossom	C	Bon	B	Bosket	B
Blount	C	Bonaccord	D	Bosler	B
Blountville	C	Bonaparte	A	Bosque	B
Blucher	C	Bond	D	Boss	D
Blue Earth	D	Bondranch	D	Boston	C
Blue Lake	A	Bondurant	B	Bostwick	B
Blue Star	B	Bone	D	Boswell	D
Bluebell	C	Bong	B	Bosworth	D
Bluejoint	B	Bonham	C	Botella	B
Bluepoint	B	Bonifay	A	Bothwell	C
Bluewing	B	Bonilla	B	Bottineau	C
Bluffdale	C	Bonita	D	Bottle	A
Bluffton	D	Bonn	D	Boulder	B
Bluford	D	Bonner	B	Boulder Lake	D
Bly	B	Bonnet	B	Boulder Point	B
Blythe	D	Bonneville	B	Boulflat	D
Boardtree	C	Bonnick	A	Bourne	C
Bobs	D	Bonnie	D	Bow	C
Bobtail	B	Bonsall	D	Bowbac	C
Bock	B	Bonta	C	Bowbells	B
Bodell	D	Bonti	C	Bowdoin	D
Bodenburg	B	Booker	D	Bowdre	C
Bodine	B	Boomer	B	Bowers	C
Boel	A	Boone	A	Bowie	B
Boelus	A	Boonesboro	B	Bowman	B/D
Boesel	B	Boonton	C	Bowmansville	C
Boettcher	C	Booth	C	Boxelder	C
Bogan	C	Boracho	C	Boxwell	C
Bogart	B	Borah	A/D	Boy	A
Bogue	D	Borda	D	Boyce	B/D
Bohannon	C	Bordeaux	B	Boyd	D
Bohemian	B	Borden	B	Boyer	B
Boistfort	C	Border	B	Boynton	
Bolar	C	Bornstedt	C	Boysag	D
Bold	B	Borrego	C	Boysen	D
Boles	C	Borup	B	Bozarth	C
Bolivar	B	Borvant	D	Boze	B
Bolivia	B	Borza	C	Bozeman	A
Bolton	B	Bosanko	D	Braceville	C

TABLE F.4 (Continued)

Bracken	D	Bremen	B	Brinkerton	D
Brackett	C	Bremer	B	Briscot	B
Brad	D	Bremo	C	Brite	C
Braddock	C	Brems	A	Britton	C
Bradenton	B/D	Brenda	C	Brizam	A
Brader	D	Brennan	B	Broad	C
Bradford	B	Brenner	C/D	Broad Canyon	B
Bradshaw	B	Brent	C	Broadalbin	C
Bradway	D	Brenton	B	Broadax	B
Brady	B	Brentwood	B	Broadbrook	C
Bradyville	C	Bresser	B	Broadhead	C
Braham	B	Brevard	B	Broadhurst	D
Brainerd	B	Brevort	B	Brock	D
Brallier	D	Brewer	C	Brockliss	C
Bram	B	Brewster	D	Brockman	C
Bramard	B	Brewton	C	Brocko	B
Bramble	C	Brickel	C	Brockport	D
Bramwell	C	Brickton	C	Brockton	D
Brand	D	Bridge	C	Brockway	B
Brandenburg	A	Bridgehampton	B	Brody	C
Brandon	B	Bridgeport	B	Broe	B
Brandywine	C	Bridger	A	Brogan	B
Branford	B	Bridgeson	B/C	Brogdon	B
Brantford	B	Bridget	B	Brolliar	D
Branyon	D	Bridgeville	B	Bromo	B
Brashear	C	Bridgport	B	Bronaugh	B
Brassfield	B	Briedwell	B	Broncho	B
Bratton	B	Brief	B	Bronson	B
Bravane	D	Briensburg		Bronte	C
Braxton	C	Briggs	A	Brooke	C
Braymill	A/D	Briggsdale	C	Brookfield	B
Brays	D	Briggsville	C	Brookings	B
Brayton	C	Brighton	A/D	Brooklyn	D
Brazito	A	Brightwood	C	Brookside	C
Brazos	A	Brill	B	Brookston	B/C
Brea	B	Brim	C	Brooksville	D
Breckenridge	D	Brimfield	C/D	Broomfield	D
Brecknock	B	Brimley	B	Broseley	B
Breece	B	Brinegar	B	Bross	B
Bregar	D	Brinkert	C	Broughton	D

TABLE F.4 (Continued)

Broward	C	Bukreek	B	Buse	B
Brownell	B	Bull Run	B	Bush	B
Brownfield	A	Bull Trail	B	Bushnell	C
Brownlee	B	Bullion	D	Bushvalley	D
Broyles	B	Bullrey	B	Buster	C
Bruce	D	Bully	B	Butano	C
Bruffy	C	Bumgard	B	Butler	D
Bruin	C	Buncombe	A	Butlertown	C
Bruneel	B/C	Bundo	B	Butte	C
Bruno	A	Bundyman	C	Butterfield	C
Brunt	C	Bunejug	C	Button	C
Brush		Bunker	D	Buxin	D
Brussett	B	Bunselmeier	C	Buxton	C
Bryan	A	Buntingville	B/C	Byars	D
Brycan	B	Bunyan	B	Bynum	C
Bryce	D	Burbank	A	Byron	A
Bucan	D	Burch	B	Caballo	B
Buchanan	C	Burchard	B	Cabarton	D
Buchenau	C	Burchell	B/C	Cabba	C
Bucher	C	Burdett	C	Cabbart	D
Buckhouse	A	Buren	C	Cabezon	D
Buckingham		Burgess	C	Cabin	C
Buckland	C	Burgi	B	Cabinet	C
Bucklebar	B	Burgin	D	Cable	D
Buckley	B/C	Burke	C	Cabo Rojo	C
Bucklon	D	Burkhardt	B	Cabot	D
Buckner	A	Burleigh	D	Cacapon	B
Buckney	A	Burleson	D	Cache	D
Bucks	B	Burlington	A	Cacique	C
Buckskin	C	Burma		Caddo	D
Bucoda	C	Burmester	D	Cadeville	D
Budd	B	Burnac	C	Cadmus	B
Bude	C	Burnette	B	Cadoma	D
Buell	B	Burnham	D	Cador	C
Buena Vista	B	Burnside	B	Cagey	C
Buff Peak	C	Burnsville	B	Caguabo	D
Buffington	B	Burnt Lake	B	Cagwin	B
Buffmeyer	B	Burris	D	Cahaba	B
Buick	C	Burt	D	Cahill	B
Buist	B	Burton	B	Cahone	C

TABLE F.4 *(Continued)*

Cahto	C	Camargo	B	Canoncito	B
Caid	B	Camarillo	B/D	Canova	B/D
Cairo	D	Camas	A	Cantala	B
Cajalco	C	Camascreek	B/D	Canton	B
Cajon	A	Cambridge	C	Cantril	B
Calabar	D	Camden	B	Cantua	B
Calabasas	B	Cameron	D	Canutio	B
Calais	C	Cameyville	C	Canyon	D
Calamine	D	Camgeny	C	Capac	B
Calapooya	C	Camillus	B	Capay	D
Calawah	B	Camp	B	Cape	D
Calco	C	Campbell	B/C	Cape Fear	D
Calder	D	Camphora	B	Capers	D
Caldwell	B	Campia	B	Capillo	C
Caleast	C	Campo	C	Caples	C
Caleb	B	Campone	B/C	Capps	B
Calera	C	Campspass	C	Capshaw	C
Calhi	A	Campus	B	Capulin	B
Calhoun	D	Camroden	C	Caputa	C
Calico	D	Cana	C	Caraco	C
Califon	D	Canaan	C/D	Caralampi	B
Calimus	B	Canadian	B	Carbo	C
Calita	B	Canadice	D	Carbol	D
Caliza	B	Canandaigua	D	Carbondale	D
Calkins	C	Canaseraga	C	Carbury	B
Callabo	C	Canaveral	C	Carcity	D
Callahan	C	Canburn	D	Cardiff	B
Calleguas	D	Cande	B	Cardington	C
Callings	C	Candelero	C	Cardon	D
Calloway	C	Cane	C	Carey	B
Calmar	B	Caneadea	D	Carey Lake	B
Calneva	C	Caneek	B	Careytown	D
Calouse	B	Canel	B	Cargill	C
Calpine	B	Canelo	D	Caribe	B
Calvert	D	Caney	C	Caribel	B
Calverton	C	Canez	B	Caribou	B
Calvin	C	Canfield	C	Carlin	D
Calvista	D	Canisted	C	Carlinton	B
Cam	B	Canninger	D	Carlisle	A/D
Camaguey	D	Cannon	B	Carlotta	B

TABLE F.4 (*Continued*)

Carlow	D	Tusel	C	Ulricher	B
Carlsbad	C	Tuskeego	C	Ulupalakua	B
Carlsborg	A	Tusler	B	Uly	B
Carlson	C	Tusquitee	B	Ulysses	B
Carlton	B	Tustin	B	Uma	A
Carmi	B	Tustumena	B	Umapine	B/C
Carnasaw	C	Tuthill	B	Umiat	D
Carnegie	C	Tutni	B	Umikoa	B
Carnero	C	Tutwiler	B	Umil	D
Carney	D	Tuxedo		Umnak	B
Caroline	C	Tuxekan	B	Umpa	B
Carr	B	Twin Creek	B	Umpqua	B
Carrisalitos	D	Twining	C	Una	D
Carrizo	A	Twisp	B	Unadilla	B
Carsitas	A	Two Dot	C	Unaweep	B
Carsley	C	Tybo	D	Uncom	B
Carso	D	Tyee	D	Uncompangre	D
Carson	D	Tygart	D	Uneeda	B
Carstairs	B	Tyler	D	Ungers	B
Carstump	C	Tyndall	B/C	Union	C
Cart	B	Tyner	A	Uniontown	B
Cartagena	D	Tyrone	C	Unionville	C
Cartecay	C	Tyson	C	Unisun	C
Caruso	C	Uana	D	Updike	D
Caruthersville	B	Ubar	C	Upsal	C
Carver	A	Ubly	B	Upsata	A
Carwile	D	Ucola	D	Upshur	C
Turnbow	C	Ucolo	C	Upton	C
Turner	B	Ucopia	B	Uracca	B
Turnerville	B	Uddlpho	C	Urbana	C
Turney	B	Udel	D	Urbo	D
Turrah	D	Uffens	D	Urich	D
Turret	B	Ugak	D	Urne	B
Turria	C	Uhland	B	Ursine	D
Turson	B/C	Uhlig	B	Urtah	C
Tuscan	D	Uinta	B	Urwil	D
Tuscarawas	C	Ukiah	C	Usal	B
Tuscarora	C	Ulen	B	Ushar	B
Tuscola	B	Ullda	B	Usine	B
Tuscumbia	D	Ulm	B	Uska	D

TABLE F.4 (Continued)

Name		Name		Name	
Utaline	B	Vandergrift	C	Venice	D
Ute	C	Vanderhoff	D	Venlo	D
Utica	A	Vanderlip	A	Venus	B
Utley	B	Vanet	D	Verboort	D
Utuado	B	Vang	B	Verde	C
Uvada	D	Vanhorn	B	Verdel	D
Uvalde	C	Vannoy	B	Verdella	D
Uwala	B	Vanoss	B	Verdico	D
Vacherie	C	Vantage	C	Verdigris	B
Vader	B	Varco	C	Verdun	D
Vado	B	Varelum	C	Vergennes	D
Vaiden	D	Varick	D	Verhalen	D
Vailton	B	Varina	C	Vermejo	D
Valby	C	Varna	C	Vernal	B
Valco	C	Varro	B	Vernalis	B
Valdez	B/C	Varysburg	B	Vernia	A
Vale	B	Vashti	C	Vernon	D
Valencia	B	Vasquez	B	Verona	C
Valent	A	Vassalboro	D	Vesser	C
Valentine	A	Vassar	B	Veston	D
Valera	C	Vastine	C	Vetal	A
Valkaria	B/D	Vaucluse	C	Veteran	B
Vallan	D	Vaughnsville	C	Veyo	D
Vallecitos	C/D	Vayas	D	Via	B
Valleono	B	Veal	B	Vian	B
Vallers	C	Veazie	B	Viboras	D
Valmont	C	Vebar	B	Viborg	B
Valmy	B	Vecont	D	Vickery	C
Valois	B	Vega	C	Vicksburg	B
Vamer	D	Vega Alta	C	Victor	A
Van Buren		Vega Baja	C	Victoria	D
Van Dusen	B	Vekol	D	Victory	B
Van Nostern	B	Velda	B	Vicu	D
Van Wagoner	D	Velma	B	Vida	B
Vanajo	D	Velva	B	Vidrine	C
Vananda	D	Vena	C	Vieja	D
Vance	C	Venango	C	Vienna	B
Vanda	D	Venator	D	Vieques	B
Vandalia	C	Veneta	C	View	C
Vanderdasson	D	Venezia	D	Vigar	C

TABLE F.4 *(Continued)*

Vigo	D	Volkmar	B	Wahtigup	B
Vigus	C	Volney	B	Wahtum	D
Viking	D	Volperie	C	Waiaha	D
Vil	D	Voltaire	D	Waiakoa	C
Vilas	A	Volumer	D	Waialeale	D
Villa Grove	B	Volusia	C	Waialua	B
Villars	B	Vona	B	Waiawa	D
Villy	D	Vore	B	Waihuna	D
Vina	B	Vrodman	B	Waikaloa	B
Vincennes	C	Vulcan	C	Waikane	B
Vincent	C	Vylach	D	Waikapu	B
Vineyard	C	Wabanica	D	Waikomo	D
Vingo	B	Wabash	D	Wailuku	B
Vining	C	Wabasha	D	Waimea	B
Vinita	C	Wabassa	B/D	Wainee	B
Vinland	C	Wabek	B	Wainola	A
Vinsad	C	Waca	C	Waipahu	C
Vint	B	Wacota	B	Waiska	B
Vinton	B	Wacousta	C	Waits	B
Vira	C	Wadams	B	Wake	D
Viraton	C	Waddell	B	Wakeen	B
Virden	C	Waddups	B	Wakefield	B
Virgil	B	Wadell	B	Wakeland	B/D
Virgin Peak	D	Wadena	B	Wakonda	C
Virgin River	D	Wadesboro	B	Wakulla	A
Virtue	C	Wadleigh	D	Walcott	B
Visalia	B	Wadmalaw	D	Waldeck	C
Vista	C	Wadsworth	C	Waldo	D
Vives	B	Wages	B	Waldron	D
Vivi	B	Wagner	D	Waldroup	D
Vlasaty	C	Wagram	A	Wales	B
Voca	C	Waha	C	Walford	C
Vodermaier	B	Wahee	D	Walke	C
Voladora	B	Wahiawa	B	Wall	B
Volco	D	Wahikuli	B	Walla Walla	B
Volente	C	Wahkeena	B	Wallace	B
Volga	D	Wahkiacus	B	Waller	B/D
Volin	B	Wahluke	B	Wallington	C
Volinia	B	Wahmonie	D	Wallis	B
Volke	C	Wahpeton	C	Wallkill	C/D

TABLE F.4 (*Continued*)

Wallman	C	Warrenton	B/D	Waukee	B
Wallowa	C	Warrior		Waukegan	B
Wallpack	C	Warsaw	B	Waukena	D
Wallrock	B/C	Warsing	B	Waukon	B
Wallsburg	D	Warwick	A	Waumbek	B
Wallson	B	Wasatch	A	Waurika	D
Walpole	C	Wasepi	B	Wauseon	B/D
Walsh	B	Washburn		Waverly	B/D
Walshville	D	Washde	C	Wawaka	C
Walters	A	Washington	B	Waycup	B
Walton	C	Washougal	B	Wayden	D
Walum	B	Washtenaw	C/D	Wayland	C/D
Walvan	B	Wasidja	B	Wayne	B
Wamba	B/C	Wasilla	D	Waynesboro	B
Wamic	B	Wassaic	B	Wayside	
Wampsville	B	Watab	C	Wea	B
Wanatah	B	Watauga	B	Weaver	C
Wanblee	D	Watchaug	B	Webb	C
Wando	A	Watchung	D	Weber	B
Wanetta	A	Waterboro		Webster	C
Wanilla	C	Waterbury	D	Wedekind	D
Wann	A	Waterino	C	Wedertz	C
Wapal	B	Waters	C	Wedge	A
Wapato	C/D	Watkins	B	Wedowee	D
Wapello	B	Watkins Ridge	B	Weed	B
Wapinitia	B	Wato	B	Weeding	A/C
Wapping	B	Watopa	B	Weedmark	B
Wapsie	B	Watrous	B	Weeksville	B/D
Warba	B	Watseka	C	Weepon	D
Ward	D	Watson	C	Wehadkee	D
Wardboro	A	Watsonia	D	Weikert	C/D
Wardell	D	Watsonville	D	Weimer	D
Warden	B	Watt	D	Weinbach	C
Wardwell	C	Watton	C	Weir	D
Ware	B	Waubay	B	Weirman	B
Wareham	C	Waubeek	B	Weiser	C
Warm Springs	C	Waubonsie	B	Weishaupt	D
Warman	D	Wauchula	B/D	Weiss	A
Warners	A/D	Waucoma	B	Weitchpec	B
Warren		Wauconda	B	Welaka	A

TABLE F.4 (*Continued*)

Welby	B	Westville	B	Whitewood	C
Welch	C	Wethersfield	C	Whitley	B
Weld	C	Wethey	B/C	Whitlock	B
Welda	C	Wetterhorn	C	Whitman	D
Weldon	D	Wetzel	D	Whitney	B
Welduna	B	Weymouth	B	Whitore	A
Weller	C	Whakana	B	Whitsol	B
Wellington	D	Whalan	B	Whitson	D
Wellman	B	Wharton	C	Whitwell	C
Wellner	B	Whatcom	C	Wholan	C
Wellsboro	C	Whately	D	Wibaux	C
Wellston	B	Wheatley	D	Wichita	C
Wellsville	B	Wheatridge	C	Wichup	D
Welring	D	Wheatville	B	Wickersham	B
Wemple	B	Wheeler	B	Wickett	C
Wenas	B/C	Wheeling	B	Wickiup	C
Wenatcher	C	Whelchel	B	Wickliffe	D
Wendel	B/C	Whellon	D	Wickman	B
Wenham		Whetstone	B	Wicksburg	B
Wenona	C	Whidbey	C	Widta	B
Wentworth	B	Whippany	C	Widtsoe	C
Werlow	C	Whipstock	C	Wiehl	C
Werner	B	Whirlo	B	Wien	D
Weso	C	Whit	B	Wiggleton	B
Wessel	B	Whitaker	C	Wigton	A
Westbrook	D	Whitcomb	C	Wilbraham	C
Westbury	C	White Bird	C	Wilbur	C
Westcreek	B	White House	C	Wilco	C
Westerville	C	White Store	D	Wilcox	D
Westfall	C	White Swan	C	Wilcoxson	C
Westfield		Whitecap	D	Wildcat	D
Westford		Whitefish	B	Wilder	B
Westland	B/D	Whiteford	B	Wilderness	C
Westminster	C/D	Whitehorse	B	Wildrose	D
Westmore	B	Whitelake	B	Wildwood	D
Westmoreland	B	Whitelaw	B	Wiley	C
Weston	D	Whiteman	D	Wilkes	C
Westphalia	B	Whiterock	D	Wilkeson	C
Westplain	C	Whitesburg	C	Wilkins	D
Westport	A	Whitewater	B	Will	D

TABLE F.4 (Continued)

Willacy	B	Winger	C	Wolfesen	C
Willakenzie	C	Wingville	B/D	Wolfeson	C
Willamar	D	Winifred	C	Wolford	B
Willamette	B	Wink	B	Wolftever	C
Willapa	C	Winkel	D	Wolverine	A
Willard	B	Winkleman	C	Wood River	D
Willette	A/D	Winkler	A	Woodbine	B
Willhand	B	Winlo	D	Woodbridge	C
Williams	B	Winlock	C	Woodburn	C
Williamsburg	B	Winn	C	Woodbury	D
Williamson	C	Winnebago	B	Woodcock	B
Willis	C	Winnemucca	B	Woodenville	C
Willits	B	Winneshiek	B	Woodglen	D
Willoughby	B	Winnett	D	Woodhall	B
Willow Creek	B	Winona	D	Woodhurst	A
Willowdale	B	Winooski	B	Woodinville	C/D
Willows	D	Winston	A	Woodly	B
Willwood	A	Winters	C	Woodlyn	C/D
Wilmer	C	Wintersburg	C	Woodmansie	B
Wilpar	D	Winterset	C	Woodmere	B
Wilson	D	Winthrop	A	Woodrock	C
Wiltshire	C	Wintoner	C	Woodrow	C
Winans	B/C	Winu	C	Woods Cross	D
Winberry	D	Winz	C	Woodsfield	C
Winchester	A	Wishard	A	Woodside	A
Winchuck	C	Wisheylu	C	Woodson	D
Wind River	B	Wishkam	C	Woodstock	C/D
Winder	B/D	Wiskam	C	Woodstown	C
Windham	B	Wisner	D	Woodward	B
Windmill	B	Witbeck	D	Woolman	B
Windom	B	Witch	D	Woolper	C
Windsor	A	Witham	D	Woolsey	C
Windthorst	C	Withee	C	Wooskow	B/C
Windy	C	Witt	B	Woosley	C
Wineg	C	Witzel	D	Wooster	C
Winema	C	Woden	B	Woostern	B
Winetti	B	Wolcottsburg		Wooten	A
Winfield	C	Woldale	C/D	Worcester	B
Wing	D	Wolf	B	Worf	D
Wingate	B	Wolf Point	D	Work	C

TABLE F.4 *(Continued)*

Worland	B	Yamhill	C	Yuba	D
Worley	C	Yampa	C	Yuko	C
Wormser	C	Yamsay	D	Yukon	D
Worock	B	Yana	B	Yunes	D
Worsham	D	Yancy	C	Yunque	C
Worth	C	Yardley	C	Zaar	D
Worthen	B	Yates	D	Zaca	D
Worthing	D	Yauco	C	Zacharias	B
Worthington	C	Yawdim	D	Zachary	D
Wortman	C	Yawkey	C	Zafra	B
Wrentham	C	Yaxon	B	Zahill	B
Wright	C	Yeary	C	Zahl	B
Wrightman	C	Yeates Hollow	C	Zaleski	C
Wrightsville	D	Yegen	B	Zalla	A
Wunjey	B	Yelm	B	Zamora	B
Wurtsboro	C	Yenrab	A	Zane	C
Wyalusing	D	Yeoman	B	Zaneis	B
Wyard	B	Yesum	B	Zanesville	C
Wyarno	B	Yetull	A	Zanone	C
Wyatt	C	Yoder	B	Zapata	C
Wyeast	C	Yokohl	D	Zavala	B
Wyeville	C	Yollabolly	D	Zavco	C
Wygant	C	Yolo	B	Zeb	B
Wykoff	B	Yologo	D	Zeesix	C
Wyman	B	Yomba	C	Zell	B
Wymoose	D	Yomont	B	Zen	C
Wymore	C	Yoncalla	C	Zenda	C
Wynn	B	Yonges	D	Zenia	B
Wyo	B	Yonna	B/D	Zeniff	B
Wyocena	B	Yordy	B	Zeona	A
Xavier	B	York	C	Ziegler	C
Yacolt	B	Yorkville	D	Zigweid	B
Yahara	B	Yost	C	Zillah	B/C
Yahola	B	Youga	B	Zim	D
Yaki	D	Youman	C	Zimmerman	A
Yakima	B	Youngston	B	Zing	C
Yakus	D	Yourame	A	Zinzer	B
Yallani	B	Yovimpa	D	Zion	C
Yalmer	B	Ysidora	D	Zipp	C/D
Yamac	B	Yturbide	A	Zita	B

TABLE F.4 *(Continued)*

Zoar	C	Zufelt	B/D	Zunhall	B/C
Zoate	D	Zukan	D	Zuni	D
Zohner	B/D	Zumbro	B	Zurich	B
Zook	C	Zumwalt	C	Zwingle	D
Zorravista	A	Zundell	B/C		

Source: U.S. Department of Agriculture (1986). *Hydrology for Small Watersheds*, TR55 USDA Soil Conservation Service, Washington, D.C.

APPENDIX G

Description of Common Stormwater Management Practices

The efficiencies of stormwater treatment practices vary greatly, and the data presented in this appendix are site-specific but useful as a guide. The reported efficiencies are average yearly efficiencies.

URBAN — NONSTRUCTURAL

1. *Street Cleaning — Broom*

    Description: Brush-type mechanical sweepers remove litter, dust, and dirt from city streets.

    Results: Broom-type sweepers are relatively inefficient for the removal of the fine solids fraction of street materials, but useful for collecting leaves and paper. Efficiency ranges are: solids, 60–80%; BOD_5, 10–20%; N, 10–25%; P; 0–30%.

2. *Street Cleaning — Broom and Vacuum (Sometimes Called Advanced Sweeping)*

    Description: Brush and vacuum mechanical sweeping.

    Results: A greater portion of the fine solids fraction of the street surface contaminants is removed. Efficiencies are approximately: solids, 70–90%; BOD_5, 20–40%; PO_4-P, 20–50%; heavy metals, 30–60%.

3. *Street and Catch Basin Flushing*

    Description: During dry weather flow, high-water-pressure cleaning to remove solids to a point-source treatment facility.

    Results: Reduces pollution loading by removing solid materials before discharge to receiving water bodies. If done on a regular basis, efficiencies can approach an average yearly value of 20 to 30% as the frequency of cleaning approaches once per day. However, street and watershed areas are not cleaned.

4. Solid Waste Management

Description: Using antilitter laws and management of the solid waste collection activity.

Results: Reduces the solid loadings and improves public health. Reduction of pollutant loadings is site specific and must be determined for each situation.

5. Infiltration / Inflow Studies

Description: By inspection of existing sewers.

Results: Infiltration is reduced; thus additional storage and treatment is available for stormwaters.

6. Deicing Control

Description: By salt or hydrophobic substances.

Results: Salt costs are lower (about 10 to 20% when compared to hydrophobic substances); however, salt causes about $3 billion/yr in damage to paved areas, highway structures, and vehicles. This adds dissolved solids to the stormwater.

URBAN — STRUCTURAL

1. Wet-Detention Ponds

Description: Attenuation of runoff with a permanent (below groundwater) storage pool.

Results: Detained runoff increases the time of concentration or reduces the maximum discharge rate of runoff from an area. Removal efficiencies for most designs are 25 to 40% BOD_5, 30 to 70% heavy metals, and about 40–90% SS, if maintained.

2. Dry-Detention Ponds

Description: Attenuation of runoff without a permanent pool of water.

Results: Peak discharges are decreased and time of concentration is increased. Efficiencies are typically half of that obtained using wet ponds.

3. Off-Line Diversion for Seepage or Recharge by Infiltration

Description: Infiltration of runoff after preliminary concentration by diversion.

Results: Allows a large percentage of runoff to recharge a valuable aquifer or prevent saltwater intrusion. Runoff is collected in various storm systems prior to being passed into the basin. Efficiencies will vary between 50 to 95%, depending on the recharge basin.

4. Off-Line Exfiltration

Description: Perforated or slotted pipe in exfiltration rock and wrapped in a permeable membrane.

Results: Efficiencies similar to diversion for seepage or recharge.

5. Physical Treatment

Description: Similar to point-source treatment, including microstrainers, high-rate filtration, and dissolved air flotation.

Results: Microstrainers at 20 gpm/ft^2 loading rate remove 40 to 50% BOD_5 and 70% SS. High-rate filtration at 24 gpm/ft^2 removes 40 to 55% BOD_5 and 80% SS, while dissolved air flotation removes 50 to 60% BOD_5 and 80% SS at 2.5 gpm/ft^2 design loading. A comparison of these physical treatment systems is shown in Table G.1.

6. Trash Catcher of Various Screen Sizes

Description: To remove debris, such as bottles, cans, leaves, and others.

Results: Device is effective in removing debris and periodic maintenance is required. Additional maintenance after storms. There is no significant chemical removal efficiencies.

7. French Drains

Description: Drainage pipe in a pervious material (usually gravel) for drainage.

Results: Used to drain an area to prevent flooding and route groundwater to a desirable location for treatment or other control measures.

8. Porous Paving — Asphalt or Concrete

Description: Infiltration of precipitation "at source" prior to concentration.

Results: Increases infiltration and may reduce flood peaks. Porous paving may also be used to reduce the need for separation of storm sewers, especially in cases where the system is already overloaded. Some removal efficiency (both particulate and dissolved) can be expected.

TABLE G.1 Comparison of Physical Treatment Systems

Physical Unit Process	Percent Reduction						Average Capital Cost ($/MGD[a])
	Suspended Solids	BOD$_5$	COD	Settleable Solids	Total Phosphorus	Total Kjeldah Nitrogen	
Sedimentation							
Without chemicals	20–60	30	34	30–90	20	38	23,000
Chemically assisted	68	68	45	—	—	—	23,000
Swirl concentrator/ flow regulator	40–60	25–60	—	50–90	—	—	4,500
Screening							
Microstrainers	50–95	10–50	35	—	20	30	19,500
Drum screens	30–55	10–40	25	60	10	17	19,300
Rotary screens	20–35	1–30	15	70–95	12	10	19,900
Disc screens	10–45	5–20	15	—	—	—	—
Static screens	5–25	0–20	13	10–60	10	8	17,600
Dissolved air flotation[b]	45–85	30–80	55	—	55	35	34,000
High-rate filtration[c]	50–80	20–55	40	55–95	50	21	58,000

[a]ENR Construction Cost Index 2000.
[b]Process efficiencies include both prescreening and dissolved air flotation with chemical addition.
[c]Includes prescreening and chemical addition.

9. Recharge of Excess Runoff via Pressure Injection Wells

Description: Recharge of groundwater.

Results: Recharges groundwater by injecting under pressure water directly into water-bearing strata. Method can prevent saltwater intrusion. Quality conditions may be a problem.

10. Swales

Description: Vegetative stabilization of drainage channels.

Results: For velocities of up to about 6 ft/s runoff is reduced by grass channels if correctly graded and stabilized. If used with pervious soils, infiltration and treatment is provided. Removal efficiencies proportional to the quantity of water infiltrated.

11. Gravel Barriers on Flat Roofs, Findams on Pitched Roofs

Description: Delay of runoff from roofs.

Results: Flood peaks are lowered by delaying runoff from roofs, either flat or sloping. In some cases the measures also impounded a small quantity of runoff (depression storage) and provided some treatment by oxidation.

12. Dutch Drains (Gravel-Filled Ditches with Optional Drainage Pipe in Base)

Description: Infiltration of precipitation "at source" prior to concentration.

Results: Reduces volume of storm runoff and reduces flood peaks by increasing ground infiltration. First flushes can be percolated, thus reducing polution.

13. Terraces, Runoff Spreaders, etc.

Description: Increase time of concentration by increasing length of overland flow.

Results: With increased time of concentration of runoff, the hydrograph peak is lowered. This can be achieved by spreading runoff or by directing it into a system of terraces.

14. Seepage Pits or Rock Wells, Pits Usually Filled with Gravel or Rubble, Sometimes Cased

Description: Infiltration of runoff after preliminary concentration.

APPENDIX G   **533**

Results: Seepage pits collect runoff and store it until runoff percolates into the soil, but unlike Dutch drains, seepage pits do not conduct water along their length when filled.

CONSTRUCTION AND FOREST

1. Temporary Mulching and Seeding of All Stripped Areas

Description: Immediate seeding and mulching.
Results: On sites that remains bare, erosion can be significantly reduced by the use of a temporary mulch and/or seeding.

2. Traffic Control on Construction Sites, Berms, and Crushed Stone on Construction Roads

Description: Detour or slow down traffic.
Results: Being in constant use, construction roads are a particular source of pollution. Where feasible, alternative routes should be made for construction traffic; one for use in dry conditions; the other incorporating measures for wet conditions.

3. Temporary Checkdams on All Waterways Draining More Than $\frac{1}{2}$ Acre of Land Under Construction

Description: Sandbags or other impervious material.
Results: Prevents gully erosion (usually during the construction periods) in unvegetated temporary channels and therefore is temporarily unable to handle design flows.

4. Roadside Swales

Description: Final grading and the establishment of protective vegetative cover.
Results: 1. Modifies the form of steep slopes to minimize the erosion potential of runoff originating on the slope.
2. Controls runoff from elsewhere so that it can pass down the slope in a protected channel. Complete retention and removal of pollutants is possible using swale blocks.

5. Streambank Protection Using Mattresses, Blankets, and Gabions

Description: Stabilization of stream channels and banks.
Results: The use of various types of mattresses and blankets for lining stream channels can give effective control of scouring.

AGRICULTURE / NONURBAN

1. No-Till Plant in Prior-Crop Residues

Description: Prevention of cropland soil erosion and control of runoff.

Results: Most effective in dormant grass and small grain; highly effective in crop residues; minimizes spring sediment surges and provides a year-round control; reduces personnel, machine, and fuel requirements, and delays soil warming and drying.

2. Conservation Tillage

Description: Methods such as till plant, sweep tillage, and wheel-track plant are used to prevent erosion.

Results: Includes a variety of no-plow systems that retain some of the residues on the surface; more widely adaptable but somewhat less effective than "no-till plant in prior crop residues"; advantages and disadvantages generally the same as "no-till" but to lesser degree.

3. Sod-Based Rotations

Description: Prevention of cropland soil erosion by sodding.

Results: Good meadow will lose virtually no soil and reduce erosion from succeeding crops. Total soil loss is greatly reduced by losses unequally distributed over rotation cycle. Substantial runoff reduction in the sod year is evident. It will aid in controlling some diseases and pests.

4. Meadowless Rotations

Description: The rotation of two types of rowcrop to prevent soil erosion and control runoff.

Results: It will aid in disease and pest control and may provide more continuous solid protection; however, it is much less effective than "sod-based rotations." The time of exposure is shortened; thus soil loss is minimized.

5. Winter Cover Crop

Description: Early planting or winter plantings for control of erosion and runoff.

Results: Reduced winter erosion where corn stover has been removed. Provided good base for soil planting of next crop. Usually, no advantage over heavy crop of nitrate. In addition, water use by winter cover may reduce yield of a cash crop.

6. Timing of Field Operations

Description: Plowing at correct times will prevent soil erosion and control of runoff.

Results: Fall plowing facilitates more timely planting in wet springs, but greatly increases winter and spring erosion hazards. Optimum timing of spring operations can reduce erosion and increase yields.

7. Plow – Plant Systems

Description: Moldboard plowing in wheeltracks to reduce erosion.

Results: The rough cloddy surface increases infiltration and reduces erosion. It is much less effective than "no-till plant" and "conservation tillage" when long rain periods occur. The seedling stands may be poor when moisture conditions are less than optimum.

8. Contouring and Graded Rows

Description: Crop rows follow field contours to prevent erosion and runoff.

Results: This method can reduce average soil loss by 50% on moderate slopes and less on steep slopes. It must be supported by terraces on long slopes. There are no soil or climatic limitations, but is it not feasible on very irregular topography.

9. Contour Strip-Cropping

Description: Strips of sod alternated with strips of two crops to prevent erosion and runoff.

Results: Row crop and hay in alternate 50 to 100-ft strips reduce soil loss to about 50% of that with the same rotation contoured only. Fall-seeded grain in lieu of meadow is about half as effective. Alternating corn and spring grain is not effective. Area must be suitable for across-slope farming and establishment of rotation meadows. Favorable and unfavorable features are similar to "sod-based rotations" and "contouring."

10. Terraces

Description: Almost level areas prevent soil erosion and control runoff.

Results: Support contouring and agronomic practices by reducing effective slope length and runoff concentration. They reduce erosion and conserve soil moisture. About 90% of the soil that is moved is deposited in the terrace channels.

APPENDIX H

Solutions to Selected Case Studies

Some of the problems at the end of each chapter were developed from actual case studies and applied research in which the authors were involved. Some case studies and research projects were reduced to present the details relative to the subject matter of this book, and in some cases substantially reduced because of irrelevant material. Approximately 30% of the problems at the end of each chapter use data from actual case studies. In some instances, names and places have been removed. It is anticipated that these solutions will help the user understand the concepts and approach to problem solving in stormwater management.

Section 2.6, Problem 2. Assume that intensity, duration, and interevent times are all independent:

(a) $i: T_r = \dfrac{1}{\Pr\{i > 0.40\} \times n} = \dfrac{1}{0.02(100)} = \dfrac{1}{2}$ or 2 per year

$D: T_r = \dfrac{1}{\Pr\{D \geq 6\text{ h}\} \times n} = \dfrac{1}{0.05(100)} = \dfrac{1}{5}$ or 5 per year

$\Delta: T_r = \dfrac{1}{\Pr\{\Delta \geq 92\text{ h}\} \times n} = \dfrac{1}{0.10(100)} = \dfrac{1}{10}$ or 10 per year

(b) $\Pr\{i \geq 0.40 \cap D \geq 6\} = 0.02(0.05) = 0.001$

no./year $= 0.001(100) = 0.1$ per year or 1 time every 10 years

$$T_r = \dfrac{1}{\Pr\{i \cap D\} \times n} = \dfrac{1}{0.001(100)} = 10 \text{ years}$$

Section 2.6, Problem 10. Using 2 months of rainfall data reported on an hourly basis, develop an empirical frequency distribution for two minimum interevent dry periods. Use $\Delta = 4$ h and $\Delta = 24$ h. Assume that the first event is preceded by a 24-h interevent dry period. Station and period used are as follows:

Apalachicola WSO ARPT
June and July 1990

$\Delta = 4$ h		$\Delta = 24$ h	
Event	P (in.)	Event	P (in.)
1	0.47	1	0.47
2	0.95	2	0.95
3	0.02	3	0.38
4	0.01	4	1.02
5	0.35	5	3.06
6	0.05	6	4.97
7	0.95	7	1.25
8	0.01	8	0.05
9	0.01	9	0.01
10	0.59		
11	0.05		
12	0.07		
13	2.00		
14	0.35		
15	0.32		
16	0.30		
17	0.80		
18	3.54		
19	0.01		
20	0.80		
21	0.20		
22	0.25		
23	0.05		
24	0.01		

For minimum interevent dry period = 4 h:

Event (Plot Position, m)	P (in.)	P sorted (in.)	Weibull $m/(n+1)$	California m/n	Foster $(2m-1)/2n$	Exceedence $(m-1)/n$
1	0.47	0.01	0.04	0.04	0.02	0.00
2	0.95	0.01	0.08	0.08	0.06	0.04
3	0.02	0.01	0.12	0.13	0.10	0.08
4	0.01	0.01	0.16	0.17	0.15	0.13
5	0.35	0.01	0.20	0.21	0.19	0.17
6	0.05	0.02	0.24	0.25	0.23	0.21
7	0.95	0.05	0.28	0.29	0.27	0.25
8	0.01	0.05	0.32	0.33	0.31	0.29
9	0.01	0.05	0.36	0.38	0.35	0.33
10	0.59	0.07	0.40	0.42	0.40	0.38
11	0.05	0.20	0.44	0.46	0.44	0.42
12	0.07	0.25	0.48	0.50	0.48	0.46
13	2.00	0.30	0.52	0.54	0.52	0.50
14	0.35	0.32	0.56	0.58	0.56	0.54
15	0.32	0.35	0.60	0.63	0.60	0.58
16	0.30	0.35	0.64	0.67	0.65	0.63
17	0.80	0.47	0.68	0.71	0.69	0.67
18	3.54	0.59	0.72	0.75	0.73	0.71
19	0.01	0.80	0.76	0.79	0.77	0.75
20	0.80	0.80	0.80	0.83	0.81	0.79
21	0.20	0.95	0.84	0.88	0.85	0.83
22	0.25	0.95	0.88	0.92	0.90	0.88
23	0.05	2.00	0.92	0.96	0.94	0.92
24	0.01	3.54	0.96	1.00	0.98	0.96

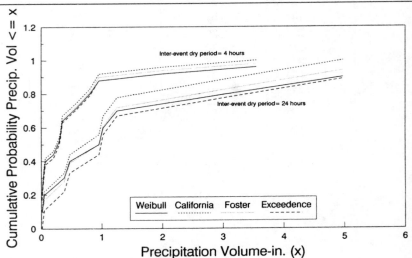

Pr{$P <= x$ given inter-event dry per = 4 or 24 hr}
Guaging Station, Apalachicola WSO ARPT
June and July 1990

APPENDIX H **539**

For interevent dry period = 24 h:

Event (Plot Position, m)	P (in.)	P sorted (in.)	Weibull $m/(n+1)$	California m/n	Foster $(2m-1)/2n$	Exceedence $(m-1)/n$
1	0.47	0.01	0.10	0.11	0.06	0.00
2	0.95	0.05	0.20	0.22	0.17	0.11
3	0.38	0.38	0.30	0.33	0.28	0.22
4	1.02	0.47	0.40	0.44	0.39	0.33
5	3.06	0.95	0.50	0.56	0.50	0.44
6	4.97	1.02	0.60	0.67	0.61	0.56
7	1.25	1.25	0.70	0.78	0.72	0.67
8	0.05	3.06	0.80	0.89	0.83	0.78
9	0.01	4.97	0.90	1.00	0.94	0.89

Original data format:

STATION	DATE	A.M. HOUR ENDING												P.M. HOUR ENDING											TOTAL
		1	2	3	4	5	6	7	8	9	10	11	12	1	2	3	4	5	6	7	8	9	10	11	12	
JUNE 90	5			.47																							.47
	10																	.84	.11							.95
	16																		.02							.02
	17													.01												.01
	18													.28	.07											.35
	22					.05												.53	.02	.07	.20	.08	.02	.02	.01	1.00
	23							.01														.01				.02
JULY 90	1									.38	.21															.59
	2										.03	.02									.01	.05	.01			.12
	3														1.56	.22	.06	.01			.02	.08	.05			2.00
	4										.05				.26	.04										.35
	12				.10	.22			.08	.22																.62
	13				.78	.02					.89	.14	1.58	.04	.01	.01	.01	.01	.01		.04	.01		.04		3.59
	14		.14	.07	.12	.11	.02	.18	.09		.02								.01							.76
	17																				.07	.73				.80
	18													.04	.10	.06										.20
	19				.01	.09	.01	.01			.08	.05														.25
	27		.05																							.05
	29	.01																								.01

Section 2.7, Problem 3

LOG PEARSON
METHOD OF MOMENTS (DIRECT)

MEAN OF X(N)	:	0.3781e+01
STANDARD DEV OF X(N)	:	0.1656E+01
VARIANCE OF X(N)	:	0.2743E+01
SKEW OF X(N)	:	0.1625E+01

ALPHA 0.3840E-01	M1	-.2883E+01
BETA 0.1076E+03	M2	0.1587E+00
GAMMA -.2885E+01	SKEW	0.1928E+00

DISTRIBUTION : CHI SQUARE VALUE IS
:0.4257E+04 TABLE VALUE IS : 16.919

METHOD OF MOMENTS (INDIRECT)

MEAN OF LN X(N)	:	0.1255E+01
STANDARD DEV OF LN X(N)	:	0.3806E+00
VARIANCE OF LN X(N)	:	0.1449E+00
SKEW OF LN X(N)	:	0.7928E+00

DISTRIBUTION : CHI SQUARE VALUE IS
:0.1684E+02 TABLE VALUE IS : 16.919

RETURN PERIOD (Yrs)	2	5	10	25
PREDICTED	0.3339E+01	0.4717E+01	0.5826E+01	0.7481E+01
STD. ERR	0.2190E+00	0.38545E+00	0.6291E+00	0.1208E+01

RETURN PERIOD (Yrs)	50	100	200	500
PREDICTED	0.8916E+01	0.1055E+02	0.1240E+02	0.1528E+02
STD. ERR	0.1883E+01	0.2810E+01	0.4048E+01	0.6303E+01

Section 3.6, Problem 12. From FID curve (Appendix C, zone 7), $i = 5$ in./h

Rational: $Q_p = CiA = (1)(5)(100) = 500 \text{ ft}^3/\text{s}$
SCS: $K = 550$ from Figure 3.2
$Q_p = KAR = 550(100/640)(5) = 430 \text{ ft}^3/\text{s}$

APPENDIX H   541

Santa Barbara: $R = \dfrac{2.5 \text{ in.} \times 100 \times 60}{30} = 500 \text{ ft}^3/\text{s}$

$K = \dfrac{\Delta t}{2t_c + \Delta t} = \dfrac{30}{2(30) + (30)} = 0.33$

$Q = 0 + 0.33[0 + 500 - 2(0)] = 165 \text{ ft}^3/\text{s}$

Volume of pond $= \tfrac{5}{2}$ in. $\times$ 100 acres $\times$ 43,560 $\times \tfrac{1}{12} = 907{,}500 \text{ ft}^3$

Section 3.6, Problem 17

$$O = \dfrac{S - 50{,}000}{11{,}350}$$

$$S = 11{,}350(O) + 50{,}000$$

$$N = 11{,}350(O) + 50{,}000 + \dfrac{O}{2}(60)(5)$$

$$N = 11{,}500(O) + 50{,}000$$

$$O = \dfrac{N - 50{,}000}{11{,}500}$$

$i = 9$ in./h from FID curve

$Q_p = ciA = (1)(9)(4) = 36 \text{ ft}^3/\text{s}$

t	I	I_{i-1}	$\bar{I}$	$\bar{I}(300)$	S_{i-1}	O_{i-1}	$O_{i-1}(150)$	N_i^a	O_i^b	S_i
5	18	0	9	2700	50,000	0	0	52,700	0.23	52,348
10	36	18	27	8100	52,348	0.23	34.5	60,414	0.91	59,100
15	18	36	27	8100	59,100	0.91	136.5	67,064	1.48	64,838
20	0	18	9	2700	64,838	1.48	222	67,316	1.51	65,100
		0	0	0[c]						

[a] $N_i = \bar{I}(300) + S_{i-1} - O_{i-1}(150)$

[b] $O_i = \dfrac{(N_i - 50{,}000)}{11{,}500}$

[c] Storage goes down because there is no more input.

Section 5.10, Problem 9. The watershed characteristics are:

Solids contaminant load from Table 5.13 using urban average of 450 kg/ha-yr

BOD_5/SS composition is the same: 0.03 kg/kg

Impervious area now 80% or 2 mi² (1280 acres)

Total area $= 2.5$ mi²

Initial solids loading assuming 10-day antecedent dry condition:

$$P_0 = \left(450\frac{\text{kg}}{\text{ha-yr}} \times \frac{1\text{ yr}}{365\text{ days}} \times 10\text{ days} \times 0.891\frac{\text{lb/acre-yr}}{\text{kg/ha-yr}} \times 1280\text{ acres}\right)$$

$$= 12{,}773\text{ lb solids}$$

Now set up table similar to text:

Δt (min)	Runoff Volume (ft³)	r_{avg} (in./h)	P_0 (lb)	ΔP (lb)	$\Sigma\Delta P$ (lb)	Total Solids Concentration (mg/L)	BOD₅ Concentration (mg/L)
0–10	15,000	0.019	12,773	186	186	199	5.97
10–20	70,000	0.090	12,587	859	1055	197	5.91
20–30	160,000	0.207	11,728	1861	2916	186	5.58
30–40	187,000	0.242	9,867	1831	4747	156	4.68
40–50	157,000	0.203	8,036	1251	5998	128	3.84
50–60	127,000	0.164	6,785	853	6851	108	3.24
60–70	87,000	0.112	5,932	509	7360	94	2.82
70–80	70,000	0.090	5,423	374	7734	86	2.58
80–90	57,000	0.074	5,049	286	8020	80	2.40
90–100	47,000	0.061	4,763	223	8243	76	2.28
100–110	42,000	0.054	4,540	188	8431	72	2.16
110–120	35,000	0.045	4,352	150	8581	69	2.07
120–130	29,000	0.037	4,202	119	8700	66	1.98
130–140	22,500	0.029	4,083	90.8	8791	65	1.95
140–150	17,500	0.023	3,992	70.4	8861	64	1.92
150–160	16,000	0.021	3,922	63.1	8924	63	1.89
160–170	12,000	0.015	3,859	44.4	8968	59	1.77
170–180	9,000	0.012	3,815	35.1	9003	62	1.86
180–190	6,500	0.008	3,780	23.2	9026	57	2.71
190–200	4,000	0.005	3,757	14.4	9040	58	1.74
200–210	1,500	0.002	3,743	5.7	9046	61	3.57

$P_0 = (6386\text{ kg})(2205\text{ lb/kg}) = 12{,}773$ lb solids; all other: $P_0 = P_{0(i-1)} - \Delta P$
Volume $= Q_{avg} \times (\Delta t)_{min} \times 60$
$r_{avg} = \text{volume(ft}^3)/[(\Delta t)_{min} \times \text{area} \times 60.5]$
$\Delta P = (4.6)(r_{avg})(P_0)(\Delta t)/60$
BOD₅ conc. $= [\text{TS}] \times (0.03\text{ kg/kg})$
conc. (mg/L) $= \Delta P \times 16019/\text{runoff volume for total impervious area}$

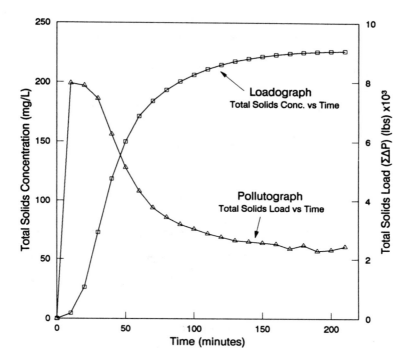

Section 5.10, Problem 13

$$\text{Loading} = \text{EMC}(2.72)(R/12); \text{ assume 50 in./yr}$$

where

$$R = 0.60(P) = 0.60(50) = 30 \text{ in./yr}$$

$$\text{and Loading} = (0.50 \text{ mg/L})(2.72)(30/12)$$

$$= 3.4 \text{ lb/acre-yr}$$

For 2 acres

$$\text{Load} = 3.4 \times 2 = 6.8 \text{ lb/yr}$$

$$\text{Urban loading} \approx 0.40 \text{ kg/ha-yr} \times 2 \text{ acres} \times 0.892 \text{ (lb/acre)/(kg/ha)}$$

$$= 0.713 \text{ lb/yr}$$

Section 5.10, Problem 14

(a) Initial load = remaining load + 2 days' accumulation

$$P_0 = 0 + \text{cattle load (Table 5.5)} + \text{urban load (Table 5.19)}$$

$$= 0.069 \frac{\text{g}}{\text{kg-day}} \times 600 \frac{\text{kg}}{\text{head}} \times 40 \text{ head}$$

$$\times 2 \text{ days} \times 0.40 \begin{pmatrix} \text{fraction} \\ \text{removed} \end{pmatrix}$$

$$+ 1800 \frac{\text{g}}{\text{ha-yr}} \times \frac{1 \text{ yr}}{365 \text{ days}} \times 2 \text{ days} \times 60 \text{ acres}$$

$$\times \frac{1 \text{ ha}}{2.47 \text{ acres}} \times 0.90 \begin{pmatrix} \text{fraction} \\ \text{removed} \end{pmatrix}$$

$$= 1325 + 216 = 1541 \text{ g}$$

$$R = \frac{1}{2} \frac{\text{in.}}{\text{h}} \times 100 \text{ acres} \times 101.6 \frac{\text{m}^3/\text{h}}{\text{acre-in.}/\text{h}}$$

$$\times 1000 \frac{\text{L}}{\text{m}^3} \times 0.4 = 2(10^6) \text{ L}$$

$$C_0 = \frac{P_0}{R} = \frac{1541 \text{ g} \times 10^3 \text{ mg/g}}{2(10^6) \text{ L}} = 0.77 \text{ mg/L}$$

(b) $$\Delta P = 4.6 \times 0.3 \times \frac{1541 \text{ g}}{454 \text{ g/lb}} \times \frac{10 \text{ min}}{60 \text{ min/h}} = 0.78 \text{ lb}$$

$$\text{Conc. (mg/L)} = \frac{0.78 \times 16{,}019}{0.4(10^6) \text{ L} \times 1 \text{ ft}^3/28.3 \text{ L}} = 0.89 \text{ mg/L}$$

Section 6.10, Problem 5. Given:

$$K_{Le} = 0.15 \text{ day}^{-1}$$
$$K_N = 0.12 \text{ day}^{-1}$$
$$K_a = 0.06 \text{ day}^{-1}$$
$$H = 6 \text{ m } (19.7 \text{ ft})$$
$$D_0 = 2 \text{ mg/L}$$
$$L_0 = 9 \text{ mg/L}$$
$$N_0 = 2 \text{ mg/L}$$
$$U = 4 \text{ km/day}$$

(*Note:* Neglect nitrogen demand.)

$$t_c = \frac{1}{K_a - K_L} \ln\left[\frac{K_a}{K_L}\left(1 - \frac{D_0(K_a - K_L)}{K_L L_0}\right)\right]$$

$$= \frac{1}{0.06 - 0.15} \ln\left[\frac{0.06}{0.15}\left(1 - \frac{2(0.06 - 0.15)}{0.15(9)}\right)\right]$$

$$= 8.79 \text{ days}$$

$$X_c = U \times t_c$$

$$= 4\frac{\text{km}}{\text{day}} \times 8.79 \text{ days}$$

$$= 35.16 \text{ km}$$

D_c, neglecting nitrogen demand:

$$D_c = 2 \exp\left[\left(\frac{-0.06(35.16)}{4}\right)\right]$$

$$+ \frac{9(0.15)}{0.06 - 0.15}\left[\exp\left(\frac{-0.15(35.16)}{4}\right) - \exp\left(\frac{-0.06(35.16)}{4}\right)\right]$$

$$= 6.02 \text{ ppm}$$

At $t = 3$ days:

$$X = 4\frac{\text{km}}{\text{day}} \times 3 \text{ days}$$

$$= 12 \text{ km}$$

$$D_{12} = 2 \exp\left[\left(\frac{-0.06(12)}{4}\right)\right]$$

$$+ \frac{9(0.15)}{0.06 - 0.15}\left[\exp\left(\frac{-0.15(12)}{4}\right) - \exp\left(\frac{-0.06(12)}{4}\right)\right]$$

$$= 4.64 \text{ ppm}$$

At $t = 6$ days:

$$X = 24 \text{ km}$$
$$D_{24} = 5.76 \text{ ppm}$$

At $t = 12$ days:
$$X = 48 \text{ km}$$
$$D_{48} = 5.80 \text{ ppm}$$

At $t = 15$ days:
$$X = 60 \text{ km}$$
$$D_{60} = 5.33 \text{ ppm}$$

At $t = 27$ days:
$$X = 108 \text{ km}$$
$$D_{108} = 3.10 \text{ ppm}$$

At $t = 39$ days:
$$X = 156 \text{ km}$$
$$D_{156} = 1.59 \text{ ppm}$$

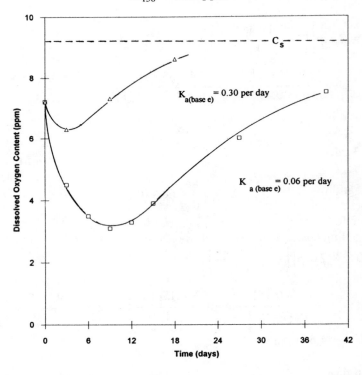

Section 7.9, Problem 7. Using Figures 7.3 and 7.5, the flow rate per unit equivalent impervious area is determined, the area of exfiltration is calculated based on the exfiltration rate. Finally, the length of trench is calculated knowing the unit exfiltration area. All these data must be available or

engineering judgment must be used and supported. For the Orlando area:

$$R_Q = 0.12 \text{ (ft}^3/\text{s)}/\text{EIA} \quad \text{and} \quad \text{EIA} = (0.60)(5) = 3.0$$

$$R_Q = 0.12(3.0) = 0.36 \text{ ft}^3/\text{s}$$

$$A = (0.36 \text{ ft}^3/\text{s})(3600 \text{ s/h})(12 \text{ in./ft})/(2 \text{ in./h}) = 7776 \text{ ft}^2$$

$$L = \frac{7776 \text{ ft}^2}{20 \text{ ft}^2/\text{ft}} = 388.8 \text{ ft}$$

Check using unit volume:

$$V = \frac{(\tfrac{1}{2} \text{ in.})(43{,}560 \text{ ft}^2/\text{acre})(3 \text{ acres})}{12 \text{ in./ft}} = 5445 \text{ ft}^3$$

$$L = \frac{5445 \text{ ft}^3}{14 \text{ ft}^2/\text{ft}} = 388.9 \text{ ft}$$

Available lengths of pipe are 20 ft; thus the trench length would be rounded to 400 ft. For the Atlanta area:

$$R_Q = 0.08 \text{ (ft}^3/\text{s)}/\text{EIA} \quad \text{and} \quad L = 260 \text{ ft}$$

Next, the data for flow rate reduction with first flush are available for the Atlanta area in Figure 7.5. The flow rate decreases to 0.06 (ft³/s)/EIA and the length of pipe is 195 ft, which rounds up to 200 ft. The cost of trench is $250 per foot or the cost savings in assuming first flush is $15,000(60)(250). The sampling cost is $10,000. Therefore, one may wish to sample and design at a lower flow rate.

Section 7.9, Problem 10. First calculate velocity:

$$A = \frac{WH}{2} \quad \text{but } W = 20H$$

$$= \frac{20H^2}{2} = 10H^2$$

$$v = \frac{Q}{A} = \frac{3}{10(\tfrac{1}{2})^2} = 1.2 \text{ ft/s}$$

548 SOLUTIONS TO SELECTED CASE STUDIES

1.2 ft/s < 2.5 ft/s (from Table 7.8). Therefore, v is OK. Next, determine an N factor:

$$vR = \left(\text{where } R = \frac{A}{P} = \frac{wH/2}{\sim w} = H/2\right)$$

$$= 1.2\left(\frac{H}{2}\right) = 1.2\left(\frac{1}{4}\right) = 0.3 \text{ ft}^2/\text{s}$$

Select $N \cong 0.08$. Calculate slope:

$$Q = \frac{1.486}{N}(R)^{2/3}(S)^{1/2}(A) \quad (\text{but } R = H/2: A = 10H^2)$$

$$3 = \frac{1.486}{0.08}\left(\frac{1}{4}\right)^{2/3}(S)^{1/2}(2.5)$$

$$S^{1/2} = \frac{3(0.08)}{1.486(0.40)(2.5)}$$

$$S = 0.026 \text{ ft/ft}$$

Sensitivity to N:

$$\text{Pick } N = 0.03; \; S = 0.0036 \text{ ft/ft}.$$

Section 7.9, Problem 12. From Table 5.17: $\text{EMC}_{Zn} = 0.135$ mg/L.

$$\text{Zinc load} = (0.135 \text{ mg/L})(2 \text{ acres} \times 1 \text{ in.} \times 0.15 \times 3630 \times 28.35 \text{ L/ft}^3)$$

$$= (0.135)(30{,}873) = 4168 \text{ mg}$$

$$\text{Mass reduction (rate)} = 1.85 \text{ mg/m}^2\text{-h}$$

$$\text{Detention time} = 150 \text{ m}/(0.02 \text{ m/s} \times 3600 \text{ s/h}) = 2.08 \text{ h}$$

Removal% = (mass reduction rate × detention time

$$\times \text{wetted area/input}) \times 100$$

let P_w = wetted perimeter

$$\frac{(60)(4168 \text{ ft})}{100} = (1.85)(2.08)(150 \text{ m} \times P_w)$$

$$P_w = 4.33 \text{ m}$$

Section 7.10, Problem 5

```
Utilities  Watershed  Rainfall  Hydrograph  Routing  Order

                         ─── Watershed Edit ───  Initial
    Drainage area (acres)                                 48
    Time of Concentration (min)                           30
    Impervious Watershed Area (acres)                     24
    Percent of Impervious Directly Connected (%)          50
    Additional Abstraction for Impervious Area (in)
    Additional Abstraction for Pervious Area (in)
    Maximum Infiltration Capacity (in)
    Input 999 if No Limit Exists                         999

    SCS Curve Number (for Pervious)                       50
    SCS Abstraction Factor                                 .2

         < Continue >         < Cancel >         < Help >
```

```
Utilities  Watershed  Rainfall  Hydrograph  Routing  Order

                         ─── Watershed Edit ─── Impervious Area Reduced
    Drainage area (acres)                                 48
    Time of Concentration (min)                           30
    Impervious Watershed Area (acres)                     12
    Percent of Impervious Directly Connected (%)          50
    Additional Abstraction for Impervious Area (in)
    Additional Abstraction for Pervious Area (in)
    Maximum Infiltration Capacity (in)
    Input 999 if No Limit Exists                         999

    SCS Curve Number (for Pervious)                       50
    SCS Abstraction Factor                                 .2

         < Continue >         < Cancel >         < Help >
```

```
Utilities   Watershed   Rainfall   Hydrograph   Routing   Order      24 Acre
======================================= HYDROGRAPH =======================  Impervious
   Time      Rainfall    Infiltration       Hydrographs in cfs
   Hours      (in)          (in)         Instantaneous      Outflow

    .5        .11          .06875          5.32224         2.612736
    .75       .12          .075            5.80608         3.793306
    1         .15          .09375          7.2576          4.88872
    1.25      .16          9.999999E-02    7.74144         5.93304
    1.5       .17          .10625          8.22528         6.753168
    1.75      .27          .16875         13.06368         8.309692
    2         .3           .1875          14.5152         10.50159
    2.25     1.08          .6228274       62.35201        21.6744
    2.5      1.14          .5146726       93.44447        44.16393
    2.75      .32          .124031        30.18562        51.22438
    3         .28          .1023841       27.60131        42.29201
    3.25      .24          8.354793E-02   24.47303        35.79007
    3.5       .24          7.992856E-02   25.1735         31.40335
    3.75      .18          5.771046E-02   19.31287        27.73928

Total ---  6                                431241.9 cf    430996.4 cf
Rational Coefficient =   .415                 Peak Flow =   51.22438 cfs
```

```
Utilities   Watershed   Rainfall   Hydrograph   Routing   Order      12 Acre
======================================= HYDROGRAPH =======================  Impervious
   Time      Rainfall    Infiltration       Hydrographs in cfs
   Hours      (in)          (in)         Instantaneous      Outflow

    .5        .11          .0928125        2.66112         1.306368
    .75       .12          .10125          2.90304         1.896653
    1         .15          .1265625        3.6288          2.44436
    1.25      .16          .135            3.87072         2.96652
    1.5       .17          .1434375        4.11264         3.376584
    1.75      .27          .2278125        6.53184         4.154846
    2         .3           .253125         7.2576          5.250796
    2.25     1.08          .8702199       34.06817        11.41563
    2.5      1.14          .7413387       70.26059        27.71513
    2.75      .32          .180553        25.05264        35.69173
    3         .28          .1496036       23.54295        31.13416
    3.25      .24          .1224591       21.29688        27.64846
    3.5       .24          .1174739       22.26169        25.30079
    3.75      .18          8.501438E-02   17.2945         23.09171

Total ---  6                                330466.8 cf    330237.3 cf
Rational Coefficient =   .325                 Peak Flow =   35.69173 cfs
```

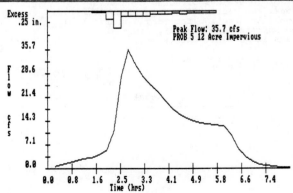

Peak Flow: 35.7 cfs
PROB 5 12 Acre Impervious

Utilities   Watershed   Rainfall   Hydrograph   Routing   Order T_c=60 minutes
================================ HYDROGRAPH =================================

Time Hours	Rainfall (in)	Infiltration (in)	Hydrographs in cfs Instantaneous	Outflow
.				
.5	.11	.0928125	2.66112	.7735466
.75	.12	.10125	2.90304	1.219887
1	.15	.1265625	3.6288	1.674561
1.25	.16	.135	3.87072	2.135716
1.5	.17	.1434375	4.11264	2.548153
1.75	.27	.2278125	6.53184	3.164617
2	.3	.253125	7.2576	3.993529
2.25	1.08	.8702199	34.06817	7.69783
2.5	1.14	.7413387	70.26059	17.57928
2.75	.32	.180553	25.05264	24.26314
3	.28	.1496036	23.54295	24.27084
3.25	.24	.1224591	21.29688	23.85952
3.5	.24	.1174739	22.26169	23.39725
3.75	.18	8.501438E-02	17.2945	22.59299

Total --- 6   330466.8 cf   330141 cf
Rational Coefficient = .325   Peak Flow = 24.27084 cfs

Utilities   Watershed   Rainfall   Hydrograph   Routing   Order
------------------------------- Rainfall Edit -------------------------------

Time min	Rain inch	Time min	Rain inch	Time min	Rain inch	Time min	Rain inch	Time min	Rain inch	Time min	Rain inch
15	.1	300	.12								
30	.11	315	.12								
45	.12	330	.12								
60	.15	345	.11								
75	.16	360	.11								
90	.17										
105	.27										
120	.3										
135	1.08										
150	1.14										
165	.32										
180	.28										
195	.24										
210	.24										
225	.18										
240	.16										
255	.14										
270	.13										
285	.13										

                         Total Rain   6 in   < Complete >

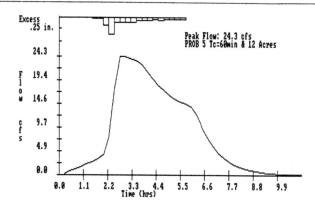

Peak Flow: 24.3 cfs
PROB 5 Tc=60min & 12 Acres

552 SOLUTIONS TO SELECTED CASE STUDIES

Section 8.5, Problem 8. Wet-detention pond design must also consider:

1. Littoral zone: 1:4 (or 6) side slopes; max. depth of 3 ft 6 in.; fluctuation of runoff storage ≤ 18 in.
2. Permanent pool < 4 to 5 ft depth.
3. Quality storage (above sediment pool) from 3 in. of rainfall

$$\text{Vol}_Q = \frac{CPA}{12} = \frac{0.7(3)(10)}{12} = 1.75 \text{ acre-ft}$$

$$= 76{,}230 \text{ ft}^3$$

4. Sediment pool volume (depends on geometry): Assume 3 ft depth × 1000 ft² = 3000 ft³. (Very shallow, would probably need more.)
5. Area needed (say 5 ft deep)

$$A = \frac{\text{sediment volume} + \text{quality volume}}{5 \text{ ft}}$$

$$= \frac{79{,}230}{5 \text{ ft}} = 15{,}846 \text{ ft}^2$$

(But only have 5000 ft², need much bigger area.)

6. Change permanent pool to 75,000 ft³ (25,000 ft² × 3 ft) and calculate T_d (based on average inflow and outflow)

$$T_d = \frac{\text{vol}}{Q_{\text{out}_{\text{avg}}}} = \frac{80{,}000}{1.5} = 53{,}333 \text{ s or } 14.81 \text{ h}$$

(*Note:* A short detention time; however, only an average.)

Section 8.5, Problem 9. To determine the partition coefficient for phosphorus in soil–water equilibrium systems, the concentration in the soil phase must be compared with the concentration in the water phase using material balance as follows: For example, the first experiment

$$P_l V_l + P_s M_s = PM_s$$

where P = phosphorus added, μg/g soil
M_s = mass of solids
P_l = phosphorus concentration in liquid, μg/mL water
P_s = phosphorus concentration in soil, μg/g soil
V_l = volume of liquid

or

$$\left(0.025\frac{\mu g\ P}{mL}\right)(1500\ mL) + P_s(300\ g) = \left(15\frac{\mu g\ P}{g\ soil}\right)(300\ g)$$

$$P_s = 14.875\ \mu g/g\ soil$$

Continue the calculation and develop the following table:

Phosphorus Distribution Between Solid and Water Phase

	Aerobic			Anaerobic	
P Added (μg P/g soil)	P in Soil (μg P/g soil)	P in Water (μg P/mL)	P Added (μg P/g soil)	P in Soil (μg P/g soil)	P in Water (μg P/mL)
15	14.90	0.025	15	14.9	0.025
30	29.85	0.05	30	29.8	0.04
45	44.4	0.12	45	44.5	0.095
60	56.0	0.80	60	58.0	0.40
180	160.0	4.0	180	155	5.0
420	367.0	10.6	420	270	30.0
780	630.0	30.0	780	330	90.0
1080	880.0	40.0	1080	579	100.2

The corresponding dimensionless partition coefficients of phosphorus between the soil and water phase can be calculated as the ratio of phosphorus in the soil to that in the water.

Partition Between Soil and Water

P Added (μg P/g soil)	Partition of Phosphorus (μg P/g soil/μg P/mL water)	
	Aerobic	Anaerobic
15	595	595
30	596	745
45	370	468
60	70	145
180	40	31
420	35	9
780	21	3.7
1080	22	5.8

The partitioning of phosphorus between soil and water indicates that phosphorus is partitioned (held in the soil) more strongly in soil under aerobic conditions. Under an anaerobic environment, the phosphorus in the soil tends to be released from the soil and go back in solution and the concentration of phosphorus in water is increased.

Section 8.5, Problem 10

Pond Location	Sedimentation Accumulation Rate (cm/yr)		
	In Situ	Core Tube	U.S. EPA Model
Silver Star	3.43	0.67	2.46
Horizon Park	1.44	0.21	1.11
New Smyrna	1.03	0.25	0.13
Ft. Myers	4.20	0.58	0.76
Ocala	2.13	0.72	2.53
Eau Gallie	3.63	0.58	0.76
Clearwater	2.20	0.22	0.78
Miami Gardens	1.08	0.20	0.26

Section 8.6, Problem 1. First, use the RETEN computer program to calculate the size of the off-line retention pond. We use the rational volume equation with a runoff volume coefficient equal to the directly connected impervious area, or 0.25. The actual computer screen is:

```
Design Precipitation =   1.00 inches
Watershed area =    48.0 acres
Rational 'C' Factor = 0.25
Retention Side Slope = 0.250 vert/horiz
-------------------------------------------
Depth         Retention Volume         Surface Area
 ft              acre-feet                acres
-------------------------------------------
  1                1.000                  1.078
  2                1.000                  0.585
  3                1.000                  0.428
  4                1.000                  0.355
  5                1.000                  0.316
-------------------------------------------
```

The off-line volume is 1.00 acre-ft and the surface area of the pond that is four ft deep is 0.355 acres.

For the first in. of rainfall, there is no runoff from the pervious area or the CN value is 50, maximum storage is 10 in., and using an initial abstraction that is 0.2 of maximum storage results in 2 in, thus no runoff until rainfall exceeds 2 in. The watershed input data are listed in the following actual computer screen print.

APPENDIX H 555

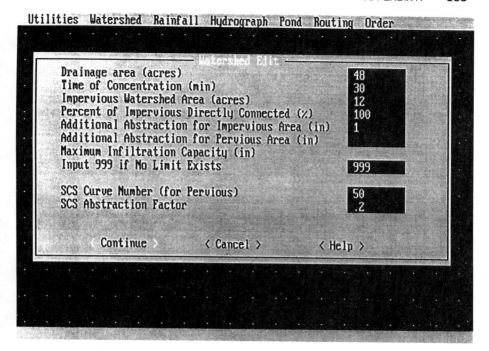

The rainfall data used for this problem is found in the following computer screen capture.

Time min	Rain inch	Time min	Rain inch	Time min	Rain inch	Time min	Rain inch	Time min	Rain inch	Time min	Rain inch
15	.1	300	.12								
30	.11	315	.12								
45	.12	330	.12								
60	.15	345	.11								
75	.16	360	.11								
90	.17										
105	.27										
120	.3										
135	1.08										
150	1.14										
165	.32										
180	.28										
195	.24										
210	.24										
225	.18										
240	.16										
255	.14										
270	.13										
285	.13										

Total Rain 6 in   < Complete >

We use the Santa Barbara urban hydrograph generation program routine to estimate the hydrograph shape. All the output data are shown in the following computer generated printer output. The peak flow is 45.17 ft^3/s. Runoff does occur from the pervious surfaces because rainfall exceeds 2.0 in. Thus, the rational coefficient is 0.351. The routing coefficient is set at 0.20 because of the time of concentration and the rainfall interval. The coefficient may be changed if needed.

```
-------------------------------------------------SMADA  PAGE 1
Chapter 8 Problem 1     Hydrograph Type :Santa Barbara Method
----------------------------------------------------------------
Time     Rain    Cumulative  Infiltration  Instantaneous  Outflow
(hr)     (in)      (in)         (in)           (cfs)       (cfs)
----------------------------------------------------------------
0.00     0.00      0.00         0.00           0.00        0.00
0.25     0.10      0.10         0.08           0.00        0.00
0.50     0.11      0.21         0.08           0.00        0.00
0.75     0.12      0.33         0.09           0.00        0.00
1.00     0.15      0.48         0.11           0.00        0.00
1.25     0.16      0.64         0.12           0.00        0.00
1.50     0.17      0.81         0.13           0.00        0.00
1.75     0.27      1.08         0.20           3.87        0.77
2.00     0.30      1.38         0.23          14.52        4.14
2.25     1.08      2.46         0.79          55.19       16.43
2.50     1.14      3.60         0.70          84.25       37.74
2.75     0.32      3.92         0.17          28.34       45.17
3.00     0.28      4.20         0.14          26.24       38.02
3.25     0.24      4.44         0.12          23.49       32.76
3.50     0.24      4.68         0.11          24.36       29.23
3.75     0.18      4.86         0.08          18.81       26.17
4.00     0.16      5.02         0.07          17.10       22.88
4.25     0.14      5.16         0.06          15.24       20.20
4.50     0.13      5.29         0.06          14.37       18.04
4.75     0.13      5.42         0.05          14.58       16.61
5.00     0.12      5.54         0.05          13.64       15.61
5.25     0.12      5.66         0.05          13.81       14.86
5.50     0.12      5.78         0.05          13.97       14.47
5.75     0.11      5.89         0.04          12.95       14.07
6.00     0.11      6.00         0.04          13.08       13.64
6.25     0.00      6.00         0.00           0.00       10.80
6.50     0.00      6.00         0.00           0.00        6.48
6.75     0.00      6.00         0.00           0.00        3.89
7.00     0.00      6.00         0.00           0.00        2.33
7.25     0.00      6.00         0.00           0.00        1.40
7.50     0.00      6.00         0.00           0.00        0.84
7.75     0.00      6.00         0.00           0.00        0.50
8.00     0.00      6.00         0.00           0.00        0.30
8.25     0.00      6.00         0.00           0.00        0.18
----------------------------------------------------------------
Total       6.00 in                          367027 cf   366782 cf
Rational Coefficient  = 0.351     Peak Flow =   45.17 cfs
Routing Coefficient = 0.200
Watershed File: CHAP8-1.SHD       Rainfall File: CHAP7-1.RNF
----------------------------------------------------------------
```

The stage storage discharge data are entered into the computer program and all files are saved. We now exit the program and re-enter going directly to routing. All files are now retrieved and the following computer print out illustrates an attenuation in the runoff hydrograph from 45.17 ft^3/s to 13.18 ft^3/s. The entire computer print-out is not shown. Also shown is a computer plot comparing the input and output hydrographs.

Pond Node:Output by Storage Discharge Relationship
Maximum Volume =   6.73 acre-feet   Maximum Stage =   5.86 ft

Stage - Storage - Discharge Relationship

Stage (ft)	Storage (Acre-ft)	Discharge (cfs)
3.0	2.00	0.00
4.0	3.00	2.00
5.0	5.00	8.00

Time(hr)	Inflow (cfs)	Outflow(cfs)	Stage (ft)
0.25	0.00	0.00	3.00
0.50	0.00	0.00	3.00
0.75	0.00	0.00	3.00
1.00	0.00	0.00	3.00
1.25	0.00	0.00	3.00
1.50	0.00	0.00	3.00
1.75	0.77	0.03	3.01
2.00	4.14	0.18	3.09
2.25	16.43	0.79	3.40
2.50	37.74	2.31	4.05
2.75	45.17	4.85	4.47
3.00	38.02	6.89	4.81
3.25	32.76	8.48	5.08
3.50	29.23	9.75	5.29
3.75	26.17	10.76	5.46
4.00	22.88	11.51	5.58
4.25	20.20	12.05	5.67
4.50	18.04	12.42	5.74
4.75	16.61	12.68	5.78
5.00	15.61	12.87	5.81
5.25	14.86	12.99	5.83
5.50	14.47	13.08	5.85
5.75	14.07	13.14	5.86
6.00	13.64	13.18	5.86
6.25	10.80	13.05	5.84
6.50	6.48	12.68	5.78
6.75	3.89	12.16	5.69
7.00	2.33	11.58	5.60
7.25	1.40	10.97	5.50
7.50	0.84	10.36	5.39
7.75	0.50	9.77	5.30

Peak Flow: 45.2 cfs
Chapter 8 Problem 1

Section 8.6, Problem 3. We wish to reduce the size of the detention pond, thus the problem statement proposes to increase the retention volume to 2.0 in. that is the rainfall at which no discharge occurs from the pervious areas. Increasing the retention volume and increasing the time of concentration will decrease the runoff rate into the pond. With a decreased runoff rate, the outflow rate will decrease and the pond size should decrease. The peak discharge decreases from 45.17 to 24.56 ft^3/s, the outflow decrease from 13.18 to 11.37 ft^3/s, and the pond volume decreases from 6.73 acre-ft to 6.13 acre-ft. The following computer printer outputs show the results.

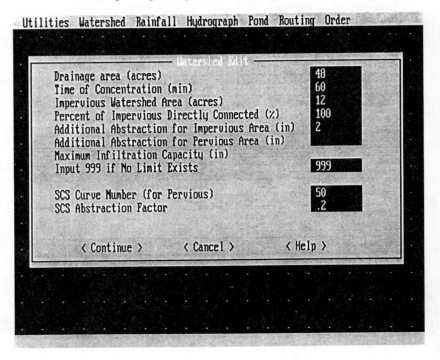

```
Node :  2  - Pond Node              Flowing To Final Outlet
                                                   SMADA PAGE 1
   Pond Node:Output by Storage Discharge Relationship
   Maximum Volume =   6.13 acre-feet  Maximum Stage =    5.56 ft

      Stage - Storage - Discharge Relationship
      Stage (ft)    Storage (Acre-ft)    Discharge (cfs)
         3.0            2.00                0.00
         4.0            3.00                2.00
         5.0            5.00                8.00
```

Node : 2 - Pond Node   SMADA PAGE 2

Time(hr)	Inflow (cfs)	Outflow(cfs)	Stage (ft)
0.25	0.00	0.00	3.00
0.50	0.00	0.00	3.00
0.75	0.00	0.00	3.00
1.00	0.00	0.00	3.00
1.25	0.00	0.00	3.00
1.50	0.00	0.00	3.00
1.75	0.00	0.00	3.00
2.00	0.00	0.00	3.00
2.25	2.80	0.10	3.05
2.50	14.34	0.63	3.32
2.75	23.66	1.53	3.77
3.00	24.47	2.68	4.11
3.25	24.56	4.00	4.33
3.50	24.42	5.23	4.54
3.75	23.79	6.36	4.73
4.00	22.49	7.34	4.89
4.25	21.09	8.17	5.03
4.50	19.69	8.88	5.15
4.75	18.53	9.46	5.24
5.00	17.55	9.96	5.33
5.25	16.70	10.37	5.39
5.50	16.07	10.72	5.45
5.75	15.49	11.01	5.50
6.00	14.94	11.25	5.54
6.25	13.07	11.37	5.56
6.50	10.17	11.31	5.55
6.75	7.91	11.12	5.52
7.00	6.15	10.83	5.47
7.25	4.78	10.48	5.41
7.50	3.72	10.08	5.35

Peak Flow: 24.6 cfs
Chapter 8 Problem 3

Section 8.6, Problem 4. This problem illustrates an increase in time of concentration and a decrease in impervious area results in a decrease in peak discharge.

Node : 2 - Pond Node   Flowing To Final Outlet
                                              SMADA PAGE 1
Pond Node:Output by Storage Discharge Relationship
Maximum Volume =   5.76 acre-feet   Maximum Stage =   5.38 ft

Stage - Storage - Discharge Relationship

Stage (ft)	Storage (Acre-ft)	Discharge (cfs)
3.0	2.00	0.00
4.0	3.00	2.00
5.0	5.00	8.00

Node : 2 - Pond Node   SMADA PAGE 2

Time(hr)	Inflow (cfs)	Outflow(cfs)	Stage (ft)
0.25	0.00	0.00	3.00
0.50	0.00	0.00	3.00
0.75	0.00	0.00	3.00
1.00	0.00	0.00	3.00
1.25	0.00	0.00	3.00
1.50	0.00	0.00	3.00
1.75	0.00	0.00	3.00
2.00	0.00	0.00	3.00
2.25	1.97	0.07	3.04
2.50	10.55	0.46	3.23
2.75	18.01	1.14	3.57
3.00	19.39	1.88	3.94
3.25	20.13	2.92	4.15
3.50	20.59	3.98	4.33
3.75	20.58	4.98	4.50
4.00	19.95	5.89	4.65
4.25	19.13	6.69	4.78
4.50	18.23	7.39	4.90
4.75	17.44	8.00	5.00
5.00	16.73	8.53	5.09
5.25	16.09	8.99	5.17
5.50	15.59	9.39	5.23
5.75	15.12	9.74	5.29
6.00	14.65	10.04	5.34
6.25	13.13	10.23	5.37
6.50	10.75	10.28	5.38
6.75	8.79	10.20	5.37
7.00	7.19	10.03	5.34
7.25	5.89	9.79	5.30
7.50	4.82	9.49	5.25

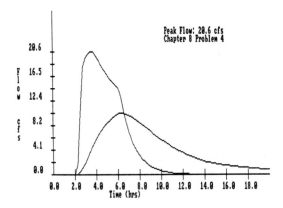

Section 9.6, Problem 11. Annual construction:

Reservoir: $$A = 120{,}000\left[\frac{0.1(1.1)^{60}}{(1.1)^{60} - 1}\right] = \$12{,}039.54$$

On-site: $$A = 160{,}000\left[\frac{0.1(1.1)^{60}}{(1.1)^{60} - 1}\right] = \$16{,}052.72$$

Annual loss and benefits:

Reservoir: Benefits = Damages before − Damages after
= $160,000 − $50,000 = $110,000/yr

On-site: Benefits = $160,000 − $25,000 = $135,000/yr

Decision based on construction cost:

Reservoir is the choice: 12,039.54 < 16,052.72

Decision based on construction and ORM cost:

Reservoir: 12,039.54 + 20,000 = $32,039.54
On-site: 16,052.72 + 10,000 = $26,052.72   (Choice)

Decision based on costs and losses:

Reservoir: 32,039.54 + 50,000 = $82,039.54
On-site: 26,052.72 + 25,000 = $51,052.72   (Choice)

Decision based on benefit and cost:

	B − C		B/C
Reservoir:	110,000 − 32,039.54 = $77,960.46		3.43
On-Site:	135,000 − 26,052.72 = $108,947.28		5.18 (Choice)

Net present worth:

Reservoir:

$$\text{NPW} = [(160{,}000 - 50{,}000) - 20{,}000]\left[\frac{(1.1)^{60} - 1}{0.1(1.1)^{60}}\right] - 120{,}000$$

$$= \$777{,}000$$

On-site: $\text{NPW} = [(160{,}000 - 25{,}000) - 10{,}000]\left[\dfrac{(1.1)^{60} - 1}{0.1(1.1)^{60}}\right] - 160{,}000$

$$= \$1{,}245{,}735$$

Pick on-site.

Section 9.6, Problem 12. Design for 20-year flow of 31 ft^3/s

Size pipe: $A = Q/V = 31/5 = 6.2$ ft^2 and

$d = (4A/\pi)^{1/2} = 2.8$ ft (say, $d = 30$ in.)

Capital cost = $3/in.-ft × 30 in. × 5280 ft = $475,200

Stage construction at year 10

Linear increase in flow = 10.8 + 10.1 = 20.9 ft^3/s

Size pipe:

$A = 20.9/5 = 4.18$ ft^2 and

$d = [4(4.18)/\pi]^{1/2} = 2.3$ ft (say, $d = 24$ in.; 27 in. not available)

Capital cost (year 0) = $3 × 24 in. × 5280 ft = $380,160

$$\Delta Q = (31 - 20.9) = 10.1 \text{ ft}^3/\text{s}$$

Size pipe: $A = 10.1/5 = 2.02$ ft^2 and $d = 18$ in.

Capital cost (year 10) = $3 × 18 in. × 5280 in. = $285,120

PW = 380,160 + 285,120(1/1.0735^{10}) = $520,444

Design and build for 20-yr flow ($475,200 < 520,444).

Section 9.6, Problem 13

Alternative A: Replace with 60-in. pipe

$$PW_A = 1000 \left[\frac{1.09^{50} - 1}{0.09(1.09^{50})} \right] + 35,000 = \$45,969.68$$

Alternative B: Add parallel pipe

$$PW_B = 800 \left[\frac{1.09^{20} - 1}{0.09(1.09)^{20}} \right] + \frac{1000}{1.09^{20} \left[\frac{1.09^{30} - 1}{0.09(1.09)^{30}} \right]}$$

$$- \frac{10,000}{1.09^{50}} + 28,000 + \frac{50,00}{1.09^2}$$

$$= 7302.84 + 1833.14 - 134.49 + 27,000 + 8921.55$$

$$= \$45,923.04$$

Strictly, $PW_B < PW_A$. However, practically there is no difference. Sensitivity analysis may reveal other distinctions. You may wish to consider the cash flow situation.

Section 10.8, Problem 9. Let

R_{1i} = water supply releases in season i, acre-ft

R_{2i} = agricultural releases in season i, acre-ft

R_{3i} = excessive releases in season i, acre-ft

S_i = storage in season i, acre-ft

Maximize

$$1.2R_{11} + 1.5R_{12} + 2.0R_{13} + 1.5R_{14} + 0.2R_{21} + 2.0R_{22} + 3.0R_{23} + 1.0R_{24}$$
$$- 0.5R_{31} - 1.0R_{32} - 1.5R_{33} - 3.0R_{34} + 0.4S_1 + 1.0S_2 + 2.5S_3 + 1.0S_4$$

subject to:

(1) $Q_1 - E_1 - R_{11} - R_{21} - R_{31} + 100{,}000 - S_1 = 0$

(2) $100{,}000 + Q_1 + Q_2 - E_1 - E_2 - R_{11} - R_{12} - R_{21} - R_{22} - R_{31} - R_{32} - S_2 = 0$

(3) $100{,}000 + \sum_{i=1}^{3} Q_i - \sum_{i=1}^{3} E_i - \sum_j \sum_i R_{ji} - S_3 = 0$

(4) $100{,}000 + \sum_{i=1}^{4} Q_i - \sum_{i=1}^{4} E_i - \sum_j \sum_i R_{ji} - S_4 = 0$

(5) $S_1, S_2, S_3, S_4 \geq 50{,}000$

(6) $S_1, S_2, S_3, S_4 \leq 150{,}000$

(7) $R_{ji} \geq 0 \quad \text{for all } j, i$

Thus a 16-variable problem.

Section 10.8, Problem 13. Calculate rainfall excess first:

$CN = 80 \quad S' = 2.5 \text{ in.} \quad R = [4 - 0.2(2.5)]^2 / [4 + 0.8(2.5)] = 2.04 \text{ in.}$
volume (acre-ft) = (4 in. × 48 acres × 2.04 in. × 52 acres)/100 × 100 acres/(12 in./ft) = 24.84 acre-ft

Let X_1 = acre-ft for golf; X_2 = acre-ft for sod.
Maximize $250 X_1 + 100 X_2$
Subject to:

(1) $X_1 + X_2 \leq 24.84$
(2) $X_2 \geq 10$
(3) $X_1 / X_2 \leq \frac{1}{2}$ or $2X_1 - X_2 \leq 0$ or $X_2 \geq 2X_1$

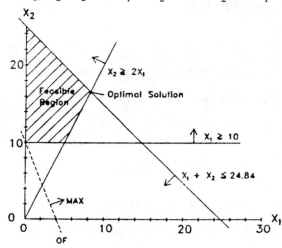

APPENDIX H   565

Point 0 is optimal point: $X_1 = 8.28$; $X_2 = 16.56$. Mathematical solution:

$$X_1 + X_2 = 24.84$$
$$2X_1 - X_2 = 0$$
$$3X_1 = 24.84$$
$$X_1 = 8.28$$

Section 10.9, Problem 3. First, construct mass balances (*note:* start reservoir with zero, end with zero):

Wet Season	Dry Season
$V_2 = V_1 + I - Y_1$	$V_2 = V_1 + I - Y_1$
$V = 0 + 80 - Y_1$	$0 = V + 25 - Y_1$
$Y_1 = 80 - V$	$Y_1 = 25 + V$
$Y_2 = 80 - V + 40$	$Y_2 = 25 + V + 15$
$Y_2 = 120 - V$	$Y_2 = 40 + V$
$Y_3 = 120 - V - 0.2X$	$Y_3 = 40 + V - 0.8X$

Now, set up as a linear programming problem:

Maximize $Z = 36X_1 + 28X_2 + 10X_3 - 6V_1 - 7.5V_2$

Subject to:

1. Cannot allocate to agriculture more than available

    Wet: $120 - V - 0.2X \geq 0$ or $V + 0.2X \leq 120$
    Dry: $40 + V - 0.8X \geq 0$ or $-V + 0.8X \leq 40$

    and

    $$0.2X_1 + 0.2X_2 + 0.2X_3 + V_1 + V_2 \leq 120 \qquad (1)$$
    $$0.8X_1 + 0.8X_2 + 0.8X_3 - V_1 - V_2 \leq 40 \qquad (2)$$

2. Maximize piecewise linear quantities

    $$X_1 \leq 20; \; X_2 \leq 45; \; X_3 \leq 15; \qquad (3),(4),(5)$$
    $$V_1 \leq 30; \; V_2 \leq 120 \qquad (6),(7)$$

3. Reservoir cannot be larger than inflow

    $$80 - Y \geq 0 \text{ or } Y \leq 80$$

and

$$V_1 + V_2 \leq 80 \tag{8}$$

Using computer (see next pages)

$$X_1 = 20(10^3) \text{ acre-ft} \quad V_1 = 24(10^3) \text{ acre-ft}$$

$$X_2 = 45(10^3) \text{ acre-ft} \quad V_2 = 0$$

$$X_3 = 15(10^3) \text{ acre-ft} \quad \text{Total cost} = \$1,986,000$$

Note: This is an opportunity cost; may wish to build reservoir larger, provided a use of stored water. Also, constraints (3), (4), and (5) show that the objective function can be increased if more water is made available to agriculture.

```
IN THE FOLLOWING PROBLEM DESCRIPTION ( V 1 ) DENOTES
DECISION
VARIABLE #1, ( V 2 ) DENOTES DECISION VARIABLE #2, ETC.
        MAXIMIZE
        OBJECTIVE FUNCTION
   Z = 36 ( V 1 ) + 28 ( V 2 ) + 10 ( V 3 ) - 6 ( V 4 )
       -7.5 ( V 5 )
   CONSTRAINTS
   1 . .2 ( V 1 ) + .2 ( V 2 ) + .2 ( V 3 )
       + 1 ( V 4 ) + 1 ( V 5 ) < = 120
   2 ..8 ( V 1 ) + .8 ( V 2 ) + .8 ( V 3 )
       -1 ( V 4 ) - 1 ( V 5 ) < = 40
   3 .1 ( V 1 ) + 0 ( V 2 ) + 0 ( V 3 ) + 0 ( V 4 ) + 0
       ( V 5 ) < = 20
   4 .0 ( V 1 ) + 1 ( V 2 ) + 0 ( V 3 ) + 0 ( V 4 ) + 0
       ( V 5 ) < = 45
   5 .0 ( V 1 ) + 0 ( V 2 ) + 1 ( V 3 ) + 0 ( V 4 ) + 0
       ( V 5 ) < = 15
   6 .0 ( V 1 ) + 0 ( V 2 ) + 0 ( V 3 ) + 1 ( V 4 ) + 0
       ( V 5 ) < = 30
   7 .0 ( V 1 ) + 0 ( V 2 ) + 0 ( V 3 ) + 0 ( V 4 ) + 1
       ( V 5 ) < = 120
   8 .0 ( V 1 ) + 0 ( V 2 ) + 0 ( V 3 ) + 1 ( V 4 ) + 1
       ( V 5 ) < = 80
```

VARIABLE NUMBER	VARIABLE TYPE	VALUE	OPPORTUNITY COST
1	DECISION	20	0
2	DECISION	45	0
3	DECISION	15	0
4	DECISION	24	0
5	DECISION	0	1.5
6	SLACK CONSTRAINT 1	80	0
7	SLACK CONSTRAINT 2	0	6
8	SLACK CONSTRAINT 3	0	31.2
9	SLACK CONSTRAINT 4	0	23.2
10	SLACK CONSTRAINT 5	0	5.2
11	SLACK CONSTRAINT 6	6	0
12	SLACK CONSTRAINT 7	120	0
13	SLACK CONSTRAINT 8	56	0

TOTAL = 1986 $1000

RANGE OF CONSTRAINING VALUES

	LOWER	PRESENT	UPPER
CONSTRAINT 1	40	120	INFINITY
CORRESP. $1000	1986	1986	1986
CONSTRAINT 2	34	40	64
CORRESP. $1000	1950	1986	2130
CONSTRAINT 3	0	20	27.5
CORRESP. $1000	1362	1986	2220
CONSTRAINT 4	15	45	52.5
CORRESP. $1000	1290	1986	2160
CONSTRAINT 5	0	15	22.5
CORRESP. $1000	1908	1986	2025
CONSTRAINT 6	24	30	INFINITY
CORRESP. $1000	1986	1986	1986
CONSTRAINT 7	0	120	INFINITY
CORRESP. $1000	1986	1986	1986
CONSTRAINT 8	24	80	INFINITY
CORRESP. $1000	1986	1986	1986

Section 10.9, Problem 9. Using SM rational method, program 5, the watershed data are (using zone 7 rainfall):

INPUT DATA FOR FILE C: CHAP

NODE #	USER LABEL	SUB AREA (ACRES)	IMPERVIOUS AREA (ACRES)	C FACTOR PERVIOUS	C FACTOR IMPERVIOUS	TIME OF CONCENTRATION (MINUTES)
1	1	3	2	.2	.9	8.21
2	2	4	1.5	.2	.9	12.55
3	3	5	2	.2	.9	16.43
4	POND	0	0	.2	.9	16.43

RESULTS FOR FILE C: CHAP

STRUCTURE LABEL	COMPOSITE C FACTOR (TOTAL AREA)	TIME OF CONCENTRATION (MINUTES)	INTENSITY ZONE 7 (IN/HR)	SUB AREA (ACRES)	TOTAL AREA (ACRES)	CUMULATIVE FLOWRATE (CFS)
1	.66	8.21	7.91	3.00	3.00	15.65
2	.55	12.55	6.75	4.00	7.00	25.98
3	.51	16.43	6.14	5.00	12.00	37.55
POND	.51	16.43	6.14	0.00	12.00	37.55

Presented are two solutions using the PRINT SCREEN key to obtain a printout. The first solution violates the 3-ft cover requirement. Use at least two pipe selections per pipe length.

```
GLOBAL OPTIMUM COST = 62600 DOLLARS

OPTIMAL CONFIGURATION :
PIPE 3 30 IN.
PIPE 2 27 IN.
PIPE 1 18 IN.

        PRESS RETURN TO CONTINUE . . .?
```

```
         PIPE # 3
Diameter (in) 30
Length (ft) 150
Velocity (fps) 7.712782
Slope 8.520674E-03
Headwater Pipe Invert (ft) 189.2781
Tailwater Pipe Invert (ft) 188
Hydraulic Grade Line Elev (ft) at Headwater 192.24
Cost of this Pipe Link 14100

PRESS RETURN TO CONTINUE . . .?
```

```
      PIPE # 2
Diameter (in) 27
Length (ft) 400
Velocity (fps) 6.483781
Slope 6.930252E-03
Headwater Pipe Invert (ft) 192.7621
Tailwater Pipe Invert (ft) 189.99
Hydraulic Grade Line Elev (ft) at Headwater 195.3385
Cost of this Pipe Link 32000

PRESS RETURN TO CONTINUE . . .?
      PIPE # 1
Diameter (in) 18
Length (ft) 300
Velocity (fps) 8.856095
Slope 2.220663E-02
Headwater Pipe Invert (ft) 200.5004
Tailwater Pipe Invert (ft) 193.8385
Hydraulic Grade Line Elev (ft) at Headwater 202.6094
Cost of this Pipe Link 16500

PRESS RETURN TO CONTINUE . . .?
```

This solution uses at least two pipe sizes per pipe length. Cover of 3 ft is maintained.

```
          GLOBAL OPTIMUM COST = 71800 DOLLARS

          OPTIMAL CONFIGURATION :
          PIPE 3  30 in.
          PIPE 2  30 in.
          PIPE 1  24 in.
               PRESS RETURN TO CONTINUE . . .?


      PIPE # 3
Diameter (in) 30
Length (ft) 150
Velocity (fps) 7.649629
Slope 8.381709E-03
Headwater Pipe Invert (ft) 189.2572
Tailwater Pipe Invert (ft) 188
Hydraulic Grade Line Elev (ft) at Headwater 192.2116
Cost of this Pipe Link 14100

PRESS RETURN TO CONTINUE . . .?
```

PIPE # 2
Diameter (in) 30
Length (ft) 400
Velocity (fps) 5.292606
Slope 4.012277E-03
Headwater Pipe Invert (ft) 191.3165
Tailwater Pipe Invert (ft) 189.7116
Hydraulic Grade Line Elev (ft) at Headwater 194.034
Cost of this Pipe Link 37600

PRESS RETURN TO CONTINUE . . .?

      PIPE # 1
Diameter (in) 24
Length (ft) 300
Velocity (fps) 4.981554
Slope 4.786955E-03
Headwater Pipe Invert (ft) 193.47
Tailwater Pipe Invert (ft) 192.034
Hydraulic Grade Line Elev (ft) at Headwater 195.6627
Cost of this Pipe Link 20100

PRESS RETURN TO CONTINUE . . .?

Section 10.9, Problem 11.
      HYDROLOGY PROGRAMS
         LINEAR PROGRAMMING
IN THE FOLLOWING PROBLEM DESCRIPTION (V 1) DENOTES DECISION
VARIABLE #1, (V 2) DENOTES DECISION VARIABLE #2, ETC.
      MINIMIZE
        OBJECTIVE FUNCTION
  Z = 50 (V 1)+ 80 (V 2)+ 60 (V 3)+ 70 (V 4)
    + 65 (V 5)+ 65 (V 6)

CONSTRAINTS
1 . 1 (V 1)+ 1 (V 2)+ 0 (V 3)+ 0 (V 4)+ 1
(V 5)+ 0 (V 6)<= 100
2 . 1 (V 1)+ 0 (V 2)+ 0 (V 3)+ 0 (V 4)+ 0
(V 5)+ 0 (V 6)<= 80
3 . 0 (V 1)+ 1 (V 2)+ 0 (V 3)+ 0 (V 4)+ 0
(V 5)+ 0 (V 6)<= 15
4 . 0 (V 1)+ 0 (V 2)+ 0 (V 3)+ 0 (V 4)+ 1
(V 5)+ 0 (V 6)<= 30
5 . 0 (V 1)+ 0 (V 2)+ 1 (V 3)+ 1 (V 4)+ 0
(V 5)+ 1 (V 6)<= 200
6 . 0 (V 1)+ 0 (V 2)+ 1 (V 3)+ 0 (V 4)+ 0

```
                ( V 5 ) + 0 ( V 6 ) <= 160
        7 . 0 ( V 1 ) + 0 ( V 2 ) + 0 ( V 3 ) + 1 ( V 4 ) + 0
                ( V 5 ) + 0 ( V 6 ) <= 30
        8 . 0 ( V 1 ) + 0 ( V 2 ) + 0 ( V 3 ) + 0 ( V 4 ) + 0
                ( V 5 ) + 1 ( V 6 ) <= 50
        9 . 1 ( V 1 ) + 1 ( V 2 ) + 1 ( V 3 ) + 1 ( V 4 ) + 1
                ( V 5 ) + 1 ( V 6 ) >= 240
```

VARIABLE NUMBER	VARIABLE TYPE	VALUE	OPPORTUNITY COST
1	DECISION	80	0
2	DECISION	0	20
3	DECISION	160	0
4	DECISION	0	10
5	DECISION	0	5
6	DECISION	0	5
7	SLACK CONSTRAINT 1	20	0
8	SLACK CONSTRAINT 2	0	10
9	SLACK CONSTRAINT 3	15	0
10	SLACK CONSTRAINT 4	30	0
11	SLACK CONSTRAINT 5	40	0
12	SLACK CONSTRAINT 6	0	0
13	SLACK CONSTRAINT 7	30	0
14	SLACK CONSTRAINT 8	50	0
15	SURPLUS CONSTRAINT 9	0	60

TOTAL = 13600 DOLLARS

RANGE OF CONSTRAINING VALUES

	LOWER	PRESENT	UPPER
CONSTRAINT 1	80	100	INFINITY
CORRESP. DOLLARS	13600	13600	13600
CONSTRAINT 2	80	80	100
CORRESP. DOLLARS	13600	13600	13400
CONSTRAINT 3	0	15	INFINITY
CORRESP. DOLLARS	13600	13600	13600
CONSTRAINT 4	0	30	INFINITY
CORRESP. DOLLARS	13600	13600	13600
CONSTRAINT 5	160	200	INFINITY
CORRESP. DOLLARS	13600	13600	13600
CONSTRAINT 6	160	160	INFINITY
CORRESP. DOLLARS	13600	13600	13600
CONSTRAINT 7	0	30	INFINITY
CORRESP. DOLLARS	13600	13600	13600
CONSTRAINT 8	0	50	INFINITY
CORRESP. DOLLARS	13600	13600	13600
CONSTRAINT 9	80	240	240
CORRESP. DOLLARS	4000	13600	13600

INDEX

Activated carbon, 237-239
Acute criteria, 163, 164
Agricultural loadings, 121
   animal comparisons, 122
   controls, 397-398
   conversions (tons to cubic yards), 409
   example problems, 123, 134, 417, 418
   feedlots, 415
   land sprinkling, 416
   NPDES, 415
   solids disposal, 414-416
   storage, 417
Agricultural runoff model (ARM), 398
Algal assay, 173-174, 207-208
Algorithm, 497
Ambient life standards, 128
Antecedent moisture condition (AMC), 508
Attenuation factor, 73-74
Average daily traffic, 175
   related to algae, 175

Bacteria, 112, 128, 154, 160, 168, 208
Benefit-cost, 320, 321, 330
   example problem, 320-321
Benefits, 317-318
   fish and wildlife, 318
   flood control, 317
   irrigation, 318
   power supply, 318
   recreation, 318
   water quality, 318
   water supply, 318
Benthic demand, 188
Berms, 248-252
Best management practices (BMP's), 371-372
   agricultural, 534
   comparative data, 379
   construction, 533
   defined, 371
   description, 528-535
   listing, 219
   non-structural, 218-219, 528-529
   structural, 218-219, 529-533

Bioaccumulation, 111
Bioassay, 163, 168, 169, 171-173
Bivariate linear regression, 496, 502
BOD, 125-126, 131-137
Broom sweeping, 372
Bulk precipitation, 91, 113, 114, 159

Capital recovery, 326
Case studies (Appendix H), 526-572
Cash flow diagram, 325
Chlorophyll, 112, 209
Classical optimization, 360
   example problems, 360, 361-362
Clean waters act, 109
Coagulation, 216, 239-241
Coliforms, 110
Combined Sewers, 167
   antecedent dry conditions, 169
Complete-mix continuous flow model, 210
   example problem, 203
Composite exponential reaction, 197
Comprehensive stormwater management, 313
Computer programs, 19, 22, 31, 34, 44, 77, 92, 102, 110, 178, 385, 464
   ARM, 398
   EPA-PTR, 91
   HEC-1, 14, 91, 96-98
   HY8, 92
   ILLUDAS, 91
   installation instructions, 465
   LEASTS (error estimation), 497
   linear programming, 496-499
   OPSEW, 365, 370, 496, 499-502
   probability distributions, 496
   QUALHYMO, 92
   RETEN, 464, 492
   SMADA, 14, 91, 98-103, 464, 466
   STORM, 91-95
   SWMM, 14, 91
   TR20, 92
   TR55, 92
Confidence limit, 39, 43, 497

573

574 INDEX

Conservation tillage, 534
Constraints, 96, 346–349, 351, 358, 360–364
  defined for models, 347
  examples for models, 348
Continuous convolution, 73, 77–79, 85, 86
  storage coefficient, 77
  travel time, 77
Contouring BMP, 535
Contributing area, 69, 70, 73
Conversion factors, 430–434
Convex cost curve, 381
Convolution (unit hydrograph), 352–354
Convolution, continuous hydrograph procedure, 73, 77–79
Cost curve, 317, 356, 364
Cost-effectiveness, 316, 371–378
  alternative comparisons, 378
  BMP's, 371
  defined, 371
  diversion systems, 373
Costs, defined, 315
  average cost, 316
  cost curve, 316–317
  economies of scale, 316
  indirect, 12
  marginal cost, 316
Council on Environmental Qualtiy, 132, 160, 406
Criteria, 216, 226, 247, 262, 298, 311–316, 330, 338, 341
Critical DO conditions, 184–185
  example problem, 205–206
  mass balance, 204
Cropping management factor, 95, 399, 405
Curve number, 71, 470, 475, 505
  agricultural, 507
  composite, 475
  composite urban, 505–506
  pervious area, 71

Dead volume, 307
Decay coefficient, 94, 181, 183
Decision space, 359–361
Deicing agents, 167, 177, 207, 529
Deposition rates, 114, 116, 132
Design criteria, 10, 11, 14, 20, 21, 32, 134, 161, 164, 171–173, 212, 214
Design storm, 235, 270, 435
  FID, 270, 435–442
  PIF, 270, 293, 494
Detention pond, flood control:
  hydrograph defined, 65
  inventory equation, 79–82
  peak attenuation, 100

Detention pond, water quality:
  design considerations, 264–274
  fate of pollutants, 283–292
  plant species, 112, 160, 161, 192, 276
  probability models, 278–283
  reuse, 292–306
  wet-detention, 262–274
Detention time, 88, 100, 192, 210, 233–239, 270–273
Dillon Model, 195, 196, 207
Dimensionless precipitation, 435–438
  Corps of Engineers, 437
  SCS type II, 435
  SCS type III, 436
  SMADA program, 471
Diskette (care for), 465
Dissolved contaminants, 239, 267
Dissolved fraction, 110, 135
Dissolved oxygen, 16, 110, 160–167, 177–181, 185–190, 284, 291, 320
  benthic demand, 188
  critical DO conditions, 185
  decay coefficient, 94, 181, 183
  DO sag curve, 185, 209
  example problems, 183, 185
  impoundment, 188, 189, 192
  reaeration coefficient, 181, 183
  solubility of oxygen, 179
Diversion systems, 225, 375
  design based on events, 221–222
  design based on volume, 222–225
  exfiltration trench, 259, 376
  off-line defined, 16
  plan and profile, 374
Diversion volume curve, 223
DO sag curve, 185, 209
Double-ring infiltrometer, 85
Drinking water, 4, 132–134, 138, 160, 379
Dry detention ponds, 529
Dustfall, 113, 114, 116, 129, 154
Dutch drains, 532
Dynamic programming, 365–372, 388, 392, 398–399
  capacity expansion, 368
  complete enumeration, 368
  example problems, 366, 368, 370
  recursive relationship, 364

Economic feasibility, 12, 313, 319, 337
Edit files, 500
Efficiency considerations, 220
Empirical distribution, 24, 29, 34–36, 41–48, 224, 268–273, 330
Emptying rate ratio, 282

INDEX 575

Environmental impact, 9, 20, 238
Erosion, 8, 12, 20, 22, 92, 95, 110, 163, 190,
    218, 242-247, 260-264, 319, 397-401,
    408-411, 421-423
  control procedures, 398
  defined, 397
Eutrophic, 190-194, 285, 314
Eutrophication, 159, 160, 167, 190-192
  eutrophic, 190-194
  oligotrophic, 190, 195
Evaporation, 2, 5-7, 10
  water budget, 5-8
Event loading, 120
Event mean concentration (EMC), 56,
    117-123, 133, 138, 158
  comparative data, 133
  definition, 117
  example problem, 119
  used in RETEN, 492
Exceedence probability, 26
Exfiltration trench, 231, 259, 262, 379, 530
  computer program, 495
  example problem, 231
Expected values, 278, 331, 333
Extreme events, 24, 26-28, 35, 36, 39, 42, 48,
    59, 60
  defined, 24
  estimation methods, 26-27

Fate of pollutants, 283-292
Feasibility tests, 1, 11-13
  comprehensive plan, 313-314
  site related, 9
Feedlot runoff, 415
FID or IDF, 26, 270, 435-442
Filtration, 216, 239-240
Financial plan, 11, 12, 20
Financial planning, 322-323, 338, 370
  example problem, 322-325
First flush, 128, 142, 164, 167, 208, 225, 231,
    258, 375
First-order complete-mix model, 202-203
First order reaction, 144, 154-158, 185,
    209-210, 227, 228, 263, 284, 307, 308,
    355, 356
Fiscal concerns, 1, 4, 313, 373
Floodplain defined, 3
Flow rates, 7, 118
  agricultural, 121
  attenuation, 78
  control objectives, 9
  discharge example, 80-82
  loadings for water quality, 119-120
  peak flow equations, 32-34

Foster, 35, 36, 61, 400, 405, 421
French drains, 530
Function space, 359-360
Future worth, 325

Grab sampling, 119
Grain size distribution, 124
Gumbel distribution, 32-33, 44

HEC-1, 14, 91, 96-98, 107-108
  default values, 97
  example application, 96-98
  optimization, 97
Highway runoff impacts, 166-169
Histogram, 28-30, 62, 222-225, 275
  example problem, 29-30
Horton equation, 66-68, 470
Hydraulic residence, 190
Hydrograph attenuation, 78
  comparisons, 7
  continuity equation, 79
  inventory equation, 79
  muskingum method, 82
  storage-outflow relationship, 80-82
Hydrograph defined, 65
Hydrograph generation, 26, 27, 73, 85, 99, 107,
    110, 159, 261
  convolution, 73, 78, 79, 85-87, 354, 355
  rational, 13, 70-77, 84-88, 90, 109, 156, 246,
    261, 372
  Santa Barbara, 73, 77, 85-89, 107, 109, 261,
    311
  SCS, 73-76, 80
Hydrologic cycle, 1, 4, 5-6, 8, 20, 24
Hydrologic soil classification, 71, 508
Hydrology, 1, 11, 19-24, 63, 90
  comprehensive plan, 314
  defined, 24
  hydrologic balance, 9
  swale hydrology, 242-244
  water budget, 3-7
Hyetograph, 24, 51, 78

IDF or FID, 26, 270, 435-442
ILLUDAS model, 91
Impacts (time scale), 161
Impoundment, 186-188, 192
Indirect costs, 12
Infiltration, 2-9, 13, 20, 66, 244
  double-ring infiltrometer, 244
  Horton, 66
  potential, 66-67
  rate, 66
  volume, 67

Infiltration inflow studies (I&I), 219, 529
Initial abstraction factor, 71, 470
Instantaneous flow, 77
Instantaneous unit hydrograph, 76
Interest rates, 323
Interevent dry period, 25, 190, 276
  defined, 25
  diversion volume curve, 222–223
  minimum interevent dry period, 25
  PIF curve, 269–270, 293–295
  probability model, 53
  wet detention, 262–263
Interval between rain mid-points, 57–58
Inventory equation, 80, 348
Irrigation, 8, 15, 295–300, 358–360, 398

Junction node, 473

Kuichling (rational formula), 69, 72

Lake, 9, 160–162
  area in peak flow equations, 33
  dissolved oxygen, 186
  trophic state, 190
Land use, 109
  event mean concentration, 133
  loadings, 137, 493
  runoff coefficient, 70
  SCS-CN, 71, 505–508
  treatment philosophy, 217
  water budget changes, 5–7
LC_{50} defined, 171
  example problem, 171–173
Leasts computer program, 497
Least-squares method, 36
Linear programming, 346–359, 362–364, 371, 381
  computer program, 497–499
  example problems, 349, 355, 497
  graphical solution, 348
  load reduction model, 354
  objective function, 347
  optimal load reduction, 354–357
  piecewise linear, 362
  regional stormwater facilities, 351–352
  unit hydrograph, 352–354
Loading rates, 120, 135
  average yearly, 137
  computer program defaults, 493
  highway related, 136
Loadograph, 118, 121, 148–150, 154–159
  example problem, 144–146
Log-normal distribution, 32, 43, 45, 129, 278
Load reduction model, 354
  decay rate, 354
  example problem, 355
Log-Pearson type III distribution, 32, 43, 47–48, 62–63, 272, 497

Manning's roughness coefficient, 245–246
  channel flow (N), 245–246
  concrete pipe (N = 0.013), 486
  pipe flow (n), 473
  swales, 243
Mass balance, 5, 220–231
  efficiencies, 220
  example problems, 18, 231
Mass loadings, 31, 109, 119, 136–146, 156, 159, 163, 164, 207
Mathematical models, 13, 20, 90, 107, 132, 144, 154, 178, 188, 207, 267
Maximum likelihood, 35, 38, 60
Metals, 4, 111, 128–140, 162–177, 252–254, 291
  EPA criteria, 171–172
  human hazards, 169–170
  LC_{50} defined, 171
  LC_{50} example, 171–173
  toxicity, 174
Metal transformation, 288
Method of moments, 35, 38, 60, 62
Municipal service taxing, 328–329
Muskingum method, 83

Nonlinear optimization, 361
  example problem, 361
  graphical solution, 361
Nonpoint sources, 3, 4, 20–22, 154, 160–167
  assessment, 159
  defined, 3
  mathematical modeling, 140
  relative magnitude, 4, 159–160
  runoff water quality, 127–136
  rural, 120–123
  separation from point, 163–165
  urban, 123–127
  water quality relationship, 109–125, 163–166
Nonstructural BMP's, 219, 528–529
Normal distribution, 32, 33, 39, 43, 401, 497
Notation, 423–429
NPDES, 414
NURP, 139, 140, 163, 164
Nutrients, 4, 111, 128, 160–163, 190–195, 252–254, 267–268, 417
  agricultural waste, 122
  dissolved, 127
  loadings and land use, 137
  transported, 148–149
  urban loads, 126

INDEX **577**

Objective function, 96–99, 345–355, 359–366, 388, 392, 394, 397
Off-line retention, 16, 109, 216, 225, 258–264, 311–317, 389, 398
  attenuation, 494
  computer program, 493
  development of designs, 225
  diversion based on number of events, 221
  diversion based on volume, 222–225, 230
  diversion structure, 374
Oligotrophic, 192, 193, 195, 209
On-line infiltration, 274–277
  example problem, 277
Operations research, 344, 388, 399
OPSEW, 366, 496, 499–502
Optimal load reduction, 354
  example problem, 355–357
Optimization, 343–351
  decision space, 359, 361
  function space, 359–360
ORM, 316–317, 340, 373–374, 381, 393
Overflow rate, 237–239, 281–284
Oxygen demand, 94, 111, 112, 124, 154, 177, 181, 284

Parameter estimation, 35
Particle sizes, 123, 234
Particulate fraction, 110
Peak attenuation, 100
Peak discharges and streamflow, 30, 66, 75, 79, 80, 86, 102
  example problems, 33, 74
  regression equations, 32
Pearson distribution, 497
Permanent pool, 262, 266, 269–275, 292–294
pH, 115–117, 162–164, 169, 175, 239
Photosynthesis, 111, 182, 184–186, 188
Physical treatment systems, 530
Piecewise linear, 356, 363, 364, 365, 384
PIF curves, 270, 293–295, 494
  example problem, 294
Planning horizon, 323
Plant species, 112, 160, 161, 192, 276
Plot Position, 35, 61, 269
  California, 35–36
  example problem, 35–36
  exceedence, 35–36
  Foster, 35–36
  Weibull, 35–36, 61, 269
Point sources, 3, 4, 10, 16, 126, 160–164, 177, 192, 207, 263, 415
Pollutant load computer program, 492
Pollutograph, 118, 148–150, 159
  example problem, 144–146
Precipitation, 1, 2, 5, 6, 13, 20–28, 60, 64, 67, 69–74, 89, 118, 131, 137, 146, 154, 159, 161, 225–231, 238, 254, 259, 267, 270, 292–299, 315, 418–421
  bulk precipitation, 91, 113, 114, 159
  dry precipitation, 91
  FID or IDF curves, 435–442
  interevent dry period, 190
  PIF curves, 270, 293–295, 494
  probability models, 51–59
  regional variability, 56
Present worth, 317, 325, 327, 337, 339, 341, 348, 370, 395
Probability distributions, 14, 27, 59, 217–220, 307, 329
  computer program, 496
  empirical, 220, 224, 237, 240, 270–275, 331–336, 362, 400, 420
  example computer applications, 44–48
  Gumble, 44–48
  independent events, 28
  log-normal, 45–48, 280
  log-Pearson type III, 46–48, 272
  normal, 43, 280, 401
Probability methods, 24
  confidence interval, 24
  independence, 28
  parameter estimation, 35
  regression, 30
  sampling, 27
  theoretical distributions, 34
Public health, 2, 111, 160–161
  benefits, 317–319
PVC, 162

Quiescent conditions, 283, 284

Rainfall dimensionless curves, 435–438
Rainfall events, 25, 51–52
  example problem, 52
  interevent dry period, 25, 51
  interval between mid-points, 57–58
  runoff producing, 52, 56–59
Rainfall excess, 5–10, 16, 21, 53–56, 65–78, 92–95, 225–231
  defined, 66
  diversion balance, 225–231
  mass balance for reuse, 297
  reuse pond calculations, 303–304
  swale design, 245–255
  unit hydrograph related, 354
Rainfall excess calculations, 66–72
  constant rate, 66–68
  example problem, 71–72
  Horton method, 66–68

578 INDEX

Rainfall excess calculations (*Continued*)
  mass balance, 66, 68-70
  SCS-CN, 66, 70-72
Rainfall histograms, 29, 220
Rainfall intensity, 57, 254
  duration, 31
  geographic variability, 57-58
Rainfall-runoff, 51
Rainfall volumes, 51-52
Rational formula, 13, 73-75, 77, 84-88, 90, 157, 246, 261, 372
Reaction kinetics, 307
  first order, 144, 154, 158, 185, 209-210, 228, 284, 308, 356
  second-order, 412
Reaeration coefficient, 181, 183
Regional stormwater facilities, 352
Regression analysis, 27, 30, 31, 59, 60, 141, 154, 156
Regulations, 10, 22, 108, 159
Respiration, 180, 196
RETEN, 464, 492
Retention pond defined, 16
  computer program, 493
  development of designs, 225
  diversion based on number of events, 221
  diversion based on volume, 222-225, 230
  diversion structure, 374
  example problem, 377
  exfiltration trench, 259, 262, 379
  minimum interevent dry period, 222-223
  off-line exfiltration, 376
  plan and profile, 374
  relation to rainfall, 220
  runoff diversion, 95, 220, 398
  yearly mass diversion curves, 228-231
Return period, 13, 26, 30, 32-35, 39, 60-62, 86-89, 233, 270, 294, 308, 333, 335, 372
  example problem, 270
Reuse, 265, 292, 294, 295-304, 305-314, 359, 390, 393
Reuse computer program, 494-495
Reuse pond, 292, 296, 299, 302, 305, 359
Reuse rate, 306
REV curve, 300, 301
  example problem, 301, 302-305
Risk, 11, 48, 63, 217, 278, 308, 316, 329, 332, 335, 343
Rock wells, 532
Runoff coefficient, 13, 69-75, 92-93, 243-244, 304
  defined, 69
Runoff diversion, 95, 220, 398
Runoff hydrograph, 25, 60, 62, 66, 87-99, 109, 145, 249, 260, 396

Rural nonpoint sources, 123, 126, 154

Sampling (rainfall), 27
  example problem, 29-30
Sampling (runoff), 116
Santa Barbara urban hydrograph, 73, 77, 85-87, 89, 107, 109, 261, 311
  example problem, 100-103
SCS-CN, 67, 70-73, 97, 470
  composite example, 377-378
  hydrograph example, 73-76
Second foot day (SFD), 432
Sediment accumulation, 160, 190, 233
Sedimentation rate, 281, 310
Sediment ponds, 232
  design storm sizing, 235
  example problems, 236, 280-283
  settling velocities, 232
Sediments, 12, 111, 163, 176, 190-196, 239-242, 264-268, 272, 283-288
Separate sewers, 167
Settling velocities, 233, 235, 272
Shannon-Brezonik, 192
Sieve analysis, 123-124
Silt, 110
Sinking fund, 324
  example problem, 327
SMADA, 14, 91, 99-102
  example application, 98-103
  example problem, 474-492
  installation, 465
  pull-down menus, 468-469
  rainfall computer files, 471
Snowmelt, 3, 28
Soil hydrologic classification, 508
Solids disposal, 415
Solid waste management, 529
Solubility of oxygen, 179
Solutions to real world problems, 536-572
Stage-storage-discharge, 474
Statistical tables, 443
  binomial, 446-455
  Gumbel, 459
  log-normal, 458
  log-Pearson, 456-457, 460
  normal, 444-445, 458
  Poisson, 463
Stick diagram, 14-16, 481
Stochastic equations, 494
STORM model, 91-95
Stormwater financing alternatives, 327
Stormwater management:
  classification, 218-219
  criteria, 10
  defined, 1

INDEX **579**

feasibility, 11–13
limitations, 9
models, 13, 90
objectives, 8
philosophies for water quality, 216–218
policies, 10
practices, 15–17, 90, 219
Stormwater management policies, 10
Stormwater sampling, 116–117
Stormwater utility, 328, 329, 343
Street surface contaminants, 125–126
Street sweeping, 123, 528
    broom sweeping, 372, 528
    comparative data, 379
Structural BMP's, 219, 529–533
Surface storage, 66
Swales, 219, 241–247, 249–258, 264, 393, 532
    berms or blocks, 248–250
    checkdams, 533
    computer program, 495
    design, 244–250
    double-ring infiltrometer, 244
    example problems, 247, 252–253
    hydrographs, 241–242
    length of swale, 245–247
    roadside, 533
    roughness, 243
    water quality, 249–253
Synthetic hydrographs, 27, 65

Temporary storage, 262, 265, 266, 292, 294, 296, 299, 302, 308, 311
Theoretical frequency distribution, 34
    incomplete records, 39
    parameter estimation, 35
Three parameter log normal, 497
Time of concentration, 11, 13, 26, 72–80, 85–88, 96–99, 103, 109, 157, 245, 260
Toxic chemicals, 110
Toxicity, 170–172, 175, 176, 211
Transport rate coefficient, 152, 153
Transport rates, 146–150
Trophic state index, 192–195
    example problems, 191, 192, 194
Truncated normal distribution, 497
Turbidity, 110
Two parameter log normal, 44, 497

Uncertainty and risk, 328–330

example problem, 334–337
expected monetary value, 328
expected values, 332
Unit hydrograph, 14, 78, 90, 96–97, 159, 353–355, 396
Urban nonpoint sources, 123–135
    dissolved fraction, 131
    example problem, 133–134
    probability distribution, 129–130
    variability, 131
USLE, 14, 398, 400, 409, 411, 421
    cropping management factor, 95, 399–405
    erosion-control practice factor, 95, 399, 408
    example problems, 408, 409
    rainfall factor, 95, 399–400
    slope-effect (length slope) factor, 95, 406–407
    soil erodibility factor, 95, 399
U.S. Water Resources Council, 3, 64, 330

Vollenweider model, 194, 207, 209

Water budget, 5, 7, 269
Water quality assessment, 160, 212, 215
    acute criteria, 163, 164
    toxicity, 170–172, 175, 176, 211
Water quality parameters, 111, 121, 131, 134, 144, 156, 208, 274, 420
    bacteria, 110, 161
    BOD, 125–126, 131–137, 142
    dissolved fraction, 110, 135
    dissolved oxygen, 110, 121, 176
    metals, 128–130, 136, 174
    nutrients, 128, 137, 174
    particulate fraction, 110
    physical appearance, 159
    sediments, 110, 188, 231
    toxic chemicals, 110
Water yield model, 92, 110, 398, 410–413
    example problem, 412
Weirs (in SMADA), 472
Wet detention ponds, 262–292, 529
    design considerations, 264–274
    fate of pollutants, 283–292
    peak attenuation, 100
    plant species, 276
    reuse from, 292–306
Wet flow effects, 164–166
Winter crop cover, 534